全球最大会客厅
深圳国际会展中心

NEW CITY
LANDMARK

SHENZHEN WORLD
EXHIBITION & CONVENTION CENTER

深圳市欧博工程设计顾问有限公司　编著
SHENZHEN AUBE ARCHITECTURAL ENGINEERING DESIGN CO., LTD.

中国建筑工业出版社

图书在版编目（CIP）数据

全球最大会客厅深圳国际会展中心 = NEW CITY LANDMARK SHENZHEN WORLD EXHIBITION & CONVENTION CENTER ／ 深圳市欧博工程设计顾问有限公司编著. —北京：中国建筑工业出版社，2020.8

ISBN 978-7-112-25285-5

Ⅰ．①全⋯ Ⅱ．①深⋯ Ⅲ．①会堂－建筑设计－深圳 Ⅳ．①TU242.1

中国版本图书馆CIP数据核字（2020）第115046号

责任编辑：戚琳琳
文字编辑：吴　尘
责任校对：赵　菲

全球最大会客厅　深圳国际会展中心
NEW CITY LANDMARK SHENZHEN WORLD EXHIBITION & CONVENTION CENTER
深圳市欧博工程设计顾问有限公司　编著
SHENZHEN AUBE ARCHITECTURAL ENGINEERING DESIGN CO., LTD.

＊

中国建筑工业出版社出版、发行（北京海淀三里河路9号）
各地新华书店、建筑书店经销
北京锋尚制版有限公司制版
北京富诚彩色印刷有限公司印刷

＊

开本：787毫米×1092毫米　1/8　印张：62　字数：1538千字
2021年1月第一版　2021年1月第一次印刷
定价：750.00元
ISBN 978-7-112-25285-5
（36055）

谨以此书

献给深圳特区成立四十周年

This book
is dedicated to the 40th Anniversary
of Shenzhen Special Economic Zone

同时致谢

Thanks to

深圳市人民政府

Shenzhen Municipal People's Government

深圳国际会展中心建设指挥部

Construction Headquarter of Shenzhen World Exhibition & Convention Center

宝安区人民政府

People's Government of Bao'an District

及相关
各级政府主管部门

And relevant
government authorities

对深圳国际会展中心建筑设计工作所给予的帮助与支持！

For the help and support of the design work of Shenzhen World Exhibition & Convention Center.

目录 CONTENTS

导言
Introduction

"启" 源
The Origin

"大" 城
Great City

"浩" 展
Grand Exhibition Center

08

24

44

92

10 前言
Foreword

12 序言
Preface

深圳国际会展中心诞生背景
The Circumstances of the Shenzhen
World's Birth

26 一个时代的宏图 —— 规划背景
The Big Blueprint
— Planning Background

33 一座新城的起点
——大空港区域规划理念
Visions of The Future — Planning
Concept of The Airport Area

超大型会展城市综合规划
The Comprehensive Planning of the
Mega Exhibition & Convention City

46 会展 EDD 与 会展经济
EDD & Exhibition Economy

62 设计一座城的未来 —— 会展中心
The Shenzhen World —— The Design
of a Future City

69 展城呼应 —— 开放提升城市空间
Exhibition Center + City—A More
Open and Advanced Space

83 规划成果表现
Planning Achievement Showcase

深圳国际会展中心设计全貌
The Overall Design of the Shenzhen World

94 会展中心的发展与演变
The Development & Evolution of the
Exhibition & Convention Center

104 会展中心总体设计
The General Design of the Shenzhen Wo

110 会展核心功能设计
The Core Functional Designs of the
Shenzhen World

184 会展配套
The Supporting Facilities

199 会展休闲带综合配套 ——特殊会展配
Special Supporting Facility—The Leisure
Zone of the Exhibition City

221 经济技术指标表
The Economic and Technical Indicators

"巨"构
Giant Stucture

230

建造结构与系统设计
Construction Structure and
System Design

232 结构设计
Structure Design

254 设备系统设计
Equipment System Design

284 消防设计
Fire Protection Design

293 构造与节点
Structure and Node

306 防水设计
Waterproof Design

"重"任
Great Mission

316

开发模式与设计管理
Construction Mode and Design
Management

318 设计总牵头管理
Design and Engineering Procurement
Construction

334 会展景观＋海绵城市
Exhibition Landscape & Sponge City

347 交通设计
Transportation and Circulation Design

378 幕墙设计
Curtain Wall Design

402 标识设计
Signage System Design

411 室内设计
Interior Design

415 垂直交通
Vertical Transportation

422 声学设计
Acoustic Design

435 绿建设计
Green Architecture Design

"远"景
Far Sight

448

新时代的起点
The Start of A New Era

450 用户体验
User Experience

458 开局之年
The Year Ahead

459 未来图景
Visions for The Future

附录
Appendix

462

462 大事记年表
Chronicle of Events

470 设计单位
Project Team

476 建设单位
Construction Team

478 联合编辑单位
Co-editorial Team

导 言

Introduction

10 前言　Foreword

12 序言　Preface

前言

深圳国际会展中心项目是深圳建市以来最大的单体建筑，也是全球第一个集大型建筑、轨道交通、水利工程、市政工程为一体的建筑开发工程，整体建成后将成为全球最大的会展中心。

秉承深圳市委、市政府提出的"一流的设计，一流的建设，一流的运营"的建设目标，为保证建设项目的高完成度和高品质，深圳国际会展中心采用设计总承包管理模式进行专业化的设计管理，由 AUBE 欧博设计担当设计总承包牵头的重任，承担方案深化、施工图设计、景观设计等多项工作，并对会展休闲带进行全程全专业设计。

自 2016 年方案中标，到 2019 年落地建成，短短三年时间，深圳国际会展中心在粤港澳大湾区的湾顶拔地而起、惊艳绽放，被誉为"全球最大会客厅"的惊世之作，让人们再次见证了"深圳速度"和"深圳质量"——一个丰富多样、活力开放的城市公共空间，一个特色鲜明的深圳地标，一个创造了八项"世界之最"的超级工程。

为了纪念这个具有划时代意义的项目，AUBE 欧博设计再一次牵头，组织深圳国际会展中心的主创团队编撰本书，向世人和同行揭秘这个世界级超大型会展综合体背后的技术难关和创新突破。

本书不仅汇集了项目在建筑、结构、给排水、暖通空调、强弱电、通信、交通、绿建等多方面的研究成果和经验总结，还整合了声学、景观、美学、运营等专业力量等，均为项目的关键技术，对于促进同类大型公共建筑技术提高和创新有积极作用。

本书编制过程中得到了深圳市政府，宝安区政府等部门以及项目相关领导的关心和指导。在此，AUBE 欧博设计谨代表全体编著人员，向所有支持单位和个人表示诚挚的谢意。

从一片滩涂，到横亘在珠江口东岸，跃跃欲飞的巨龙，深圳国际会展中心凝聚着近 3 万名管理、施工人员两年多的辛劳、智慧与汗水。AUBE 欧博设计亦以此书，向深圳国际会展中心的所有参与者和贡献者致敬。

本书涉及领域广泛，内容浩瀚，如有不足之处，欢迎读者监督指正。

Foreword

The Shenzhen World Exhibition & Convention Center (known as Shenzhen World for short) is the largest single building ever in Shenzhen, also the first construction project in the world that involved simultaneous developments of large-scale construction, rail transit system, water engineering and municipal engineering. It will be the largest convention and exhibition center in the world when completed.

Adhering to the "the first-class design, first-class construction, first-class operation" principle proposed by the Shenzhen Municipal Committee and Shenzhen Municipal government, and to guarantee the high quality and full completion, the design construction contract management mode was conducted as a professional design management way led by AUBE, who has undertaken multiple responsibilities such as detailed scheme, construction drawing design, landscape design, etc., and the whole leisure belt professional design as well.

From the winning of bid in 2016 to the completion in 2019, it only took three years to build such a grand city landmark, which once again witnessed the Shenzhen speed and the Shenzhen quality. The Shenzhen World is not only a diverse, dynamic urban public space, but also a super project with distinct feature, which will be famous for its eight world records ever.

In order to commemorate this epoch-making project, AUBE once again took the lead in organizing a team for this publication, trying to reveal all the technical difficulties and innovative breakthroughs behind this world-class ultra-large exhibition complex.

This book has not only shown the various research results and experiences in construction, structure, water supply and drainage, heating ventilation air conditioning, strong and weak electricity, communications, transportation, green building, etc., but also the professional skills of landscape, aesthetics, operation, etc., which are all key technologies of the project, and will be a good reference for the technology improvement and innovation of similar large-scale public construction.

The preparation of this book has been guided by the Shenzhen Municipal Government, Bao'an District Government and other departments as well as the project leaders. On behalf of the whole Editorial Team, AUBE would like to express its sincere gratitude to all the supporting units and individuals.

Built from nothing, now looks like a flying dragon on the east bank of the Pearl River Estuary, The Shenzhen World embodies the efforts, wisdom and sweat of nearly 30,000 management and construction personnel for more than two years. Here with this publication, AUBE would also like to pay a tribute to all the participants and contributors of the Shenzhen World.

This book covers a wide range of areas and is rich in contents, any letters from readers for advices and discussions will be welcomed and appreciated.

序言

赋能未来·城市向新力

—— 打造全球最大会展中心

深圳国际会展中心，被誉为"全球最大会客厅"，位于粤港澳大湾区纵深地带，珠三角广深科技创新走廊，狮子洋与内伶仃洋交汇处的会展新城片区，是深圳市委市政府布局粤港澳大湾区的重要引擎。

2016 年 2 月，深圳市面向全球范围进行深圳国际会展中心设计竞赛，法国 VP+AUBE 欧博设计联合体的方案中标。在项目实施过程中，AUBE 欧博设计作为设计总牵头单位，协同 30 余家设计顾问单位，采取集成一体化的设计方法，对标国际领先的行业标准，攻克了如大跨度钢结构、建筑消防、机电设计等一系列技术难题，实现了深圳市委市政府提出"一流的设计、一流的建设、一流的运营"的会展建设目标。

在深圳市政府的鼎力支持以及代建、参建各方的辛勤努力下，深圳国际会展中心于 2019 年 9 月竣工。

"深圳距离世界级城市还差一个国际会展中心"的瓶颈被突破，深圳会展业将迎来"一城两馆"全新发展机遇。

作为深圳国际会展中心的设计总牵头单位，AUBE 欧博设计参与了深圳国际会展中心这座超级城市地标建设的全过程，并将见证这个全球最大会展中心承载一个新区诞生与发展的梦想，成为驱动粤港澳大湾区发展的新生力量。

城市力：从会展中心到会展新城

最近十年，会展经济概念在中国经历了快速成长，逐渐形成围绕北京、上海、广州为中心的热点板块，同时影响力向周边区域扩散。深圳国际会展中心本身与其他国际大型会展场馆有何不同？其对城市乃至周边区域的发展有什么特别的意义？

冯越强（AUBE 欧博设计董事长）：会展经济起源于 19 世纪中期，1851 年伦敦万国工业产品大博览会是历史上首次全球性的会展，向全世界展示了各国经济、科学、文化、交通运输等技术的进步成果，同时作为交流平台，极大地促进了贸易发展。高级别会展中心可以带动会展上下游和相关科创产业发展，助力本地经济腾飞。中国的会展产业近年虽然发展迅猛，出现了部分世界级展览和会议，如夏季达沃斯论坛、博鳌亚洲论坛。但是，如北京上海等城市仍尚未形成世界级的会展产业集群，换言之，中国目前真正意义上产城融合、展城一体的会展综合发展区十分匮乏。

深圳期待通过全球最大的国际会展中心造一座"会展新城"。未来，新展馆周边将逐步规划为"三城一港"——"三城"分别为国际会展城、海洋新城、会展田园城，"一港"为综合大空港区。

作为全球单体建筑中最大的会展中心，深圳国际会展中心从城市设计方案中就提出了"会展综合发展区"的概念，会展核心功能带、会展休闲带、西岸、东岸和南岸三个片区的区域规划结构，以会展为核心辐射周边，带动综合配套功能区域的发展。

大型国际性的展览和会议将直接带来高品质会展上下游产业的集集，其所吸引的高端人流将带动本地消费升级。同时创新型人才的聚落和大量信息流、技术流的汇聚，形成相关尖端科创产业集群，进而构建起完善的生态系统，形成一座"会展新城"。

从经济特区到"先行示范区"，深圳在新时期被赋予了新的使命。作为粤港澳大湾区的城市"超级地标"，深圳国际会展中心的面世将开启粤港澳大湾区会展经济新时代。

设计力：一座地标的诞生

作为地标性的城市公建项目，深圳国际会展中心如何与深圳市的城市规划形成互融互动？在整体的设计理念上有什么亮点？

杨光伟（AUBE 欧博设计董事、ARC 建筑创意中心主持人）：关于会展综合发展区的整体设计是 AUBE 欧博设计会展产品的核心设计内容，其内容涵盖会展片区的规划设计、城市设计、建筑设计以及景观设计等。

从 2002 年竣工的安徽国际会展中心、西南片区第一个获得 LEED 铂金奖项目的贵阳生态国际会议展览中心，到 2016 年作为全球最大单一会展工程深圳国际会展中心的设计总牵头单位，再到 2018 年绍兴国际会展中心的中标，充分展现了 AUBE 欧博设计在会展设计领域的核心竞争力。

2016 年的深圳国际会展中心从投标阶段开始，AUBE 欧博设计就尝试以会展促进片区经济、提升城市竞争力的、宏观层面的观念来审视设计，关注大型会展对城市的积极作用，通过有效的设计方式消除会展对片区带来的交通重负等不利影响，让会展更好地融入城市。

在城市设计层面，将会展中心沿西侧道路布局，将东侧沿河界面留给城市，作为会展中心各类外部配套。同时在会展中心与沿河地块之间创造性地引入一座高线公园，高度 8 米，公园下部布置了服务于会展中心的各类公交与市政设施，地下空间接驳轨道线路与地铁商业，并布置了近 4000 辆的社会停车位，为会展中心和周边用地服务。这座复合型的高线公园在合理解决了会展中心所需的各类交通配套、商业配套和市政服务设施的同时，大大降低了会展中心对周边地块日常运行的不利影响。

建筑设计层面，设计秉持高效、融合与弹性的三大原则。高效是指从展厅的功能出发，选择最高效清晰的货运与人流组织模式；融合是指标准展厅及配套设施的便捷关系，使每一个标准展厅能自成一体，功能完备，除了能满足单厅使用外，人性化的就近服务也体现了融合的原则；弹性是指每座展厅的灵活分割与多功能使用的兼容设计，旨在降低运营压力，更好地适应各类不同需求。上述三大原则使得深圳国际会展中心能够更好地满足展会运营的各类诉求。

使命力：团队凝聚的荣光

面对这样一个巨大且复杂的项目，AUBE 欧博深圳设计会展中心团队作为设计总牵头单位该如何整合各个专业，又要如何进行分工和协作？

林建军（AUBE 欧博设计董事总经理）：深圳国际会展中心项目本身最具价值的是其城市设计策略理念更好地平衡了会展与城市的关系，这其中包括综合交通组织与停车策略、建设周期与造价、片区景观资源的利用与梳理、开发策略与经济平衡等。

在深圳国际会展中心的前期投标阶段中，AUBE 欧博设计团队在设计联合体的工作中承担了片区城市设计、选址方案研究、会展中心周边配套用地的建筑概念设计和片区景观概念设计等工作，同时负责外聘并组织会展中心片区交通顾问、运营顾问、经济顾问、绿色建筑及海绵城市顾问、造价咨询等部分的工作组织与协调工作。深圳国际会展中心设计与建设过程中，涉及到各类政府部门 20 多个，设计总牵头单位需协调的各类专项设计单位 30 余个，清晰而严密的组织架构是深圳会展项目得以高效顺利推进的前提和保障。深圳市政府以市领导牵头成立的会展建设指挥部，为会展中心建设提供了强有力的支撑，进行了大量关键而细致的统筹协调工作。回溯整个前期设计工作，多专业协作、集成一体化的思维方式与工作方法是 AUBE 欧博设计在整个项目取得成功的关键。从城市设计的角度切入问题，广泛组织各类内外部团队介入，合力提出更科学合理的会展片区与会展中心建筑综合解决方案。面对此等规模尺度的巨大项目，AUBE 欧博设计要从全公司统筹调配资源，组成技术和管理能力兼备的优秀设计团队才能完成此艰巨的设计任务。团队构成分三个层级：决策层由项目总负责人、设计总负责人、大项目经理组成；核心层由各专业总工程师、总建筑师、各功能分区的副设计总负责人，以及运营中心和项目助理组成；执行层由各功能分区的多专业项目组成员构成。

在该项目的管理中，对海量项目信息的有效梳理与协调指令精确下达是最为关键的环节。AUBE 欧博设计以项目总负责人牵头、项目经理团队与各设计总负责人形成了核心设计管理层，该管理层对于项目质量控制、计划协调、信息管理进行了全方位地管理把控，确保整个设计总牵头单位队伍集团作战方向的一致性和快速反应机制的建立。

创造力：实践最佳提案

AUBE 欧博设计一直秉承"国际化经验，地域化实践，设计提升价值"的设计理念和"集成一体化"的设计方法。该原则在深圳国际会展中心项目中是如何体现并付诸实践的？

丁荣（AUBE 欧博设计董事、总建筑师）：欧博拥有许多来自世界不同国家的设计师，多元的文化背景和对设计的不同视角相互碰撞交融，为同一项目提供了更为丰富多样的解决思路和方法。

深圳国际会展中心参考了欧洲、美国等会展发达地区和国家的会展建设标准和国际通行的常规做法，结合场地条件、国内运营特点和操作方式，经过充分调研、考察、论证，力求创造符合深圳使用特性且国际一流的会展中心。

欧博在设计不同阶段遇到复杂技术问题时，不是追求单专业最佳解决方式，而是强调通过专业间整合，在满足本专业特性要求、发挥本专业优势的同时，为其他专业创造发挥的条件，寻求多专业综合最优解决方案。如项目设计初期，为配合地下室先行施工，地下空间方案的合理性决定了施工组织、施工周期及其工程造价。AUBE 欧博设计团队在对标国内外多个会展项目之后，发现地下综合管廊并不适用本项目的场地条件和气候条件，于是提出了地上综合管廊的解决方案。此方案不仅为项目节省工期，更利于后期运营维护，为项目减少了挖方 21 万立方米，和 3 万平方米地下管廊的混凝土工程量。

在项目运营团队 SMG 公司进入该项目后，新的运营标准带来新的空间需求。AUBE 欧博设计团队通过：增加展厅与展厅之间会议空间（约 7000 ㎡）；增加国际报告厅席至 1920 座；改善餐饮空间水准和增设地下中央厨房等措施以提升会展会议和餐饮服务水准。尤其是在北区 17 号、18 号和 20 号展厅全部植入了会议、餐饮或体育赛事等多功能。并科学地创造了超大空间的消防疏散和施救条件，运用水幕和可开启屋面对超出规范规模的展厅进行有效的防火分隔，保证消防设计不留死角，并使得展厅展位使用率接近 50%，确保了北区展厅空间的功能转换，以及使用的合规性和合理性，最终，在已基本稳定的规划结构条件下，实现了运营团队的各项要求和运营目标。

专业力：成就超级工程

作为功能性很强的超大型综合体，深圳国际会展中心的景观休闲空间是如何规划设计的？

祝捷（AUBE 欧博设计董事、总景观师）：深圳国际会展中心的景观设计作为城市空间形态和建筑空间形态的交织体，在营造功能合理、空间怡人的物理环境的同时，需要平衡超大尺度与人性化体验、大面积硬化与生态性，以及功能性与休闲性之间的矛盾，并结合周边环境、建筑、场地功能及使用人群进行综合考量，打造专业、高效、复合、人性、可持续的会展景观。

设计层面，深圳会展户外空间，有极强的专业功能性和交通需求，从布展、参展、运营管理等各方面都进行了深入地剖析。未来的会展已经不仅仅是单一的展览和会议模式，因此景观要有很大的弹性，通过各类专项的研究、流线的分析、各种可使用的模拟，以适应各种需求的变化。在此基础上，进一步思考如何在大尺度开放空间里，营造"近人尺度"的宜人体验，通过大空间节奏化、界面模糊化、设施人性化、引入自然，将公园融入会展等手段，构建一个可持续的生态和有人情味的会展景观。实施层面，项目虽然工期非常紧张，景观也基本是最后一步施工，现场的很多问题往往在最后需要景观综合处理解决，但欧博在集成一体化的工作模式下，景观在建筑设计初期便已介入，各专业的密切沟通协调，让很多问题在设计初期便得到了解决，并给景观预留了充分的条件，以此才能整体呈现出一个相对完好的作品。

对于这样一个超大尺度的项目，如何在从确保结构的安全性基础上赋予建筑美感？

黄用军（AUBE 欧博设计董事、总工程师）：深圳国际会展中心的整体设计风格以尊重功能使用为出发点，拒绝夸张造型。钢结构的表现力在会展建筑中是不可或缺的，几乎成为了建筑美的"主角"。结构处处展现出张力之美，韵律之美。

本项目主体采用钢框架结构 - 屋盖为钢三角桁架及空间网格结构，利用钢结构的施工速度快、精度高、工程机械化程度高、自动化程度高等优点，缩短了建造周期。采用的铸钢节点都是整体浇铸成形，并进行了精整处理，实现了节点光滑连贯的最佳建筑效果。
其中，标准展厅屋盖结构采用完全拟合建筑外形的空间三角桁架，99m 跨度的主桁架跨中矢高 7m，桁架间距 18m，展厅间货运通道跨度 42m，其屋面采用由展厅延伸而出的变截面空间三角桁架，采用连续的空间三角桁架可顺利的实现展厅屋盖悬挑，满足屋盖挑檐的要求，并将两个展厅与其间的货运通道完整地连为一体。这样由连续屋盖而形成的连续桁架，其受力合理、形态美观，实现了屋盖形体与结构形式的完美统一，赋予了建筑结构美感。
17 号超大展厅是全球规模最大，配套最齐全，科技含量最高的展厅。展厅屋盖平面投影尺寸为 224mX257m，为减小其结构跨度，同时兼顾展厅使用的要求，在展厅中央增设了 4X4 共 16 个承重柱。建筑结构一体化协调及精细化设计，成就了深圳国际会展中心这座地标性建筑。

深圳国际会展中心是目前唯一通过三认证体系（绿色二星，LEED，BREEAM）的会展中心。欧博在设计过程中运营了哪些保障低碳节能的专业技术？

黄煜（AUBE 欧博设计机电设计中心总经理）： "绿色会展和智慧会展"是全球最大单体会展建筑：深圳国际会展中心建设的重要目标，项目因地制宜地应用了 52 项全球领先的绿色建筑技术和智慧会展技术，全力打造规模宏大、节能环保的一流绿色智慧展馆和全球首个通过三个绿色认证的展馆。

绿色建筑技术应用，智能化控制管理技术运用，供水供电现场服务及时性的保障，在本项目中都得到了合理科学的采纳。AUBE 欧博设计团队通过节地、节能、节水、节材、保护环境五方面实现了绿色建筑技术在深圳国际会展中心中的应用。欧博机电团队建模的 BIM 模型，为设计各阶段各专业配合解决了提供方案（如展厅空调机排烟管的方案等）、方案验证（如综合管廊和管沟管线布置等）、三维可视化（即所见所得）、错漏碰缺等问题，对机电方案、机房、线路设计以及建筑结构等起到很大作用。为实现节能环保绿色会展目标，深圳国际会展中心的展厅还为吊装布展方式预留了大量的吊点，有效地保证了展位布展的重复使用，大量降低了展装垃圾的产生。同时，设计团队还对有限的展装垃圾和生活垃圾的运输流线进行了科学规划，垃圾就近分类压缩处理，并与周边垃圾填埋物和焚烧厂资源整合。此外，防线设计、票证管理系统、视频监控技术、人车安检系统、人车导航系统、门禁系统、面相识别技术、人数控制技术、消防控制系统、通讯保障、公安安全指挥系统、城市联防联动体系、应急保障系统、对外信息发布管理中心以及会展大数据运用和手机会展 APP 等等，会展的智能化设计是深圳国际会展中心国际领先一流定位最重要的体现。

未来可期：与城市共生长

AUBE 欧博对于深圳国际会展中心的未来有着怎样的期待？

冯越强（AUBE 欧博设计董事长）： 深圳国际会展中心是一个伟大的工程，也是 AUBE 欧博设计以集成一体化方式在会展综合发展区领域建成的当之无愧的明星作品。深圳国际会展中心的建成与运营，会对整个地区、深圳乃至粤港澳地区产生非常大的影响，因为它不单纯只是一个尺度超规模的或者是一个巨大的建筑物，它的影响范围是多重的，甚至会叠加无数的可能性。

会展本就是城市的组成部分，除去会展的专业性、安全性以及科学性，这个超级地标在建成后，将会更关注其与周围的城市居民的关系，这是 AUBE 欧博设计从最初竞标中标到后来的深化研究一直在探索和追求的。未来，我们希望这座地标性建筑可以成为城市友好街区的一部分。

Preface
Shenzhen World: The Urban Innovation

The Shenzhen World is known as "the world's largest meeting room", it is located in the core area of the GBA, the Guangzhou-Shenzhen technology innovation corridor of the pearl river delta, and the exhibition new city area at the intersection of the Lion ocean and Neilingding river. It is an important engine for the Shenzhen municipal government to planning the layout of the GBA.

In February 2016, the VP+AUBE AUBE design consortium won the bidding in the global design competition of the Shenzhen World. In the process of the implementation, AUBE, as the Design EPC, has led and cooperated with more than 39 design consulting companies. By an integrated design method, AUBE overcame Several technical problems such as large-span steel structure, building fire protection, mechanical and electrical design and eventually achieved the goal of "first-class design, first-class construction, first-class operation" proposed by the Shenzhen government. With the full support of Shenzhen municipal government and the hard work of all parties involved in the construction, Shenzhen World was completed in Sep. 2019, which means the period of " Shenzhen is lack of an international exhibition center to be the world-class city" came to an end, and Shenzhen is entering a new era with the development opportunity of "one city with two exhibition center".

As the general contractor of Shenzhen World, AUBE has participated the whole process of the development of Such a great city landmark, and will witness this world largest exhibition & convention center carrying the dream of the birth and development of a new district, which would become a new innovation force, to left the whole city to a new level. Years on, even some world-class exhibitions and conferences have emerged, such as the Summer Davos and the Boao Forum for Asia. However, cities like Beijing and Shanghai Still do not have world-class exhibition industrial clusters. In other words, there lack of comprehensive exhibition development areas that truly integrate the exhibitions with the cities in China.

Shenzhen is looking forward to establish an "exhibition city" through such a world largest international exhibition & convention center. In the future, the surrounding area of the new exhibition center will be gradually planned as "three cities and one port" -- the "three cities" are the international exhibition city, the ocean new city and the exhibition garden city, While the "one port" indicates the comprehensive port area.
The design of such a world largest exhibition & convention center put forward the concept of "comprehensive development area", which means a planning area of exhibition core function belt including exhibition leisure belt, west, east and south bank, taking the exhibition center as its core area, and promoting the development of the comprehensive function area.
Large international exhibitions and conferences would directly gather high-quality upstream and downstream industries and attract top talents to promote local consumption upgrading. Furthermore, innovative talents would gather a large information and technology flow to form a related cutting-edge science and innovation industry cluster, and thus build an "ecosystem " to form a "new exhibition city".
From special economic zones to "pilot demonstration zones", Shenzhen has been given a new mission in the new era. As a "super landmark" in GBA, the completion of Shenzhen World has marked the new start of GBA exhibition economy.

City-driven: From Exhibition Center to Exhibition New City

In the past decade, the exhibition economy has experienced rapid growth in China. Beijing, Shanghai and Guangzhou have become the major city for exhibition and begin to influence the surrounding areas gradually. As the current largest exhibition center in China as well as the world, what are the differences between the Shenzhen World and other large international exhibition centers? What particular significance it will bring to Shenzhen and surrounding areas?

Feng Yueqiang (President of AUBE): The exhibition economy originated in the middle 19th century. The first global exhibition in the history is the Great Exhibition of the Works of Industry of all Nations in London 1851, within which many countries has shown their progress and achievements of economy, science, culture, and transportation technology, and which also promoted the trade development greatly. The high-level exhibition center can promote the development of the upstream and downstream industries , and therefore other related science and innovation industries and help the local economy to rise rapidly. Although the exhibition industry in China has developed rapidly in recent years, even some world-class exhibitions and conferences were held , such as the Summer Davos and the Boao Forum for Asia. However, cities like Beijing and Shanghai do not have formed world-class exhibition industrial clusters. In other words, there is a lack of comprehensive exhibition development areas that truly integrate the exhibitions with the cities of China.

Shenzhen is looking forward to establish an "exhibition city" through such a world largest international exhibition & convention center. In the future, the surrounding area of the new exhibition center will be gradually planned as "three cities and one port" -- the "three cities" are the international exhibition city, the ocean new city and the exhibition garden city, while the "one port" indicates the comprehensive port area.
The design of such a world largest exhibition & convention center put forward the concept of "comprehensive development area", which means a planning area of exhibition core function belt including exhibition leisure belt, west, east and south bank, taking the exhibition center as its core area, and promoting the development of the comprehensive function area.
Large international exhibitions and conferences would directly gather high-quality upstream and downstream industries and attract top talents to promote local consumption upgrading. Furthermore, innovative talents would gather a large information and technology flow to form a related cutting-edge science and innovation industry cluster, and thus build an "ecosystem" to form a "new exhibition city".
From special economic zones to "pilot demonstration zones", Shenzhen has been given a new mission in the new era. As a "super landmark" in GBA, the completion of Shenzhen World has marked the new start of GBA exhibition economy.

Design: The Birth of a Landmark

As a landmark of urban public construction project, how would the Shenzhen World interact with the Shenzhen urban planning? What are the highlights of the overall design concept?

Yang Guangwei (Director of AUBE, chief director of architectural creativity center): the overall design of the comprehensive exhibition development area is the core design content of AUBE for exhibition projects, which includes planning, urban design, architectural design and landscape design of the exhibition area.

Completing the Anhui international convention and exhibition center in 2002, winning the first LEED platinum award in the southwest area for Guiyang ecological international convention and exhibition center, becoming the general contractor of the world's largest exhibition center ever - the Shenzhen World in 2016, and winning the bidding of the Shaoxing international conference and exhibition center in 2018, AUBE has shown the core competitiveness in the field of exhibition & convention center design.

From the bidding stage, AUBE has already tried to look at the overall design from a macro perspective , to promote area economy and strengthen the urban competitiveness and focus on the positive influence of exhibition on the city, and through effective design, the AUBE also tried to eliminate bad influence which the exhibition could bring to the citys such as traffic burden, while ensure the exhibition center fits properly in the city.

For the urban design, the exhibition & convention center is located on the west side of the road, and the east side which along the river, is for the urban supporting facilities. Between the exhibition center and the area along the river, there will be a high line park with the height of 8 meters, under which all kinds of public transportation and municipal infrastructures are arranged. This underground space is also connected to the rail line , the underground business, and a parking lot for nearly 4000 vehicles. This complex high line park serves all kinds of transportation facilities, commercial facilities and municipal service facilities according to the need of the exhibition & convention center, while greatly reducing the possible negative impact of the exhibition & convention center on the surrounding area.

The architectural design adopts three principles : efficiency, integration and flexibility. Efficiency means the most efficient and clear organization of freight and people according to the function of the exhibition hall. Integration refers to the convenient connection between the standard exhibition hall and supporting facilities, so that each standard exhibition hall can be self-contained with complete functions. moreover, the nearby service also embodies this principle. Flexibility refers to the flexible partition of each exhibition hall and the compatible design of multi functional use, aiming at reducing the operating pressure and better adaption for different needs. These three principles guarantee the Shenzhen World to better meet the needs of all kinds of exhibition operations.

Mission:The Glory of Teamwork

For such a huge and complicated project, How AUBE, the general contractor, lead and integrate various specialties and how to distribute the work and cooperate?

Lin Jianjun (Managing Director of AUBE) : The greatest value of the Shenzhen World is how the urban design strategy concept balance the relationship between exhibitions and the city, which includes the comprehensive traffic organization and parking strategy, construction period and cost, area landscape resource use and arranging, development strategy and the economic balance and so on.

In the early bidding stage, AUBE design team ,as a part of the design consortium, had undertake the responsibility of urban design of the exhibition area, the site selection study, the conceptual design of the building in the surrounding area of the exhibition & convention center, and the conceptual design of landscape. At the same time, AUBE is also responsible for selecting and organizing consulting expertise of other territorial such as the traffic, operational, economic consultancy, the green buildings and sponge city and cost consultation.

During the design and construction process, there are more than 20 government departments involved, and more than 30 special design units need to be coordinated for the overall design contract. A clear and tight organizational structure is the premise and guarantee for the efficient and smooth progress of this project. The headquarters of exhibition construction led by the Shenzhen municipal government has provided a strong support for the construction of the exhibition & convention center and carried out a large number of key and meticulous coordinations.

Looking back on the whole preliminary design work, the thinking mode and working method of multi-professional cooperation and integration are the key to the success of the whole project. From the perspective of urban design, various internal and external teams are extensively organized to intervene and jointly propose a more scientific and reasonable comprehensive solution for exhibition area and exhibition center building.

For such a huge project, AUBE has managed to accomplish this arduous design task by coordinating and allocating resources from the whole company and thus forming an excellent design team with both technologies and capabilities of management. The team consists of three layers: the "decision-making layer" consists of the project leader, the design leader and the project manager; The "core layer" includes chief engineer, chief architect, deputy chief designer of each functional area, operation center and project assistants; The "executive layer" is composed of multi-professional project team members of each functional area.

In the management of this project, the most critical section is to effectively sort out the massive project information and to issue the coordination instructions accurately. AUBE has formed a core design management team led by the project leader, this team consists of a project management team and various design leaders, whom are responsible and have full control for the project quality control, project coordination and information management, so as to guarantee the group consistency of design direction and a rapid response mechanism.

Creativity: Best Solutions

AUBE has been adhering to the design concept of "international experience, regional practice, design value" and the design method of "integration". How is this principle embodied and put into practice in the project of the Shenzhen World ?

Ding Rong (Director of AUBE, Executive Chief Architect) : In AUBE, there are many designers from different countries around the world. Diverse cultural backgrounds and different perspectives collide and blend with each other, providing more diversified solutions and methods for the project.

The Shenzhen World has referred to the construction standards of Europe and the United States, as well as the international prevailing standard practice. By combining with the field conditions, characteristics and operation mode in China, and after comprehensive research, investigation and argumentation, we strived to create a world-class exhibition & convention center that fits Shenzhen itself.

When faced with complex technical problems in different stages of design, we did not just pursue the best solution for one specific problem, but also emphasized the integration among different specialties to create more possibilities and conditions for other specialties while meeting the characteristic requirements of the specialty and giving play to the advantages of the specialty, so as to seek the optimal solution for multi-specialty integration.

For example, in the early stage of the project design, in order to comply with the construction of the basement, the rationality of the underground space plan determines the construction organization, construction period and project cost. After bidding for a number of exhibition projects at home and abroad, the AUBE design team found that the underground comprehensive pipe corridor was not suitable for the site and the climatic conditions of the project, so we proposed the solution of an aboveground comprehensive pipe corridor. This solution not only saves time, but also facilitates the later operation and maintenance, and reduces the work load of 210,000 cubic meters of concrete excavation and 30,000 square meters of underground pipe corridors.

After the project operation team SMG entered the project, the new operation standard brought new space requirements. AUBE team thus increased the meeting room space between exhibition halls (about 7000 ㎡); increased the number of seats in the international lecture hall to 1,920; improved the catering space as well as the conference and catering services by adding underground central kitchen and other measures. In particular, NO. 17, NO. 18 and NO. 20 exhibition halls in the north district are all equipped with multi-functional facilities such as meeting, catering or sports events. In addition, the AUBE create large space for fire evacuation and rescue, and uses water curtain and open roof for the ultra-large exhibition halls for effective fire compartmentation, make sure there is no dead angle of fire control design, and make the utilization rate of exhibition hall rate nearly 50%, to guarantee the compliance and rationality of function transformation in the north exhibition halls. Finally, under the condition of the stable planning structure, we achieved the operation goal of SMG.

Professionalism: Super Engineering

For such a super-large complex with powerful professional functions, how do you plan and design its leisure space landscape?

Zhu Jie (Director of AUBE, Executive Chief Landscape Architects): The landscape design of the Shenzhen World is a mixed object of city space form and architectural space form. When creating a pleasant physical environment with reasonable functions, it needs to balance the ultra-scale and human experiences, stiffness and ecosystem, function and leisure. With a comprehensive consideration for the surrounding environment, building & site function and people who use it, we created a professional, efficient, and sustainable landscape with complex function and humanity.

The outdoor space of the Shenzhen World demands professional functions and transportation, which need to be deeply analyzed from the aspects of arrangement, participation, operation and management of exhibitions. The exhibition & convention center In the future is not just for exhibitions and conferences, therefore the landscape should be more flexible. Through various special studies, streamline analysis, and use of various possible simulations, the landscape was designed to adapt the changes of various needs.

On this basis, we think further about how to create a pleasant experience within approachable scale in such a large open space. With large space rhythm, fuzzy interface, humanized facilities, nature elements, we managed to integrate the park into the exhibition & convention center area, and thus to construct a sustainable ecological and humanistic landscape.

For the implementation of the whole process, due to a tight schedule, and the landscape is the last step of the construction and therefore there were many problems left about the landscape to be solved comprehensively. However, because of the integration work mode created by AUBE, the landscape work has engaged at the beginning of the architectural design, and with the close communication and coordination of each professional team, many problems has been resolved at the beginning, and thus provided sufficient conditions for landscape, and guaratee the display of a rather compelete result.

How to maintain the beauty of the building while guarantee the structure safty of such a large project?

Huang Yongjun (Director of AUBE, Chief Engineer): The design principle of the Shenzhen World is "function first" and "avoiding exaggeration." The steel structure is indispensable in exhibition architecture, almost the "leading role" of architectural beauty. Therefore, the beauty of tension and rhythm lies in the Shenzhen World's structure.

This project adopts steel frame structure, steel triangle truss roof and space grid structure, which have the advantages of fast construction, high precision, high mechanization and high automation which could shortened the construction period. All the steel joints are cast onsite into one piece and futher processed to achieve the best result of smooth and continuous construction appearancel.

Among them, the roof structure of the standard exhibition hall uses a spatial triangle truss that fits the facade of the building, a 99m-span main truss with a 7m midspan and a 18m truss spacing. The freight channel of the exhibition hall has the span of 42m, and its roofing uses a variable cross-section triangle truss that extends from the exhibition hall. And the continuous spatial triangle truss provides a cantilevered roof which meets the requirements of roof overhangs, and connects two exhibition halls with the freight channel in between as a whole. This continuous truss formed by the continuous roof is both beautiful and rational, guaratees a perfect unification of the roof shape and the structure, and endows the whole architectural structure with a sense of aesthetic.

No. 17th super large exhibition hall is the world's largest exhibition hall with the most complete equipment and highest technology. The projection size of the roof is 224mX257m. In order to reduce the span of the structure and meet the requirements of the exhibition hall, 16 load-bearing columns of 4X4 were added in the center of the exhibition hall. With integrated structure and refined design, we have accomplished this landmark building.

The Shenzhen World is currently the only exhibition & convention center with the certification of three Green Buliding evaluation systems (green two-star, LEED, BREEAM). What are the low-carbon and energy-saving technologies that AUBE has applied in this project?

Huang Yu (General Manager of AUBE Mechanical and Electrical Design): Green and smart exhibition center are important objectives of the construction of The Shenzhen World. It has applied 52 world's leading green building technologies and smart exhibition technologies, strive to build as a world-class green exhibition & convention center with large scale, energy efficiency and environmental protection, as well as the first exhibition center to acquire three green building certifications.

The application of green building technology, intelligent control and management technology, the guarantee of water supply and power supply on site and in time, have all been reasonably and scientifically adopted in this project. AUBE design team has accomplished the application of green building technology in the Shenzhen World from the following five aspects: land saving, energy saving, water saving, material saving and environment protection.

In different stages of the design, AUBE mechanical and electrical team's BIM model has provided various specialties with solutions(such as exhibition hall air-conditioning exhaust pipe solution, etc.), program verification (such as the layout of comprehensive pipe corridor and the pipe trench and pipeline, etc.), 3D visualization, etc.

In order to achieve the green exhibition goal of energy-saving and environmental protection, the exhibition hall of the Shenzhen World also reserved a large number of suspension points for exhibition, effectively guarantee the repeated use of booth layout and greatly reduce exhibition garbage. At the same time, the design team also made a scientific plan for the transportation lines of exhibition garbage and domestic waste, which can be sorted and compressed nearby, and treated by the surrounding landfill and incinerator.

In addition, the smart design of the Shenzhen World includes the line design, ticket management system, video monitoring technology, security system, car navigation system, entrance guard system, facial recognition technology, public control technology, fire control system, communications security, public security command system, city co-prevention and co-management system, emergency guarantee system, external information release and management center, big data use and exhibition center APP on the mobile phone, etc., all of these can make the Shenzhen World a first-class smart exhibition & convention center.

Future: Evolue with the City

What is AUBE's expectation for the future of the Shenzhen World?

Feng Yueqiang (President of AUBE): The Shenzhen World is a great project, and also a well-deserved star work built by AUBE with an integrated method within the exhibition comprehensive development area. Its completion and operation will have a great impact on the whole area, on Shenzhen and on the GBA. It is not just a super scale or a huge building, it has multiple marginal influence and countless possibilities.

Exhibition is a part of a city. Apart from the professionalism, security and scientificity of exhibition, we pay more attention to the relationship between this super landmark and its surrounding city residents. This is what AUBE has been exploring and pursuing since we won the bidding and deepened its research later. In the future, we hope that this landmark could be part of a city-friendly neighborhood.

"启"源

The Origin

深圳国际会展中心诞生背景　The Circumstances of the Shenzhen World's Birth

26 一个时代的宏图 —— 规划背景
The Big Blueprint — Planning Background

33 一座新城的起点 ——大空港区域规划理念
Visions of The Future — Planning Concept of The Airport Area

一个时代的宏图

—— 规划背景

The Big Blueprint — Planning Backgroun

粤港澳大湾区是海上丝绸之路的起点之一，以其特殊的历史文化、区位
势背景，承担着我们国家"一带一路"扩大对外开放的战略意图，是新
政策形势下，中国经济进一步与世界互联的战略枢纽之一。2017 年，
家政府工作报告提出"制定粤港澳大湾区城市群发展规划"，至此，粤港
大湾区的建设已正式上升为国家战略。

在此背景下，深圳国际会展中心所在的空港新城作为深圳大空港核心地区
地处粤港澳大湾区和广深港经济带核心位置，是深圳未来经济和城市发
的重点区域，对深圳实现有质量、内涵式的发展，提升城市功能、经济
能和长远竞争力具有重要意义，未来更将成为粤港澳大湾区引领发展、
务全国、连通世界的重要经济发展制高点。

The GBA is one of the starting points of the Maritime Silk Roa
With its special historical, cultural and geographical advantage
it bears the Chinese strategic intention of "The Belt and Road"
expand its opening to the outside world. It is one of the strateg
hubs for China's economy to further connect with the world und
the new policy. In 2017, the national government work report p
forward a proposal - "formulate a plan for the development
urban agglomeration in the GBA", and therefore, the constructi
of GBA has been officially promoted as a national strategy.
In this context, the Shenzhen World Convention and Exhibiti
Center's location — the Airport New Town, which is the co
part of the Shenzhen Airport Area, is located in the GBA and t
central position of Guangzhou-Shenzhen-Hongkong econom
belt,which is the key area for Shenzhen's future economy a
urban development that will have great significance in high-qual
and profound development, the enhancement of urban functio
economic functions and long-term competitiveness of Shenzhe
In the future, it will also be an important economic highland
leading the development of the GBA, serving the whole count
and connecting to the world.

区域背景：天时地利，这是最好的时代
Regional Background: The Best of Times

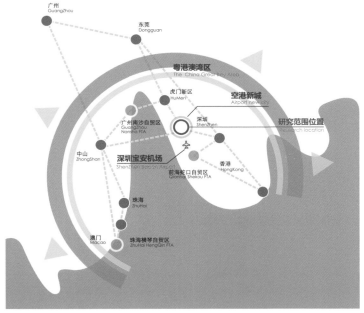

图 1-1　项目在粤港澳大湾区的位置
Figure 1-1　Location of the Project in the GBA

The Shenzhen World is located in the Airport New Town, which is one of the 17 major development areas in Shenzhen

As the development of the Pearl River Delta area enters the construction stage of the Airport New Town in the GBA, Guangzhou Nansha New District, Shenzhen Qianhai - Airport Area, Zhuhai Hengqin New Area, Dongguan Changan New Area will all focus on the development of this world-class bay area around the Pearl River Estuary, where the core functions such as shipping, trade, finance, scientific and technological innovation and international culture, etc, will gather together and generate enormous influences.

In this context, Shenzhen Urban Planning and Land Resources Committee, together with Baoan District Government, drew up the Comprehensive Development Plan of Shenzhen Airport Area, indicating the direction of the development of this area. According to the Airport Area's strategic positioning as the economic central region of the GBA, the world-leading airport metropolitan area, the Airport North - International Exhibition Center Area, where the Shenzhen World Exhibition & Convention Center situated, will focus on the constructions of two Core Projects: the Airport North transport hub and the Shenzhen World Exhibition & Convention Center, and thus promote the developments of convention and exhibition services, e-commerce, creative designs, as well as related logistics, tourism, design and other industries.

深圳国际会展中心落户空港新城，深圳 17 大重点发展区域之一

随着珠三角地区发展进入湾区新城建设的阶段，广州南沙新区、深圳前海 - 大空港地区、珠海横琴新区、东莞长安新区等都将发展重点瞄准环珠江口这一世界级湾区。湾区的核心功能将主要体现在航运、贸易、金融、科技创新、国际文化等集聚辐射功能。

2013 年底，深圳市委第五届十八次全会上明确提出发展 "湾区经济"。围绕发展湾区经济，充分发挥深圳区位、产业等方面的优势，打造高质量的滨海产业集群，同时增强深圳湾区经济的国际影响力；抓住深圳机场扩建和港口转型升级的机遇，开辟更多国际航运、航空航线，增强海港、空港辐射能力。

《深圳市大空港地区综合发展规划》在此背景下由市规划国土委及宝安区政府牵头编制，为片区空间发展指明了方向。根据大空港地区建设粤港澳大湾区经济核心区、国际一流空港都市区的战略定位，深圳国际会展中心所在的机场北 - 国际会展区，将重点围绕机场北交通枢纽、深圳国际会展中心核心项目建设，促进会展服务、电子商务、创意设计以及相关展会物流、展会旅游、展会设计等产业发展。在截流河以西、凤塘大道以南、沿江高速公路以东规划建设近期室内展览面积 30 万平方米大型展馆，远期预留展览面积 20 万平方米，与会展场馆一体化配套建设各类星级及商务酒店、餐饮服务、商务办公等功能设施。在机场北规划建设 T4 航站楼，与机场北轨道交通站组合形成综合交通枢纽（图 1-1）。

深圳国际会展中心作为空港新城启动区建设的有力抓手，将与南侧的机场北枢纽综合交通枢纽及周边道路交通、给排水等市政基础设施，穗莞深城际轨道、片区多条城市轨道线、外环高速等交通基础设施同步规划建设，建成后将拉开空港都市区发展框架（图 1-2）。

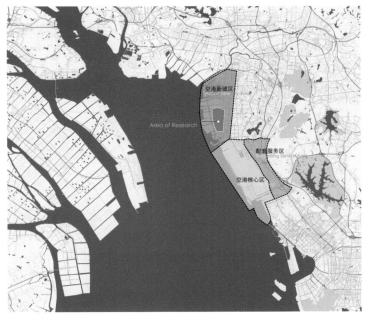

图 1-2　项目在空港新城的位置
Figure 1-2　Location of the Project in the Airport New Town

生态活力海岸带，深圳滨海城市未来新名片

2014年，宝安区开展了"宝安西部活力海岸带概念城市设计国际咨询"，着手规划和建设宝安区西部的海岸带，强力推进滨海城市建设。

空港新城片区位于西部海岸带，将依托以茅洲河口指状生态湿地为核心的自然生态资源，和以机场为依托的会展中心及其相关功能的落位与布局（图1-3），与机场片区及宝安中心片区形成合力、相得益彰，共同串联成深圳西部生产、生活、生态深度融合的亲海休闲空间，充分展现滨海印象、滨海特色，打造滨海内涵、滨海品牌，成为深圳滨海城市的又一张亮丽名片（图1-4）。

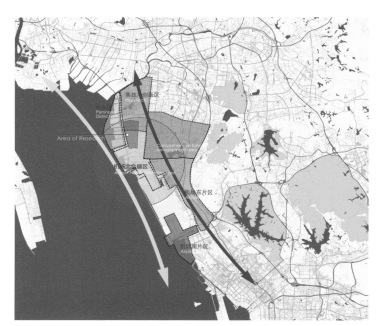

图1-3 项目在空港片区功能结构中的位置
Figure 1-3 The Location of the Project in the Functional Distribution of the Airport New Town

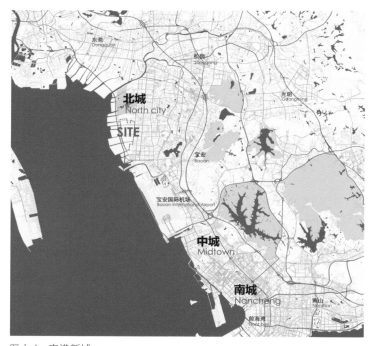

图1-4 空港新城
Figure 1-4 The Airport New Town

Ecological Vibrant Coastal Zone, the future name card of Shenzhen coastal city

With the development of the pearl river delta entering the stage of the construction of the new bay area, the Guangzhou Nansha new area, Shenzhen Qianhai-Big Airport Area, Zhuhai Hengqin new area and Dongguan Changan new area will all focus on the world-class bay area around the pearl river estuary. The core functions of the bay area are mainly the functions of shipping, trade, finance, scientific and technological innovation, international culture and so on.

At the end of 2013, in the 18th plenary session of the fifth Shenzhen Municipal Committee, it was clearly proposed to develop the "bay area economy". Focusing on the bay area economy, to give full play to the advantages of location and industry, build a high-quality coastal industrial cluster, and enhance the international influence of Shenzhen bay area economy; Seize the opportunities brought by the expansion of Shenzhen airport and the port transformation and upgrading, open up more international shipping and air routes, and enhance the radiation capacity of the port and airport.

图 1-5　空港北城　摘自中国城市规划设计研究院深圳分院：《深圳国际会展中心（一期）配套商业用地详细蓝图》
Figure 1-5　The north district of the Airport New Town - From the Supporting Commercial Facilities Blueprint of the Shenzhen World Exhibition and Convention Center (Phase I)
by the China Academy of Urban Planning & Design, Shenzhen

图 1-6　空港中城　摘自中国城市规划设计研究院深圳分院：《深圳国际会展中心（一期）配套商业用地详细蓝图》
Figure 1-6　The middle district of the Airport New Town - From the Supporting Commercial Facilities Blueprint of the Shenzhen World Exhibition and Convention Center (Phase I)
by the China Academy of Urban Planning & Design, Shenzhen

图 1-7　空港南城　摘自中国城市规划设计研究院深圳分院：《深圳国际会展中心（一期）配套商业用地详细蓝图》
Figure 1-7　The south district of the Airport New Town - From the Supporting Commercial Facilities Blueprint of the ShenzhenWorld Exhibition and Convention Center (Phase I)
by the China Academy of Urban Planning & Design, Shenzhen

区位优势：空港新城资源禀赋，得天独厚
Regional Advantage: The Airport New Town is Endowed with Unique Resources

核心区域位置，湾区战略支点

深圳国际会展中心所在的空港新城位于深圳宝安机场以北，距深圳宝安国际机场 T3 航站楼 7 公里，距规划 T4 枢纽 3 公里，是粤港澳大湾区、穗莞深港经济走廊的核心部位，珠三角的地理中心和广东自贸区的中心（图 1-8）。

Core Area: the strategic pivot of the GBA

The Shenzhen World is located in the Airport New Town, which is in the north side of Bao'an airport, 7 km away from the Terminal 3, 3 km to the Terminal 4 under planning. It is within the core area of the GBA and Guangzhou-Dongguan-Shenzhen-Hongkong Economic corridor, as well as the geographic center of Pearl River Delta and the center of Guangdong Free Trade Zone (Figure 1-8).

图 1-8　项目在空港新城的区位
Figure 1-8　The location of the project in the Airport New Town

交通便捷，快速融入湾区、抵达空港枢纽

从交通条件来看，深茂铁路（规划）以及穗莞深城际线分别从研究范围东西两侧经过，使片区可快速对接广州、东莞、香港，融入湾区；规划地铁 20 号线、12 号线均在范围内设有站点，其中 20 号线纳入会展一期工程同步开展，向北可衔接东莞市轨道 R2 线，向南可达空港枢纽，进而向东达福田，12 号线向南经福永片区与 20 号线于空港枢纽交汇，进而向南抵达前海中心；同时，沿江高速公路于研究范围西侧通过，并设有 2 个高速公路出入口，其中福洲大道福海出入口已建成通车（图 1-9）。

图 1-9　研究范围及周边交通条件分析
Figure 1-9　P1 The Analysis of the research area and its surrounding traffic conditions

Convenient Transportation, quick entry to the GBA, directly to the Airport Hub.

According to the transportation conditions, Shenmao railway (under planning) and Guangzhou-Dongguan-Shenzhen intercity line will pass through the east and west sides of this area respectively, which makes it connect to Guangzhou, Dongguang, Hong Kong faster and incorporate into the GBA.

The under planning metro line 20, line12 will also set up stations here; line 20 will be built synchronously as the Shenzhen World phase I is in progress, which connects to line R2 of Dongguan in the north, the Airport Hub in the south, and the center of Futian in the east; line 12 will pass Fuyong area to the south, and meet line 20 in the Airport Hub, further towards south to the center of Qianhai.

Meantime, expressway along the river will pass through the west side of the area, setting up two expressway entrances, of which the Fuyong entrance in the Fuzhou avenue has been completed and opened to traffic.(Figure 1-9)

发展机遇：深圳会展经济进入快车道
Development opportunities: Shenzhen exhibition economy enters the fast track

会展经济作为一种新兴经济形态，拥有巨大的潜能与广阔的发展空间。

1999 年，首届"中国国际高新技术成果交易会"在深圳举办，就此拉开深圳会展业发展序幕。为满足展会对场馆设施的需要，深圳市政府斥资 3 亿元在福田中心区投资建设"中国国际高新技术成果交易会展览中心"（简称"高交会馆"）。到 2003 年，深圳全年举办展会数量 60 多个，四年间展会增幅达 37.7%（图 1-10）。

在"高交会"的带动下，深圳会展业快速发展，展会数量及规模逐年递增，"高交会馆"的使用率趋于饱和。

As an emerging economic form, exhibition economy has great potential and vast space for development.

In 1999, the first "China Hi Tech Fair" (CHTF) was held in Shenzhen, which was seen as the prelude for the development of Shenzhen exhibition industry. In order to meet the demand of the venue facilities for the exhibition industry, Shenzhen Municipal Government invested 300 million yuan in the construction of CHTF Exhibition & convention Center (Figure 1-10) in the central area of Futian. Until 2003, there were more than 60 exhibitions were held in Shenzhen, with the increase of 37.7% within four years.

Driven by CHTF, the exhibition industry in Shenzhen has developed rapidly, with the number and scale increased progressively year after year. Therefore, the usage rate of CHTF Exhibition Center tends to be saturated.

图 1-10 1999 年 深圳高新技术成果交易会馆
Figure 1-10 The Venue of the China Hi Tech Fair, 1999

2004 年新建成深圳会展中心启用后，深圳市举办展会数量、规模迅速增涨，品牌展会的国际化、市场化、专业化程度不断提升。2017 年的统计数据显示，深圳场馆使用率高达 57.48%，全国排名第三，但展览数量与面积仅为全国第九（图 1-11）。

随着全球会展业发展重心向中国转移，作为国内最重要的会展中心城市之一的深圳，展馆数量和展厅面积不足的问题严重制约了会展业发展。

面向大湾区巨量人口、资本、信息、机遇涌入的未来，深圳急需一个全新的国际会展中心来承载其会展产业为核心的、千亿级别的经济体量，助力驱动湾区经济腾飞。

Since the new Shenzhen Convention and Exhibition Center opened in 2004, the number and scale of exhibitions held in Shenzhen have increased rapidly, and the internationalization, marketization and specialization of branded exhibitions have been improved continuously. According to the statistical data in 2017, the usage rate of Shenzhen venues was 57.48%, which ranked the 3rd in China. However, the numbers and areas of the exhibitions just ranked the 9th across the nation (Figure 1-11).

As the global exhibition industry shifted its development focus to China, Shenzhen, which is one of the most important exhibition center cities in China, is facing the problem of Shortage of both numbers and areas of exhibitions spaces,which seriously restricting the development of the the exhibition industry with in the area.

Facing a future with enormous population, capital, information and opportunities flooding to the GBA, there is an urgent need for a brand new international convention and exhibition center in Shen zhen in order to carry a grand billion-level economic volume with exhibition industry as its core, and help drive the GBA economic to growth.

图 1-11 2004 年 深圳会展中心
Figure 1-11 The Venue of the China Hi Tech Fair, 1999

2017年，深圳场馆使用率为57.48%，全国排名第三，展览数量与面积为全国第九（表1-1）。深圳会展中心的规模对深圳会展业发展存在制约。

In 2017, the utilization rate of Shenzhen exhibition centers was 57.48%, ranking the third in China, and the number and area of exhibitions ranked ninth in China (Table 1-1). The scale of the Shenzhen Convention and Exhibition Center has restricted the development of Shenzhen exhibition industry.

2017年全国城市展览数量和展览面积统计
Statistics on the number of national city exhibitions and exhibition area in 2017

城市 city	展览数量（场） Number of exhibitions (field)	展览数量全国占比 National share of exhibitions	展览面积（万平方米） Exhibition area	展览面积全国占比 National share of exhibition area	平均展览面积（万平方米） National share of exhibition area
上海 Shanghai	767	7.4%	1689	11.82%	2.20
广州 Guangzhou	662	6.39%	976	6.83%	1.47
重庆 Chongqing	496	4.79%	876.5	6.14%	1.77
北京 Beijing	365	3.52%	595.5	4.17%	1.63
南京 Nnanjing	509	4.91%	487.35	3.41%	0.96
沈阳 Shenyang	405	3.91%	410.3	2.87%	1.01
成都 Chengdu	207	2%	366	2.56%	1.77
青岛 Qingdao	239	2.31%	345	2.42%	1.44
深圳 Shenzhen	114	1.1%	325.44	2.28%	2.85

场馆名称 Venue name	使用面积（万平方米） Use area (10,000 square meters)	利用率 Utilization rate
广东现代国际展览中心（东莞） Guangdong Modern International Exhibition Center (Dongguan)	312.77	61.63%
中国国际展览中心（北京朝阳馆） China International Exhibition Center (Beijing Chaoyang Hall)	173.96	58.75%
深圳会展中心 Shenzhen Convention and Exhibition Center	398.72	57.48%
上海国际博览中心 Shanghai International Expo Center	643.83	55.96%
云南国际会展中心 Yunnan International Convention and Exhibition Center	220.99	52.41%
新疆国际会展中心 Xinjiang International Convention and Exhibition Center	177.24	52.19%
成都世纪新城国际会展中心 Chengdu Century New City International Convention and Exhibition Center	385.64	50.13%
海南国际会议展览中心 Hainan International Conference and Exhibition Center	151.21	49.34%
上海世博展览中心 Shanghai World Expo Exhibition Center	227.90	48.27%

表1-1 2017 年全国会展现状表格
Table1-1 Table of domestic EXPO situation, 2017

一座新城的起点
—— 大空港区域规划理念
Visions of The Future —— Planning Concept of The
Airport Area

粤港澳大湾区发展的时代背景下，深圳市政府于 2010 年就对沙井西部
海（大空港片区）提出了战略规划。

2015 年《深圳市大空港地区综合规划》确立了深圳大空港地区作为粤港
大湾区经济核心的地位，机场北交通枢纽、深圳国际会展中心等项目的
划建设拉开了序幕。

the development context of the GBA, Shenzhen government has
oposed a strategic plan for the west coast of Shajing (big airport
ea) in 2010.

2015, the Comprehensive Planning of Shenzhen Airport Area has
anted the Shenzhen Airport Area as the economic core of the
BA, and the planning and construction transportation hub in the
orth of the airport. The Shenzhen World project kicked off.

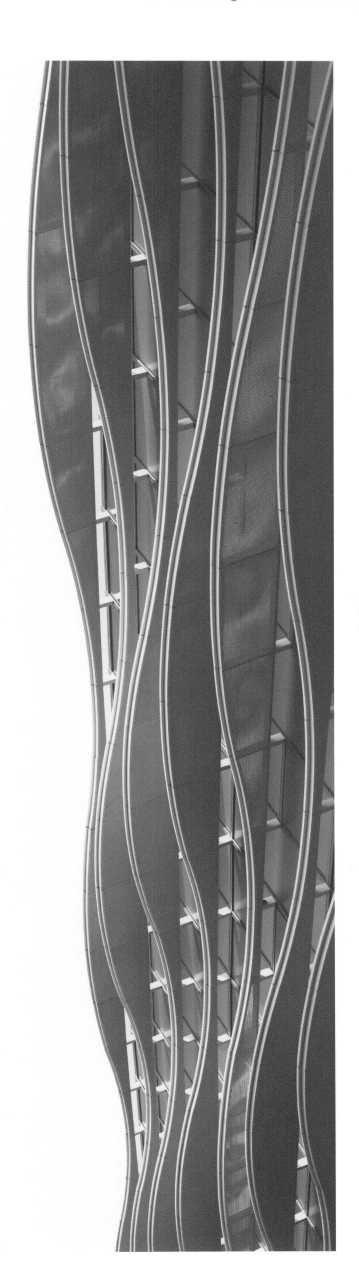

前期规划事记年表
The Chronology of Early Planning

自 20 世纪 80 年代起，大空港地区的水域经过逐步填海，形成现今陆地，在深圳城市总体规划中属于城市重点储备地区，拥有宝贵的用地资源。
随着深圳市的城市发展，大空港地区战略性的地位逐渐凸显，深圳国际会展中心选址于此，该片区的发展目标、定位及功能也逐步明晰（图表 1-1）。

Since 1980s, the Airport Area has been taking in the land from the sea which form the present region.It is the key reserved city area according to the Shenzhen Overall Urban Planning, possess valuable land resources.
With the urban development of Shenzhen, the strategic position of the airport area has been highlighted. When the site for the Shenzhen World was chosen, the development goal, its positioning and functions gradually revealed (Chart 1-1).

图 1-12　2015.06 宝安西部活力海岸带概念城市设计
Figure 1-12　The Conceptual Urban Design of Bao' an West Vibrant Coastal Zone (June, 2015)

图 1-13　2016 .01 深圳市会展中心建筑工程设计国际咨询
Figure 1-13　The International Consultancy for the Architectural al Engineering Design of the Shenzhen World Exhibition & Conventio Center (January, 2016)

深圳总规
The 2010 General Urban Planning of Shenzhen

隶属于沙井西部沿江战略发展地区、高薪技术产业制造基地，是城市重点储备控制地区。

It attaches to the Shajing west coastal strategic development area, a manufacturing base of high-tech industry, and the key reserved and controlled city area.

2010

大空港地区综合发展规划
The 2014 Comprehensive Planning of the Shenzhen Airport Area

隶属机场北 - 国际会展区，空港新城的启动区，将启动国际会展中心的展馆及其配套设施建设。

It belongs to the Airport North - International Exhibition & Convention Center Area, the starting area of the Airport New Town which begin with the construction of the Shenzhen World and its supporting facilities.

2014

2013

宝安综合发展规划
The 2013 Comprehensive Development Plan of Bao'an District

隶属空港新城，打造深莞城镇集聚区的核心、国际化科技创新基地、区域性中枢控制区以及"宜居、宜业、宜游"的珠江三角城产融合创新区。

It attatches to the Airport New Town, will be built as the core of Shenzhen-Dongguan conurbation, the international scientific and technological innovation base, the regional central control area, and a city and industry integrated innovation area in the pearl river delta with livable community, good working environment, and tourist destinations.

2015

宝安西部活力海岸带概念城市设计
The 2015 Conceptual Urban Design of Bao'an West Vibrant Coas

隶属湾区海城之北城，将启动截流河、国际会展中心及其配套设施的建设。

Attaches to the north section of the west coas it will start with the construction of the Shenzhe World and its supporting facilities.

图表 1-1　前期规划纪事年表
Chart 1-1　The Chronology of Early Planning

图 1-14　2016.01 大空港启动区城市设计
Figure 1-14　The Urban Design of the Starting Area in the Airport Area
(January, 2016)

图 1-15　2016 .04 深圳国际会展中心及配套用地城市设计
Figure 1-15　The Urban Design of the Shenzhen World Exhibition &
Convention Center and its Supporting Land (April, 2016)

空港启动区城市设计
.01 The Urban Design of the Starting Area within the Airport Area

化会展新城，片区整体空间风貌初步确定。

ernational exhibition new city, preliminarily determined
overall outlook of the area.

016.01

深圳国际会展中心及配套用地城市设计
2016.04 The Urban Design of the Shenzhen World Exhibition & Convention Center and its Supporting Land

规划设计条件初步明确。

The planning and design conditions has been preliminarily clarified.

2016.04

2016.01

深圳国际会展中心建筑工程设计国际咨询
2016.01 The International Consultancy for the Architectural and Engineering Design
of the Shenzhen World Exhibition & Convention Center

会展主体场馆建筑设计初步方案确定。

Preliminarily determined the design solution for the main venue
building of the Shenzhen World.

2016.07

福永西片区法定图则
2016.07 The Statutory Plan of Fuyong West Area

宜居宜业的空港新城，划定规划控制单元，具体要求
交由详细蓝图细化。

To build an Airport New Town that is suitable for living and
working, the planning control districts need to be allocated,
with specific requirements to be refined in the future process

上位规划解读
Interpretation of the Upper Region Planning

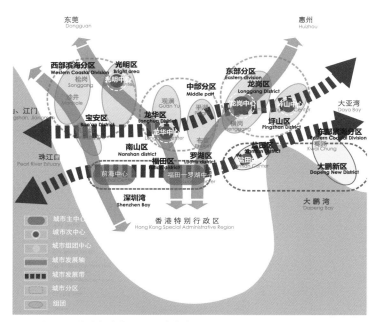

图 1-16 《深圳城市总体规划》城市布局规划结构图 摘自中国城市规划设计研究院深圳分院：《深圳国际会展中心（一期）配套商业用地详细蓝图》
Figure 1-16 The planning of urban layout map in Shenzhen Overall Urban Planning

《深圳市城市总体规划（2010-2020）》

规划确定深圳国际会展中心项目（以下建成"项目"）所在的沙井西部滨海片区将与前海地区一同被作为城市两大战略发展地区，根据发展需要进行重点建设；未来该区域将依托沿江高速公路、机场扩建、深港机场快线、穗深港城际线等重点交通基础设施，通过西部发展轴，向南对接前海中心联系香港，向北联系东莞西部并通往广州（图 1-17）。

这将是深圳提升国际化城市职能、实现深港合作、促进珠三角城镇群协调发展的战略性走廊。

The General Urban Planning of Shenzhen (2010-2020)

According to the General Urban Planning, the Shajing west coastal area where the Shenzhen World is located, along with the Qianhai area will be taken as two major strategic development areas. In the future, relying on the key construction of transportation infrastructure: the expressway along the river, the expansion of the airport, the Shenzhen-Hongkong airport express, the Guangzhou-Shenzhen-Hongkong intercity line, along with western development axis, the region will connect the Qianhai center and Hong Kong in the south, and the Dongguan west and Guangzhou in the North.(Figure 1-18)

This will be a strategic corridor for Shenzhen to enhance its functions as an international city, achieve cooperation with Hong Kong, and promote the coordinated development of pearl river delta city clusters.

《深圳市大空港地区综合规划》（2015）（简称"《大空港综合规划》"

规划将大空港地区定位为粤港澳大湾区经济核心区、国际一流空港都市区。项目属于机场北 - 国际会展区，该片区位于空港新城南部、T4 航站楼以北的空港新城启动区，也是大空港地区的重点功能区（图 1-18）。该片区重点围绕机场北交通枢纽、深圳国际会展中心核心项目建设，促进会展服务、电子商务、创意设计以及相关展会物流、展会旅游、展会设计等产业发展，带动福永旧工业区转型升级（图 1-19）。

The Comprehensive Planning of the Shenzhen Airport Area (2015)(CPSAA for short)

According to the CPSAA, the Shenzhen Airport Area is positioned to be the economic core region of the GBA, and the world-leading airport metropolitan area. The Shenzhen World is located in the Airport North - International Exhibition Center Area, which is situated in the south of the Airport New Town, and the starting area of the Airport New Town in the north of the terminal T4, and also an important functional part of the Airport area (Figure 1-18). Here, it will focus on the constructions of two Core Projects: the Airport North transport hub and the Shenzhen World Exhibition & Convention Center, and promote the developments of convention and exhibition services, e-commerce, creative designs, as well as related logistics, tourism, design and other industries, and finally to the upgrade of the industrial districts in Fuyong.(Figure 1-17)

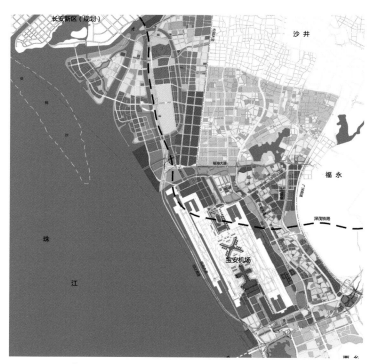

图 1-17 《大空港综合规划》用地规划图 摘自中规院深圳分院《深圳国际会展中心（一期）配套商业用地详细蓝图》
Figure 1-17 The land use map in the Comprehensive Planning

《宝安区综合发展规划 (2013-2020)》

该规划将宝安区的战略地位定位为国际化滨海名城和现代化产业强区，将把宝安建设为亚太地区最具活力和国际竞争力的滨海城区，以及多元、创新、转型的现代化产业强区为目标。以宝安中心区和空港新城为核心，将会形成"三带、两心、一谷"的城区空间结构（图 1-18）。

项目所在的空港新城被定位为深莞城镇集聚区的核心、国际化科技创新基地、区域性中枢控制区以及"宜居、宜业、宜游"的珠三角城产融合创新区。同时，规划提出需在明确会展中心选址和规模的同时，同步开展前期立项、规划、方案设计工作。

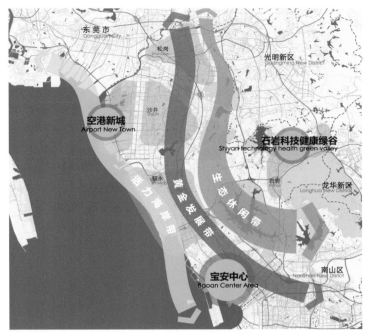

图 1-18　《宝安区综合发展规划》规划结构图
Figure 1-18　The Planning structure map in the Comprehensive Development Plan

The Comprehensive Development Plan of Bao'an District (2013-2020)

According to the comprehensive plan, Bao'an district is strategically positioned as an international coastal city and an advantaged area of modern industries, and will be built in the most vibrant coastal city with international competitiveness among the Asian-Pacific region, as well as an advantaged area of modern industries features diversification, innovation and transformation. With the Bao'an center and the Airport New town as its core, it will form an urban spatial structure of "three belts, two centers, and one valley".(Figure 1-18)

The Airport New City, where the Shenzhen World is located, is positioned as the core of Shenzhen-Dongguan conurbation, the international scientific and technological innovation base, the regional central control area and a city and industry integrated innovation area in pearl river delta with livable communities, good working environments, and tourist destinations.

Meanwhile, this plan also emphasis that when determin the site selection and the scale of the Shenzhen World, it has to carried out the preliminary works including project approval, planning, and scheme design at the same time.

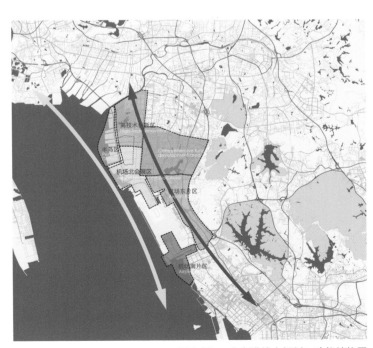

图 1-19　《大空港综合规划》功能结构图
Figure 1-19　P4 The functional structure map in the Comprehensive Planning

相关规划解读
Interpretation of Correlative Planning

《宝安西部活力海岸带概念城市设计国际咨询》（2015 年 6 月）

该次国际咨询从城市设计的角度全面谋划了空港新城的空间和功能组织，指出了研究范围内以机场为依托的会展中心及其相关功能的落位与布局将作为新城的核心功能区，且也将于近期启动实施，而其空间组织也应与东莞和宝安内陆地区的功能保持连接（图 1-20）。

项目区位及功能——位于"三城四湾"西海岸的北城区段，是组成宝安城市空间结构的重要部分。北城将以湾区公共服务和现代产业为主导，发展战略性新兴产业、海洋经济、区域服务枢纽、会展经济等功能，在复合功能上，结合湿地桑田等海岸生态景观本底，塑造以田园湿地观光休闲、城市水岸观光休闲为特征的生态城市体验区（图 1-21）。

项目的主要建设内容——2015 年启动的国际会展中心展馆，配套酒店、餐饮、商场、银行以及停车场等功能设施的建设。此外还有同步规划的展馆与机场、轨道以及城市道路交通的便利衔接，保证会展人流和物流的高效集散。

图 1-20　2015.06 宝安西部活力海岸带概念城市设计
Figure 1-20　The Conceptual Urban Design of Bao'an West Vibrant Coastal Zon (June, 2015)

International Consultancy for the Conceptual Urban Design of Bao'an West Vibrant Coastal Zone (June, 2015)

From the perspective of urban design, this proposal carried out a plan of the spatial and functional organizations for the Airport New Town, points out that the placement and layout of the Shenzhen World and its related functions are going to be the core area of the Airport New Town, and it will start implementing this strategical plan recently, and the spatial organization of the core area should keep its connection with Dongguan and the inland area of Bao'ao. (Figure 1-20)

The location and function of the project — at the north section of the "three cities and four bays" on the west bank, which is an important part of the spatial structure of Bao'an. Driven by the GBA public service and modern industries, This area will develop strategic emerging industry, marine economy, regional services hub, and exhibition economy, etc,and as regardvte the compound functions, it will combine the coastal ecological landscape such as the wetland landscape, to build ecological experience area which features sightseeing and leisure of gardens, wetland, and the city bank.(Figure 1-21)

The main construction of the project— in 2015, begin with the construction of the Shenzhen World's exhibition hall and other supporting facilities such as hotels, restaurants, shopping malls, banks, and parking lots and etc, and simultaneously plans the convenient connection between the exhibition hall and the airport, the railway, and city road traffic, so as to guarantee the efficient distribution of people and logistics.

图 1-21　2015.06 宝安西部活力海岸带概念城市设计总平面图
Figure 1-21　The Masterplan of the International Consultancy for the Conceptual Urba Design of Bao'an West Vibrant Coastal Zone (June, 2015)

图 1-22 2016 .01 深圳市会展中心建筑工程设计国际咨询中标方案
ure 1-22 The International Consultancy for the Architectural and Engineering Design of the Shenzhen World Exhibition & Convention Center (January, 2016)

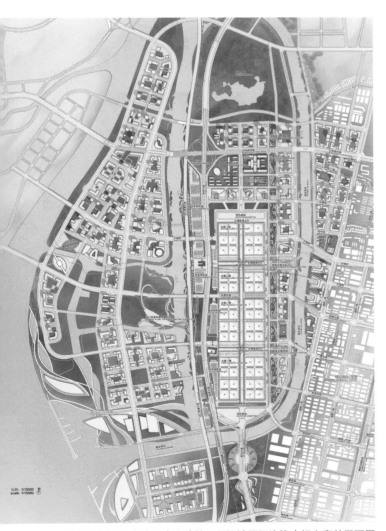

图 1-23 深圳市会展中心建筑工程设计国际咨询中标方案总平面图
Figure 1-23 The Masterplan of the International Consultancy for the Conceptual Urban Design of Bao'an West Vibrant Coastal Zone (June, 2015)

《深圳国际会展中心建筑工程设计国际咨询》（2016 年 1 月）

在该次国际咨询中，中标方案明确了会展中心片区的整体空间结构，并初步确定了会展主体场馆建筑的设计方案（图 1-22）。

项目规划被定位为一座有机结合的"城"与"市"；一座极具深圳滨海特征、充满活力与亲和力的公共空间的复合功能岛；一组运作良好、能融入城市生活的会展中心建筑群。

项目规划结构——"一岛、一轴、一带、四城"，其中"一岛"即会展岛，回应大片区岛群意向，打造功能复合的会展岛；"一轴"为会展 —— 海上田园功能轴，延伸会展人行廊道南至 T4 空港，北通海上田园；"一带"为复合型的会展中央公园带，完善片区城市绿脉，同时辐射两侧城市功能带；"四城"即由三条东西向景观水带将用地划分成的、自南向北的四个复合城区（图 1-23）。

The International Consultancy for the Architectural and Engineering Design of the Shenzhen World Exhibition & Convention Center (January, 2016)

This proposal clarified the overall spatial structure of the exhibition center area and preliminarily determined the design solution of the Shenzhen World's main venue building. (Figure 1-22)

The planning positioning of the project — a well integrated composition of "city" and "business", an island of multifunction with Shenzhen's characteristics, vigor and appeal, a cluster of exhibition related buildings which is well operated and integrated into the city life.

The planning structure of the project — one island, one axis, one belt and four comprehensive urban areas.
One island: an island for exhibition and convention, an island of integrated functions, echoing the image of the islands;
One axis: from the exhibition and convention area to the Shenzhen Waterland Resort, Stretching the corridor of exhibition and convention from the south to T4 airport, and from north to the Shenzhen Waterlands Resort;
One belt: a comprehensive central park belt in the exhibition and convention area, to improve the urban green land, and provides transportation convenience for the urban function areas on either side of the park;
Four comprehensive urban areas: three east-west water landscape belts will separate the land into four comprehensive urban areas from south to the north. (Figure 1-23)

相关规划解读
Interpretation of Correlative Planning

图 1-24　2016.01 大空港启动区城市设计
Figure 1-24　The Urban Design of the Starting Area in the Airport Area (January, 2016)

图 1-25　2016.01 大空港启动区城市设计总平面图
Figure 1-25　The Masterplan of the Urban Design of the Starting Area in the Airport Area (January, 2016)

《大空港启动区城市设计》（2016 年 1 月）

此次城市设计对大空港启动区的整体风貌及空间布局进行了统筹，并在《深圳国际会展中心建筑工程设计国际咨询》中标方案的基础上，完善了会展中心及其配套商业用地的设计（图 1-24）。

城市设计范围——为北至茅洲河、南至福永河、西邻沿江高速、东接锦程路，规划总用地面积为 15.53 平方公里。

城市设计目标——国际会展岛及科技创新城，并在此基础上提出国际化会展新城、高端型产业平台及复合化生态城区的三个分目标。

城市设计空间结构——以十字主轴为骨架，建立东西向多廊道，南北向多核心串联的设计构架。

项目定位及目标——在会展中心片区的主要功能为一期、会展二期，以及办公、酒店、公寓等会展配套服务设施，旨在建设具有片区影响力的区域性地标建筑群，以体现大空港地区的高端形象（图 1-25）。

The Urban Design of the Starting Area in the Airport Area (January, 2016)

This urban design carried out the overall layout of the outlook and spatial layout of the Starting Area in the Airport New Town, and also improved and refined the Design of the Shenzhen World Exhibition & Convention Center and the Supporting Land based on the International Consultancy for the Architectural and Engineering Design of the Shenzhen World Exhibition & Convention Center.(Figure 1-24)

The Scope of the Urban Design — extends to the Maozhou river in the north, and Fuyong river in the south, next to the riverine expressway in the west, and adjoins to the Jincheng road in the east, with a total planning area of 15.53 square kilometers.

The Goal of the Urban Design — to build an international exhibition island and a science and technology innovation city. Then on this basis, put forward three sub-goals: international exhibition new city, high-end industrial platform and composite ecological city.

The Spatial Structure of the Urban Design — a design framework with a cross principal axis as its backbone, Set up multiple east-west corridors and south-north cascades.

The Positioning and Objective of the project—in the exhibition center area, phase i, phase ii, and supporting services such as office buildings, hotels, and apartments, are aim to build a regional architectural complex landmark with influences, in order to manifests the high-end image of the Airport Area. (Figure 1-25).

图 1-26　2016 .04 深圳国际会展中心及配套用地城市设计总平面图
Figure 1-26　The Masterplan of The Urban Design of the Shenzhen World Exhibition & Convention Center and its Supporting Land (April, 2016)

《深圳国际会展中心及配套用地城市设计》（2016 年 4 月）

该城市设计在《深圳国际会展中心建筑工程设计国际咨询》中标方案的基础上，落实了《福永西片区法定图则》的要求，并对会展中心及其综合配套区域的功能、空间进行了系统地规划，明确了会展中心一期用地及其综合配套用地的总体布局、建筑退红线要求、市政设施要求等主要规划设计条件，并最终形成了《深圳国际会展中心综合配套用地规划设计条件》（图 1-26）。

项目规划目标——"打造国际一流会展中心"，"建设融入城市生活的综合体，""塑造特色鲜明的城市地标"（图 1-27）。

The Urban Design of the Shenzhen World Exhibition & Convention Center and its Supporting Land (April 2016)

Based on *the International Consultancy for the Architectural and Engineering Design of the Shenzhen World Exhibition & Convention Center*, The Urban Design implemented the requirement of the *Statutory plan of Fuyong west area*, and systematically carried on the planning of the functions and the spaces of the Shenzhen World and its comprehensive supporting lands, made clear its overall layout, the building lines, municipal facilities requirements, and other main planning and design conditions, and eventually composed *the Planning and Design Conditions of the Shenzhen World Exhibition & Convention Center and it Supporting Land* (Figure 1-26).

The Planning Goal of the project—— to build a world-class exhibition center, a urban lifestyle complex, and a city landmark with distinctive features (Figure 1-27).

图 1-27　2016 .04 深圳国际会展中心及配套用地城市设计
Figure 1-27　The Urban Design of the Shenzhen World Exhibition & Convention Center and its Supporting Land

"三城一港"，未来之城
Three Cities and One Port — The City of the Future

《宝安西部活力海岸带概念城市设计国际咨询》(2015 年 6 月)

未来，深圳国际会展中心周边将规划成为"三城一港"，包括国际会展城、海洋新城、会展田园城和综合港区，打造世界级会展经济集群和大湾区创产业中心（图 1-28 ）（图 1-29）。

其中，最核心的部分——国际会展城，规划面积约 10 平方公里，将形成"会展＋科技"的世界产业高地，同时也是深圳向世界展示城市形象的新名片。

图 1-28 "三城一港"功能规划结构图
Figure 1-28 Three Cities and One Port : Function Distribution Map

图 1-29　"空港新城" 效果图　摘自中国城市规划设计研究院深圳分院：《深圳国际会展中心（一期）配套商业用地详细蓝图 》
Figure 1-29　The Rendering of the Airport New Town
From the Supporting Commercial Facilities Blueprint of the Shenzhen World Convention and Exhibition Center (Phase I) by the China Academy

Three Cities and One Port — The City of the Future
The International Consultancy for the Conceptual Urban Design of Bao'an West (June, 2015)

In the future, there will be an international convention and exhibition city, a new marine city, an exhibition garden city and a comprehensive port area around the Shenzhen World Exhibition & Convention Center. The area and the cities will be built into world-class exhibition economy clusters, and a science and innovation industrial center of the GBA (Figure 1-28, 1-29).

Among them, the core part — the international exhibition city, with a general planning area of 10 square kilometers, will reach a global industrial leading position of "exhibition and convention + science and technology", and therefore become Shenzhen's new image.

"大" 城

Great City

超大型会展城市综合规划
The Comprehensive Planning of the Mega Exhibition & Convention City

46 会展 EDD 与 会展经济
EDD & Exhibition Economy

62 设计一座城的未来 —— 会展中心
The Shenzhen World — The Design of a Future City

69 展城呼应 —— 开放提升城市空间
Exhibition Center + City — A More Open and Advanced Space

83 规划成果表现
Planning Achievement Showcase

会展 EDD 与会展经济

EDD & Exhibition Economy

会展业与旅游业、房地产业并称为"三大无烟产业",而"会展兴市"也已经成为国内许多城市实现经济增长的"城市面包"。据 2017 年中国会展经济研究会数据统计:自 2011 年以来,全国总展览面积的年增长率为 9.75%,展会数量也以 13.24% 的年增长率递增。深圳国际会展中心及天津国家会展中心的相继建成,将促成珠三角、长三角和环渤海的三足鼎立之势,并将共同成就中国会展业的大格局。与此同时,郑州、济南、厦门、武汉、宁波等地也都将兴建大型、超大型的会展场馆。

面对新一轮的会展场馆建设热潮,结合近年来在会展设计领域的实践与思考,AUBE 欧博设计认为:大型、超大型会展场馆相比其他类型的城市公共设施,其巨大的聚集效应已经远远超出会展场馆自身的范围,其对城市尤其是周边片区域的影响更加强烈且深远,如何在项目前期策划与规划设计时充分考虑和利用这一效应,提出更具在地性、前瞻性与系统性的规划策略与理念,对于实现会展场馆与城市发展的互融互动、发挥其经济增长的引擎作用意义重大。

Exhibition industry, tourism and real estate are known as the "three smoke-free industries", while "to prosper city by exhibition industry " has become the "urban bread" for many cities to achieve economic growth in China. According to the 2017 data statistics from China Convention and Exhibition Society, since 2011, the annual growth rate of the total exhibition area has been 9.75%, and the number of exhibitions has increased by 13.24% progressively. In the next few years, after the completion of the Shenzhen World Exhibition & Convention Center, and the Shanghai National Exhibition and Convention Center, there will form three important economic trends : the Pearl River Delta, the Yangtze River Delta and the Circum-Bohai, together they will become the great backbone structure of Chinese exhibition industry. Meantime, Zhengzhou, Jinan, Xiamen, Wuhan, Ningbo and other cities will also build large, or super large exhibition venues.

Facing a new round of construction booms of exhibition venues, and considering the recent years of experiences and thinking in exhibition design, we AUBE believe that large and super large exhibition venues has much more aggregation effects than solely themselves as venues compared to other urban public facilities, and will have more intensive and profound influences on its surrounding areas particularly. Therefore, to think over and take full advantage of this kind of effect during the preliminary planning and design of the project is essential, put forward a planning strategy and concept of indigenization, foresight, and systematization, which will help achieve the merging and interactive development between the exhibition industry and the city, and will also have a great significance as engine on the economic growth is of important signality.

培育"会展+"生态圈，实现展城一体
Develop an "Exhibition Plus" Ecosystem, and Build a City of Exhibition

从会展中心迈向活力都会，会展新城预计在 3-5 年内形成 700 万平米的展览规模。而在会展中心本身的量级、会展业作为深圳市重要支柱产业的地位，以及会展业的溢出效应等多重因素的综合作用下，将会形成"会展+"的生态圈（图 2-1），留住会展产业要素，让会展业、创新、文化、休闲、配套互相催化，带动整个宝安的产业升级。

From an exhibition center to a vibrant city, the Exhibition New City as the most important exhibition center in the South of China is expected to build up to a 7 million square meters scale. with 3-5 year period According to this scale, the exhibition industry will become a mainstay industry in Shenzhen, playing an important role on multiple effects, and forming an "Exhibition Plus" ecosystem(Figure 2-1). with these elements, the exhibition industry, innovation, culture, leisure, and supporting services will catalyze one another, and help upgrade the whole industry of Bao'an.

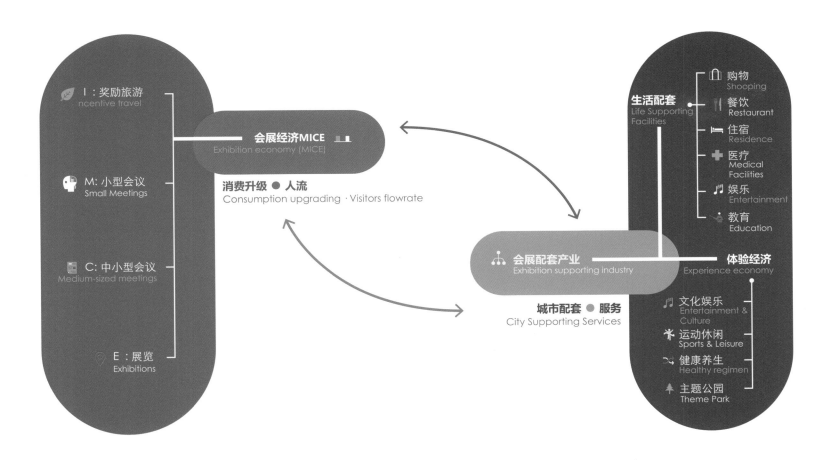

图 2-1 "会展+"生态圈分析图
Figure 2-1 Analysis of the "Meeting +" Ecosphere

事件 / 节日

规律化的城市生活使人们感到单调乏味，在会展中植入各种文化事件，如酒吧、艺术节等，形成跳跃的文化脉动，与会展区日常的文化商业活动相互补充（图 2-2）。

Events / festivals

Regular urban life will make people feel monotonous and boring. Implanting various cultural events in the exhibition, such as bars, art festivals, and etc., forming a beating cultural pulse and thus completing the daily cultural and commercial activities in the exhibition area (Figure 2-2).

一天/one day

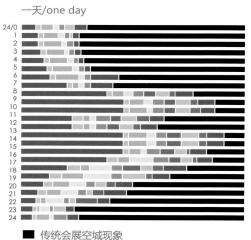

■ 传统会展空城现象
Empty city phenomenon of traditional exhibition

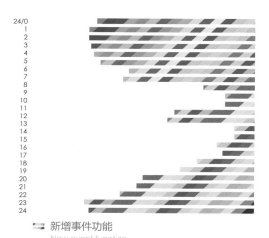

新增事件功能
New event function

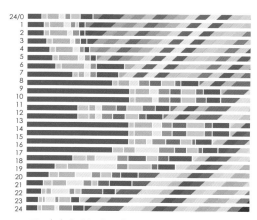

丰富充实泛会展生活
Enrich pan exhibition life

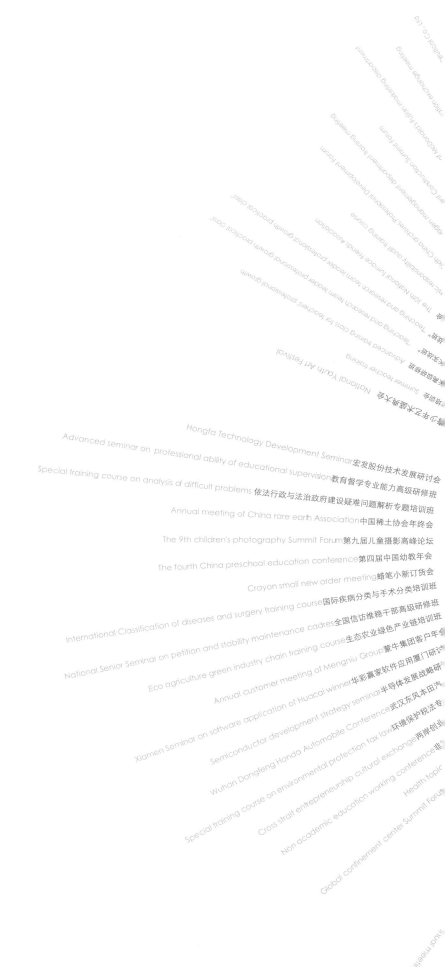

“泛会展”

补充持续活力，填补时间空洞

泛会展为传统会展区注入书城、博物馆、餐饮、KTV、购物商城、创意工作室、等全天候的功能业态，丰富展馆服务配套的推送，也为区域市民提供丰富的24小时娱乐生活目的地。把深圳新会展中心打造成一个全年充满活力的会展生活城。

For the traditional single function Exhibition Center, pan exhibition injects book city, museum, catering, KTV, shopping mall, creative studio and other all-weather functional formats into the traditional exhibition area, enriches the supporting Services of the exhibition mall , and also provides rich 24-hour entertainment and life destinations for citizens. We will build the Shenzhen world Exhibition and Convention Center into an exhibition life city full of vitality throughout the year.

国际海工论坛 International Marine industry Forum
第七届深基础工程发展论坛 Shenzhen Foundation Engineering Development Fau
危险废物环境管理与风险防控专题培训班 Special training course
广东天阳神集团第27届工作室室长班 Guangdong Sun Goa Group Studi
大学生工程训练综合能力竞赛 Training Competition for college students in F
"JM中脉"超级动力训练营 JM midvein" super power training camp
食品药品规范化执法案例培训班 Case training course
《和谐之道》培训班 Harmonious way training course
蚂蚁金服 (深圳) 信息科技有限公司年会 Top forum of life insurance industry
"现代新计量管理实务"高级培训班
达安集团2016年度营销峰会
山东齐鲁漆业营销峰会 Shandong Qilu lacquer industry Marketing Summit
火山鸣泉生态科技年终总结会 Year end meeting of ont golden clothing
深圳国际建筑设计交流会 Do'an Group Marketing Center annual meeting
Advanced training course
Shenzhen international architectural design exchange

深圳文化展业交流会 Shenzhen cultural exhibition industry exchange
中国当代绘画展 Contemporary Chinese Painting Exhibition
西安中高级人才招聘会 Xi'an middle and senior talents recruitment fair
房地产交流会 Real estate exchange meeting

《二十四城记》电影文献会 A Tale of twenty-tour Cities film Documentary
西安爵士音乐工作坊讲座 Xi'an Jazz workshop lecture
WCG魔兽争霸3：冰封皇座 WCG world of Warcraft 3: Frozen Throne
澳洲大学独立话剧《候鸟》 University of Australia independent drama migratory bird
西安独立电影播放会 Xi'an independent film show

创意的左右 “公寓派对” Creative "apartment party"
活动 "一起去看展览" Activity "go to the exhibition together"
活动 "我的老婆是设计师" Activity "my wife is a designer"

瑞士巧克力的发展 The development of chocolate in Switzerland
亲手做法国菜 Hand made French food
素食餐厅 Vegetarian Restaurant
汤姆餐厅 Tom Restaurant
吃三明治大赛 Sandwich eating competition
昆虫餐厅 Insect Restaurant
冰淇淋火锅 Ice cream hot Pot
西安中高级人才招聘会 Contemporary Chinese Painting Exhibition
房地产交流会 Xi'an middle and senior talents recruitment fair
Real estate exchange meeting
转折点健康产业高峰论坛 Turning point Health Industry Summit Forum
著名商学院"总裁执行之旅" 4.0时代 "Tour of famous schools" president executive mode 4.0
PPP项目案例专题培训 Special training course on PPP project
施耐德电气创新峰会 Schneider Electric Innovation Summit
民营企业国际合作论坛 International Cooperation Forum of private enterprises
中国长三角古玩商会第九届理事会 China Yangtze River Delta antique chamber of Commerce
昆明地区各国会员代表大会The 10th National Congress of Insect Association

China Ping An Youcai high end Entrepreneur Forum
Annual training course
Aspo beauty competition
Modern international quality training course
China Ping An

图 2-2-1 泛会展分析图
Figure 2-2-1 Pan-Convention Analysis Chart

从基础设施、生活服务两个角度，为新城配备优质足量的特色服务，满足生活需求。
From the perspective of infrastructure and life service, the new town is equipped with high-quality and sufficient characterized services to meet the needs of life.

图 2-2-2　泛会展分析图
Figure 2-2-2　Pan-Convention

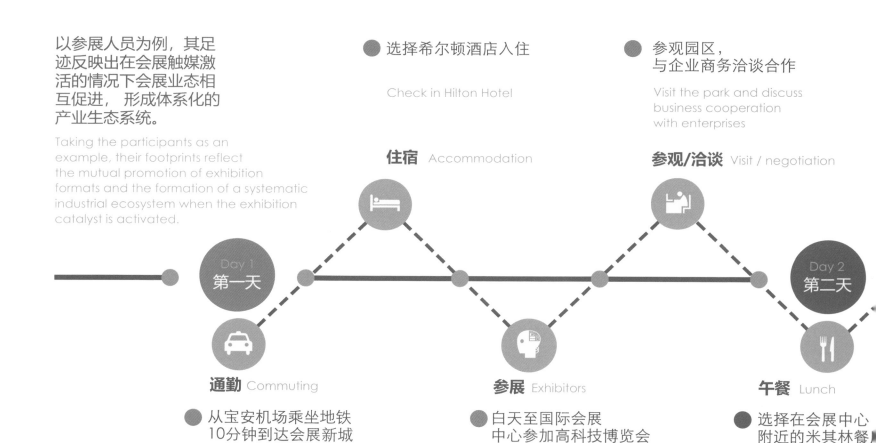

以参展人员为例，其足迹反映出在会展触媒激活的情况下会展业态相互促进，形成体系化的产业生态系统。

Taking the participants as an example, their footprints reflect the mutual promotion of exhibition formats and the formation of a systematic industrial ecosystem when the exhibition catalyst is activated.

选择希尔顿酒店入住
Check in Hilton Hotel

参观园区，与企业商务洽谈合作
Visit the park and discuss business cooperation with enterprises

住宿 Accommodation

参观/洽谈 Visit / negotiation

Day 1 第一天

Day 2 第二天

通勤 Commuting

参展 Exhibitors

午餐 Lunch

从宝安机场乘坐地铁10分钟到达会展新城
Take the subway from Bao'an airport to the exhibition new town in 10 minutes

白天至国际会展中心参加高科技博览会
Take part in the Hi Tech Expo at the International Convention and Exhibition Center during the daytime

选择在会展中心附近的米其林餐
Choose Michelin restaurant near the exhibition center

泛会展概念，会展生活不落幕
An Extensive Concept of Exhibition Creats a Non-stop Alive City

会展对相关产业有 1:9 的拉动作用，会展中心的影响力远大于会展中心本身（图 2-2-2）。所以，不同以往传统单一功能的会展中心，深圳国际会展中心将按照泛会展的理念，致力于打造一个超大型城市地标性生活服务综合体，在为传统会展区注入书城、博物馆、餐饮、KTV、购物商城、创艺工作室等全天候的功能业态、丰富展馆服务配套的同时，也为区域市民提供了丰富的 24 小时娱乐生活目的地，从而把会展中心对城市的影响扩张到最大，把价值发挥到最大，塑造国际会展活力的都会目的地（图 2-2-3）。

Exhibition Industry would help promoting other industries with a effect of 1: 9, whose influences are much bigger than the exhibition center itself.(Figure 2-2-2) Therefore, different from previous traditional exhibition centers with single function, the Shenzhen World Exhibition & Convention Center will be built according to an extensive exhibition concept, and become a super-giant city landmark complex of life and service. It will introduce series of 24 hours non-stop commercial facilities such as book stores, museums, restaurants, KTV, shopping mals, art studios and etc., making it a destination of various entertainment and recreation for local people, and thereby outspread its influences on the city to its maximum, and develop its biggest value, making it a metropolitan destination with a vibrant exhibition industry. (Figure 2-2-3)

图 2-2-3 泛会展分析图
Figure 2-2-3 Pan-Convention Analysis

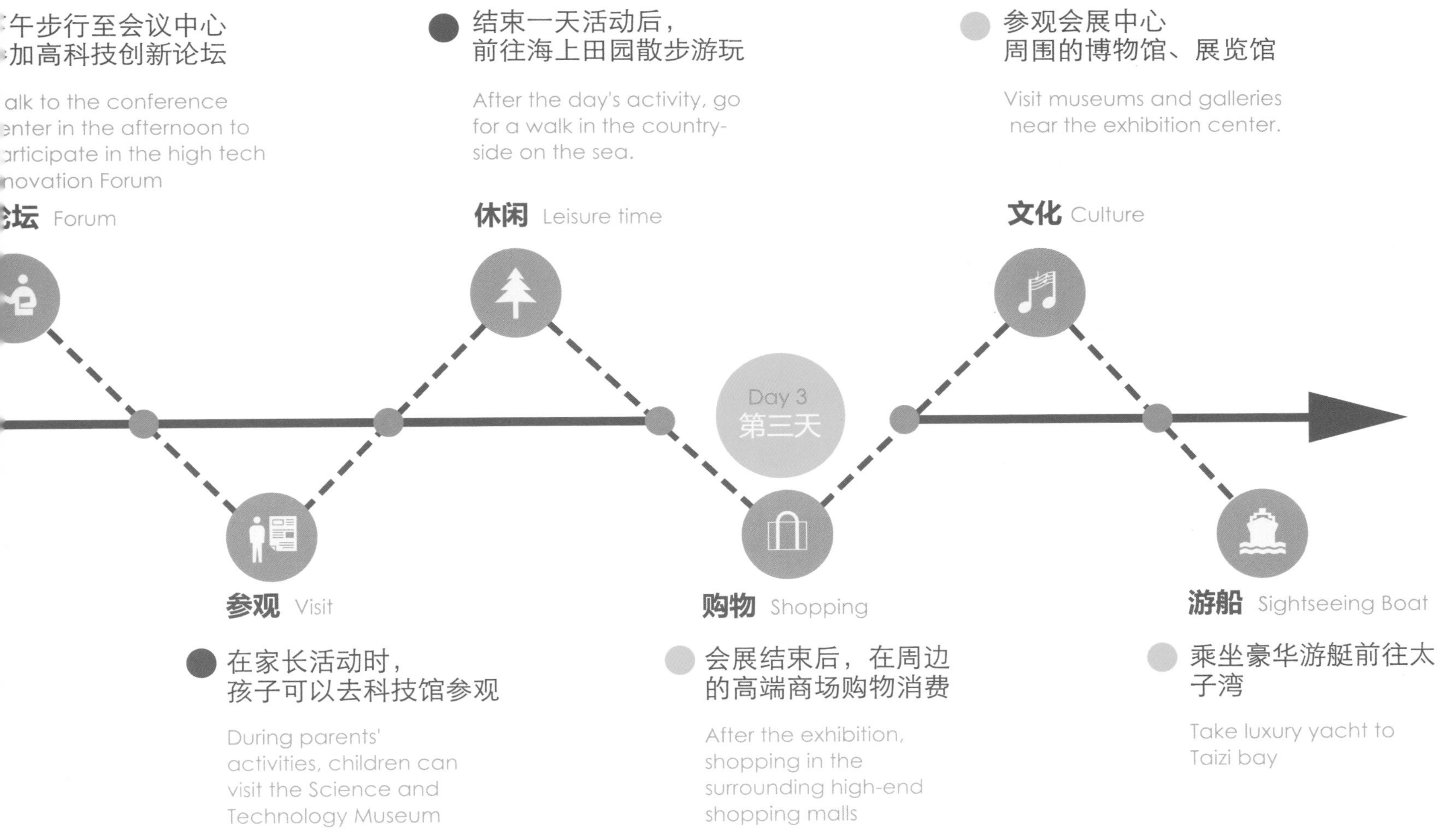

午步行至会议中心
加高科技创新论坛

alk to the conference
enter in the afternoon to
articipate in the high tech
nnovation Forum

坛 Forum

● 结束一天活动后，
前往海上田园散步游玩

After the day's activity, go for a walk in the country-side on the sea.

休闲 Leisure time

● 参观会展中心
周围的博物馆、展览馆

Visit museums and galleries near the exhibition center.

文化 Culture

Day 3
第三天

参观 Visit

● 在家长活动时，
孩子可以去科技馆参观

During parents' activities, children can visit the Science and Technology Museum

购物 Shopping

● 会展结束后，在周边
的高端商场购物消费

After the exhibition, shopping in the surrounding high-end shopping malls

游船 Sightseeing Boat

● 乘坐豪华游艇前往太子湾

Take luxury yacht to Taizi bay

会展综合发展区规划，以展兴城
The Concept of Exhibition Development District

基于上述理念，AUBE 欧博设计提出了会展综合发展区（Exhibition Development District，简称：EDD）这一规划概念：

会展综合发展区（EDD）是指在城市地区内，以会展业为核心产业，以大型、超大型会展场馆为核心建筑，聚集了与会展产业相关的各类功能用地的城市综合发展片区（图 2-3-1）。

Therefore, AUBE put forward the planning concept of Exhibition Development District (EDD) :

Taking the exhibition industry as a core industry of the city, and the large and super large exhibition venues as core buildings, So as to develop an urban comprehensive development district that gathers all kinds of functional lands related to the exhibition industry(Figure 2-3-1).

图 2-3-1   会展 EDD 概念图
Figure 2-3-1   EDD Concept

城市发展因素
Urban Develpoment

会展产业发展因素
Industry Develpoment

交通因素
Transportation

选址
Site selection

选址

会展综合发展区（EDD）选择的场地在满足会展场馆建设需求的同时，应用可供会展产业链发展的空间，用以构建会展经济圈（图 2-3-2），从而形成城市产业结构与功能上的重要节点。影响其选址的因素有以上三点：

Site selection

The site of EDD should not only meet the requirements of the construction of exhibition venues, but also provide enough space for the development of exhibition industry chains, which will form the exhibition economic circle (Figure 2-3-2), and become an important part of the urban industrial structure and function. There are three factors for the site selection as shown in the Chart 2-3-1:

图表 2-3-2   会展 EDD 概念图
Chart 2-3-2   EDD Concept

CONVENTION & EXHIBITION

Exhibition Development District

城市发展因素
Urban Develpoment

超级城市群、经济发展带及湾区经济区等理念的提出也大大影响了会展综合发展区的选址理念；例如深圳国际会展中心的选址在多轮研判后，最终选在了粤港澳大湾区的核心位置（图 2-4），意在辐射香港、澳门、广州、珠海等周边多座城市，并与广州琶洲会展中心、香港国际会展中心等多座会展中心共同打造超级会展产业群。

The concepts of super city agglomeration, economic development zone and bay area economic zone have greatly influenced the concept of the EDD's site selection. For example, after several rounds of research and evaluation, the Shenzhen World has been located in the core area of the GBA(Figure 2-4), aiming at making great impact on Hongkong, Macao, Guangzhou, Zhuhai and many other surrounding cities, and building up a super scale exhibition industries group together with exhibition centers like Guangzhou Pazhou Exhibition Center, Hong Kong Convention & Exhibition Centre and other centers.

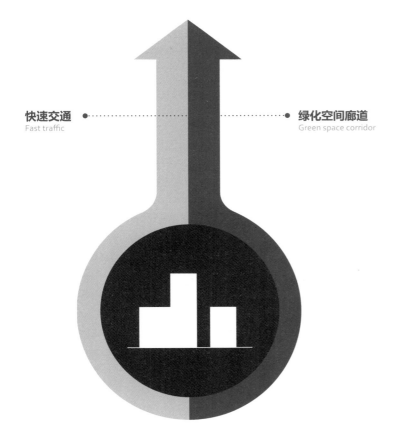

图 2-3-3　会展 EDD 概念图
Figure 2-3-3　Concept EDD

国内多数城市已经从原有的单中心城市向多中心城市发展，会展综合发展区（EDD）应作为城市发展的一个产业次中心，并通过快速交通、绿化空间廊道与现有城市中心或其他功能板块形成紧密联系（图 2-3-3）。

Most cities in China have developed from the original single-centered city to multi-centered city. The EDD should be regarded as an industrial sub-center in urban development, and form a close connection with the existing city center and other functional areas through rapid transportation and green corridors.(Figure 2-3-3)

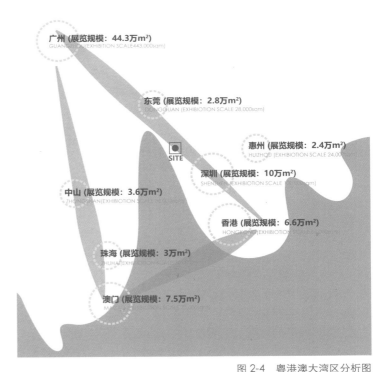

图 2-4　粤港澳大湾区分析图
Figure 2-4　Analysis map of Guangdong-Hong Kong-Macao Greater Bay Area

会展产业发展因素
Industry Develpoment

会展综合发展区（EDD）应选在所在制造业集中的城市地区或区域，减少长距离运输成本的同时形成产业聚集效应，从而增强该地区或城市的核心竞争力（图 2-5）。例如浙江绍兴国际会展中心的选址就位于"亚洲布市"之称的绍兴柯桥区；义乌国际会展中心也是因其多年发达的小商品产业，成为国内会展场馆的一个成功案例。

Exhibition Industry Developments Factors

The site of the EDD should be an area within a region or city where the manufacturing industry is concentrated, so as to reduce the long-distance transportation cost and form the industrial agglomeration effect and enhance the core competitiveness of the region or city (Figure 2-5). For example, the site of Shaoxing International Convention and Exhibition Center in Zhejiang is located in Keqiao District, which is known as "The Cloth City of Asia". Moreover, the Yiwu International Convention and Exhibition Center is a very successful case of domestic exhibition centers which benefits from the local prosperous small-commodities industry for years.

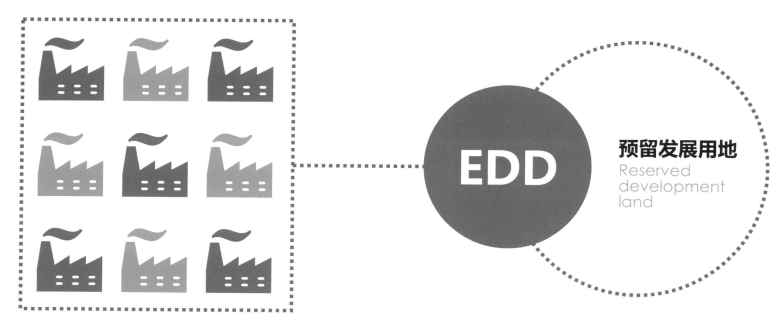

图 2-5　会展产业发展因素分析图
Figure 2-5　Analysis of Development Factors of the Exhibition Industry

交通因素
Transportation

EDD 内外部的交通应当发达便捷，且应与空港、高速公路、城市快速路、港口、火车站、汽车站、轨道等交通设施有便捷的联接；近年来国内许多城市的会展中心与传统的临空经济区相结合，正是利用其坐拥更加便捷外部交通的同时对城市主城区影响小的优势（图2-6）。国内外知名的大型、超大型会展中心多数也遵循这一理念，选择在同时拥有空港、高速公路及多条轨道线等交通设施的区域。

Transportation Factors

Both the internal and external transportation conditions of the EDD should be convenient. It should be easy and fast connection to the airport, expressway, urban expressway, port, railway station, bus station, rail and other transportation facilities.

In recent years, the convention and exhibition centers of many cities in China have combined the traditional airport with economic zone, which is to take advantage of its more convenient external transportation and reduce its influence on main urban areas (Figure 2-6).

Most of the well-known large and super large convention and exhibition centers at home and abroad also follow this concept, and choose the site near airport, expressway, multiple rail lines and other transportation facilities at the same time.

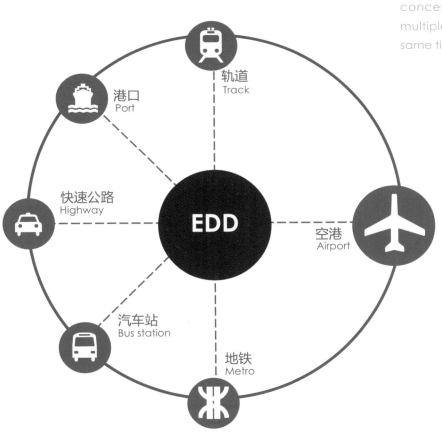

图 2-6 交通因素分析图
Figure 2-6 Analysis of Traffic Factors

功能圈层与规划布局模式

EDD 因其规模与辐射力，往往是所在城市的一个产业次中心，其内部产业功能链自我完善、自成一体。从产业链各功能之间关系的角度进行分类，会展综合发展区（EDD）自内而外可分为：会展核心功能区（图 2-7-1）、会展综合配套区、会展影响辐射区三大部分。

The Function Areas and the Layout

Due to its scale and influence, the EDD is often an industrial sub-center of the city, and the functions of different industries within should be self-improved and self-contained. According to the relationship of all kinds of functions of the industrial chain, the EDD is divided into three areas: the core function area (Figure 2-7-1), the comprehensive supporting area, and the influence area.

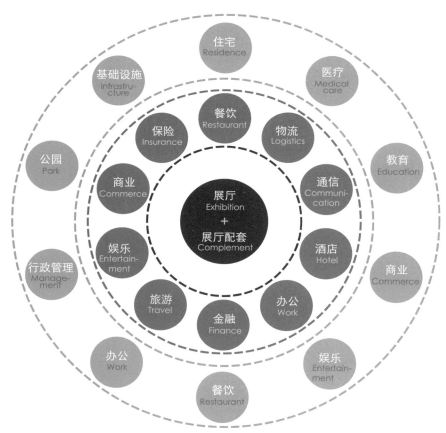

● **会展核心功能**　Core Functions

由会展中心主体及其配套场地或设施构成。主要包括：会展中心主体建筑、室外展场、货车场地、海关国检用房及场馆运营用房等；大型、超大型会展中心出于安全性考虑对上述内容进行封闭管理，地块独立，周边路网完善。

This consists of the main body of the exhibition center and its supporting sites or facilities. Mainly includes: the main building of the convention and exhibition center, outdoor exhibition field, truck field, national customs inspection room and venue operation room, etc.; large and super large convention and exhibition centers carry out closed management of the above content for safety consideration, with independent plot and complete surrounding road network.

● **会展综合配套圈**　Comprehensive Supporting Circle

由与会展中心紧密相关的功能板块构成。主要包括：酒店、办公、餐饮、商业、金融、保险、通信、物流、文化娱乐等功能。

This composed of functional plates closely related to the exhibition center. Mainly including: Hotel, office, catering, business, finance, insurance, communication, logistics, cultural entertainment and other functions.

● **会展影响圈**　Exhibition Influence Circle

由参与展会的人、货流所引发并带动的相关功能构成。主要包括：住宅、商业、办公以及相应配套的医疗、教育、餐饮、娱乐、公园、行政管理、基础设施等功能。

Formed by the related functions caused and driven by the flow of people and goods participating in the exhibition. Mainly including: residence, commercial area, office and corresponding supporting medical facilities, educational facilities, restaurant, entertainment facilities, park, administrative management facilities, infrastructure and other functional facilities.

图 2-7-1　会展结构功能分析图
Figure 2-7-1　Function Analysis of Exhibition Structure

组织模式
Organizing Mode

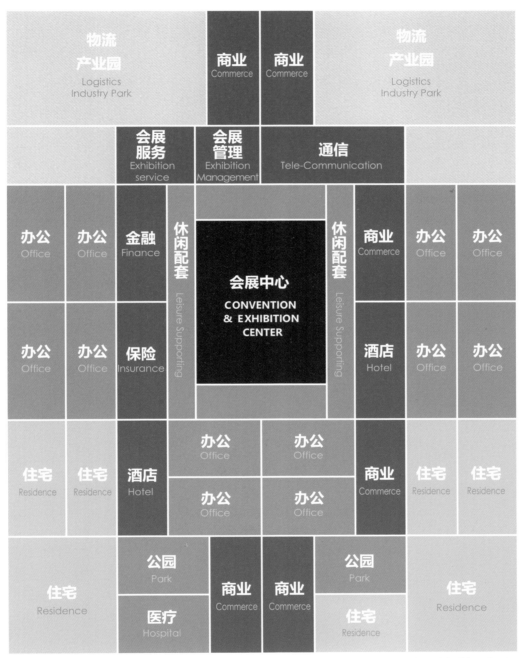

图 2-7-2　会展结构功能分析图
Figure 2-7-2　Function Analysis of Exhibition Structure

EDD 因其规模与重要程度，往往是所在城市的一个产业次中心，其内部产业功能链自我完善、自成一体（图 2-7-2）。

Because of its scale and importance, the EDD is often an industrial sub center of the city where it is located, and its internal industrial function chain is self-improvement and self integration(Figure 2-7-2).

会展核心功能区

会展核心功能区由会展场馆及其配套场地或设施构成。主要包含会展中心主体建筑、室外展场、货车场地、海关国检用房及场馆运营用房等；大型、超大型会展场馆往往出于安全性的考虑进行封闭管理，地块独立，用地四界路网完善方便组织各类交通。现代会展场馆通常常有如下三种布局模式：

对于大型、特大型会展场馆，双边式布局（模式二）因其流线明晰、货运高效及运营灵活等优势更受使用者欢迎，也是近年来国内外新建会展场馆的首选模式。在深圳国际会展中心的设计中，结合用地条件及主要的交通组织方式，明确了会展中心呈南北向展开布局，考虑到展厅整体规模已达到 50 万平方米的配置要求，最简洁明确的直线型方案对于来访者而言方向感最佳；同时双边式布局的货运场地空间更充足，相比其他布局模式能够更高效地组织布展、撤展车辆，提高"翻台"效率，进而提高场馆的使用频率；同时中廊串联各展厅的布局能够为各类不同规模的展会提供更灵活的组织方式，建成后的深圳国际会展中心一期可根据需求向主办方提供 1 万至 40 万平方米的不同展厅规模（图 2-8）。

The Core Function Area

This area is made up by the exhibition venues and its supporting venues or facilities, mainly including the main building of the exhibition center, outdoor area, truck parking area, "builds for the customs", CIQ and daily operations. For security reasons, modern large and super large exhibition venues often adopt the "closed" management method by using independent areas which has perfect road network for convenient organizations of all kinds of transportations. There are usually three modes of layout for modern exhibition centers: For large and super-large exhibition venues, the bilateral layout (mode 2) is more popular within users because of its clear streamline, efficient freight transportation and flexible operation. It is also the preferred mode for the exhibition venues newly built both at home and abroad in recent years.

According to the land conditions and the main organization of the transport, the Shenzhen World determined to adopt the south-north layout. Considering the whole scale has reached a configuration requirement of 500,000 square meters, the most simplest and clearest linear plan is best for the visitors' sense of direction.

Meanwhile, this bilateral layout had more sufficient space for the freight management. Compared to other layout, it's more efficient to organize vehicles for the assemble and removal of exhibitions, so as to improve the frequency of use. Moreover, the central corridor of this layout which connects all the exhibition halls can provide a more flexible way to organize all kinds of exhibitions in different sizes.

The phase strengthens of the Shenzhen World can provide organizers with different exhibition scales from 10,000 to 400,000 square meters according to their needs (Figure 2-8).

● 模式一：单边式布局
Mode 1: One-side Layout

● 模式二：双边式布局
Mode 2: Bilateral Layout

● 模式三：其他布局
Mode 3: Other Layouts

图 2-8　场馆布局模式分析图
Figure 2-8　Analysis of Venue Layout Mode

会展影响辐射区

会展影响辐射区（图 2-9）由会展场馆带动的产业需求相关的功能构成。主要包括办公、物流、产业园、住宅、商业及其相应配套的医疗、教育、餐饮、娱乐、公园、行政管理、基础设施等功能。这一圈层与上述两圈层共同形成整个会展产业链。在深圳国际会展中心的设计中，在解决好上述两圈层的整体布局后，深圳市政府以会展为核心，对周边 30 平方公里范围进行了规划研究，并提出"三城一港"（三城指国际会展城、会展田园城、海洋新城，一港指引领湾区的海洋产业新城的宝安综合港区）的城中城新格局。力图打造面向全球的高端功能聚集环（图 2-9）。

The Influence Area

The Influence Area (Figure 2-9) is composed of the functions driven by the exhibition venue, including offices, logistics, industrial park, residence, business, as well as supporting facilities such as medical care, education, restaurants, entertainment, parks, administrative management, and infrastructure, etc. This circle, together with the other two above, form the whole exhibition industry chain.

In the design of the Shenzhen World, after solving the overall layout of the above two areas, the Shenzhen government takes the Shenzhen World as the core, conducted planning research on its 29 hectares surrounding area, and put forward a new city structure of "Three Cities and One Port", which is an international convention and exhibition city, a new marine city, an exhibition garden city and comprehensive port area, and strive to build a global-oriented aggregation circle of high-end functions.(Figure 2-9)

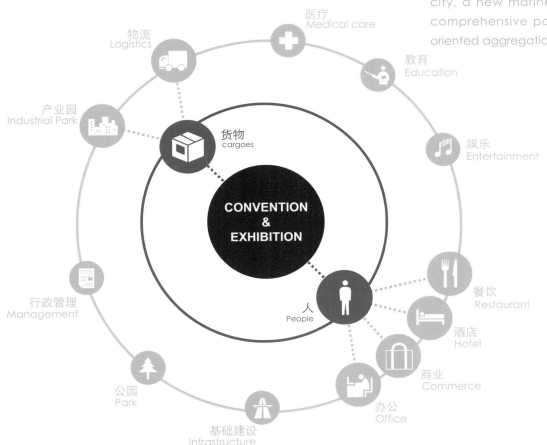

图 2-9　会展影响辐射区分析图
Figure 2-9　Analysis of the Radiation Zone Affected by the Exhibition

会展配套休闲区

沿会展核心功能区外围设置会展配套休闲区，主要由社会停车场、各类公交场站（含轨道站点）、配套商业、休闲公园等功能组成。对于整个会展片区而言，会展配套休闲区兼容了各类公共交通系统、会展外围集散、社会停车场、配套商业区及会展主题休闲公园等五大功能；在深圳国际会展中心的设计中，它自下而上地整合了地下多层车库、轨道站点、地铁商业、地面各类公共交通、市政配套设施、高线公园及休闲商业。同时，会展配套休闲区紧邻会展中心，大大减少了会展中心在使用中对于周边配套的交通与环境品质等方面的负面影响，同时也为来访者与周边市民提供了一处有特色、有规模的休闲运动场所。

会展综合配套区

会展综合配套区是由为展会聚集的人、货、信息提供服务的各类功能构成，是会展场馆的强配套功能圈层，其功能配置及服务水平将直接影响展会活动的质量。主要包含配套酒店、办公、餐饮、商业、金融、保险、通信、物流、文化娱乐等各类功能。各部分功能容量应与场馆规模比例适宜。但在实际操作中，考虑到展会活动的季节性及波动性特征，可做相应调整。例如可以通过配置相应比例的公寓类产品来减少星级酒店的配置，在降低酒店的投资压力的同时，增加开发收益，进而平衡场馆的建设支出（图 2-10）（图 2-11）。

图 2-10　会展综合配套区分析图
Figure 2-10　Analysis of Comprehensive Exhibition Area

Supporting Leisure Area

The periphery of the core function area is the supporting leisure area which including social parking lots, various bus stations (contains metro stations), supporting businesses, leisure parks and other relative functions.

For the whole area, the supporting leisure area is compatible with all kinds of public transport hubs, peripheral distributions, social parking lots, supporting business districts and theme leisure parks.

The design of the Shenzhen World, integrated the underground multi-storey garage, metro stations, subway business, ground public transportation, municipal supporting facilities, high line park and leisure business from "the bottom to the top".

Meanwhile, this area is close to the Shenzhen World, which could reduces the negative impact of the exhibition center to the surrounding transportation and the quality of the environment during the exhibitions, and provides a unique and large-scale leisure and sports place for visitors and the residents nearby at the same time.

Comprehensive Supporting Area

The comprehensive supporting area consists of various functions which provide services for people, goods and information that gathered in the exhibitions. It is a strong supporting functional circle for the exhibition venues, whose configuration and service level will directly affect the quality of exhibitions and activities. It mainly consists of hotels, offices, restaurants , business, finance, insurance, communications, logistics, entertainments and other functions.

The capacities of different functions are proportioned according to the size of the venues. In practical use, it could also be adjusted according to seasonal changes. For example, it can increase the proportion of apartments and reduce the proportion of star-rated hotels, so as to reduce the investment pressure of hotels and increase the income, and thus also, balance the construction expenditure of exhibitions venues (Figure 2-10, 2-11).

	模式一：单边式布局 Mode 1: One-side layout	模式二：双边式布局 Mode 2: bilateral layout	模式三：其他布局 Mode 3: other layouts
Small Exhibition 小型展会 ▶	优秀 Excellent	良好 Good	良好 Good
Medium Exhibition 中型展会 ▶	良好 Good	优秀 Excellent	优秀 Excellent
Logistics Organization 大型展会物流组织 ▶	--	良好 Good	较差 Not Good
Sense of Direction 步行距离及方向感 ▶	--	良好 Good	良好 Good
Multiple Exhibitions 多个展会 ▶	优秀 Excellent	优秀 Excellent	良好 Good
Land Use Efficiency 用地利用效率 ▶	良好 Good	优秀 Excellent	较差 Not Good
daptability of Land Shape 用地形状适应性 ▶	良好 Good	良好 Good	较差 Not Good

图 2-11 场馆布局模式分析图
Figure 2-11 Analysis of Venue Layout Pattern

设计一座城的未来
—— 会展中心
The Shenzhen World — The Design of a Future City

从 1851 年首届世界博览会开幕之日起，现代会展业就开始与城市发生紧密的联系，向城市展现自己独特的魅力，并深刻地影响着城市的面貌。这种会展与城市紧密的内在关联被称为会展的城市性。特别在当今中国，会展中心往往被作为城市顶层决策的重大战略部署，也因此其城市性更显重要。城市空间是为人而设计的，因而城市空间的大小尺寸，除了满足基本功能需求之外，更应将人的感受作为衡量标准。会展中心建筑体量巨大，且专业性强，往往会对城市日常生活造成巨大的压力。然而作为城市的一部分，会展中心不应该与城市互相隔离，而是应当更好地适应并融入城市生活。

Since the opening of the first world expo in 1851, modern exhibition industry has been closely connected with cities, showing their unique charm and affecting the appearance of cities profoundly. We call this close inner relation between exhibition and city "the urbanism of exhibition". Especially in China today, the urbanism is much more important because the exhibition center is often regarded as a major strategic deployment with in urban top decision-making.

Urban space is designed for human beings. Therefore, except meeting the basic needs, the humanity should be one of the determine factor of the size of urban space.
The exhibition center has a large volume and its use alway has strong specialties, which could causes great pressure on people's daily life. However, as part of the city, the exhibition center should not be isolated from, but better fit into the city life.

空间结构，一岛一轴一带，四个复合城区
Spatial Confisuration, One Island, One Axis, One Belt, and Four Comprehensive Urban Areas

用**城市设计**的方法对用地周边的**公共景观资源**进行梳理，有针对性的应对研究范围内的各类问题。

Use the method of urban design to sort out the public landscape resources of surrounding lands, in order to tackle with all sorts of problems presents within the research project.

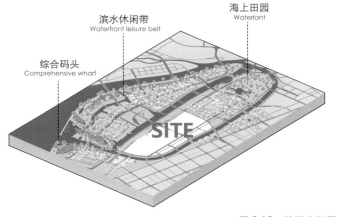

图 2-12　梳理分析图
Figure 2-12　Site Sorting Analysis Map

1. 梳理（如图 2-12）　　Summarize (Figure 2-12)

用地北侧与**海上田园**休闲公园隔路相望；
用地西南方向为福永海河**入海口**，沿河岸为极具滨海特色的**公共景观休闲带**。
用地的东西两侧均为**河岸景观**，河道宽 120-160m。

The north side of the land and the waterland leisure park is located on either side of the road.The southwest of the land is the Fuyong river estuary which has leisure and characteristic public landscaping along its coast. Both sides of the land have river landscape,and the channel is 120-160m wide.

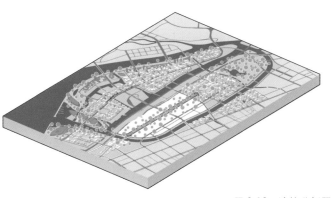

图 2-13　连接分析图
Figure 2-13　Connection Analysis Map

2. 连接（如图 2-13）　　Connection (Figure 2-13)

延续滨海带的走向，让滨海景观向陆地方向渗入是设计的出发点，沿着用地中部营造**中央公园**，向北联系海上田园休闲公园，向西南沿河连接入海口海滨带。

The purpose of this design is to let the coastal landscape infiltrate into the inner land along the coast, creates a central park at the middle of the land. The north part of the land connects the leisure park of the waterland, and the southwest along the riverside connects the estuary and littoral zone.

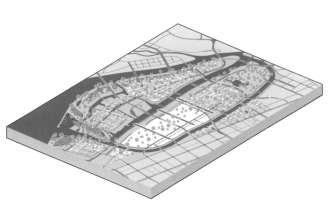

图 2-14　理水分析图
Figure 2-14　Analysis of Water Treatment

3. 理水（如图 2-14）　　Water Treatment (Figure 2-14)

理水，用地临近海湾，逢恶劣天气常遇**海水倒灌**，海洋安全是理水的主要原因，用地内设置**东西向水道**，联系两侧河流，有利于降低海水危害。

Inwelling could happen during bad weather since the land is near the gulf. Therefore the main reason for the water treatment is marine satefy, thereupon it setseast-west oriented waterway to connect the rivers on the both sides, to reduce the water damage.

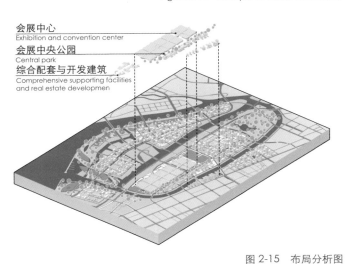

图 2-15　布局分析图
Figure 2-15　Function Layout Analysis

4. 布局（如图 2-15）　　Arrangement (Figure 2-15)

用地内部的各类功能建筑**按需配置**，西侧临路布置**会展建筑**，东侧临河布置各类**城市功能建筑**，中央公园处两者之间，**平衡**两侧悬殊的尺度，地上公园、地下交通与服务为城市服务，公园又可作为雨水花园增加场地的**透水性**。

Different functional buildings within the land were allocated according to needs. Exhibition and convection buildings were located in the west along the road, and various functional buildings were located in the east along the river, the central park is in between to balance the scale disparity, with the park on the ground and the underground traffic serve the city simultaneously.The park can also incerasing site water permeability as a rain garden at the same time.

一岛一轴一带，四个复合城区
One Island, One Axis, One Belt, and Four Comprehensive Urban Areas

一岛：会展岛

回应大片区岛群意向，打造功能复合的会展岛（图 2-16）。

One Island: an island for exhibition and convention

An island with integrated functions, echoing the image of islands group (Figure 2-16).

图 2-16　会展岛定位图
Figure 2-16　Exhibition Island Location Map

一轴：会展—海上田园功能轴

延伸会展人行廊道南至 T4 空港，北通海上田园（图 2-17）。

One axis: from exhibition and convention area to the Shenzhen Waterlands Resort

The axis can stretch the corridor of exhibition and convention from the south to T4 airport, and from the north to the Shenzhen Waterlands Resort.(Figure 2-17)

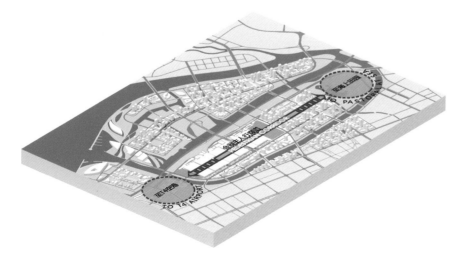

图 2-17　功能轴分析图
Figure 2-17　Function Axis Analysis Diagram

四个复合城区

三条东西向景观水带将用地划分为自南向北的四个复合城区，原用地内存在多条东西向河涌，整体规划设计结合了主要的人行广场与公共空间，并还原了三条河涌，加强了东侧城市片区与会展中心在景观与公共空间的联系与渗透，形成自南向北的四个复合城区，从城市肌理、东西向城市联系及开发时序等各方面给予回应（图2-18）。

Four Comprehensive Urban Areas: three east-west water landscape belts will separate the land into four comprehensive urban areas from north to the south.

There used to be several east-west, oriented rivers within the land. Therefore, in the general plan, three rivers will be restored in accordance with the main pedestrian square and public space. This will enhance the connection and interaction between urban areas in the east and the Exhibition and Convention Center via landscape and public space, and thus bring about the four comprehensive urban areas, provide answers relate to urban texture, relation between east and west urban areas and development phases. (Figure 2-18)

图 2-18 四个复核区分析图
Figure 2-18 Analysis of Four Comprehensive Urban Areas

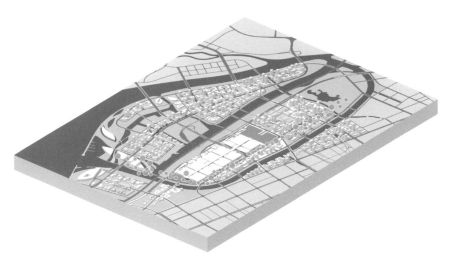

图 2-19 中央公园带分析图
Figure 2-19 Analysis Map of the Central Park Belt

一带：复合型的会展中央公园带

中央公园的引入除了能完善片区城市绿脉，提供更优质丰富的城市空间外，下部设置轨道交通，能更便利地辐射两侧城市功能带，同时利用轨道交通的深基开挖，设置社会停车场与地铁商业，达到土地资源的立体复合化利用。

在一二期会展东侧入口大厅处通过地下层直接联系地铁站厅层，并打造集地铁、商务旅游大巴、公交与出租车上下客等于一体的复合公交体系的地下交通中心，大大降低了会展期地表道路的交通压力。同时，整体中央公园可用做雨水花园，对周边区域的雨水收集及增强片区透水性提供可能性（图2-19）。

One Belt: a comprehensive central park belt in the exhibition and convention area.

The Central park belt can not only increase the urban green land and offer abundant urban space, but also provide underground space for railway transportation. With the railway transportation, the urban function areas at either sides of the park could provide convenience transportation. Moreover, with underground railway transportation, underground space can be used for public parking lot and metro business, and thus achieve three-dimensional and comprehensive use of land resources.

East entry hall of the exhibition and convention area will be directly linked with metro station hall through the underground floor. An underground transportation center will be instored to integrate metro, tourism bus, public bus and taxi. With help of the transportation center, pressure for ground transportation will be released. Meanwhile, the central park could also act as a rain garden, provide possibilities for rain recycle and water permeability within the surrounding areas. (Figure 2-21)

控制原则与要素
Control Principles and Important Factors

一岛：会展岛
One Island: an island for exhibition and convention

图 2-20 会展岛分析图
Figure 2-20 Exhibition & Convention Island

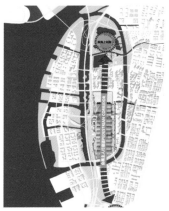

一轴：会展 - 海上田园功能轴
One Axis: from exhibition and convention area to Shenzhen Waterlands Resort

图 2-21 功能轴分析图
Figure 2-21 Function Axis Analysis Diagram

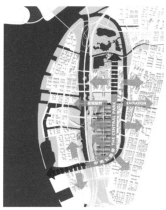

一带：复合型的会展中央公园带
One Belt: a comprehensive central pork belt in the exhibition and convention area.

图 2-22 中央公园带分析图
Figure 2-22 Analysis of Central Park

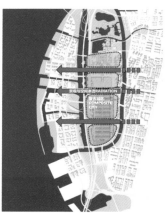

四个复合城区
Four Comprehensive Urban Areas

图 2-23 四个复合区分析图
Figure 2-23 Analysis of Four Composite Areas

控制原则

功能角度：实现功能的精准服务
结合会展中心功能、服务需求及人群动线，组织清晰有序的功能单元，建立精准、有效、多元的配套服务系统。

系统角度：组织网络化的开放空间
延续上位空间骨架，优化垂直交通，组织多水平面、多层次的立体复合开放空间网络，构建高密度、特色化的开放空间系统。

形态角度：构建识别性强的建筑场所
结合片区空间秩序，协调福永海河、会展中心、配套建筑的空间关系，塑造特色化空间及建筑，引领大空港地区的建设。

实施角度：可操作的弹性规划
应对开发实施的经济可操作性及方案可变性，在落实规划要求的基础上，预留充分的弹性。

Control Principles

Function: Accurate Service
Combining the function and service of the exhibition center and the moving line of the crowd, organize clear and orderly function unites, and establish an accurate, effective and diversified supporting service system.

System: Open Space with Networked Arrangement
Continue to use the upper space framework, optimize vertical transportation, organize Multi-Leveled open space network, and build a high-density and featured open space system.

Form: Buildings with Strong Identification
Coordinate with the spatial order of the area, to coordinate the spatial relationship among the Fuyong River, the exhibition and convention center and the supporting buildings, the built space and buildings with characteristic, leading the construction of the Airport Area.

Implementation: Flexible Operational Planning
The development and implementation of the ecnomical operability and alterable plan, guarantee the sufficient flexibility based on the requirements of the implementation.

控制要素　　　**Control Factors**

深圳国际会展中心规划及建筑设计重要控制要素（表2-1）。

表2-1　控制要素表格
Table 2-1　Control Element Table

控制分项 Control sub-item		控制要素 Control element		B类 Class B	
				优先级 priority	选择 selection
功能分析 Function Analysis	A.地块细分和开发强度控制 A. Parcel subdivision and development intensity control	A1 地块边界及面积 A1 plot boundary and area		▲	✓
		A2 用地性质及相容性 A2 Land use properties and compatibility		▲	✓
		A3 建筑总量及分类建筑面积 A3 total construction and classified building area		▲	✓
		A4 配套设施（包括公共配套及市政配套） A4 supporting facilities (including public facilities and municipal facilities)		▲	✓
		A5 容积率 A5 floor area ratio		▲	✓
		A6 绿地率（和人均公共绿地面积） A6 Greenland proportion (and per capita public green area)		▲	✓
		A7 居住人口/户数 A7 Resident population / number of households		△	
	B.道路交通及竖向控制 B. Road traffic and vertical control	B1 道路系统设计（功能、等级、密度、红线、控制点坐标、断面、重要交叉口形式） B1 Road system design (function, level, density, red line, control point coordinates, section, important intersection form)		△	✓
		B2 交通组织（公交/地铁线路及站点布局、平面/立体人行系统组织）交通分析 B2 Traffic organization (transit/metro line and site layout, plane/three-dimensional pedestrian system organization) traffic analysis		△	
		B3 停车场及配建车位 B3 Parking lot and construction parking space		▲	✓
		B4 其他交通设施 B4 Other transportation facilities		○	✓
		B5 道路及场地竖向设计（控制点标高、道路纵坡、场地基准标高） B5 Vertical design of roads and sites (control point elevation, road longitudinal slope, site reference elevation)		▲	✓
	C.市政工程规划 C. Municipal engineering planning	C1 给水工程（负荷、设施、管网） C1 water supply Engineering(load, facility, pipe network)		○	✓
		C2 排水工程（负荷、设施、管网） C2 drainage Engineering (load, facility, pipe network)		○	✓
		C3 电力工程（负荷、设施、管网） C3 Power Engineering (Load, Facilities, Pipe Network)		○	✓
		C4 电信工程（负荷、设施、管网） C4 Telecommunications Engineering (Load, Facilities, Pipe Network)		○	✓
		C5 燃气工程（负荷、设施、管网） C5 Gas Engineering (Load, Facilities, Pipe Network)		○	✓
		C6 工程管线综合（各项市政工程负荷、设施、管网及其相互关系） C6 Engineering pipeline integration (various municipal engineering loads, facilities, pipeline networks and their interrelationships)		○	✓
功能分析 Function Analysis	D.地区空间组织 D. Regional Space Organization	D1 空间结构关系 D1 spatial structure relationship		△	✓
		D2 街道空间模式（平面几何形态及空间尺度关系） D2 street space mode (plane geometry and spatial scale relationship)		△	
		D3 公共开放空间（绿地、广场和水域）布局、步行可达范围覆盖率和人均公共空间面积指标 D3 Public open space (green space, square, water area) layout, walking reach and per capita public space area		△	
	E.建筑形态控制 E. Architectural form control	E1 建筑退线 E1 Building exit line	一级 First level	▲	✓
			二级 Secondary level	△	✓
		E1 建筑贴线 E1 Building line	一级 First level	△	✓
			二级 Secondary level	△	
		E3 建筑高度 E3 Building height		▲	✓
		E4 建筑前遮蔽高度 E4 Front height of the building		△	✓
	F.公共开放空间 F. Public open space	F1 位置 F1 Position		△	✓
		F2 面积 F2 Area		△	✓
		F3 尺度 F3 Scale		△	✓
		F4 形式（绿色空间/广场空间/运动空间） F4 Form (green space / square space / sports space)		○	✓
	G.公共通道 G. Public access	G1 步行通道位置（空中/地面/地下、平面位置） G1 Walking path position (air/ground/underground, plane position)			
		G2 步行通道尺度 G2 Walking channel scale		△	✓
		G3 服务性车行通道位置 G3 Service car lane location		▲	✓
	H.地块出入口 H. Plot entrance	H1 车行出入口位置 H1 Car entrance and exit location		▲	✓
		H2 人行出入口位置 H2 Pedestrian exit location		▲	✓
其他控制要求 Other control requirements		1.无障碍设计 1.Accessible Design		○	
		2.历史遗产保护 2.Historical heritage protection		○	
		3.自然生态保护 3.Ecology protection		○	
		4.特殊活动支持（如民俗活动等） 4. Special event support (such as folk activities, etc.)		○	
		5.灯光夜景 5.Night lights		○	
		6.户外广告和标识系统 6.Outdoor advertising and signage system		○	
		7.街道设施 7.Street facilities		○	
		8.实施操作建议及分期实施计划等 8.Implementation of operational recommendations and phased implementation plans, etc.		○	✓

（注：表中优先级符号：▲——比选要素，及必须选择的基本要素；△——宜选要素，即建议选择的要素；○——可选要素，即可以自由选择的相关要素。）

(Note: Priority symbols in the table:▲——Comparison factor,And the basic elements that must be chosen；△——Optional elements, ie the elements of the proposed choice；○——Optional elements are related elements that can be freely selected.)

项目规划指标
Planning Indicators

按照规划指标，深圳国际会展中心项目用地面积 148 万平方米，规划建设室内展厅面积 50 万平方米，整体建成后将是全球最大的会展中心。

会展一期用地面积 121.4 万平方米，总建筑面积 160 万平方米，其中展厅面积 40 万平方米，会议、办公、餐饮、仓储等配套面积 64 万平方米，地下车库及设备用房 56 万平方米（图 2-24-1）。

会展周边配套用地指标为：商业用地总面积 52.8 万平方米，总建筑面积 154.3 万平方米。
其中酒店 25 万平方米；办公 26.2 万平方米；商业 32 万平方米；公寓 69.7 万平方米（含 10 万平方米人才公寓）；公共服务配套 1.4 万平方米（图 2-24-2）。

图 2-24-1　规划指标分析图 1
Figure 2-24-1　Analysis of Planning Indicators

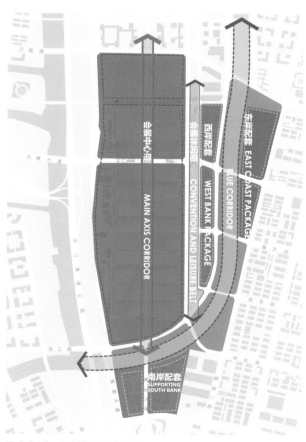

图 2-24-2　规划指标分析图 2
Figure 2-24-2　Analysis of Planning Indicators

According to the planning indicators, the Shenzhen World covers an area of 1.48 million square meters, and the planned indoor exhibition halls cover an area of 500,000 square meters, which will be the largest exhibitions center in the world when completed.

The first phase of the project covers an area of 1.214 million square meters, with a total construction area of 1.6 million square meters, including 400,000 square meters of exhibition halls, 640,000 square meters of conference rooms, office, catering, storage and other supporting facilities, and 560,000 square meters of underground parking and equipment rooms (Figure 2-24-1).

The indicator of supporting land around the exhibition center: commercial land covers an area of 528,000 square meters, with a total construction area of 1.543 million square meters, including 250,000 square meters of hotels; 262,000 square meters of offices; 320,000 square meters of business; 697,000 square meters of apartments (contain 100,000 square meters of talent apartments), and 14,000 square meters of public service area (Figure 2-24-2).

展城呼应

——开放提升城市空间

Exhibition Center + City — A More Open and Advanced Space

深圳会展中心展厅建筑的布局在满足会展功能的基础上,将展厅的短边朝向主要城市界面,在人的尺度上化整为零,从一定程度上削弱过大的建筑体量对人产生的压迫感;同时充分利用展厅之间通廊,加强东西两侧联系。展厅之间的通廊和连廊与外围的城市道路和空间通廊之间形成对位关系;主入口登录大厅前的广场也和城市滨水广场直接相连,打通人流和视线廊道。

通过积极呼应城市肌理,会展中心的布局模式留下部分空间,为未来与城市互通、对城市开放并将城市生活空间引入形成街道提供了可能性。

On the basis of meeting the requirement of building functions, the shorter edge of the exhibition halls is designed to face the city according to the layout, to reduce the pressure of such a large building forced on human beings. Meanwhile, Strengthen the east - west connection by making full use of the corridor between the exhibition halls. The corridors and connecting corridors formed a counterpoint to the peripheral city streets and the outdoor road system. The square in the front of the Arrival Hall at the main entrance is connected to the city waterfront plaza, forming a corridor for both the crowd and the sight.

By embracing the urban space texture, the layout leaves enough spaces for the possibility of connecting with the city, open to the city and introduce urban living space in the form of streets into the area in the future.

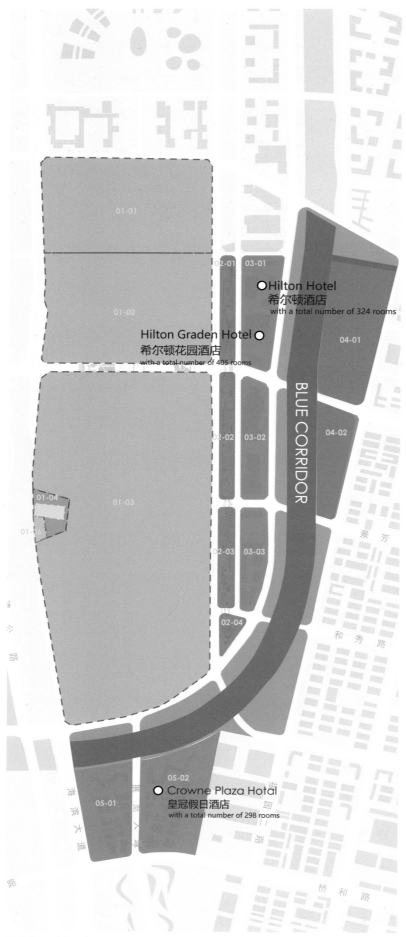

城市配套
Urban Supporting Facilities

会展配套用地总面积 52.8 万平方米，总建筑面积 154 万平方米，包含城市地标、产业总部、商业中心、国际酒店群、精品公寓、生态公园、交通枢纽等。此处将成为城市新中心，满足全球会展、参展人员的一站式需求（图 2-25）。

共 4 块酒店用地，总建筑面积为 250,000 平方米。规划建设 7 个高端酒店，酒店房间总计将超过 4,000 间，满足各种会展和商务需求。目前已确定入驻的酒店有：希尔顿酒店 (324 间房)、希尔顿花园酒店 (405 间房)、皇冠假日酒店 (298 间房)。

Exhibition bay covers a total supporting land area of 1.54 million square meters, including city landmarks, industrial headquarters, business centers, international hotels, boutique apartments, ecological parks, transportation hubs, etc., which will become a new city center to meet the one-stop needs of exhibition organizers and exhibitors around the world (Figure 2-25).

There are tatal 4 hotel-use lands with area of 250,000 square meters, with a plan to build 7 high-end hotels, with more than 4,000, hotel rooms in total to meet the needs of various exhibition and business. Hotels that have confirmed to settle in inlcude: Hilton Hotel (324 rooms), Hilton Garden Hotel (405 rooms) and Crowne Plaza Hotel (298 rooms).

图 2-25　城市配套分析图
Figure 2-25　Analysis of Urban Supporting Facilities

表2-2 会展综合配套规划指标表
Table 2-2 Comprehensive Supporting Facilities Planning Control Index Table

地块编号 Parcel number	用地性质代码 Site Property Code	用地性质 Land use	用地面积(m²) Land area (m²)	规定总建筑面积(m²) Required total construction area (m²)	容积率 Volume rate	地上规定建筑面积(m²) Ground regulations construction area (m²)	地上分项建筑面积(m²) 会展 Exhibition	酒店 Hotel	办公 Office	地上商业 Above ground business	公寓 Apartment	配套设施 Supporting facilities	地下规定建筑面积 Site Property Code	建筑覆盖率(%) 一级 First level	二级 Secon dary	绿化覆盖率(%) Green cover Cover rate(%)	配建车位(个) Build Parking space (A)	建筑限高(m) building Limit high(M)
01 01-01	GIC2	文体设施用地 Land for cultural and sports facilities	266342	199757	0.75	199757	199757	0	0	0	0	0	0	60	—	5	1998	40
01-02	GIC2	文体设施用地 Land for cultural and sports facilities	288064	216048	0.75	216048	216048	0	0	0	0	0	0	60	—	5	3000	40
01-03	GIC2	文体设施用地 Land for cultural and sports facilities	926140	694605	0.75	694605	694605	0	0	0	0	0	0	60	—	5	8000	40
01-04	G1	公园绿地 Park green	6566	—	—	—	—	—	—	—	—	—	—	—	—	—	—	—
01-05	U1	供应设施用地 Supply facility land	7529	—	—	—	—	—	—	—	—	—	—	—	—	—	—	—
01-06	S9	其他交通设施用地 Land for other transportation facilities	1029	—	—	—	—	—	—	—	—	—	—	—	—	—	—	—
02 02-01	C1	商业用地 Commercial land	23438	20023	0.26	6007	0	0	0	6007	0	0	14016	80	—	40	1000	14
02-02	C1	商业用地 Commercial land	27560	23544	0.26	7063	0	0	0	7063	0	0	16481	80	—	40	1000	14
02-03	C1	商业用地 Commercial land	22009	18801	0.26	5640	0	0	0	5640	0	0	13161	80	—	40	1000	14
02-04	C1	商业用地 Commercial land	11274	9632	0.26	2890	0	0	0	2890	0	0	6742	80	—	40	1000	14
03 03-01	C1	商业用地 Commercial land	64387	284000	4.16	268600	0	100400	144600	23000	0	0	16000	65	50	20	1949	116
03-02	C1	商业用地 Commercial land	51124	268600	5.02	256600	0	54000	35000	14000	153000	600	12000	65	50	20	1491	96
03-03	C1	商业用地 Commercial land		112200	2.43	96200	0	0	82000	14100	0	100	16000	65	50	20	957	72
04 04-01	C1	商业用地 Commercial land	110246	454680	4.12	454680	0	30600	0	25400	396000	2680	0	50	40	30	2278	112
04-01	C1	商业用地 Commercial land	42160	168800	4.00	168800	0	0	0	14000	149800	5000	0	50	40	30	860	84
05 05-01	C1	商业用地 Commercial land	65492	70800	0.92	60000	0	0	0	60000	0	0	10800	60	40	20	708	45
05-02	C1	商业用地 Commercial land	83178	112200	1.26	105000	0	65000	0	40000	0	0	7200	60	40	20	732	45

图 2-26 会展配套效果图 来自深圳国际会展城综合规划
Figure 2-26 Exhibition Supporting Facilities Rendering
Comprehensive Planning of Shenzhen International Exhibition City

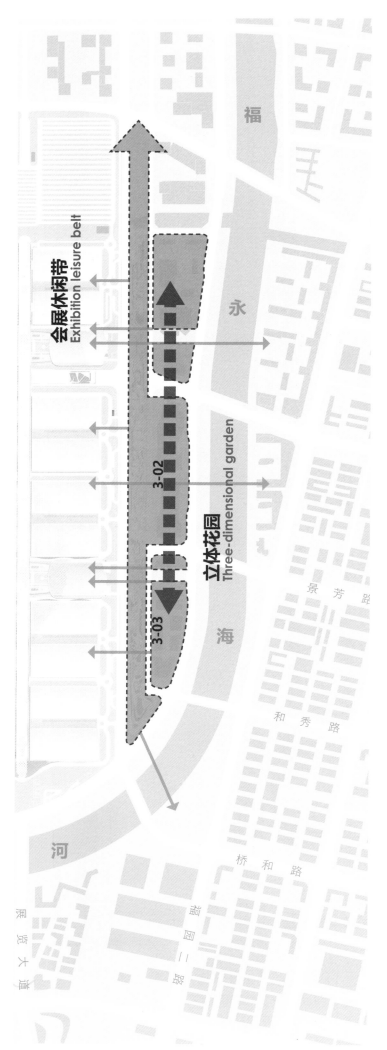

图 2-27　立体花园系统分析图
Figure 2-27　Analysis of the Three-dimensional Garden System

开放空间系统
Open Space System

立体花园系统

规划连续的二层平台系统，连接会展休闲带与福永海河，同时与周边建筑及地面开放空间形成良好的互动关系，构成一个完整的立体步行网络系统（图 2-27）。

地块内联系：于 02、03 街坊建立二层步行平台（立体花园）系统，在建筑二层设置集中的公共开放空间场所作为会展休闲带的延伸，通过空中连廊进行连接，利于不同人群的分流及休闲活动需求。03 街坊内部跨越凤塘大道、景芳路的空中联系宜连接裙房的多个楼层，形成紧密的商业氛围（图 2-28）。

地块间联系：通过跨街立体花园、空中连廊延伸会展休闲带至 03 街坊，将 02、03 街坊进行整体连接，提升会展休闲带的服务功能。

会展休闲带：为保证休闲带平台上人流动线的连续性，平台上建筑的东西向宽度不宜超过平台宽度的 2/5，局部根据具体功能可适当突破，但不应超过平台宽度的 4/5。

图例
Legend

 立体花园
Three-dimensional Garden

 03 街坊空中连廊
03 Street Square Sky Gallery

Three-dimensional Garden System

A sequential two-storey platform system is planned to connect the exhibition leisure belt with the Fuyong river, and form a good interactive relationship between the leisure belt and the surrounding buildings and open space on the ground, forming a complete three-dimensional walking network (Figure 2-27).

Connection within the plot: Building a two-storey walking platform (three-dimensional garden) system within the 02 and 03 blocks, and set up a concentrated public open space as the extension of the leisure belt and' provides different leisure activities. Inside the 03 block, there are multiple floors for intensive commercial activities (Figure 2-28).

Connection between plots: extend the exhibition leisure belt to block 03 through a cross-street three-dimensional garden and a overhang connection corridor, which connects block 02 and 03 as a whole, and enhance the service function of the exhibition leisure belt.

Exhibition leisure belt: in order to guarantee the continuity of crowd's moving line on the platform of the leisure belt, the width of east-west spaning building on the platform should not exceed 2/5 of the platform width. Some exceptions could be made according to specific function needs, but the width still should not exceed 4/5 of the platform width.

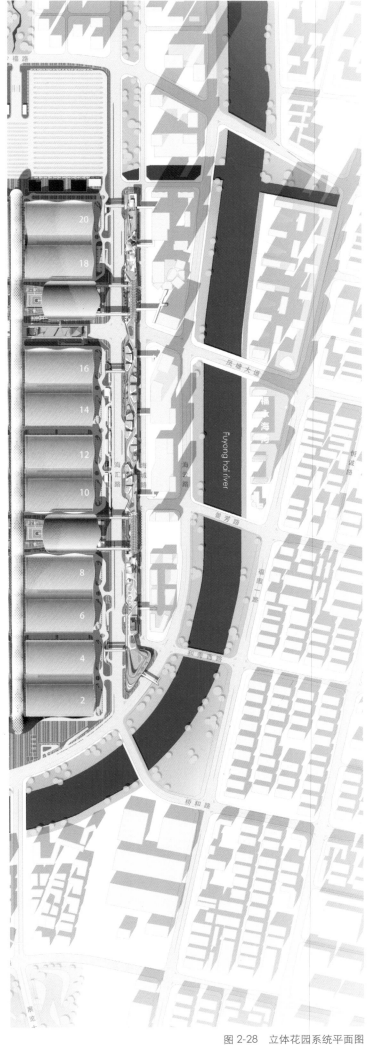

图 2-28　立体花园系统平面图
Figure 2-28　Plan View of a Three-dimensional Garden System

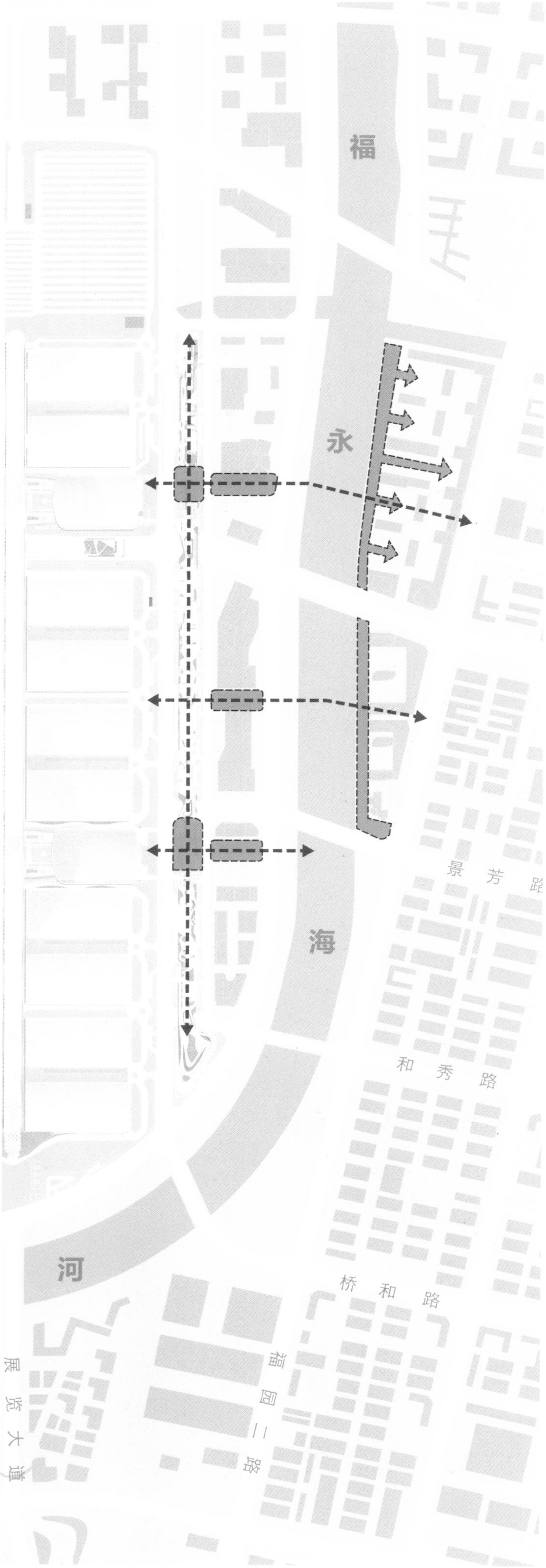

景观广场系统

利用会展的南北登陆大厅和景观商业广场形成三条横向的景观廊道，连接各个主要的功能区块（图 2-29）。

会展登录大厅面对的景观廊道应保证福园二路 - 福永海河 - 会展登录大厅的视线通畅，中部宜保持开敞，以大草坪、疏林草地为主导景观意向，广场两侧宜提供树荫、风雨廊、休闲广场等休憩空间，并布置适量商业空间（图 2-30-1，图 2-30-2，图 2-30-3）。

Landscape Square System

Three horizontal landscape corridors are formed by the north-south Arrival Hall and the landscape commercial plaza, connecting the main functional blocks.(Figure 2-29)

The corridor opposite the exhibition landing hall should guarantee a clear sight through the Fuyuan second road - Fuyong river - exhibition Arrival Hall, and the middle part should be kept open with the large lawn and sparse forest and grass land as its main landscape features. Each side of the square should provide shade, wind and rain corridors, leisure square and other public open space, and appropriate commercial space is neccessary as well.(Figure 2-30-1, 2-30-2, 2-30-3)

图例
Legend

 景观广场
Landscape Square

 开放空间轴带
Open Space Passage

图 2-29　景观广场系统分析图
Figure 2-29　Analysis of Landscape Square System

图 2-30-1 景观广场效果图
Figure 2-30-1 Rendering of Landscape Square

图 2-30-2 景观广场效果图
Figure 2-30-2 Rendering of Landscape Square

图 2-30-3 景观广场效果图
Figure 2-30-3 Rendering of Landscape Square

空间形态控制
Spatial Form Control

视线廊道控制

04 街坊应严格控制建筑面宽，每隔 75 m~150 m 设置一条垂直于福永海河的视线通廊（位置与公共空间、公共通道结合），视线通廊两侧建筑高度 24 m 以上部分的宽度不宜小于 25 m，24 m 以下部分不宜小于 15 m，其中与会展登录大厅正对处的视线应通畅且和慢行联系，其余通道具体位置可结合详细设计进行确定。

天际线与建筑高度控制

配套商业用地限高为 48.7～118 m。主体建筑会展中心建筑高度为 42.8m，会展登陆大厅为 40.1 m。

03、04 街坊较为狭长，用地布局紧凑，塔楼布局宜尽量错落，并在临海河侧适当降低高度，形成峡谷型天际线（图 2-31）。

Sight Corridor Control

The width of the construction site within Block 04 should be strictly controlled. Set up a perpendicular sight corridor to the Fuyong river every 75 to 150 meters (connecting the public space and public channel). The width of the upper parts of the buildings over 24 meters on both sides of the sight corridor should not be less than 25 meters, and no less than 15 meters for those who are below 24 meters.The space facing the exhibition hall should preserve the sight and the slow connection while the location of the rest of the channels can be determined according to the detailed design.

Skyline and Building Height control

The height limit of the supporting commercial land is 48.7 to 118 meters. The height of the main exhibition building is 42.8m, and the Arrival Hall is 40.1m.
Block 03, 04 are relatively long and narrow, the land use is relatively compacted, therefore the layout of the tower should be as scattered as possible, and reduce its height appropriatly on its side facing the river to form a canyon skyline.(Figure 2-31)

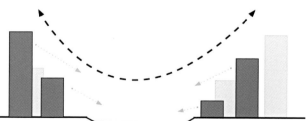
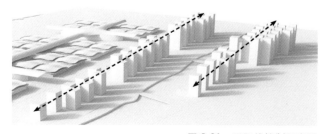

图 2-31　天际线控制示意图
Figure 2-31　Skyline Control Diagram

次要视线通廊
Secondary Sight Corridor

主要视线通廊
Main Sight Corridor

会展登录大厅
Exhibition Hall

会展休闲带 Exhibition leisure belt

主要视线通廊
Main Sight Corridor

次要视线通廊
Secondary Sight Corridor

会展登录大厅
Exhibition Hall

主要视线通廊
Main Sight Corridor

次要视线通廊
Secondary Sight Corridor

主要视线通廊
Main Sight Corridor

次要视线通廊
Secondary Sight Corridor

图例
Legend

主要视线通廊
Main Sight Corridor

次要视线通廊
Secondary Sight Corridor

图 2-32　空间形态控制分析图
Figure 2-32　Analysis of Spatial Morphology Control

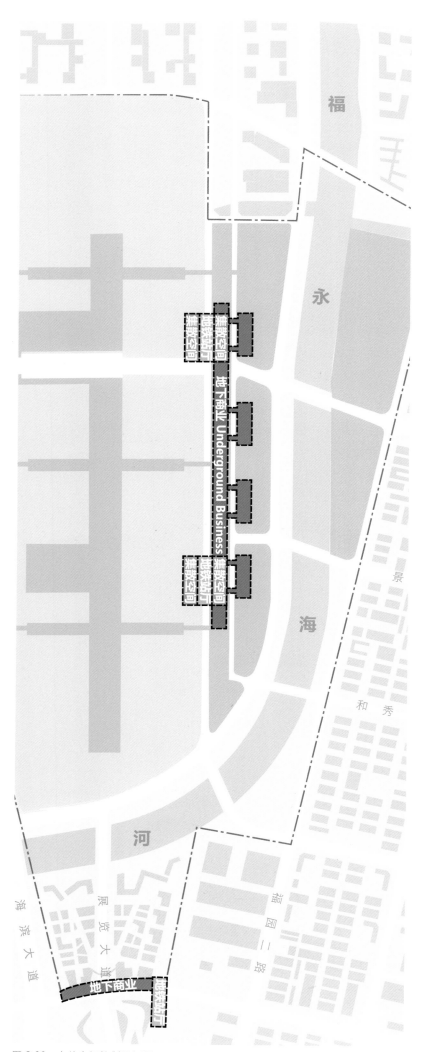

图 2-33 立体空间控制分析图
Figure 2-33 Analysis of Three - dimentional Space Control

立体空间控制
Three - dimentional Spacial Control

依据会展交通流量的模拟与预测计算，公共交通设置的需求较大，共需要 3 组公共交通枢纽，每一组包含一个公共交通站台，可同时容纳 9 辆公交车停靠，还包含一个出租车轮候上下客站，当中设置了三排轮候场地。将这些设施分设在休闲带三个地块的架空层，临近会展，整体分布，又能有效地服务整个会展中心，运营高效且释放了大量会展中心用地，可用作货运，提高会展本身的运行效率（图 2-33）。

According to the simulation and prediction calculation of traffic flow, public transportation is in huge demand. Three groups of public transportation hubs are needed. Each group includes a bus station which could accommodate 9 bus lines to pull over at the same time, a taxi station for pick-up and drop-off was set as well, with three lines of waiting and queueing areas.These facilities are set in the empty spaces of the three parcels of the leisure belt, which is close to the exhibition center and can serve it effectively. With highly efficient operation, this arrangement could save a large amount of land use for freight, so as to improve the operation efficiency of the exhibition itself. (Figure 2-33)

图例
Legend

地下停车空间
Underground Parking Space

地下商业空间
Underground Commercial Space

地铁站厅
Subway Station Hall

集散空间
Distributing Space

规划范围
Planning Scope

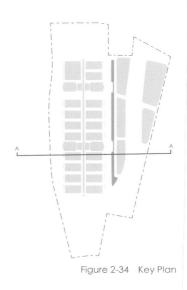

Figure 2-34 Key Plan

基于会展公众人流对于轨道交通的庞大需求，在会展覆盖的范围内，沿会展中心南北向设置了双线 4 站的地铁线路，及会展南站、会展北站两个地铁站。休闲带紧邻两条地铁线路，其地下空间也与地铁展厅空间产生了紧密的连接（图 2-35）。

Based on the huge demand for rail traffic, within the coverage of the exhibition and convention center, a south-north double subway lines and 4 stations has been set up, as well as exhibition north and exhibition south two stations. The leisure zone is adjacent to two subway lines, and its underground space is also closely connected to the subway space. (Figure 2-35)

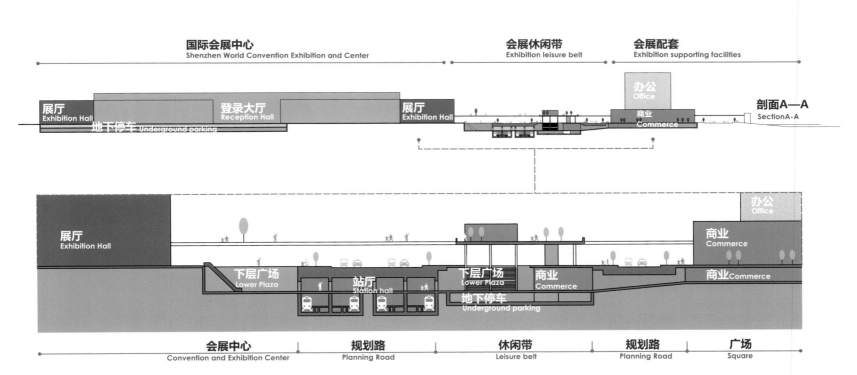

图 2-35　立体空间控制剖面示意图
Figure 2-35　Section of Three-dimensional Space Control

海绵城市策略
Rainwater collection and purification system

1．建立完善的雨水收集及净化系统：本区属于南亚热带海洋性气候，全年温湿多雨，平均年降水量1242.7mm。可收集雨水用于打造景观水景或储存于地下（图2-36）。

2．收集建筑用水，通过水净化系统，达到景观用水标准，储存于地下。

3．通过设计人工湿地，净化水质（图2-37）。

1. This area has a subtropical maritime climate. It is warm, humid and rainy over most of the year. With an average annual rainfall of 1242.7 mm, storm water is proposed to be collected and stored for landscape use, either within landscaped ponds or other open water features, or stored underground.(Figure 2-36)

2. Run-off water from buildings may also be collected, diverting that to constructed wetland filtration ponds for purifying prior to storage underground.

3. Purify water through artificial wetlands. (Figure 2-37)

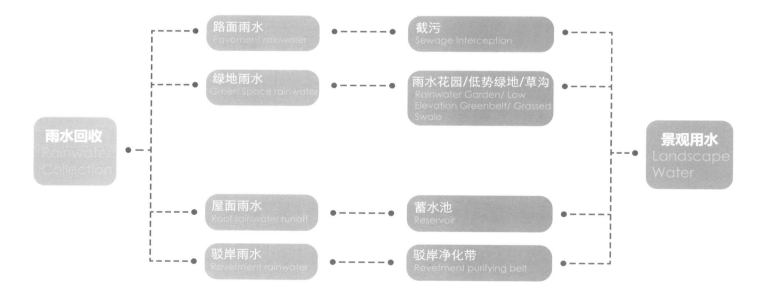

图2-36　海绵城市策略分析图
Figure 2-36　Analysis of Sponge City Strategy

图 2-37　海绵城市策略示意图
Figure 2-37　Sponge City Strategy Analysis Map

图例
Legend

 洪水公园
Flood park

 净化花园
Purification garden

 雨水花园
Rain garden

 地表径流
Runoff

 生态草沟
Bio-swale

 水系
Water System

 地下储水罐
Underground Water Storage Tank

 净化街道
Purified Street

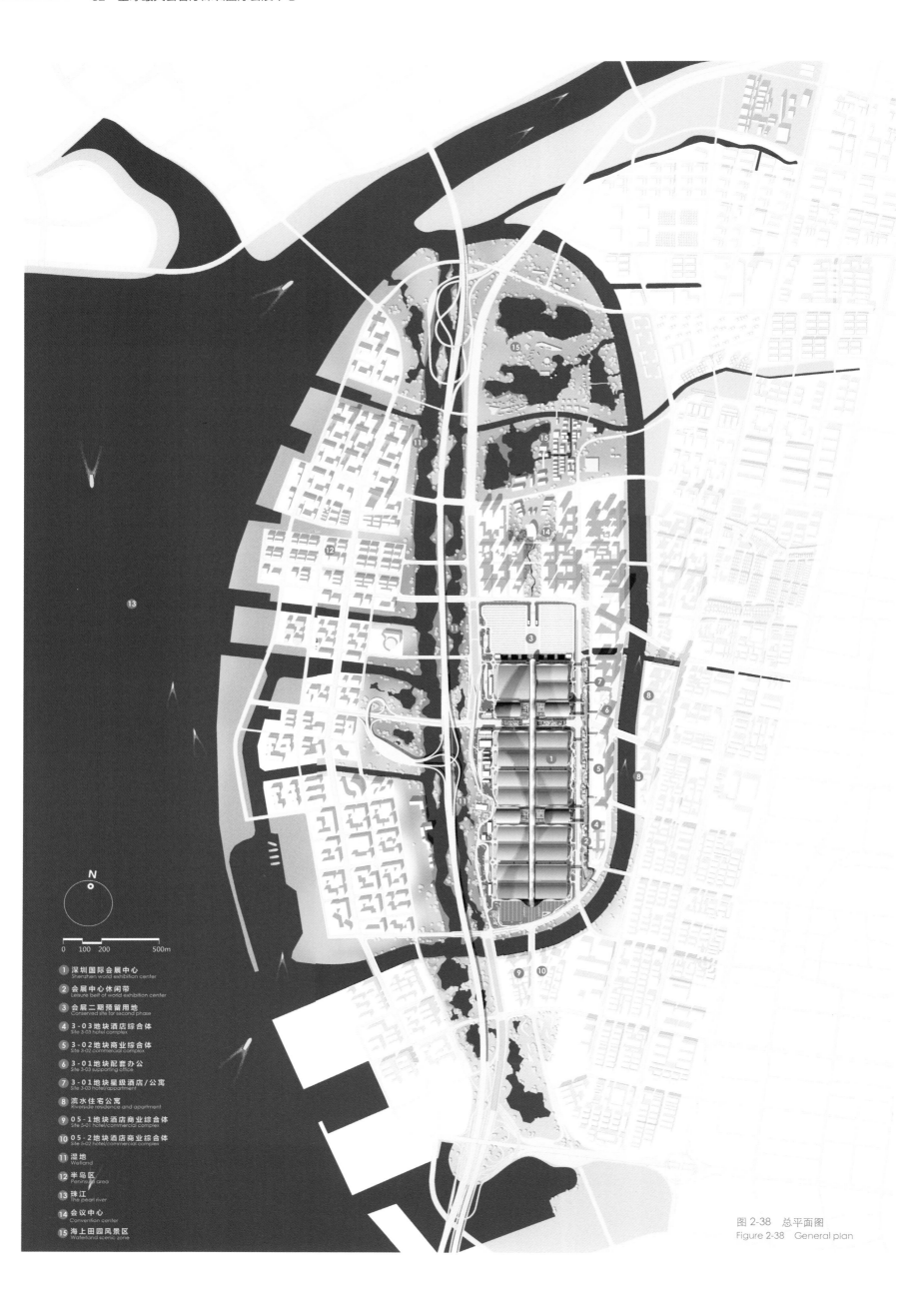

N

0   100   200   500m

① 深圳国际会展中心
   Shenzhen world exhibition center
② 会展中心休闲带
   Leisure belt of world exhibition center
③ 会展二期预留用地
   Conserved site for second phase
④ 3-03地块酒店综合体
   Site 3-03 hotel complex
⑤ 3-02地块商业综合体
   Site 3-02 commercial complex
⑥ 3-01地块配套办公
   Site 3-03 supporting office
⑦ 3-01地块星级酒店/公寓
   Site 3-03 hotel/apartment
⑧ 流水住宅公寓
   Riverside residence and apartment
⑨ 05-1地块酒店商业综合体
   Site 5-01 hotel/commercial complex
⑩ 05-2地块酒店商业综合体
   Site 5-02 hotel/commercial complex
⑪ 湿地
   Wetland
⑫ 半岛区
   Peninsula area
⑬ 珠江
   The pearl river
⑭ 会议中心
   Convention center
⑮ 海上田园风景区
   Waterland scenic zone

图 2-38 总平面图
Figure 2-38 General plan

规划成果表现

Planning Achievement Showcase

国际会展都市文旅目的地

大湾区科技创新产业新引擎

展城融合深圳西部城市中心极核

Cultural Tourism Destination as an International exhibition city.

A new engine of science and technology innovation industry in the GBA.

Shenzhen western city center core and exhibition-city integration.

图 2-39　整体鸟瞰图　来自深圳国际会展城综合规划
Figure 2-39　Bird's Eye View from the Comprehensive Planning of Shenzhen International Exhibition City

图 2-40　整体鸟瞰图　来自深圳国际会展城综合规划
Figure 2-40　Bird's Eye View From the comprehensive planning of Shenzhen International Exhibition City

图 2-41 整体鸟瞰图   来自深圳国际会展城综合规划
Figure 2-41   Bird's Eye View From the Comprehensive Planning of Shenzhen International Exhibition City

深圳国际会展中心设计全貌
The Overall Design of the Shenzhen World

———————————— 深圳国际会展中心设计全貌
The Overall Design of the Shenzhen World

94　会展中心的发展与演变
　　The Development & Evolution of the Exhibition & Convention Center

104　会展中心总体设计
　　The General Design of the Shenzhen World

110　会展核心功能设计
　　The Core Functional Designs of the Shenzhen World

184　会展配套
　　The Supporting Facilities

199　会展休闲带综合配套 —— 特殊会展配套
　　Special Supporting Facility——The Leisure Zone of the Exhibition

221　经济技术指标表
　　The Economic and Technical Indicators

展

Grand Exhibition Center

会展中心的发展与演变

The Development & Evolution of the Exhibition &
Convention Center

深圳国际会展中心既吸收了以大型专业高效工业展会为导向的德国会展类型经验，又融合了集专业展会与商务旅游、休闲、会议活动等多功能为一体的复合的美国会展类型的优势，同时还注入了深圳国际领先智慧会展的最新技术，并成就了一个全新的国际一流会展类型。

Learning from the experience of the German Exhibition Centers, which is oriented large professional and efficient industrial exhibitions, and blending the advantages from the American Exhibition centers, which combined professional exhibitions and business tourism, leisure, conferences and events, etc., and finally applying the latest advanced Shenzhen smart exhibitions technology, the Shenzhen World Exhibition and Convention Center is determined to be the latest upgraded, world-class exhibition center.

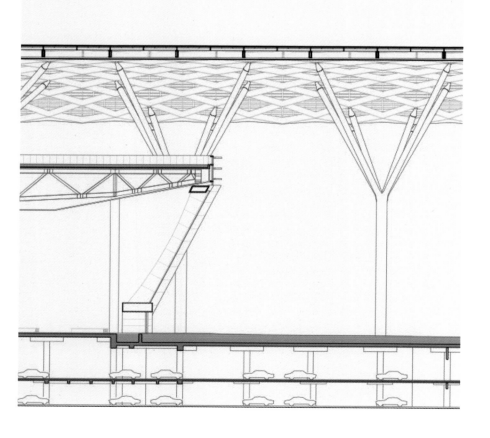

会展历史发展年表
Chronology of the Exhibition Development History

会展行为并非从现代才开始,它可以追溯到人类文明早期的集市活动,由那时以简单的商品交易为特征的活动发展而来的。而从近现代展会业的发展历史来看,会展则是欧洲工业革命的产物。自1851年诞生于欧洲的第一个国际博览会"万国工业大展览会"至今,现代展览业经历了近一个半世纪的发展历程,不仅规模、体量在不断扩大,其专业性、功能性、综合性也在不断增强,对相关产业乃至城市、全球经济的综合影响力也日益增强(图3-1)。

Exhibition does not start from modern times. It can be traced back to the bazaar activities in the early stage of human civilization, which was developed from simple commodity trading.
From a historical perspective to see the modern exhibition industry, the exhibition is the result of European industrial revolution. From 1851, when the first international exposition "the Great Exhibition of the Works of Industry of all Nations" was held in Europe, till now, modern exhibition industry has been developing for one and a half century, not only the scale, size and volume has been expending, its professionalism, functionality and comprehensiveness has also been improved continues to increase the comprehensive influence on related industries, cites and global economy throughout time.(Figure 3-1)

早期——集市活动
特征:简单的商品交易。

Early time - Bazaar / Fairs
Features: simple commodity trading.

近代——展览会的雏形
特征:商品的大量交易;展示建筑新技术,新材料,新结构。

Modern times - the prototype of exhibition
Features: large trade of commodities; exhibiting new technologies, new materials, and new structures.

现代——展览会的正式诞生
特征:专业性增强,会展建筑规模逐渐增大,但功能单一。

Contemporary - the birth of modern exhibition
Features: more specialized, the scale of exhibition building gradually increased, but still remained single functioned.

当代——展览会的发展
特征:综合性、高端化会展活动产生:产生建筑群,且有复合业态与多功能;作为城市功能部分的综合影响力日益增强。

Nowadays - the development of modern exhibition
Features: comprehensive and high-end exhibition activities; building complex with integrated business and multi functions; the comprehensive influence increases steadily as a part of the urban function .

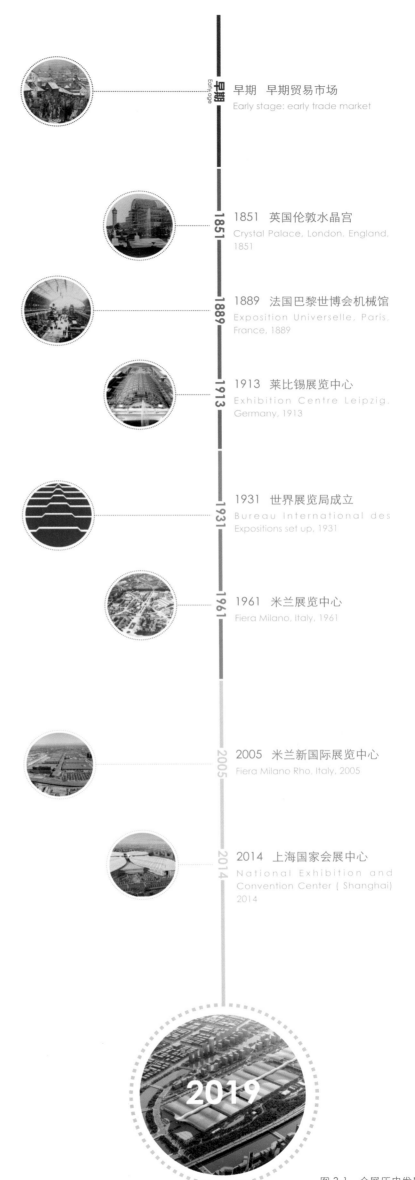

早期 早期贸易市场
Early stage: early trade market

1851 英国伦敦水晶宫
Crystal Palace, London, England, 1851

1889 法国巴黎世博会机械馆
Exposition Universelle, Paris, France, 1889

1913 莱比锡展览中心
Exhibition Centre Leipzig, Germany, 1913

1931 世界展览局成立
Bureau International des Expositions set up, 1931

1961 米兰展览中心
Fiera Milano, Italy, 1961

2005 米兰新国际展览中心
Fiera Milano Rho, Italy, 2005

2014 上海国家会展中心
National Exhibition and Convention Center (Shanghai) 2014

2019

图3-1 会展历史发展年表
Figure 3-1 Chronology of Exhibition Development history

图 3-2 米兰会展图
Figure 3-2 Milan Exhibition

鱼骨式　Fishbone Pattern

图 3-2 特点：平面形式和体量大小基本一致，可分期建设，便于扩建与改建，交通流线简洁明了。

图 3-2 鱼骨式布局图
Fishbone Pattern Layout

Characteristic (Figure 3-2) : the plane form and dimension are identical, it can be constructed phase by phase, and is more convinient for future extension and reconstruction. Its traffice circulation is simple and clear.

图 3-3 琶洲会展图
Figure 3-3 Pazhou Exhibition

集中式　Centralized Pattern

如图 3-3 特点：节省用地，体量完整，容易塑造出宏伟的体量，但一次性投入大，不利于分期建设和改扩建。

图 3-3 集中式布局图
Centralized Pattern Layout

Characteristic (Figure 3-3) : this pattern can save more land, and it's easy to generate the huge volume. But it needs more investment at the beginning, and it is not easy to construct in phases or extention.

图 3-4 上海新国际博览中心图
Figure 3-4 Shanghai New International Expo Center

围合式群落式　Enclosed Community Pattern

如图 3-4 特点：采光通风好，有利于分期建设和改扩建，但流线过长，需要增加入口和服务设施的数量。

图 3-4 围合式群落式布局图
Peri-harmony Community Layout

Characteristic (Figure 3-4) : This pattern provide good ventilation and lighting, easy for construction in phase and extention. But the circulationline is too long, which needs multiple entrances and service facilities.

图 3-5 汉诺威展览中心图
Figure 3-5 Hanover Exhibition Center

群落式　Community pattern

如图 3-5 特点：空间变化丰富，便于改扩建，但人流流线不够明确，建筑造型较难处理。

图 3-5 群落式布局图
Community Pattern Layout

Characteristic (Figure 3-5) : this pattern create space varieties, it's easy for extentions and reconstructions, but the pedestrain circulation is not clear and the appearance is difficult to define.

多层建筑属于功能性建筑，需承受高密度人流，以及复杂的交通网。这当中包括行人流通系统，机动车 / 非机动车通道系统，以及货运通道系统。

Exhibition architecture are functional buidings, who have high density of passenger flow and complex traffic flow which includ pedestrian circulation, vehicle circulation and freight circulation.

如图 3-6，大型、特大型或者有特殊要求的会展建筑应设置贵宾系统。参展商与管理方凭有效证件出入会展现场及办公空间。

As Shown in the Figure 3-6, some large, oversize, or exhibition and convention buildings with special requirments should set up vip system.
The exhibitors and staff members identity card can use the to enter and exit the exhibition site and office space.

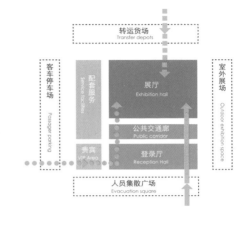

图 3-6　人行流线图
Figure 3-6　Pedestrian Circulation Diagram

如图 3-7 所示，公共交通包括公交巴士、轨道交通、轮船等通过城市道路、水路到达会展周边的落客地点。

As shown in the Figure 3-7, the public transportation includes bus, railway transport and ferries, which could approach the exhibition and convention area through the city road and waterways.

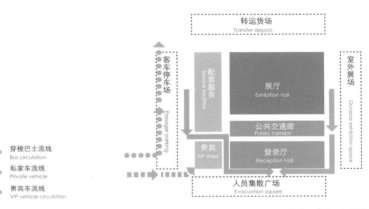

图 3-7　车行流线图
Figure 3-7　Vehicle Circulation Diagram

如图 3-8，货车从城市道路进入会展用地，通过场地内道路进入转运货场或直接进入展厅，卸货或者装货完成后返回。

As shown in the Figure 3-8, trucks enter the exhibition site through the city road, and enter the exhibition hall through the road within the area to load and unload, and they leave the same way back.

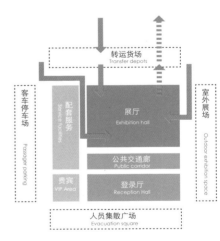

图 3-8　货运流线图
Figure 3-8　Freight Circulation Diagram

室内展览面积与停车位数量图
Indoor exhibition area and parking place diagram

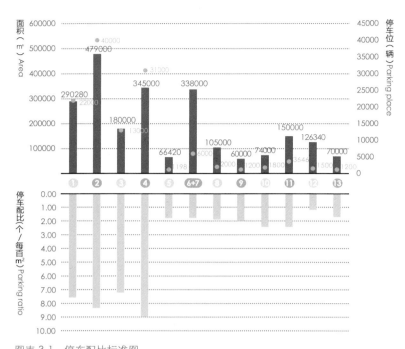

图表 3-1   停车配比标准图
Chart 3-1   Standard Diagram of Parking Ratio

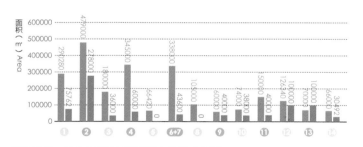

图表 3-2   室内与室外展览面积与配比关系图
Chart 3-2   Relationship Between Indoor and Outdoor Exhibition Area and ratio

会展中心各类交通与停车配比研究
The Research of transportation variety and supporting parking area ratio

国外会展建筑停车位配比量相对较大，高于 7.0；国内会展建筑停车位配比量相对较小，在 1.0-2.0 之间。国外会展通常会设置一个铁路站点，增加交通选择（图表 3-1）。

The supporting parking area ratio of foreign exhibition building are relatively high, usually higher than 7.0; The same ratio of Chinese exhibition buildings are relatively low, usually between 1.0 - 2.0. Foreign exhibition usually set up a railway station to provide move choice of (Chart 3-1) .

室内与室外展览面积配比研究
Research of indoor and outdoor exhibition area ratio

如图表 3-2，对室内与室外展览面积的配比研究，对确定本项目的面积配比有一定的参考意义。

The indoor and outdoor exhibition area ratio research can be useful to define the area ratio of this project , as Figure 3-10 shown.

■ 室内展览面积（m²）
  Indoor exhibition area

■ 室外展览面积（m²）
  Outdoor exhibition area

● 停车位
  Parking lot

■ 室内展览面积（m²）
  Indoor exhibition area

  停车配比(个/每百m²展厅)
  Parking ratio (per m²)

会展中心研究序号
Number of research for Expo

① 法兰克福国际展览中心   Messe Frankfurt

② 汉诺威展览中心   Hannover Exhibition Center

③ 慕尼黑新国际博览中心   Munich International Expo Center

④ 米兰新国际会展中心   Milano International Expo Center

⑤ 香港亚洲国际博览馆   HongKong Asia World-Expo Arena

⑥ 广州会议展览中心一期   Guangzhou Expo Phase I

⑦ 广州会议展览中心二期   Guangzhou Expo Phase II

⑧ 深圳会议展览中心   Shenzhen Convention and Exhibition Center

展厅高度（净高）
Height of Exhibition Hall (Net Height)

对国内外会展中心标准展厅室内净高数据做正态分布统计分析可得，会展建筑展厅室内净高取值在 9-23m 之间即可满足绝大多数展览需要，特殊展品（如帆船）需要更高的净高，可将会展建筑中 1-2个展厅的高度适当加高（图表 3-3）。

The result of the normal distribution statistical analysis of the height of domestic and oversea exhibition standard hall data shows that the height of indoor exhibition space between 9m to 23m could usually satisfy most of the exhibition activities. Special items (e. g. sailing boat) may demand higher height, which could be dealt with by increasing the height of several exhibition halls. (Chart 3-3).

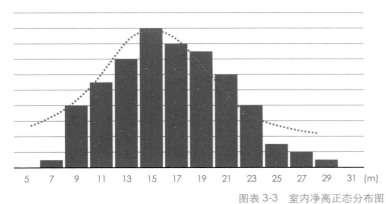

图表 3-3 室内净高正态分布图
Chart 3-3 Normal Distribution of Net Height of Indoor Space

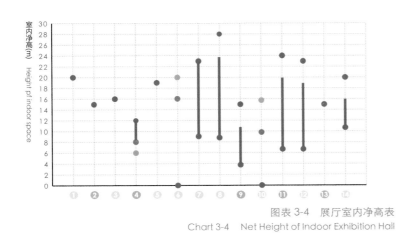

图表 3-4 展厅室内净高表
Chart 3-4 Net Height of Indoor Exhibition Hall

	轨道交通站 Subway station	铁路站 Railway station
①	0	①
②	②	①
③	②	①
④	①	①
⑤	①	0
⑥+⑦	②	0
⑧	②	0
⑨	0	0
⑩	①	0
⑪	②	0
⑫	②	0
⑬	0	0

图表 3-5 轨道交通图
Chart 3-5 Railway Transportation

⑨ 厦门国际会议展览中心　Xiamen Convention & Exhibition Center

⑩ 郑州国际会议展览中心　Zhengzhou Convention & Exhibition Center

⑪ 武汉新城国际博览中心　Wuhan Convention & Exhibition Center

⑫ 上海新国际博览中心　Shanghai New International Expo Center

⑬ 哈尔滨国际会展体育中心 I 号工程　Harbin International Exhibition & Sports Center

⑭ 西安曲江新会议展览中心　Xi 'an Qujiang New EXPO

规模尺度对比分析
The Contrastive Analysis of the Scale

落成年份：2004
Year of completion
占地面积：220000 m²
Occupied area
室内展览面积：105000 m²
Floor area

图 3-9 深圳会展中心
Figure 3-9 Shenzhen Convention & Exhibition Center

落成年份：1947
Year of completion
占地面积：1000000m²
Occupied area
室内展览面积：498000m²
Floor area

图 3-10 汉诺威会展中心
Figure 3-10 Hannover Exhibition Center

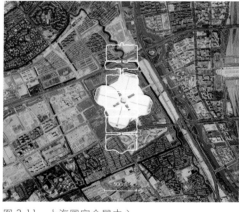

落成年份：2014
Year of completion
占地面积：860000m²
Occupied area
室内展览面积：500000m²
Floor area

图 3-11 上海国家会展中心
Figure 3-11 National Convention & Exhibition Center (Shanghai)

落成年份：2005
Year of completion
占地面积：2000000m²
Occupied area
室内展览面积：1400000m²
Floor area

图 3-12 米兰会展中心
Figure 3-12 Fiera Milano

会展规划设计的关键因素
The Key Factors of the Planning and Design of Exhibition Center

超大型会展中心是资源集约型建筑，一流的会展中心需要城市竞争力、会展经济、城市规划等多方支撑。就会展中心本身而言，满足5大客户群体不同层面的需求至关重要（图3-16）。

Super large exhibition center is a kind of building with intensive resources. The first-class exhibition center needs support from urban competitiveness, exhibition economy, urban planning and many others. Regarding the exhibition center itself, it is very important to meet the needs of the five major customer groups at different levels (Figure 3-16).

图 3-13、3-14、3-15　会展规划设计的关键因素意向图
Picture 3-13, 3-14, 3-15　The Key Factors of the Planning and Design of Exhibition Center

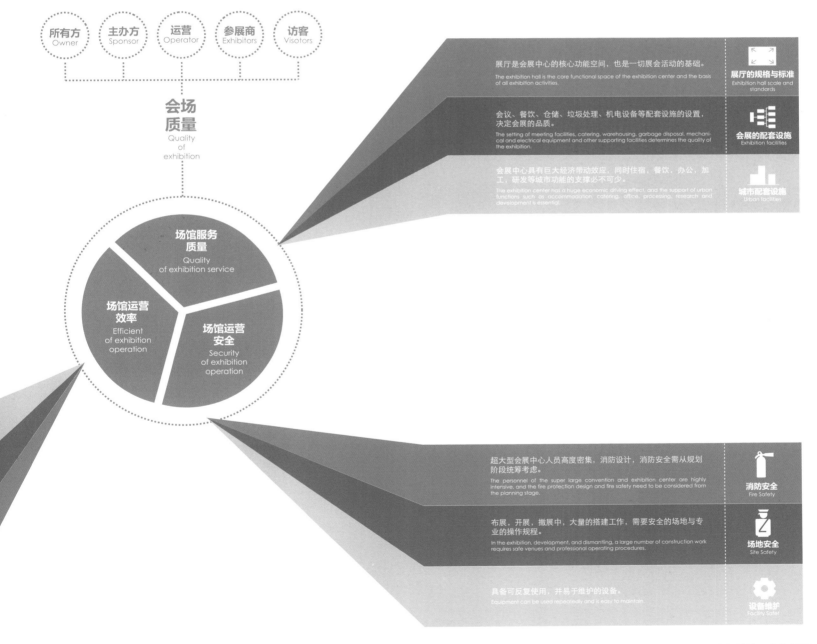

图 3-16　会展规划设计的关键因素分析图
Figure 3-16　Analysis Diagram of Key Factors in Exhibition Planning and Design

模式一：单边式布局
Pattern 1: One-Side Layout

模式：双边式布局
Pattern 2: Double-side Layout

模式三：其他类型布局
Pattern 3: Other Layout

图 3-17　会展规划布局原则分析图
Figure 3-17　Diagram of Layout Principles of Exhibition Planning

会展规划布局原则
The Planning and Layout Principles

会展规划布局模式的选择，取决于会展的展览规模、展览类型以及用地条件的影响。其中更为关键的是会展规模的影响。

大型会展（展览规模大于 10 万平方米），展厅的布局模式应满足多种展览同时举办的人流与货运需求（图 3-17）。

The layout pattern of an exhibition center depends on its scale, type and the land conditions. And the most critical factor is the scale itself. The layout pattern of large-scale exhibition center (the scale is more than 100,000 square metes) should satisfy the needs of people and freight when a variety of exhibitions are held at the same time (Figure 3-17).

● 模式一：单边式布局 Pattern 1: One - side layout	● 模式二：双边式布局 Pattern 2: double - side layout	● 模式三：其他布局 Pattern 3. other layouts	
优秀 Excellent	良好 Good	良好 Good	◀ 小型展会 Small scale exhibition
良好 Good	优秀 Excellent	优秀 Excellent	◀ 中型展会 Middle scale exhibition
---	良好 Good	较差 Bad	◀ 大型展会物流组织 Large exhibition logistics organization
---	良好 Good	距离长，方向感良好 Long distance, good oreintation	◀ 大型展会步行距离/方向感 Large exhibition orientation/walk distance
优秀 Excellent	优秀 Excellent	良好 Good	◀ 多个展会 Multiple exhibition
良好 Good	优秀 Excellent	较差 Bad	◀ 用地利用效率 Land usage
良好 Good	良好 Good	较差 Bad	◀ 用地形状适应性 Land shape adaptation

● 模式一：单边式布局
Pattern one: one-side layout

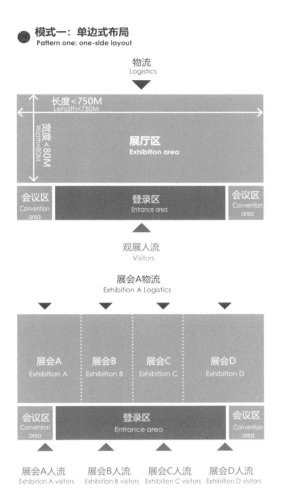

● 模式三：其他类型布局
Pattern 3: Other - layout

● 模式二：双边式布局
Pattern 2: Double side layout

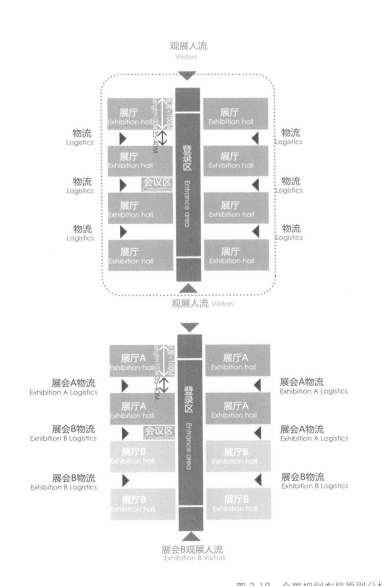

图 3-18 会展规划布局原则分析图
Figure 3-18 Diagram of Layout Principles of Exhibition Planning

会展中心总体设计
The General Design of The Shenzhen World

深圳国际会展中心项目展馆采用鱼骨状布局，一条长约 1750m 的中央廊道连接两个登录大厅和 19 个展厅，包括凤塘大道以南区域 16 个 2 万平方米的标准展厅、凤塘大道以北区域 2 个 2 万平方米的多功能展厅和 1 个 5 万平方米的超大展厅，是集展览、会议、活动（赛事、娱乐）、餐饮、办公、服务配套于一体的超大型会展综合体。

Designed with a herringbone pattern layout, the building of the Shenzhen World were given a 1750m long central concourse to connect the two regesteration halls and 19 exhibition halls, including 16 standard exhibition halls with 20000m^2 in the south, and 2 multi-function exhibition halls with 20000 m^2 and 1 super large exhibition hall with 50000 m^2 in the north. The building itself is a super huge exhibition complex that integrates the functions of exhibitions, conferences, activities (events, entertainment), catering, office, and service as a whole.

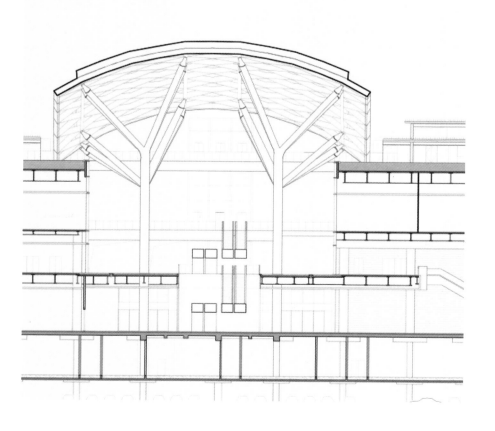

图 3-19-1　总平面图
Figure 3-19-1　General Plan

会展中心及休闲带
Exhibition center & leisure belt

① 南入口广场
South entrance plaza

② 南登入大厅
South entrance hall

③ 中央廊道
Central concourse

④ 北登入大厅
North arrival hall

⑤ 北入口
North entrance

⑥ 预留二期用地
Reserved site for phase2

⑦ 地铁会展南站/休闲带叠水广场
Exhibition South tube station
Water plaza of leisure belt

⑧ 地铁会展北站/休闲带叠绿广场
South entrance hall
Green plaza of leisure belt

⑨ 休闲带音乐岛
Music island of leisure belt

⑩ 休闲带浪漫岛
Romance island of leisure belt

⑪ 休闲带活力岛
Dynamic island of leisure belt

⑫ 休闲带运动岛
Sport island of leisure belt

西侧配套
Westside supporting

⑬ 2栋/仓库
Unit 2/ warehouse

⑭ 3栋/垃圾用房
Unit 3/ Garbage room

⑮ 4栋/仓库
Unit 4/ warehouse

⑯ 5栋/海关国检
Unit 5/ customs sffice

⑰ 6栋/安保办公
Unit 6/ Security office

⑱ 7栋/制服洗衣
Unit 7/ Uniform & laundry

⑲ 8栋/配套综合
Unit 8/ Comprehensive supporting

⑳ 9栋/行政办公
Unit 9/ Managing office

㉑ 10栋/办公
Unit 10/ Office

㉒ 11栋/仓库、车检
Unit 11/ Warehouse & vehicle check

其他 Others

㉓ 沿江高速出口
Exit of riverside highway

㉔ 变电站
Transformer substation

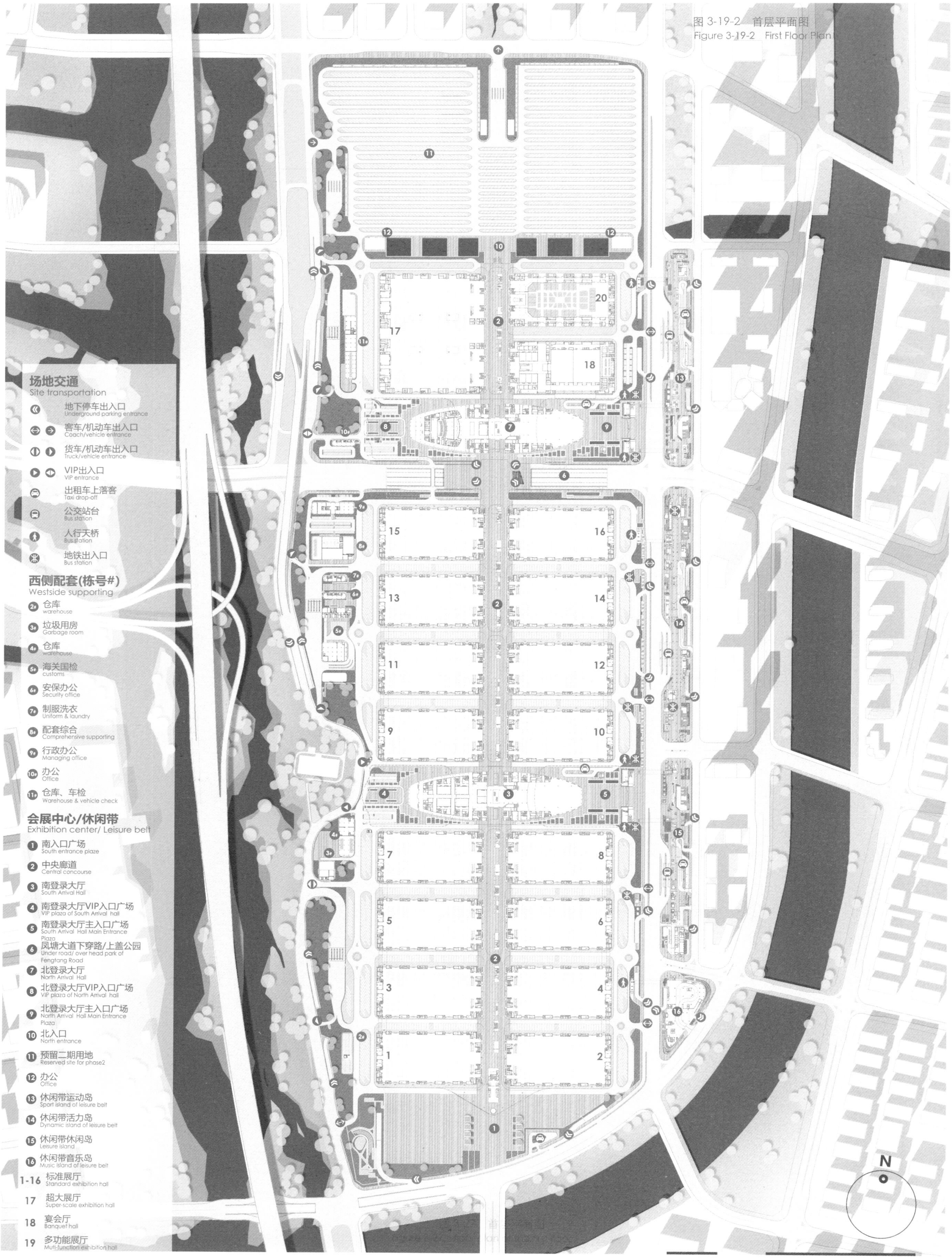

图 3-19-2　首层平面图
Figure 3-19-2　First Floor Plan

场地交通
Site transportation

- Ⓚ 地下停车出入口
  Underground parking entrance
- ⇄→ 客车/机动车出入口
  Coach/vehicle entrance
- ⓘ ↕ 货车/机动车出入口
  Truck/vehicle entrance
- ▶◀ VIP出入口
  VIP entrance
- 🚕 出租车上落客
  Taxi drop-off
- 🚌 公交站台
  Bus station
- 🚶 人行天桥
  Bus station
- 🚇 地铁出入口
  Bus station

西侧配套(栋号#)
Westside supporting

- 2# 仓库
  warehouse
- 3# 垃圾用房
  Garbage room
- 4# 仓库
  warehouse
- 5# 海关国检
  customs
- 6# 安保办公
  Security office
- 7# 制服洗衣
  Uniform & laundry
- 8# 配套综合
  Comprehensive supporting
- 9# 行政办公
  Managing office
- 10# 办公
  Office
- 11# 仓库、车检
  Warehouse & vehicle check

会展中心/休闲带
Exhibition center/ Leisure belt

- ❶ 南入口广场
  South entrance plaza
- ❷ 中央廊道
  Central concourse
- ❸ 南登录大厅
  South Arrival Hall
- ❹ 南登录大厅VIP入口广场
  VIP plaza of South Arrival hall
- ❺ 南登录大厅主入口广场
  South Arrival Hall Main Entrance Plaza
- ❻ 凤塘大道下穿路/上盖公园
  Under road/ over head park of Fengtang Road
- ❼ 北登录大厅
  North Arrival Hall
- ❽ 北登录大厅VIP入口广场
  ViP plaza of North Arrival hall
- ❾ 北登录大厅主入口广场
  North Arrival Hall Main Entrance Plaza
- ❿ 北入口
  North entrance
- ⓫ 预留二期用地
  Reserved site for phase2
- ⓬ 办公
  Office
- ⓭ 休闲带运动岛
  Sport island of leisure belt
- ⓮ 休闲带活力岛
  Dynamic island of leisure belt
- ⓯ 休闲带休闲岛
  Leisure island
- ⓰ 休闲带音乐岛
  Music island of leisure belt
- 1-16 标准展厅
  Standard exhibition hall
- 17 超大展厅
  Super-scale exhibition hall
- 18 宴会厅
  Banquet hall
- 19 多功能展厅
  Multi-function exhibition hall

会展规划结构
The Planning and Layout Principles

如图 3-20，为了解决不同功能和体量之间的过渡问题，深圳国际会展中心片区规划了一条特殊的"会展休闲带"。它是人流密集的功能体量之间的重要开放区域，也是在多个层面都起到重要作用的复合空间。会展休闲带与截流河景观带，将会在展 - 园 - 城之间形成和谐的转换与衔接。

• 在会展支撑层面，休闲带集合了公交枢纽与人流集散功能；

• 在城市生活层面，休闲带集合了休闲活动与商业服务功能；

• 在城市空间层面，休闲带则作为立体绿色开放空间，使会展中心与城市能顺畅的衔接与过渡。

As Figure 3-20 shown, In order to solve the transitional problem between different functions and volumes, a special "Exhibition Leisure Zone" is planned in the Shenzhen World. It is an important open area between densely populated functional volumes, as well as a composite space that plays an important role. The exhibition leisure zone and the intercepting river landscape zone would become a harmonious transition and connection between the exhibition, garden and the city.

· For exhibition supporting, the leisure zone integrates the functions of bus hub and crowd distribution;

· For urban life, the leisure zone integrates leisure activities and commercial services;

· For urban space, the leisure zone serves as a three-dimensional green open space, enabling the connection and transition between the exhibition center and the city smoothly.

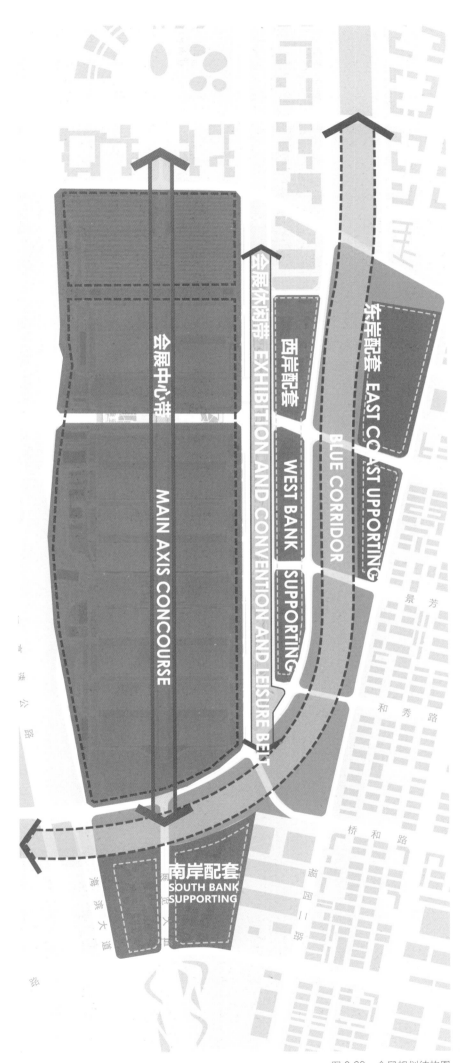

图 3-20　会展规划结构图
Figure 3-20　Exhibition Planning Structure Diagram

功能布局
Function Layout

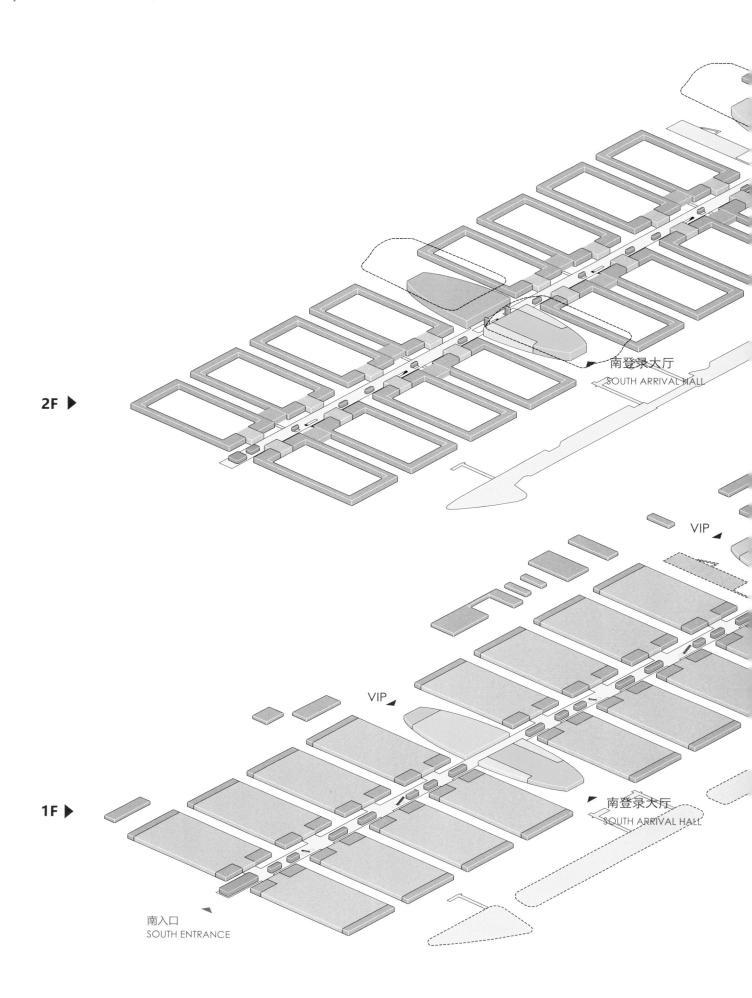

2F ▶

▶ 南登录大厅
SOUTH ARRIVAL HALL

VIP ◢

VIP ◢

1F ▶

▶ 南登录大厅
SOUTH ARRIVAL HALL

南入口
SOUTH ENTRANCE

图 3-21 会展中心功能分析图

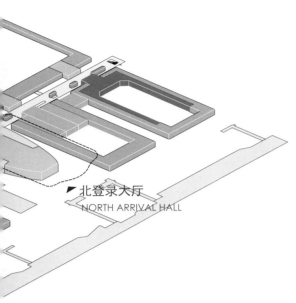

▲ 北登录大厅
NORTH ARRIVAL HALL

中心休闲带
OGICAL LEISURE GREENBELT

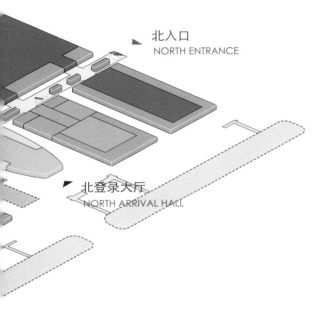

▲ 北入口
NORTH ENTRANCE

▲ 北登录大厅
NORTH ARRIVAL HALL

图 3-21　会展中心功能分析图
Figure 3-21　Functional Analysis Diagram of Exhibition and Convention Center

区域划分
Area Division

● 登录大厅
　Arrival Hall

○ 休闲带
　Leisure Belt

● 标准展厅
　Standard Exhibition hall

● 展厅配套
　Supporting Facilities

● 会议区
　Conference Area

○ 中央廊道
　Central Concourse

● 中廊配套
　Central Concourse Supporting

● 宴会厅
　Banquet Hall

○ 登录大厅配套
　Arrival Hall Supporting

● 超大展厅
　Large Exhibition Hall

● 多功能展厅
　Multi-functional Exhibition Hall

● 展厅会议区
　Exhibition Meeting Area

会展核心功能设计
The Core Functional Designs of the Shenzhen World

深圳国际会展中心一期建筑总面积达 160 万 m^2，室内展览面积约 40 万 m^2，是集展览、会议、活动（赛事、娱乐）、餐饮、购物、办公、服务于一体的超大型会展综合体。

The first phase of the Shenzhen World covers a total area of 1.6 million square meters,with an indoor exhibition area about 400,000 square meters. It is a super huge exhibition complex including conference, activities (events, entertainment), catering, shopping, office, services, and many other functions.

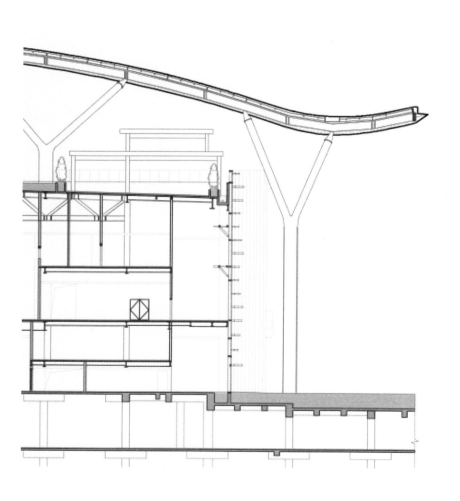

鱼骨式规划
Fish-Bone Pattern Planning

如图 3-22

- 19 个展厅沿中央廊道东西对称布局，呈鱼骨式排列
- 南入口和南北登录大厅由南至北均匀分布
- 凤塘大道将展馆划分为南区和北区
- 2 - 11 栋综合配套建筑位于西侧
- 总共 19 个展厅
  南区：16 个 2 万 m² 标准展厅
  北区：1 个 5 万 m² 超大展厅；2 个特殊展厅
- 大跨度桁架结构，全部展厅采用单层、无柱结构（超大展厅除外）
- 南区标准展厅可灵活组合，北区展厅可多功能灵活互换使用
- 展厅悬挂吊点：全馆覆盖

- As Figure 3-22 shown, 19 exhibition halls are symmetrically arranged along the central concourse as a fishbone pattern;
- the south entrance and the north and south arrival halls are evenly arranged from south to north;
- Fengtang avenue divides the exhibition center into the south and the north;
- 2 to 11 comprehensive supporting buildings are located on the west side
- 19 exhibition halls in total
  Southern district :16 standard exhibition halls with area of 20,000 square meters; North district: a large exhibition hall with 50,000 square meters exhibition area; 2 special exhibition halls.
- Large-span truss structure, all the exhibition halls adopt the structure of single-story and column-free (except for super-large exhibition halls) character;
- The space of the standard exhibition halls in the south are flexible. The multi functions of the exhibition halls in the north are interchangeable.
- Hanging points are covered in all the exhibition halls.

图例
Legend

5万m²超大展厅
50000 Square Meters Super Large Exhibition Hall

2万m²标准展厅
20000 Square Meters Standard Exhibition Hall

2万m²特殊展厅
20000 Square Meters Special Exhibition Hall

登录大厅
Arrival Hall

中央廊道
Central Concourse

综合配套建筑
Comprehensive Supporting Building

南入口
South Entrance

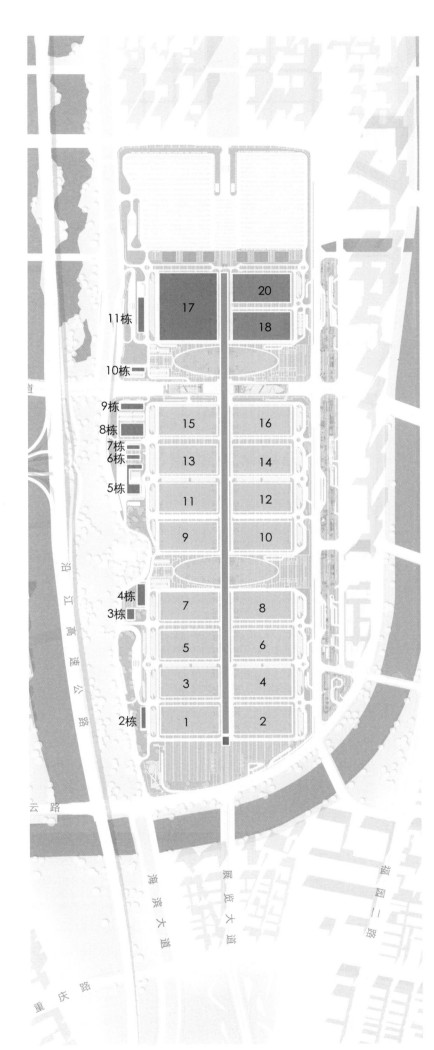

图 3-22　展厅组织模式图
Figure 3-22　Arrangement of Exhibition Hall

图 3-23   中央廊道设计（交通枢纽）图
Figure 3-23   Central Concourse Design (Circulation Hub)

中央廊道设计——交通枢纽
Central Concourse Design — The Circulation Hub

交通原则

会展主体建筑南北长 1750m，东西宽约 500m，根据鱼骨式布局会
在会展建筑一条中央公共通廊衔接两侧展厅的特点，同时为做到人
货分流，布展、观展、撤展组织高效清晰，也为了减少地上地下建
设工序之间相互交叉影响（图 3-23），中央廊道将以地上二层两个
不同标高的通廊解决人货分流和同期不同展会分流登录入场的问题。

Circulation Principle

The main building of the Shenzhen World is 1750m long
from south to north, and 500m wide from east to west,
connecting all the exhibition halls with a public center
concourse throughout the whole building mimicking the
pattern of fishbone. In order to separate the moving lines
of people and cargos, guarantee the exhibition's organization
of move-in and move-out, and the efficiency and lucidity
of the exhibition visit to the visitors, as well as to reduce the
cross impact between the ground and the underground
construction processes (Figure 3-23), it's necessary for the
center concourse to adopt 2 corridors with different heights
on the second floor to achieve what's mentioned above.

图例
Legend

● 南入口（含入口大厅1868m²和VIP接待厅）
South Entrance (including entrance hall 1868m² and
VIP reception hall)

● 北登录大厅（含东侧入口大厅7400m²和西侧1920座国际报告厅）
North Arrival Hall (including 7400m² entrance hall and 1920 international lecture hall)

● 南登录大厅（含东侧入口大厅8000m²，含会议中心和西侧3400m²功能厅）
South Arrival Hall (including 8000m² entrance hall, including conference center
and 3400m² function hall)

○ 中央廊道（长：1750m,含商业和服务配套）
Central Concourse (1750m long, including commercial and service facilities)

登录流线

深圳国际会展中心登录均需经过安检后进入中央廊道二层，安检设施布置主要设置于南北登陆大厅和南入口集中安检区、地下车库集中安检区、展馆前布展临时安检区，地下车库在中央廊道下方均匀布置了 5 个交通核，设置了自动扶梯和无障碍电梯，保证地上地下的人流快速便捷集散，从地下可通过 5 个交通核前的安检区直达中央廊道二层再进入各展厅，散场时可通过地下或者一层、二层直接离场（图 3-24）。

Circulation after Arrival

People shall go through the security check point and enter the second floor of the central concourse after arriving at the Shenzhen World. The security check point was set at the arrival halls on the South and the North, the security area of the south entrance, the underground parking, and the temporary security area in front of the exhibition halls.

Under the center concourse, there set up escalators,elevators within the 5 traffic switching centers in the underground parking lot, which guarantee the fast and convenient transit of the crowd both on the ground and underground. Through the security areas in front of the 5 traffic switching centers, people can go directly to the second floor of the center concourse, and enter the exhibition halls. When exhibitions are close, people can choose to leave from the second floor, the ground floor, or from the underground exit. (Figure 3-24)

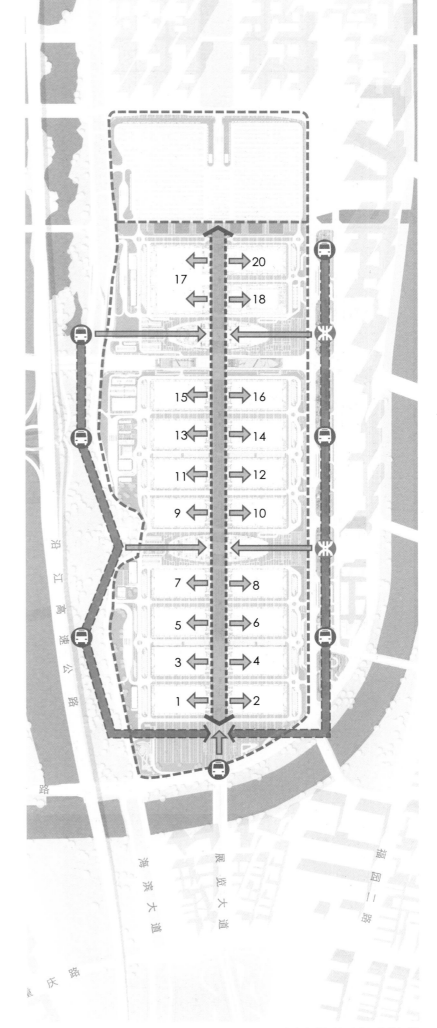

图例
Legend

公共交通流线
Public Transport Circulation

人行流线
Pedestrian Circulation

巴士站点
Bus Station

地铁站点 🚇
Subway Station

图 3-24　中央廊道设计（登录流线）图
Figure 3-24　Central Concourse Circulation Design (Crculation Arrival)

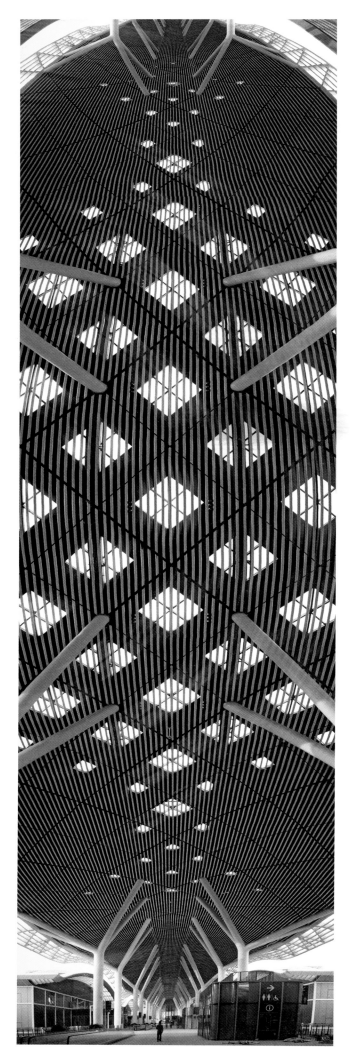

图 3-25   中央廊道实景图
Figure 3-25   Central Concourse Reality Image

展厅使用模式
Exhibition and Convention hall — Usage Pattern

模式一：作为一个大展馆使用
Pattern 1: As a large exhibition hall

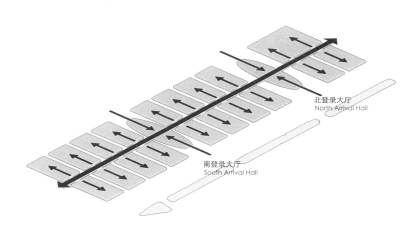

模式二：两个展览同时进行时
Pattern 2: When two exhibitions are in progress at the same time

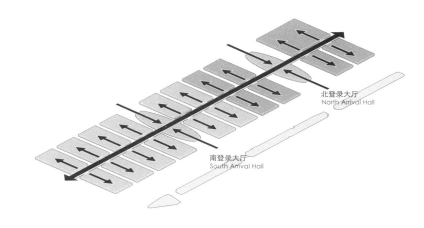

模式三：多个展览同时进行时
Pattern 3: When multiple exhibitions are in progress at the same time

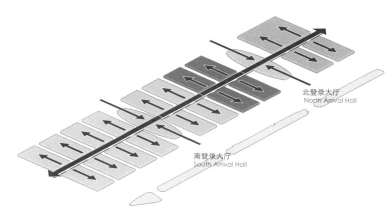

图 3-26   展厅使用模式图
Figure 3-26   Exhibition Hall Usage Pattern Diagram

图 3-27　中央廊道实景图
Figure 3-27　Central Concourse Reality Image

图 3-28　中央廊道效果图
Figure 3-28　Rendering of the Central Concourse

1栋 – 地下室组合平面图
Building 1   Basement plan

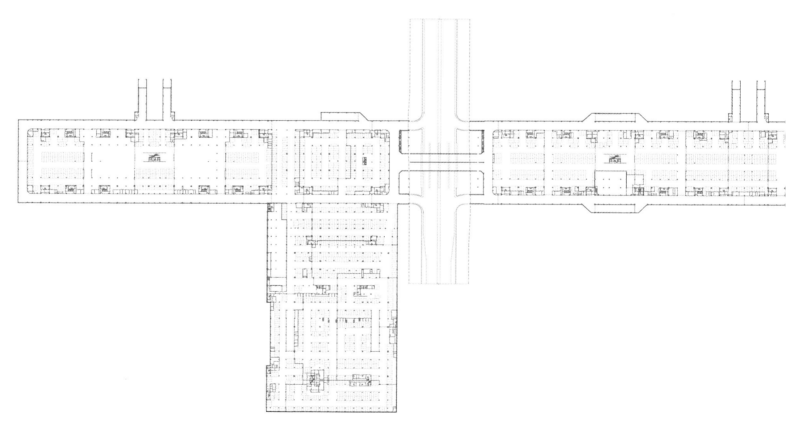

图 3-29   负 2 层地下室平面图
Figure 3-29   Basement Plan B2

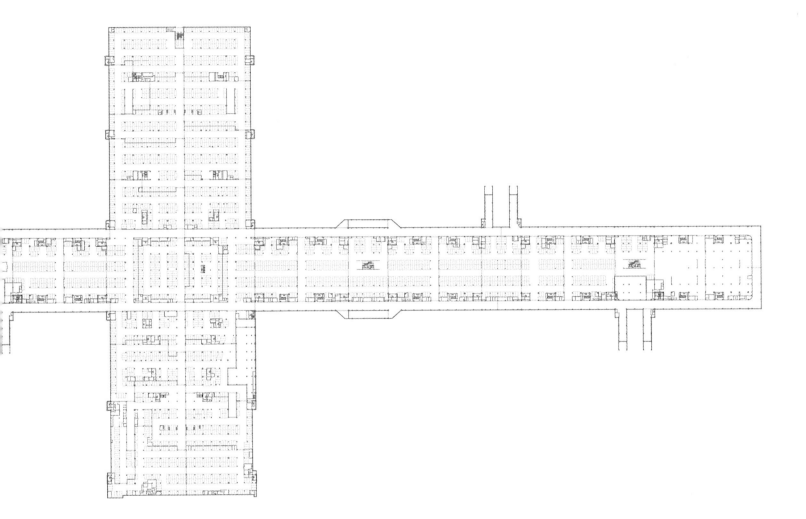

图 3-29-1　组合平面索引图
Figure 3-29-1　Key Plan

0 10 20　　50　　　100

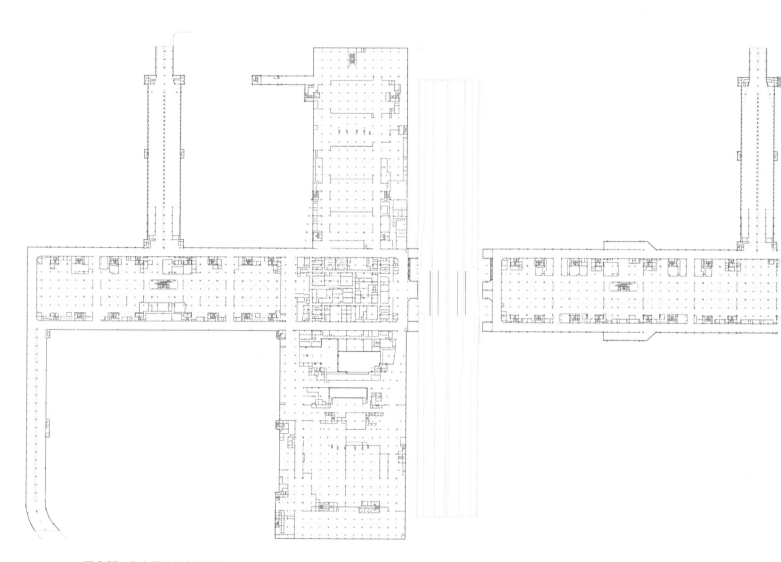

图 3-30   负 1 层地下室平面图
Figure 3-30   Basement Plan B1

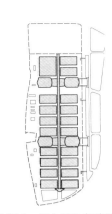

图 3-30-1　组合平面索引图
Figure 3-30-1　Key Plan

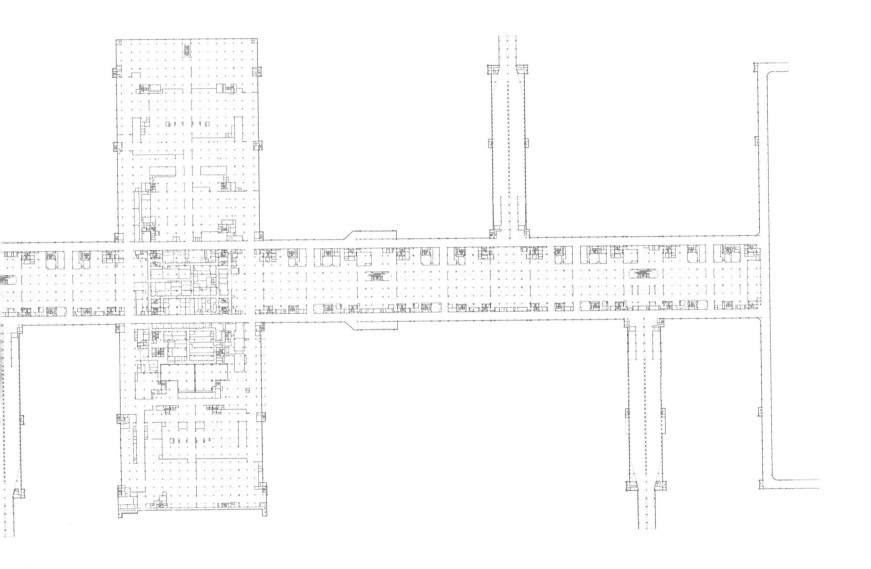

0 10 20　50　100

1栋 – 组合平面图
Building 1　General plan

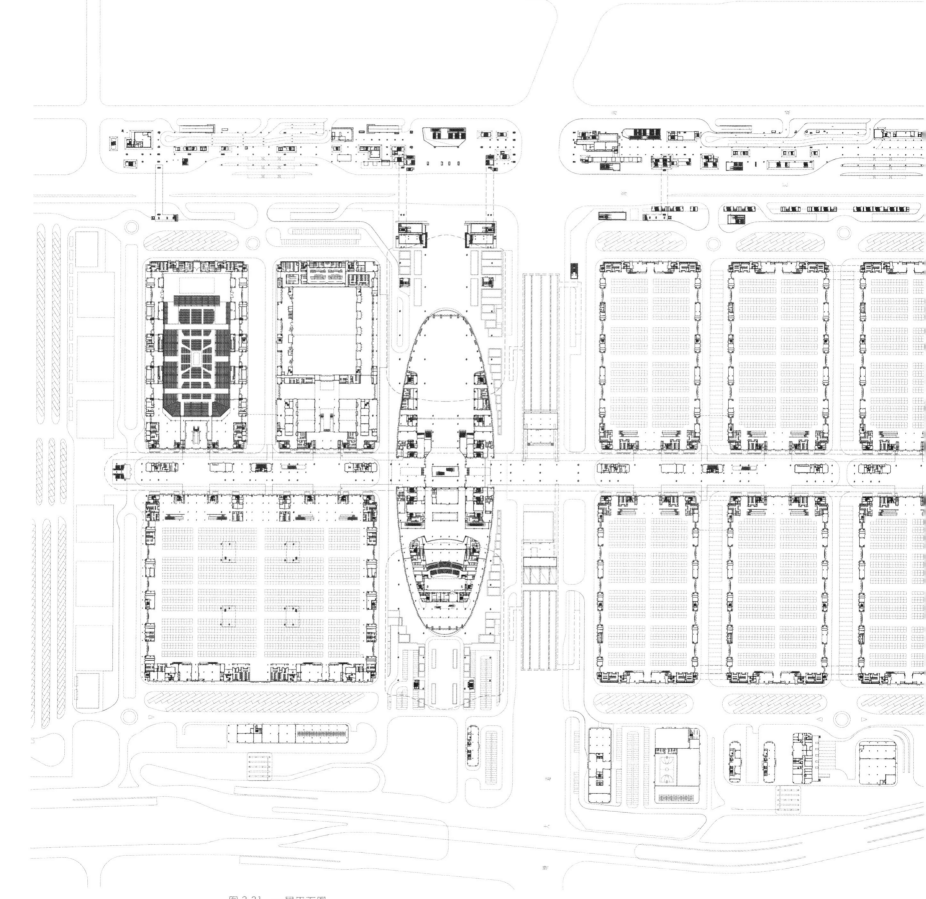

图 3-31　一层平面图
Figure 3-31　First Floor Plan

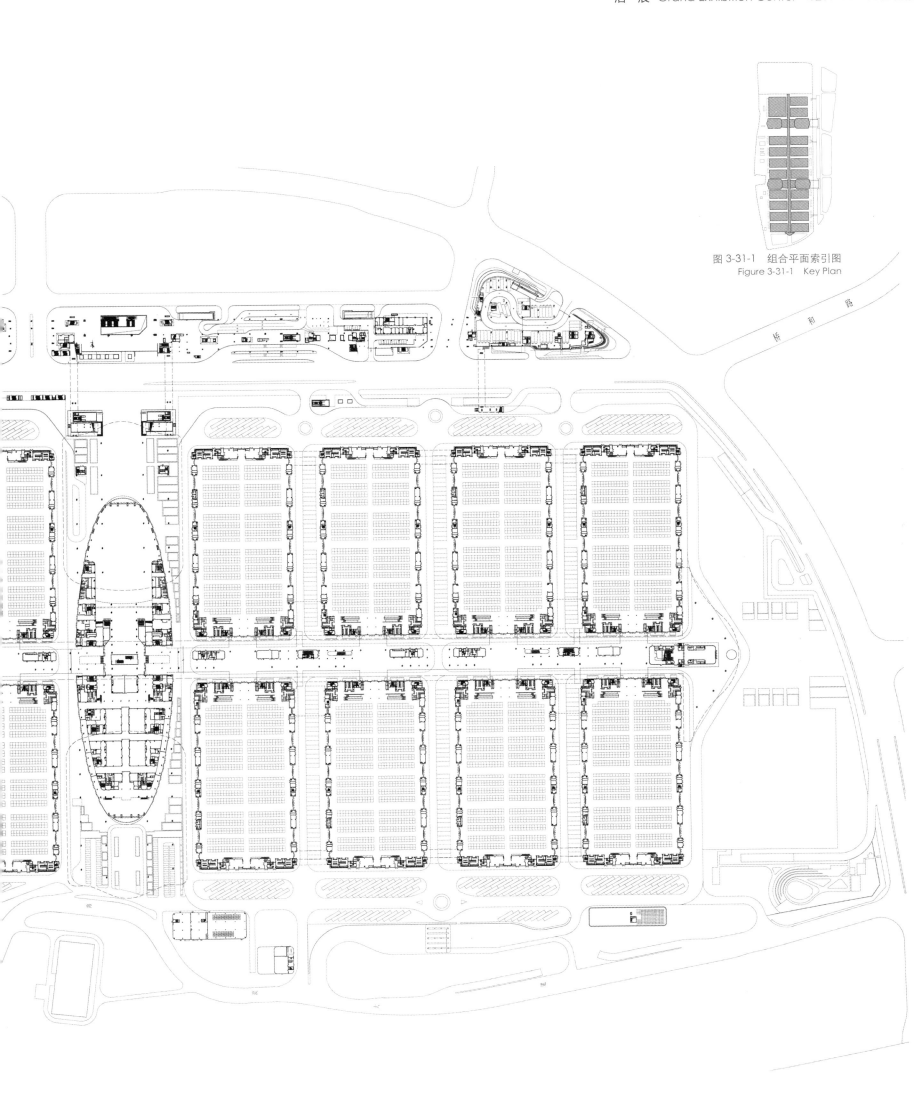

图 3-31-1　组合平面索引图
Figure 3-31-1　Key Plan

0 10 20　50　100

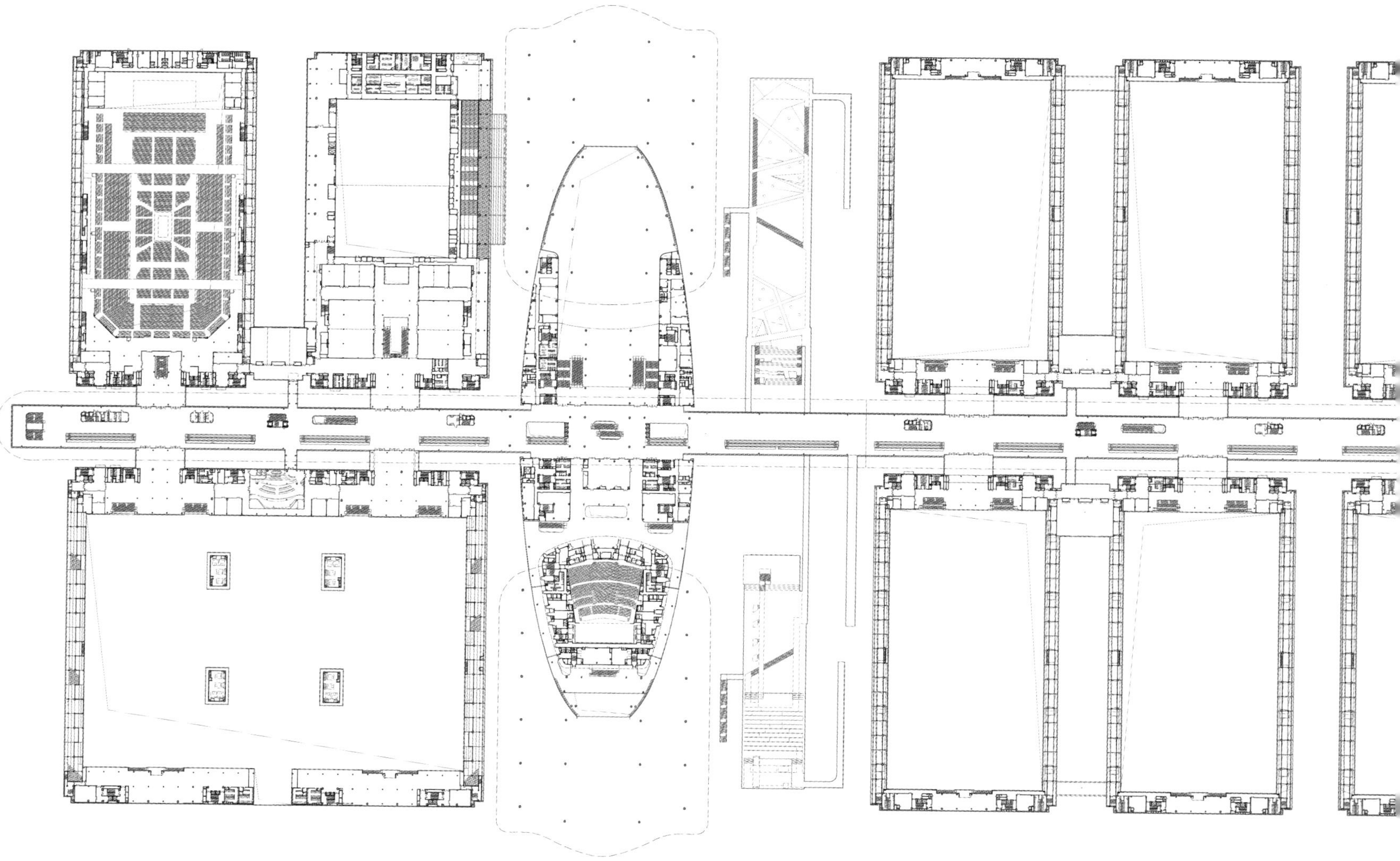

图 3-32　二层平面图
Figure 3-32　Second Floor Plan

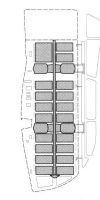

图 3-32-1 组合平面索引图
Figure 3-32-1 Key Plan

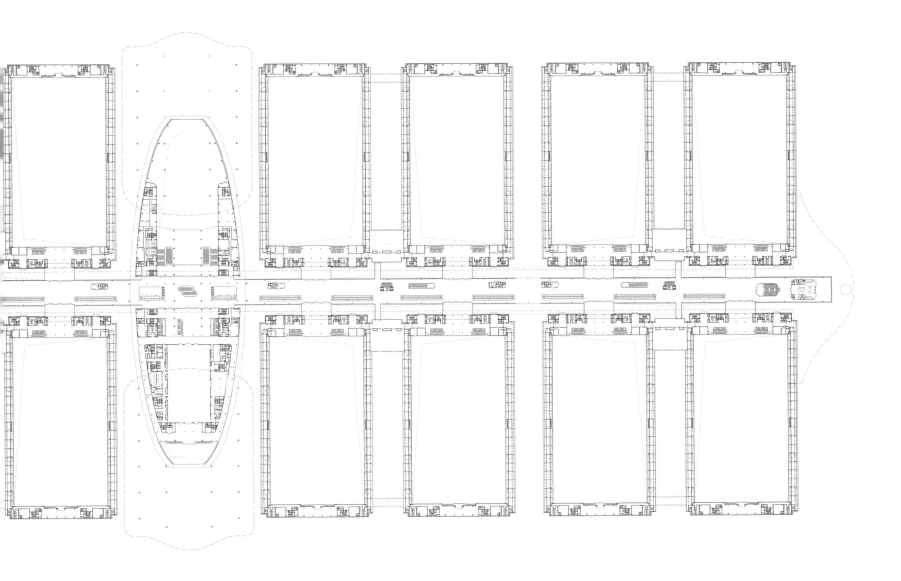

0 10 20　　50　　　　100

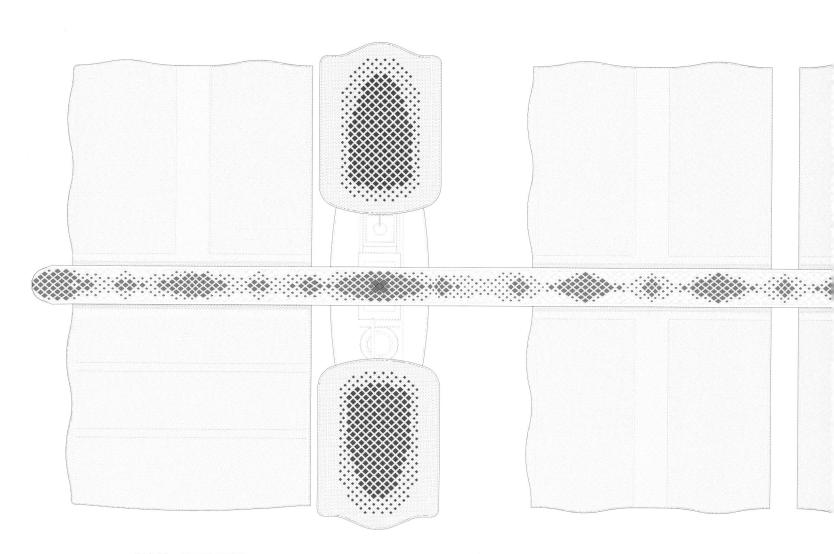

图 3-33　屋顶层平面图
Figure 3-33　Roof Plan

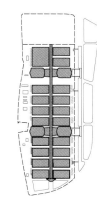

图 3-33-1　组合平面索引图
Figure 3-33-1　Key Plan

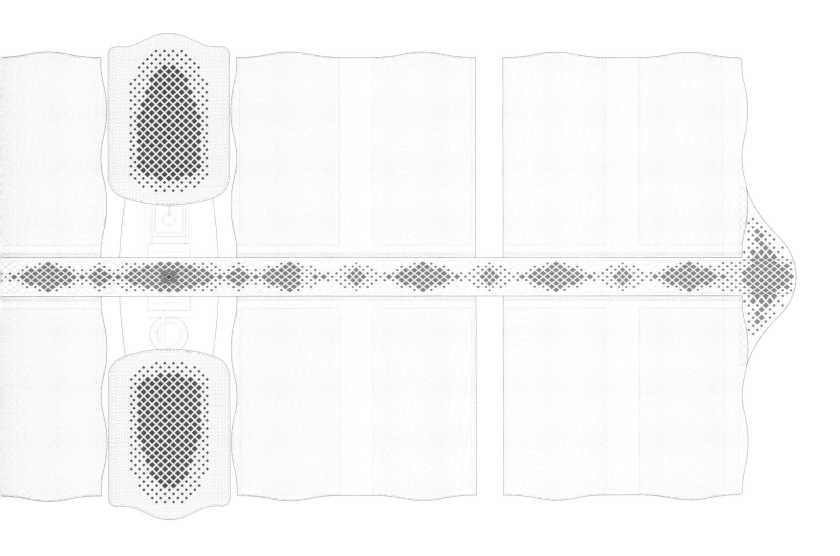

0 10 20　50　100

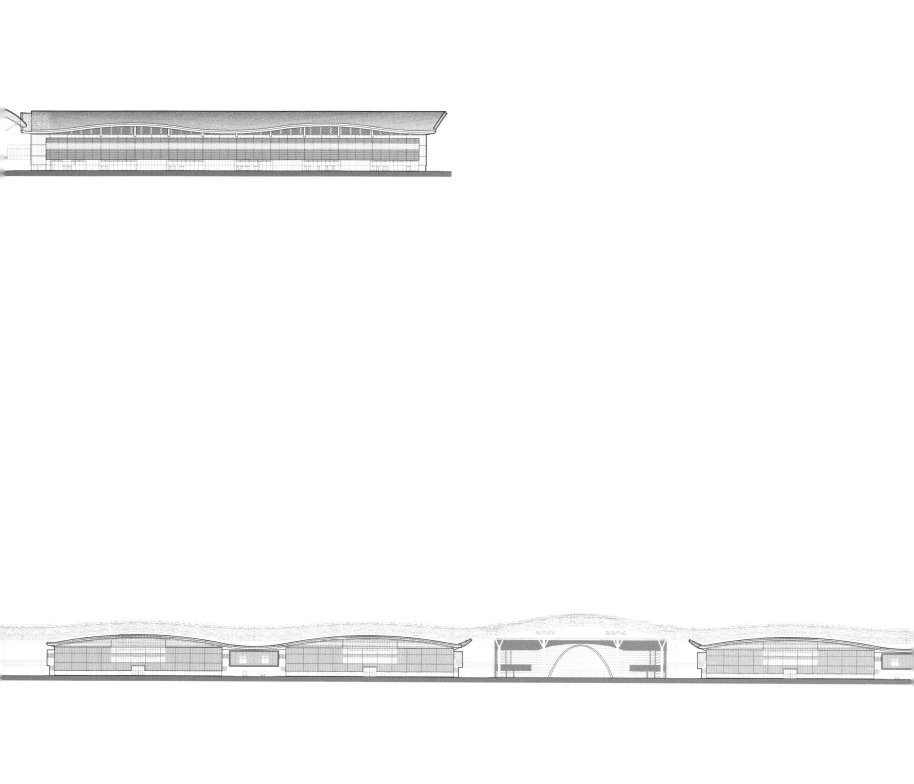

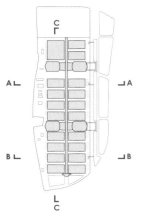

图 3-34-1 立面索引图
Figure 3-34-1 Elevation Key Plan

1栋 – 立面图
Building 1- Elevation

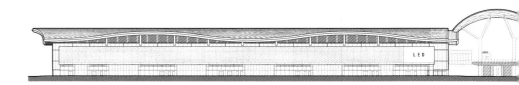

图 3-34 南立面图
Figure 3-34 Elevation B-B

图 3-35 西侧组合立面图
Figure 3-35 Elevation C-C

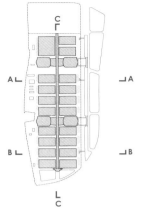

图 3-36-1　剖面索引图
Figure 3-36-1　Sectional Key Plan

1栋 – 剖面图
Building 1 - Section

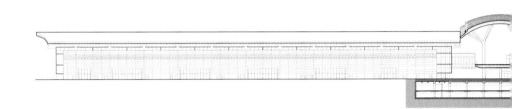

图 3-37　A-A 剖面图
Figure 3-37　Section A-A

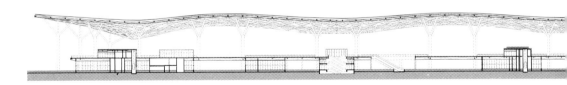

图 3-38　C-C 剖面图
Figure 3-38　Section C-C

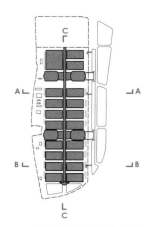

图 3-36-1　立面索引图
Figure 3-36-1　Elevation Key Plan

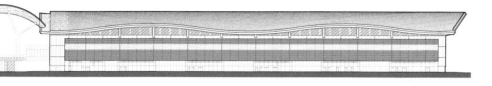

图 3-36　北立面图
Figure 3-36　B-B North Elevation

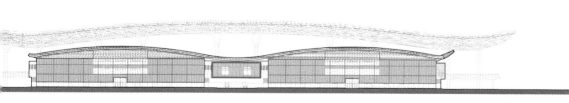

图 3-35　西侧组合立面图
Figure 3-35　C-C Section

0　10　20　　50　　100

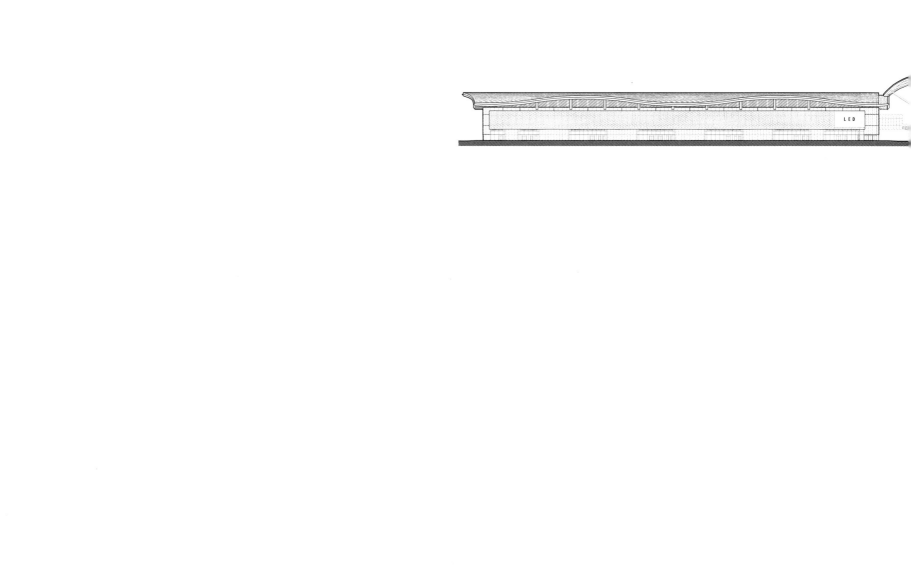

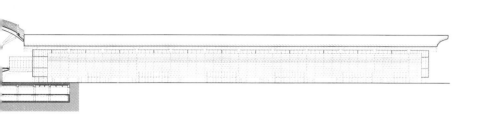

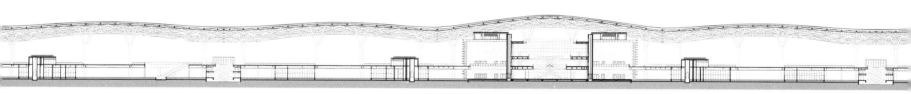

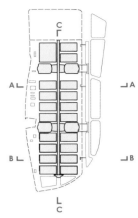

图 3-39-1　剖面索引图
Figure 3-39-1　Sectional Key Plan

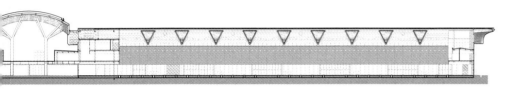

图 3-39　B-B 剖面图
Figure 3-39　Section B-B

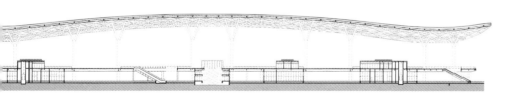

图 3-38　C-C 剖面图
Figure 3-38　Section C-C

0　10　20　　50　　　　　100

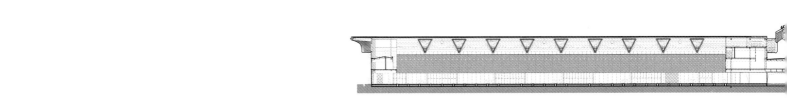
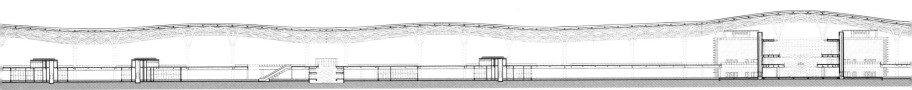

登录大厅设计
The Design of Arrival Halls

南登录大厅功能布局

东侧为单层登录大厅，配套设置有行李寄存登记，VIP 出入口等，主要解决展览建筑安检登录功能。

西侧首层设有四个大会议室及媒体发布室（图 3-40）。二层为可容纳三千人左右的多功能厅，能满足展览、餐饮、会议等多种需求（图 3-41）。三层靠近中廊部分设有中餐厅，西餐厅及美食广场（图 3-42）。

Function layout of South Arrival Hall

The east side is a single-floor arrival hall, equipped with luggage deposit, VIP entrance and exit, etc., which mainly deal with the security check and registration.

The first floor on the west side is equipped with four large conference rooms and press rooms (Figure 3-40). It is a multi-functional hall with a capacity around 3,000, which can meet various needs such as exhibition, catering and conference (Figure 3-41). There are Chinese restaurant, western restaurant and food court on the third floor near the center of the concourse.(Figure 3-42)

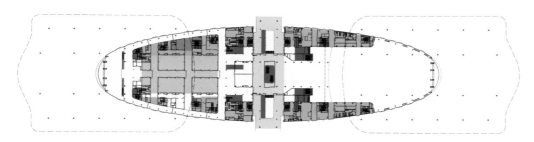

图 3-40　南登录大厅一层平面图
Figure 3-40　Floor Plan of South Arrival Hall - Level 1

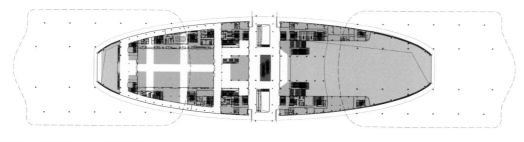

图 3-41　南登录大厅二层平面图
Figure 3-41　Floor Plan of South Arrival Hall - Level 2

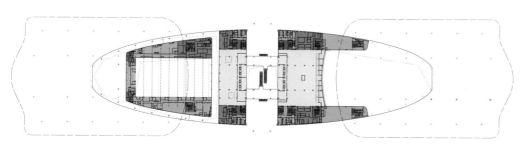

图 3-42　南登录大厅三层平面图
Figure 3-42　Floor Plan of South Arrival Hall - Level 3

图例
Legend

● 垂直交通 Vertical Circulation

○ 多功能厅 Multifunction Room

○ 中央廊道 Central Concourse

○ 会议 Meeting

○ 展厅配套 Exhibition Hall Matching

○ 登录大厅 Arrival Hall

● 厨房 Kitchen

○ 餐厅 Restaurant

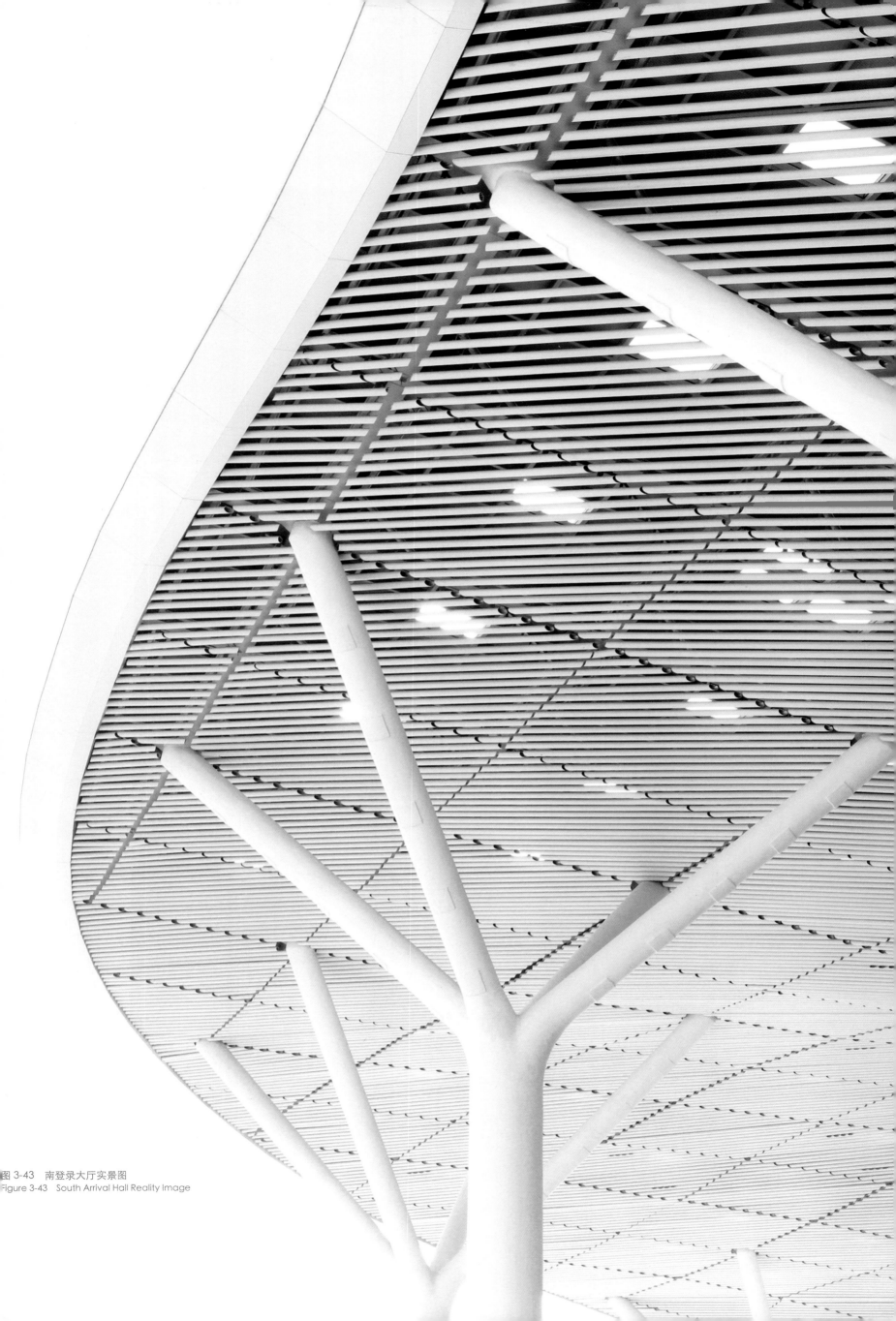

图 3-43　南登录大厅实景图
Figure 3-43　South Arrival Hall Reality Image

图 3-44　登录大厅效果图
Figure 3-44　Arrival Hall Rendering

图 3-45　登录大厅效果图
Figure 3-45　Arrival Hall Rendering

图 3-46　登录大厅南入口效果图
Figure 3-46　Arrival Hall South Entrance Rendering

登录大厅设计
The Design of the Arrival Halls

北登录大厅功能布局

北登东侧空间基本与南登一致，为登录空间。西侧为容纳 1500-1920 人的国际报告厅，内设可升降座椅，提供不同等级的座椅布置方式。三层靠近中廊的区域设有美食广场及中西餐厅（图 3-47，图 3-48，图 3-49）。

Function layout of the North Arrival Hall

The space on the east side of the north arrival hall is basically the same as the south arrival hall. On the west side, there is an international lecture hall, which can host 1500-1920 people. The seats can be raised and lowered to provide different seating arrangements. There are food court has both Chinese and western restaurants near the center of the concourse.(Figure 3-47, 3-48, 3-49)

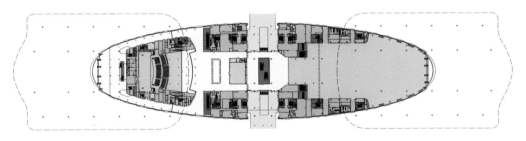

图 3-47　北登录大厅一层平面图
Figure 3-47　Floor Plan of Nouth Arrival Hall - Level 1

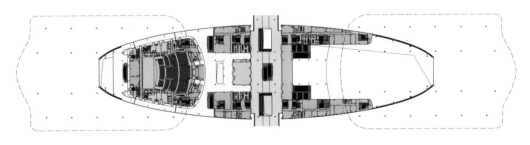

图 3-48　北登录大厅二层平面图
Figure 3-48　Floor Plan of Nouth Arrival Halll - Level 2

图例
legend

 垂直交通
Vertical traffic

国际报告厅
International Reporting Office

中央廊道
Central corridor

会议
Meeting

展厅配套
Exhibition hall matching

登录大厅
Logon Hall

厨房
Kitchen

餐厅
Restaurant

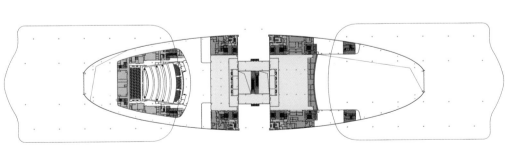

图 3-49　北登录大厅三层平面图
Figure 3-49　Floor Plan of Nouth Arrival Halll - Level 3

1栋 – 南登录大厅平面图
South Arrival Hall Building 1 - Plan

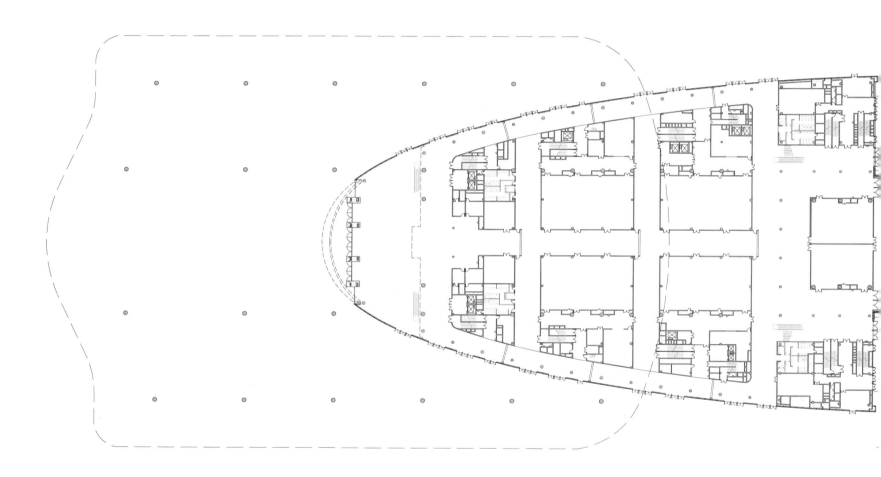

图 3-50 南登录大厅一层平面图
Figure 3-50 Floor Plan of the South Arrival Hall - Level 1

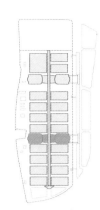

图 3-50-1 南登录大厅索引图
Figure 3-50-1 South Arrival Hall Key Plan

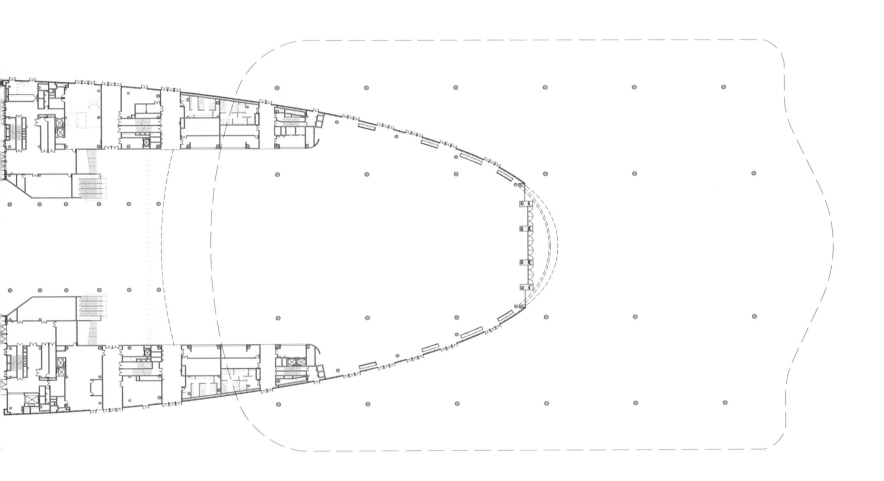

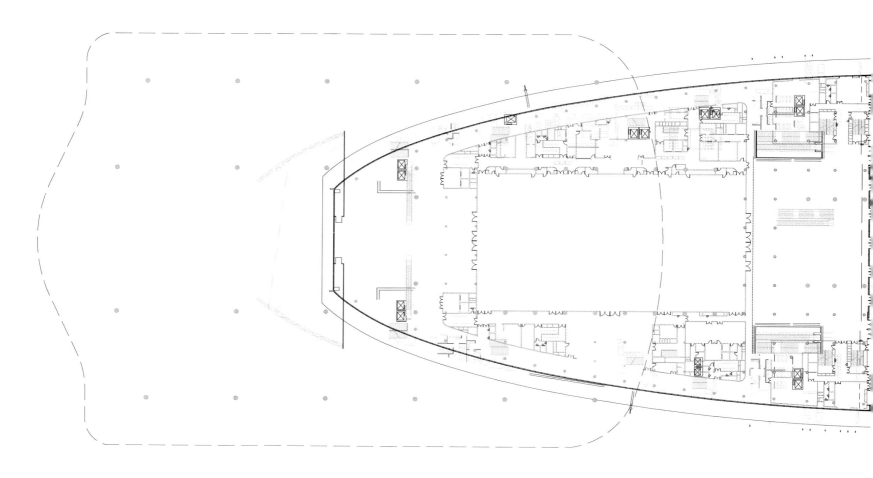

图 3-51　南登录大厅二层平面图
Figure 3-51　Floor Plan of the South Arrival Hall -Level 2

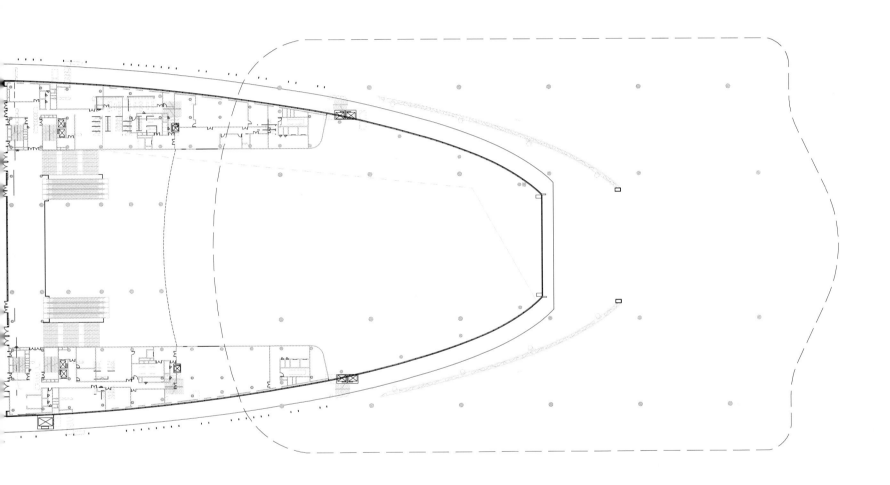

图 3-51-1　南登录大厅索引图
Figure 3-51-1　South Arrival Hall Key Plan

0　10　20　　　50　　　　　　100

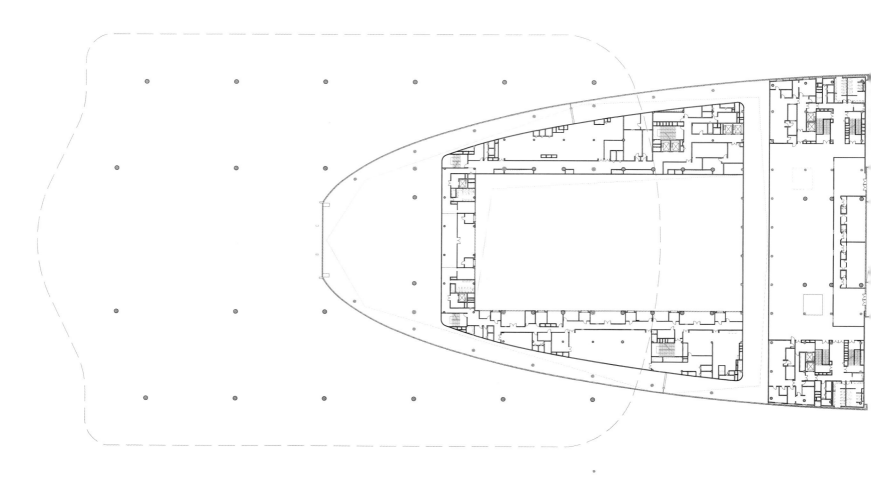

图 3-52 南登录大厅三层平面图
Figure 3-52 Floor Plan of the South Arrival Hall - Level 3

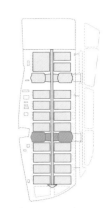

图 3-52-1 南登录大厅索引图
Figure 3-52-1 South Arrival Hall Key Plan

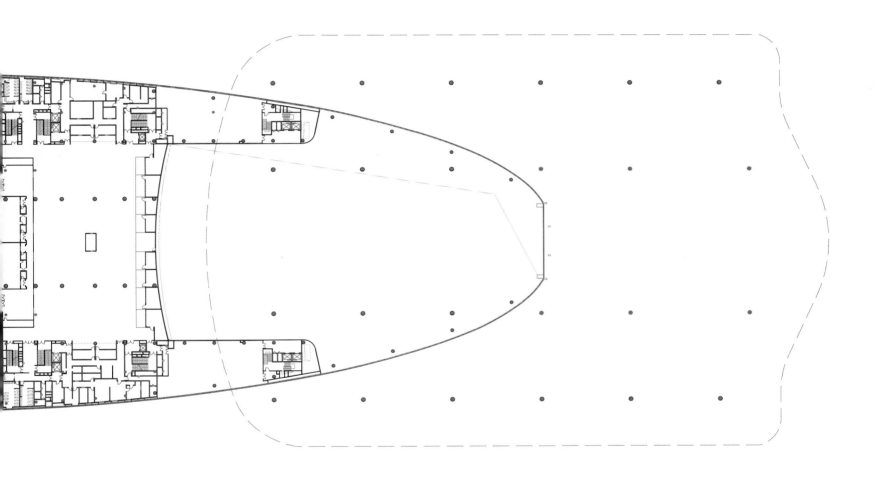

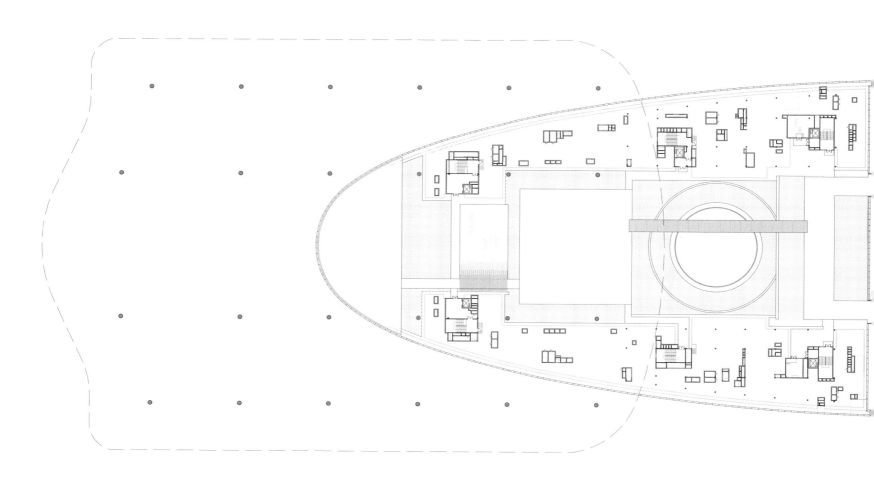

图 3-53 南登录大厅四层平面图
Figure 3-53 Floor Plan of the South Arrival Hall - Level 4

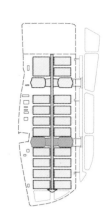

图 3-53-1　南登录大厅索引图
Figure 3-53-1　South Arrival Hall Key Plan

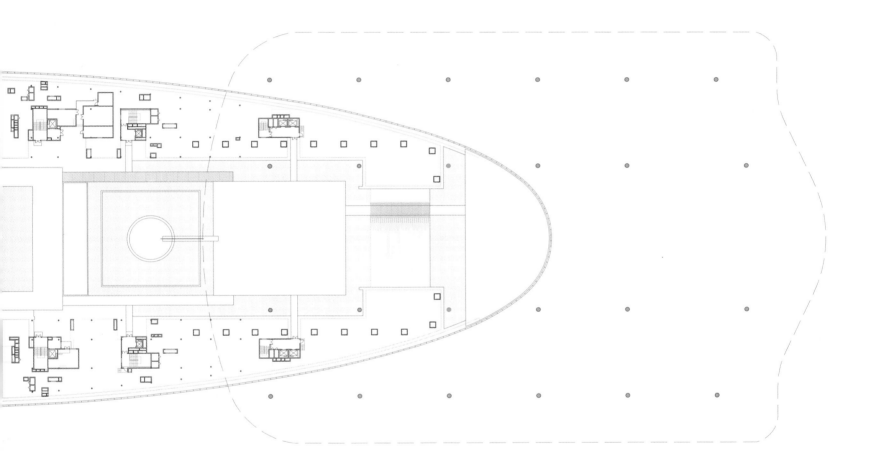

0　10　20　　　50　　　　　　100

1栋 – 南登录大厅立面图、剖面图
South Arrival Hall Building 1 - Elevation and Section

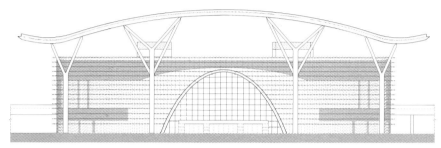

图 3-54 南登录大厅东立面图
Figure 3-54 South Arrival Hall East Elevation

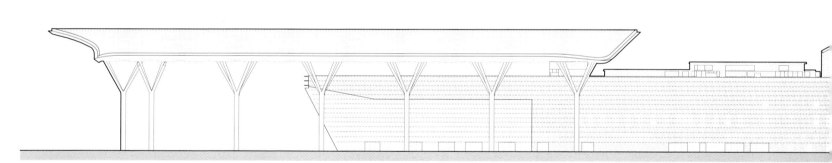

图 3-55 南登录大厅南立面图
Figure 3-55 South Arrival Hall South Elevation

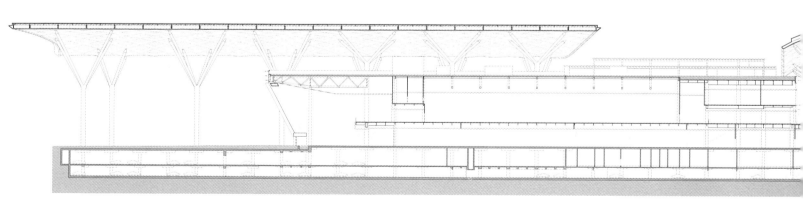

图 3-56 南登录大厅 B-B 剖面图
Figure 3-56 South Arrival Hall Section B-B

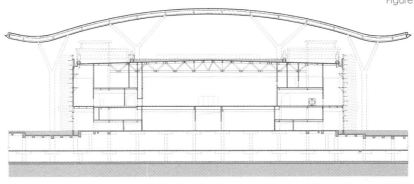

图 3-57-1　南登录大厅剖立面索引图
Figure 3-57-1　South Arrival Hall Key Plan

图 3-57　南登录大厅 A-A 剖面图
Figure 3-57　South Arrival Hall Section A-A

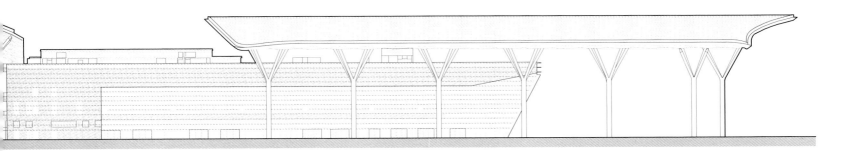

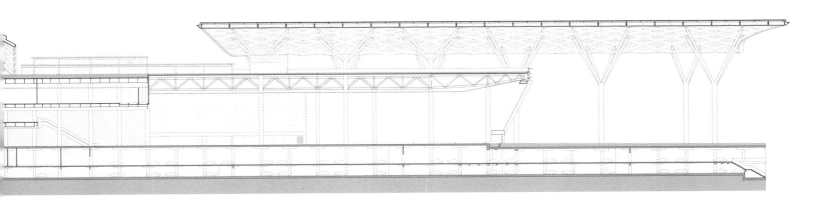

0　10　20　　　　　50　　　　　　　　　　　100

1栋 – 北登录大厅平面图
North Arrival Hall Building 1 - Plan

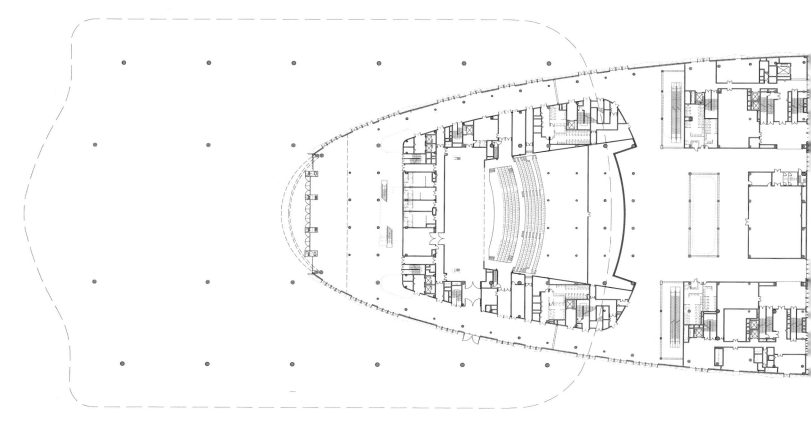

图 3-58　北登录大厅一层平面图
Figure 3-58　Floor Plan of the North Arrival Hall - Level 1

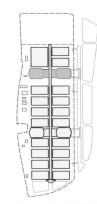

图 3-58-1 北登录大厅索引图
Figure 3-58-1 North Arrival Hall Key Plan

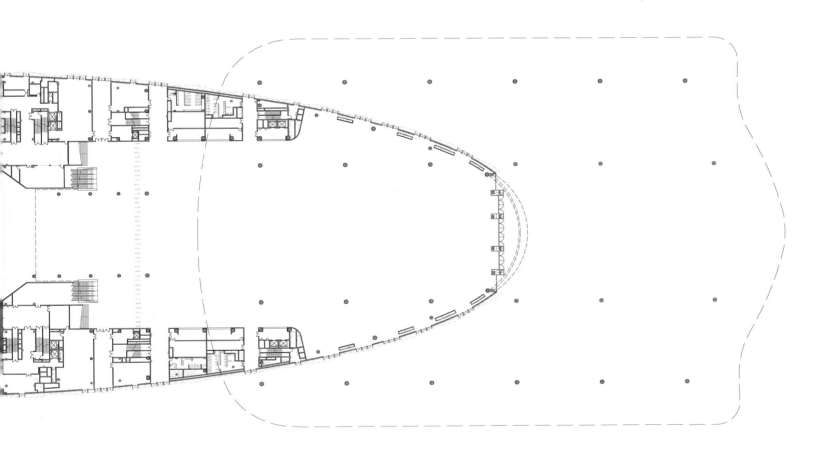

0　10　20　　　　50　　　　　　　　　100

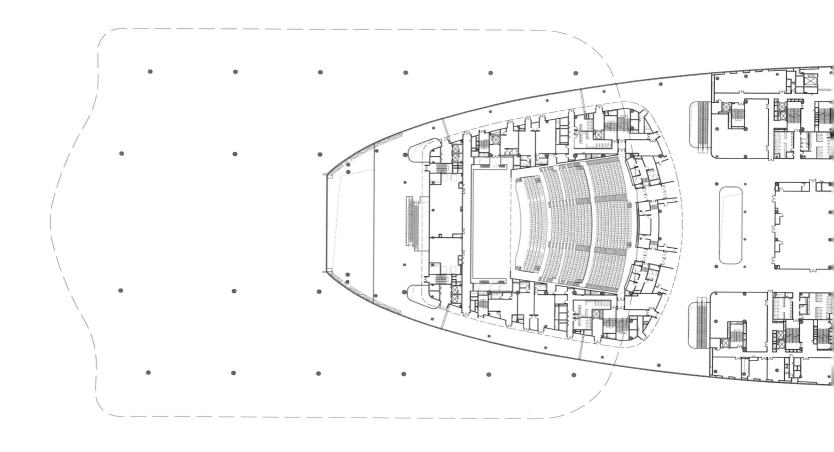

图 3-59　北登录大厅二层平面图
Figure 3-59　Floor Plan of the North Arrival Hall - Level 2

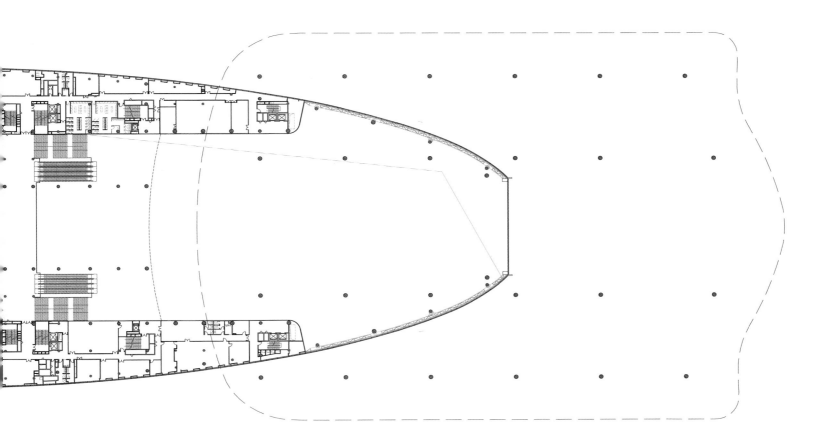

图 3-59-1　北登录大厅索引图
Figure 3-59-1　North Arrival Hall Key Plan

0　10　20　　　50　　　　　　　　100

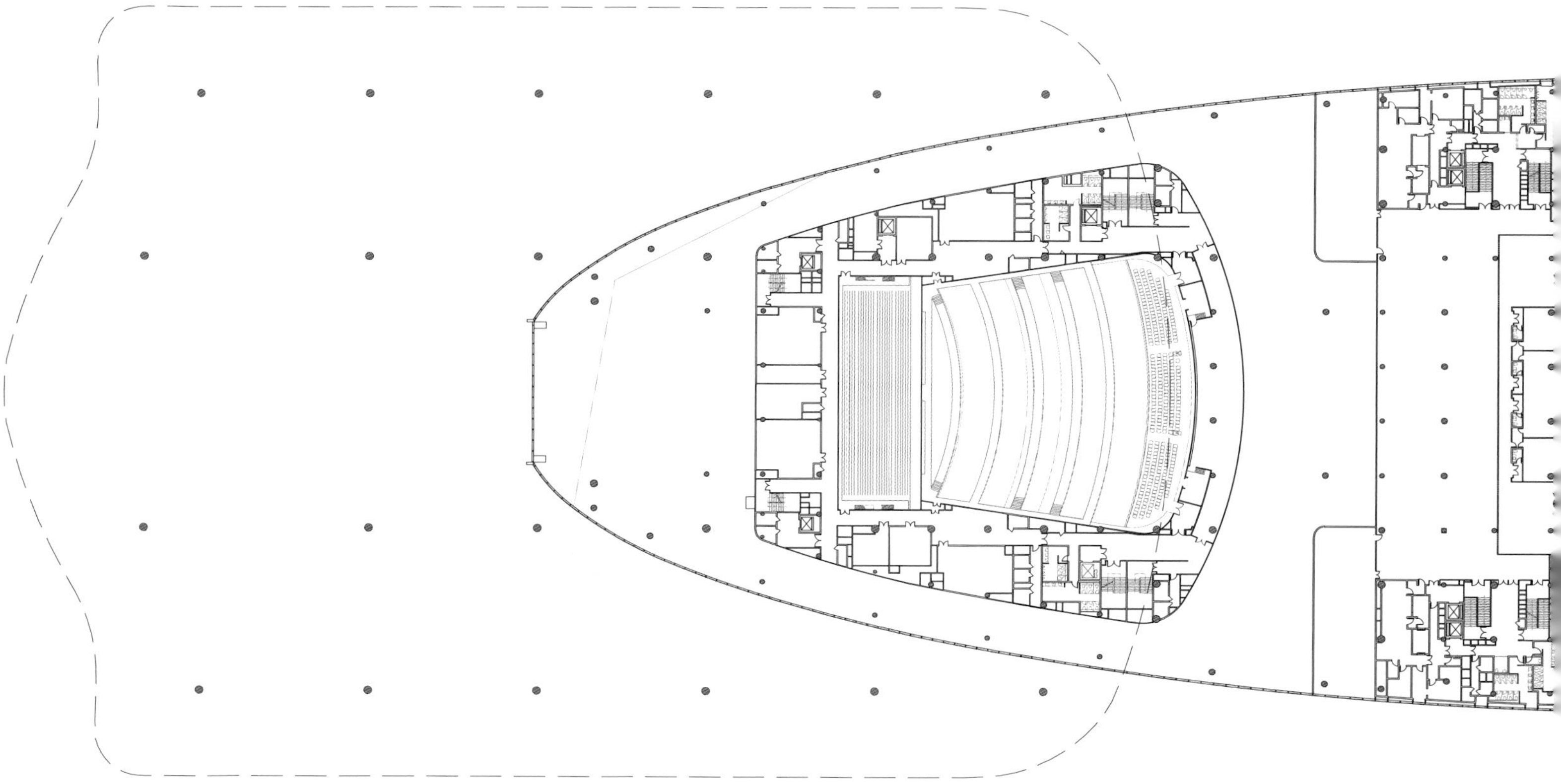

图 3-60　北登录大厅三层平面图
Figure 3-60　Floor Plan of the North Arrival Hall - Level 3

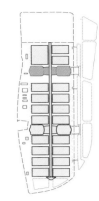

图 3-60-1　北登录大厅索引图
Figure 3-60-1　North Arrival Hall Key Plan

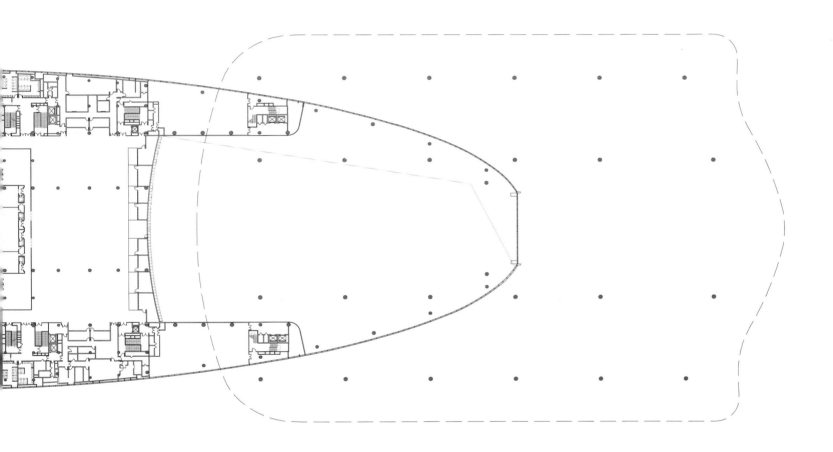

0   10   20   50   100

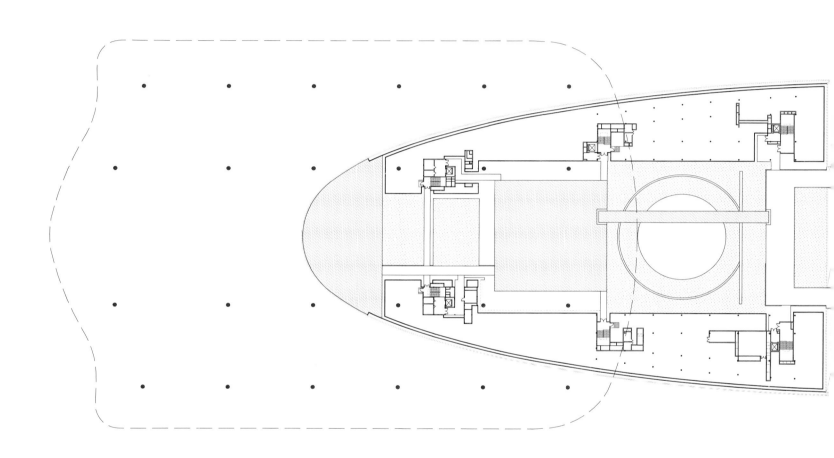

图 3-61 北登录大厅四层平面图
Figure 3-61 Floor Plan of the North Arrival Hall - Level 4

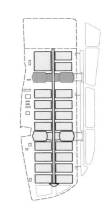

图 3-61-1　北登录大厅索引图
Figure 3-61-1　North Arrival Hall Key Plan

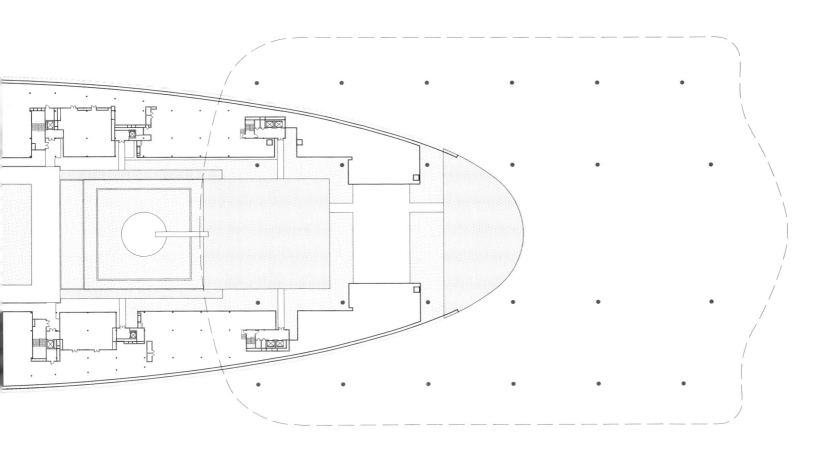

0   10   20   50   100

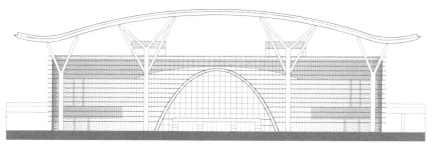

图 3-62　北登录大厅东立面图
Figure 3-62　North Arrival Hall East Elevation

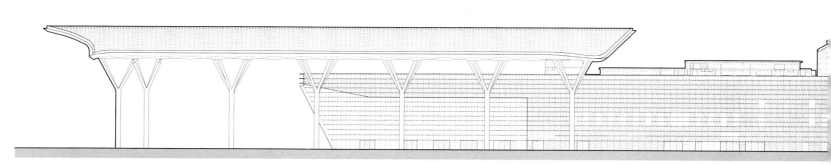

图 3-63　北登录大厅南立面图
Figure 3-63　North Arrival Hall South Elevation

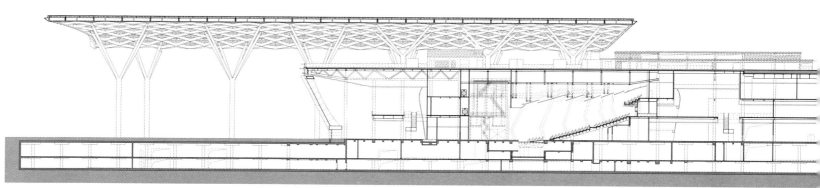

图 3-64　北登录大厅 B-B 剖面图
Figure 3-64　North Arrival Hall Section B-B

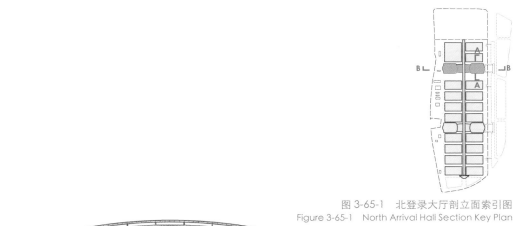

图 3-65-1　北登录大厅剖立面索引图
Figure 3-65-1　North Arrival Hall Section Key Plan

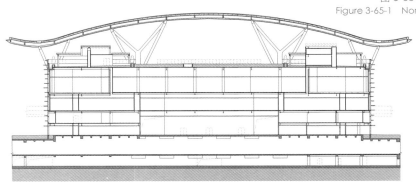

图 3-65　北登录大厅 A-A 剖面图
Figure 3-65　North Arrival Hall Section A-A

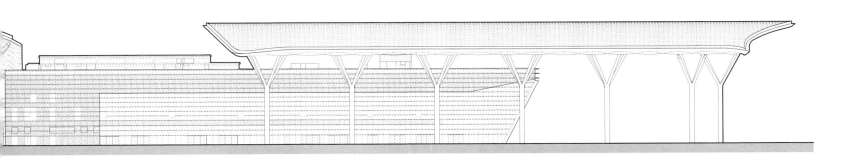

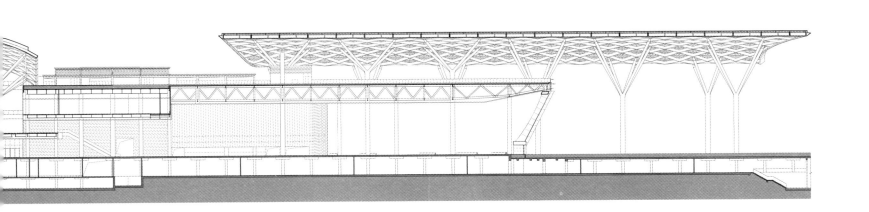

标准展厅设计
The Design of Standard Exhibition Hall

16 个 2 万 m² 标准展厅为单层大空间，展览面积约 20380m²，可布置标准展位 976 个（图 3-66），展厅的东西两侧为展览配套功能（图 3-67），其中靠近中廊一侧的二层为会议办公功能、三层为配套餐饮功能。远离中廊一侧的二层为配套餐饮功能。南北两侧设有地上综合管廊和设备用房。

The 20,000 square meters 16 standard exhibition halls are all large single-storey space with total exhibiting area around 20,380 square meters, and 976 standard exhibition booths that can be arranged (Figure 3-66) according to need. Both east and west sides are equipped with exhibition supporting facilities (Figure 3-67), the space on the second floor near the side of the central concourse can be used for meeting or as office, and the third floor is for catering. The space on the second floor away from the side of the central concourse is also for catering. The north and south sides are equipped with above ground utility tunnel and equipment rooms.

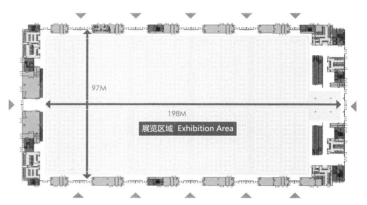

图 3-66　标准展厅平面图
Figure 3-66　Standard Exhibition Hall Plan

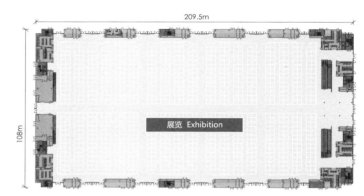

图 3-67　标准展厅平面图
Figure 3-76　Standard Exhibition Hall Plan

展厅配置参数
Exhibition Hall Configuration Parameters

标准展厅尺寸：198m×97m
Standard hall size:198m×97m

展览面积：2万m²
Exhibition area: 20000 square meters

展厅净高：15m
Clear height of exhibition hall: 15m

展厅地面荷载：50kN/m²
Floor load of exhibition hall: 50kN / m²

展厅吊点距离：9m×9m
Exhibition hall lifting point distance: 9m×9m

展厅吊点荷载：2t
Exhibition hall lifting point load: 2t

展厅管沟间距：6m
Distance between pipe trenches in the exhibition hall: 6m

展厅灯光间距：9m×9m
Lighting space of exhibition hall: 9m×9m

空调形式：侧送风
Air conditioning type: side air supply

图例
Legend

○ 展厅
EXhibition

○ 卫生间
Toilet

○ 展厅配套
Exhibition hall matching

● 楼梯
Stairs

标准展厅配套功能
Supporting functions of standard exhibition hall
单个展厅展览区域约100MX200M，展览面积2万平方米。
The exhibition area of a single exhibition hall is about 100mx200m, with an exhibition area of 20000 square meters.

深圳会展单个标准展厅面积统计表															
层数 Layer number		展厅 Exhibition hall	会议 Meeting	餐饮 Restaurant	办公 Meeting	接待 Reception	储藏 Store	安防 Security	交通 traffic	卫生间 Toilet	设备 equipment	服务中心 Service Centre	外卖区 take-out Area	其他 Other	总计 Total	总计连接处 ToTotal connectionstal
一层 1st floor		19029			176.16		685.52	93.35	1017.61	839.01	548.89	167.13			22556.67	
二层 2nd Floor	二层 2nd Floor	1247.13	296.43		244.99	63.59	224.7		751.98	304.76	258.41		697.69		4089.68	
	展厅之间 Between		759.79						309.72		42.96					1112.47
三层 3rd Floor				1373.49					576.35	152.46	2659.38				4761.68	
屋顶 Roof									116.89					970.59	1087.48	
总计 Total		20276.13	1056.2	1373.49	421.15	63.59	910.22	93.35	2772.55	1296.23	3509.64	167.13	697.69	970.59	32495.51	33607.98

表 3-1　标准展厅面积统计表
Table 3-1　statistical table of standard exhibition area

图 3-68　标准展厅效果图
Figure 3-68　Standard Exhibition Hall Rendering

图 3-69　标准展厅效果图
Figure 3-69　Standard Exhibition Hall Rendering

展厅的多功能利用图
The Multifunction of the Exhibition and Convention Hall

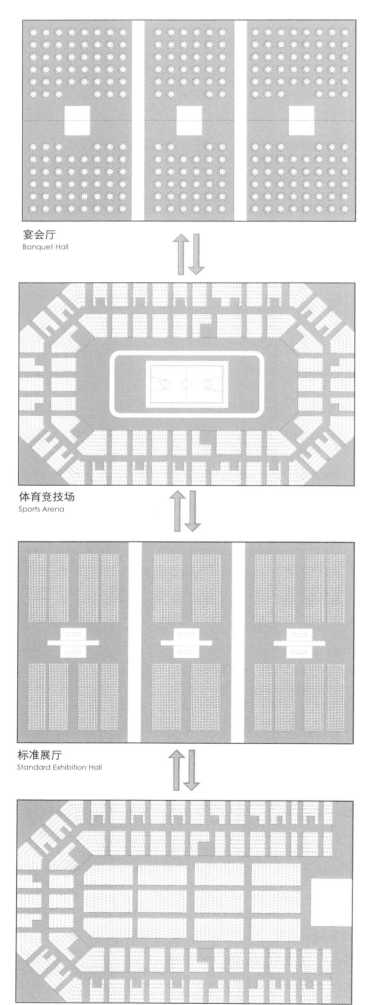

宴会厅
Banquet Hall

体育竞技场
Sports Arena

标准展厅
Standard Exhibition Hall

表演
Performance

图 3-70   展厅的多功能利用图
Figure 3-70   The Multifunction use of the Exhibition Hall

展厅作为每一座会展建筑中最核心的功能空间，其组织方式、流线安排以及配备的各类服务设施直接决定了展会的品质与观展的体验。场馆全年还应具备展期与非展期的经营灵活性（图 3-70）。由此 AUBE 欧博设计引入会展综合体模式，使展厅使用模式演变为会议、展览、宴会、演艺、运动等兼容多种功能的会展综合体，为城市及周边片区的市民提供更多室内公共活动空间。

As the core functional space in every exhibition building, the exhibition hall's organization, circulation arrangement and various service facilities directly determine the quality of the exhibition and the relative experience. The exhibition hall should also be flexible for different use both in exhibition and non-exhibition periods (Figure 3-70). Therefore, AUBE introduced the exhibition complex mode, making the exhibition hall an exhibition complex with multiple function like meeting, exhibition, banquet, arts performing, sports and etc. providing more indoor public activity space for the citizens of the city and surrounding areas.

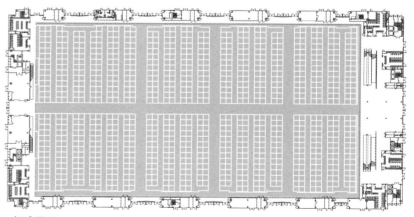

标准展厅
Standard Exhibition Hall

机电管沟和设备管廊
Mechanical Pipe Trench and Equipment Pipe Gallery

深圳国际会展中心展厅管沟间距为 6m，沿着管沟方向，接驳井间隔 9m（图 3-72），这种分布方式能更灵活地满足不同布展平面的需求。此外，由于展厅地面荷载的要求，管沟盖板需要满足与展厅地面相同的荷载要求。

地上设备管廊设计可满足机电管线布置与扩增管线的可能性，增加地下停车数量，而且施工周期短、造价低，设备管廊也可自然采光、通风，方便检修。

The gap between the pipe trench is 6m, along the trench, the space between the cross-well is 9m (Figure 3-72), in which way can flexibly meet different exhibition layout needs . In addition, due to the requirement of the ground loading capacity, the cover plate of the pipe ditch shall have the same loading capacity as the ground. The design of the above ground equipment pipe gallery can meet the needs of electromechanical pipeline layouts and the possibilities of increasing the pipelines and the number of underground parking. Its construction period is short with a low cost, and there also has natural lighting and ventilation in the equipment pipe gallery, which is convenient for future maintenance.

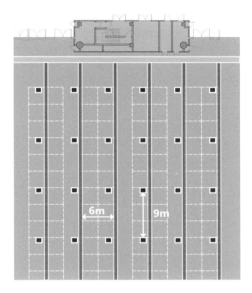

图 3-71　机电管沟和设备管廊图
Figure 3-71　Mechanical and Electrical Pipe Trench and Equipment Pipe Gallery

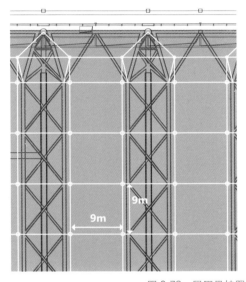

图 3-72　展厅吊挂图
Figure 3-72　The Suspention Diagram of Exhibition Hall

吊挂点规划
Suspension Point Arrangement

展厅吊挂点采用钢拉索结构，吊挂点布置密度为 9m×9m，会议室设置吊挂点密度为 3m×3m。展厅单个吊挂点容许荷载为 2t，会议室单个吊挂点容许荷载为 250kg（图 3-72）。

The Suspension points apply the steel tight wire structure form, with the density of 9m×9m in the exhibition halls, and 3m×3m in the conference rooms. The load of each Suspension point is 2t in the exhibition halls, and 250kg in the conference rooms. (Figure 3-72)

强电
1. 展厅地面每个接驳井内设置 16A/ 单相、32A/ 三相、63A/ 三相工业插座各一个；
2. 标准展厅四周均匀分布的共 10 处电井内，设置 125A/ 三相、200A/ 三相、250A/ 三相、400A/ 三相大电流工业插座，展览临时电缆可通过地沟引至需要的位置。
弱电
接驳井弱电箱内 4 个数据口和一个光纤口。
给水
主展沟内设有给水管道，排水管道，消火栓管道，主展沟尽头的集水坑里设有潜水泵，次展沟内设有消火栓管道和埋地式消火栓。
压缩空气
展厅主管沟内设置 DN32/DN80 阀门，每个阀门接一个快速接头，每个快速接头可接 10mm/19mm 口径的管线。

Strong Current
1. one 16A / single-phase, 32A / three-phase and 63A / three-phase industrial socket shall be set in each connecting well on the ground of the exhibition hall;
2. a total of 10 electric wells evenly distributed around the standard exhibition hall are equipped with 125A / three-phase, 200A / three-phase, 250A / three-phase, 400A / three-phase high current industrial sockets. The temporary cables for exhibition can be led to the required positions through the trench.
Weak Current
Four data ports and one optical fiber port in the weak current box of the connecting well.
Water Supply
The main exhibition ditch is provided with water supply pipeline, drainage pipeline, fire hydrant pipeline, the sump at the end of the main exhibition ditch is provided with submersible pump, and the secondary exhibition ditch is provided with fire hydrant pipeline and buried fire hydrant.
Compressed Air
DN32/ DN80 valves are set in the main-trench of the exhibition hall, each valve is connected with a quick connector, and each quick connector can be connected with a pipeline with a diameter of 10 mm / 19 mm.

图例
Legend

 主管道
Main Pipe

 次管道
Secondary Pipeline

 接驳井
Connecting Well

 展厅吊点
Exhibition hall Suspension Point

1栋 – 标准展厅平面图
Standard Exhibition Hall　Building 1 – Plan

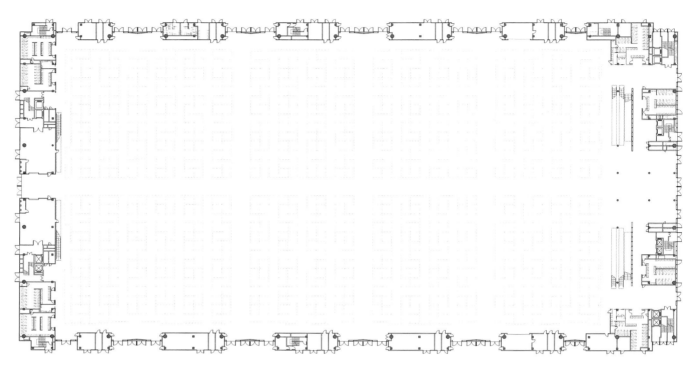

图 3-73　一层平面图
Figure 3-73　Floor Plan - Level 1

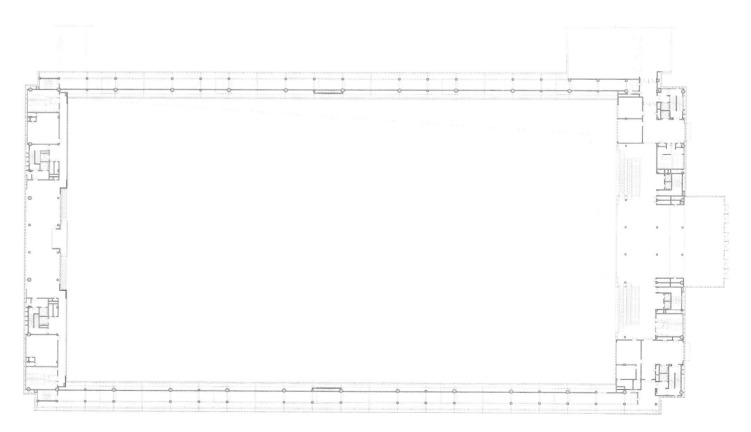

图 3-74　二层平面图
Figure 3-74　Floor Plan - Level 2

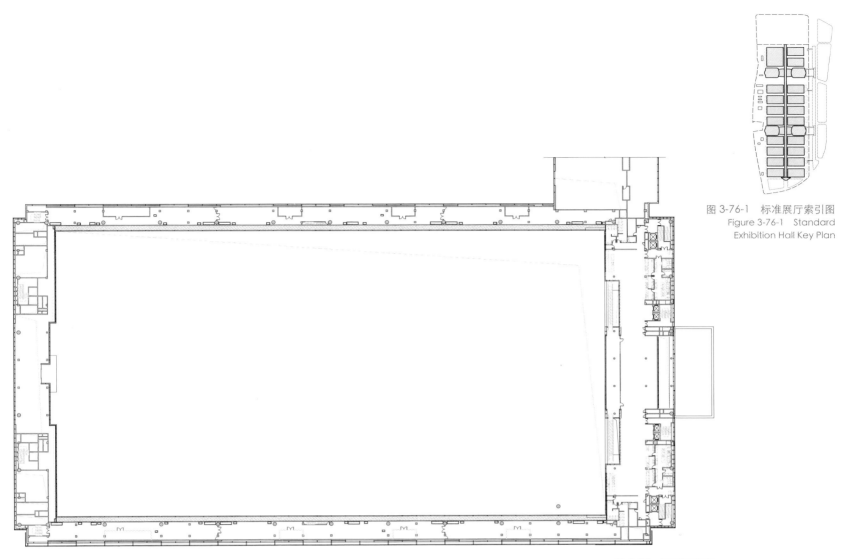

图 3-76-1　标准展厅索引图
Figure 3-76-1　Standard
Exhibition Hall Key Plan

图 3-75　三层平面图
Figure 3-75　Floor Plan - Level 3

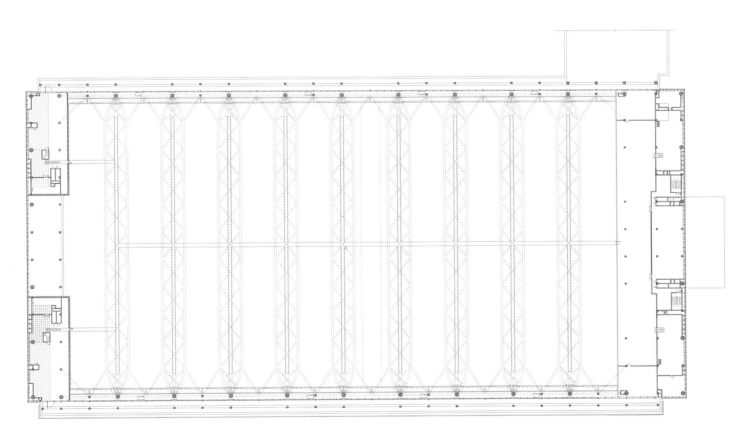

图 3-76　四层平面图
Figure 3-76　Floor Plan - Level 4

0   10   20   50   100

1栋－标准展厅立面图、剖面图
Standard Exhibition Hall Building 1 — Elevation and Section

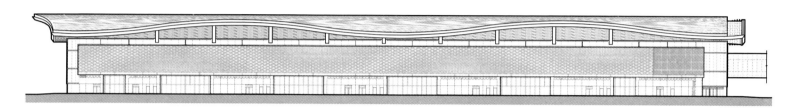

图 3-77　南立面图
Figure 3-77　South Elevation

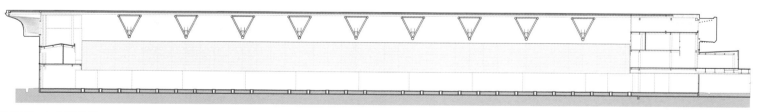

图 3-78　A-A 剖面图
Figure 3-78　Section A-A

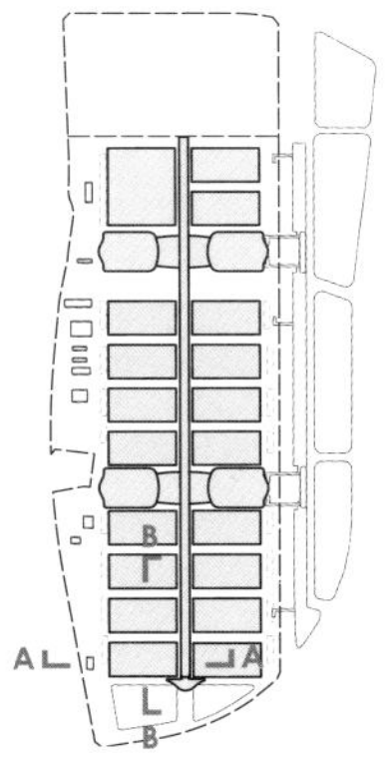

图 3-80-1 标准展厅剖立面索引图
Figure 3-80-1 Standard Exhibition Hall Key Plan

图 3-79 西立面图
Figure 3-79 West Elevation

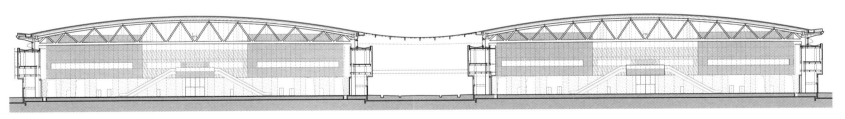

图 3-80 B-B 剖面图
Figure 3-80 Section B-B

0   10   20   50   100

特殊展厅设计
The Design of the Special Exhibition Hall

北区超大展厅和两个多功能展厅可以满足会议、宴会、体育赛事、演艺活动等功能使用要求（图 3-81）。当把一个展厅定义为多功能展厅时，其平面尺度和净高便是影响到空间适应能力和灵活程度的主要决定性因素之一。应当按照其具体的功能需要，根据实际情况来确定净空（图 3-82）。

展厅的空间模块里可以包含展览、会议、宴会、文体和商务活动空间，实现了功能互换的灵活性，这种功能复合的展览空间将会为未来会展运营带来极大的利润增长机会，也将成为国内首创的展厅多功能复合模式（图 3-83）。

The super large exhibition hall and two multi-functional exhibition halls in the north can meet the requirements of meeting, banquet, sports events, performing arts, activities and many other functions (Figure 3-81). When an exhibition hall is defined as multi-functional, its plane scale and net height are one of the main decisive factors that affect the adaptability and flexibility of the space. Therefore, the space shall be determined according to its specific functional needs and actual conditions (Figure 3-82).

The space module of the exhibition hall can be used for exhibitions, conferences, banquets, cultural events and business activities, all kinds of functions can be held and switched with flexibility. This kind of exhibition hall with compound functions will bring with the opportunities of profit growth for the exhibition operator, and it will also become the first multi-function compound model of exhibition halls in China (Figure 3-83).

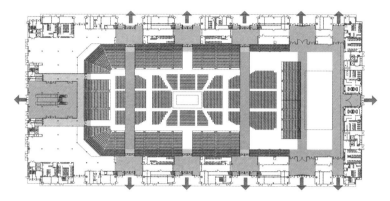

图 3-81　特殊展厅多功能利用图
Figure 3-81　Multi-functional Utilization of Special Exhibition Hall

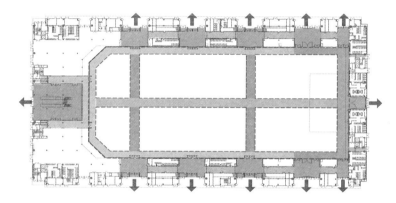

图 3-82　特殊展厅多功能利图
Figure 3-82　Multi-functional Utilization of Special Exhibition Hall

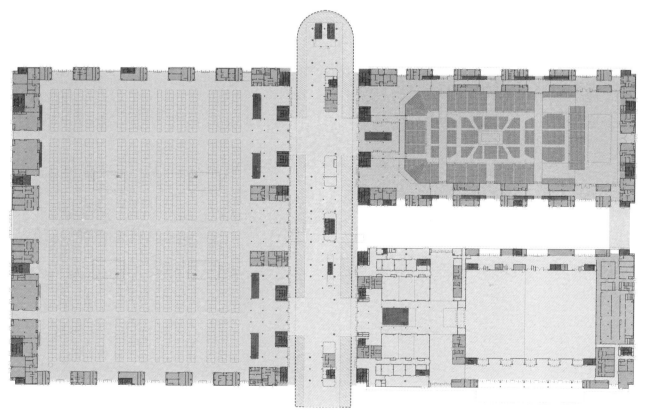

图 3-83　特殊展厅设计图
Figure 3-83　The Design Drawing of Special Exhibition Hall

图例
Legend

● 垂直交通
　Vertical Circulation

○ 体育展厅
　Sports Exhibition Hall

○ 展厅配套
　Exhibition Hall Supporting

○ 厨房
　Kitchen

○ 中央廊道
　Central Concourse

○ 消防通道
　Fire Engine Access

○ 展厅
　Exhibition Hall

○ 会议
　Meeting Room

○ 宴会厅
　Banquet Hall

图 3-84　室内交通效果图
Figure 3-84　In door Circulation Rendering

图 3-85　宴会厅效果图
Figure 3-85　Banquet Hall Rendering

图 3-86　零售店效果图
Figure 3-86　Retail Stores Rendering

图 3-87　吧台效果图
Figure 3-87　Bar Counter Rendering

17 号展厅 – 超大展厅
No. 17 Exhibition Hall- Super Large Exhibition Hall - Plan

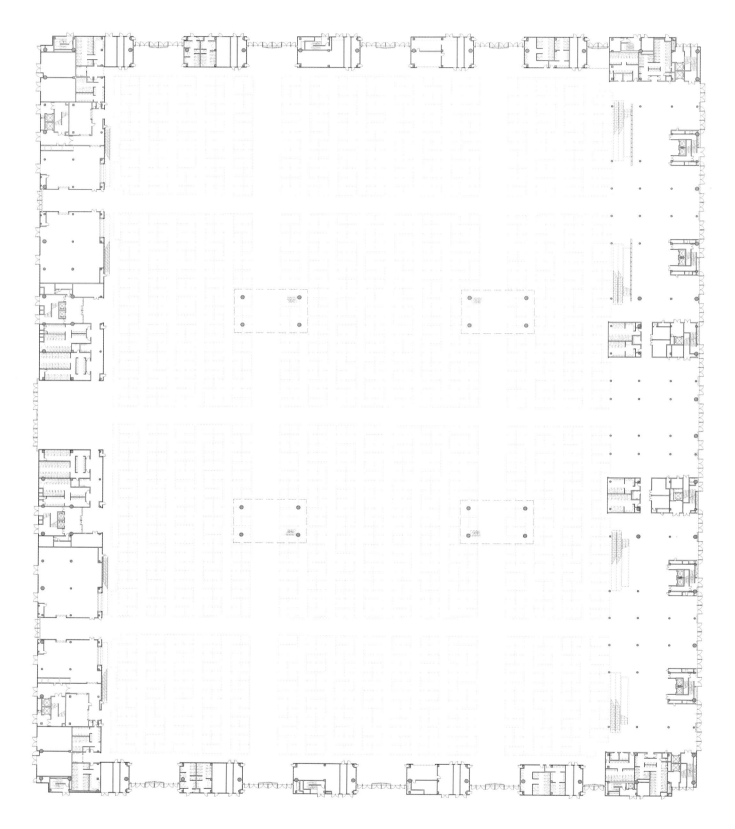

图 3-88 一层平面图
Figure 3-88 Floor Plan - Level 1

图 3-89-1 17 号展厅索引图
Figure 3-89-1 Hall 17 Key Plan

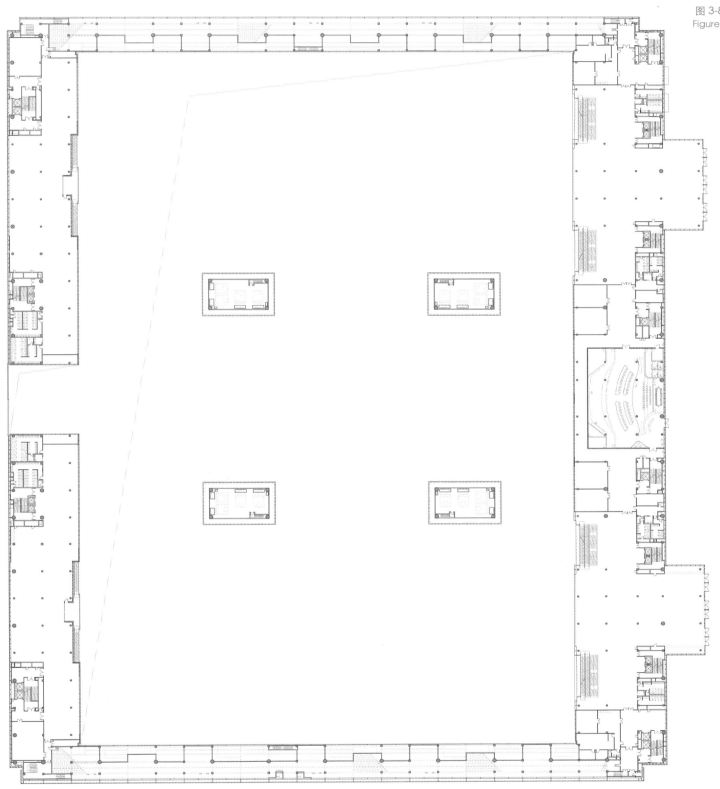

图 3-89 二层平面图
Figure 3-89 Floor Plan - Level 2

0   10   20   50   100

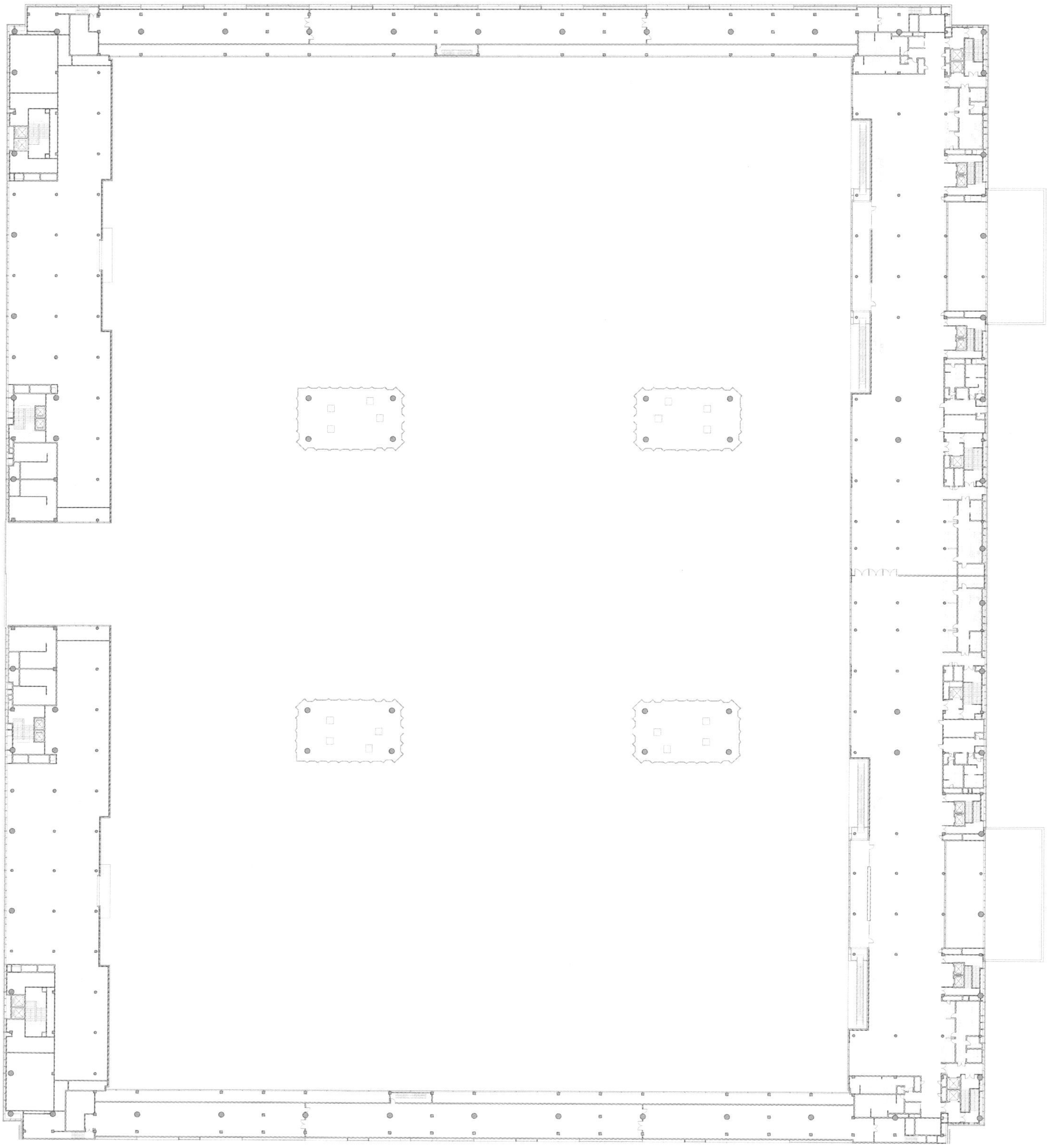

图 3-90 三层平面图
Figure 3-90 Floor Plan - Level 3

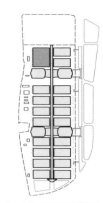

图 3-91-1　17 号展厅索引图
Figure 3-91-1　Hall 17 Key Plan

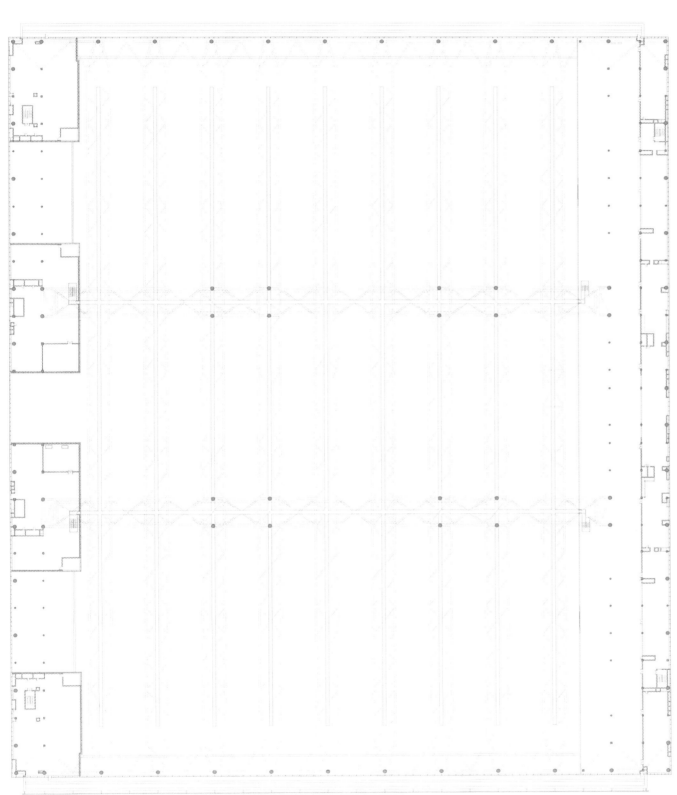

图 3-91　四层平面图
Figure 3-91　Floor Plan - Level 4

0　　10　　20　　　　　50　　　　　　　　100

17 号展厅 – 超大展厅
No. 17 Exhibition Hall - Super Large Exhibition Hall - Elevation and Section

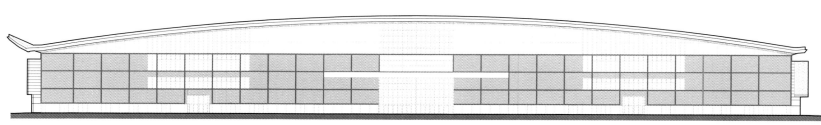

图 3-92 西立面图
Figure 3-92 West Elevation

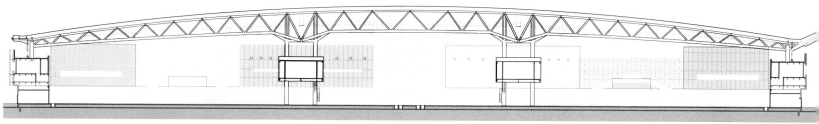

图 3-93 A-A 剖面图
Figure 3-93 Section A-A

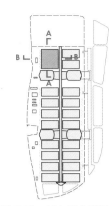

图 3-95-1　17 号展厅剖立面索引图
Figure 3-95-1　Exhibition Hall 17 Key Plan

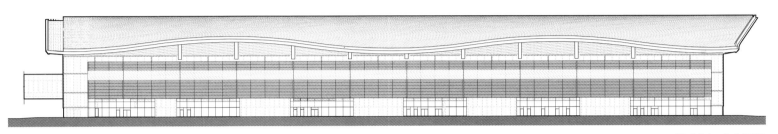

图 3-94　北立面图
Figure 3-94　North Elevation

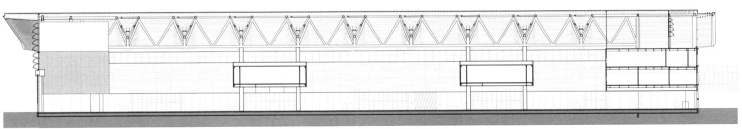

图 3-95　B-B 剖面图
Figure 3-95　Section B-B

0   10   20   50   100

18 号展厅 – 会议中心 / 宴会厅
No.18 Exhibition Hall - Conference Center/Banquet Hall - Plan

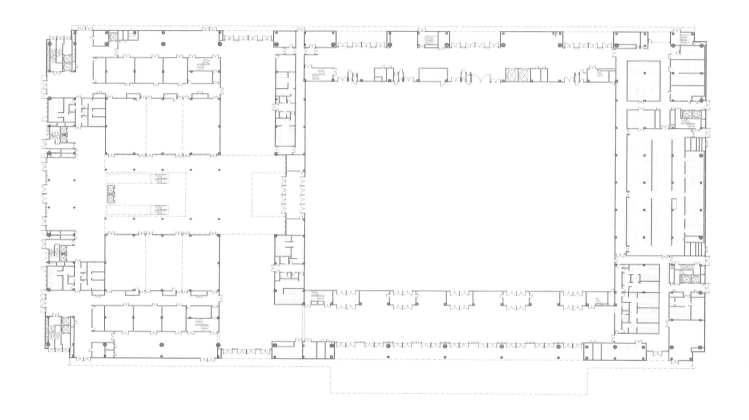

图 3-96　一层平面图
Figure 3-96　Floor Plan - Level 1

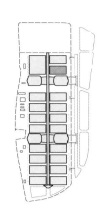

图 3-97-1   18 号展厅索引图
Picture 3-97-1   Hall 18 Key Plan

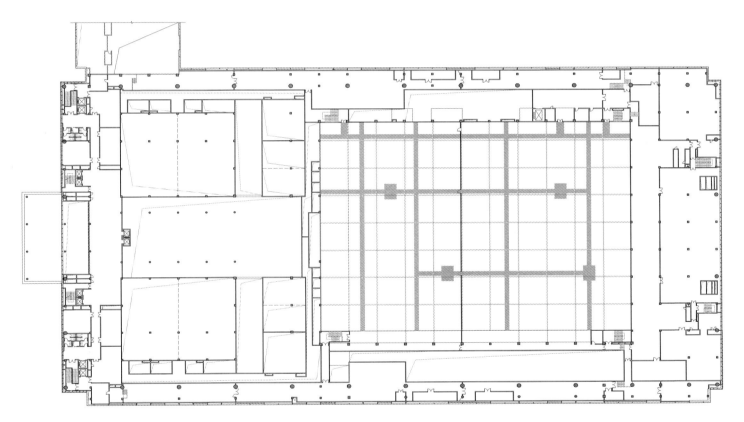

图 3-97   二层平面图
Figure 3-97   Floor Plan - Level 2

0   10   20   50   100

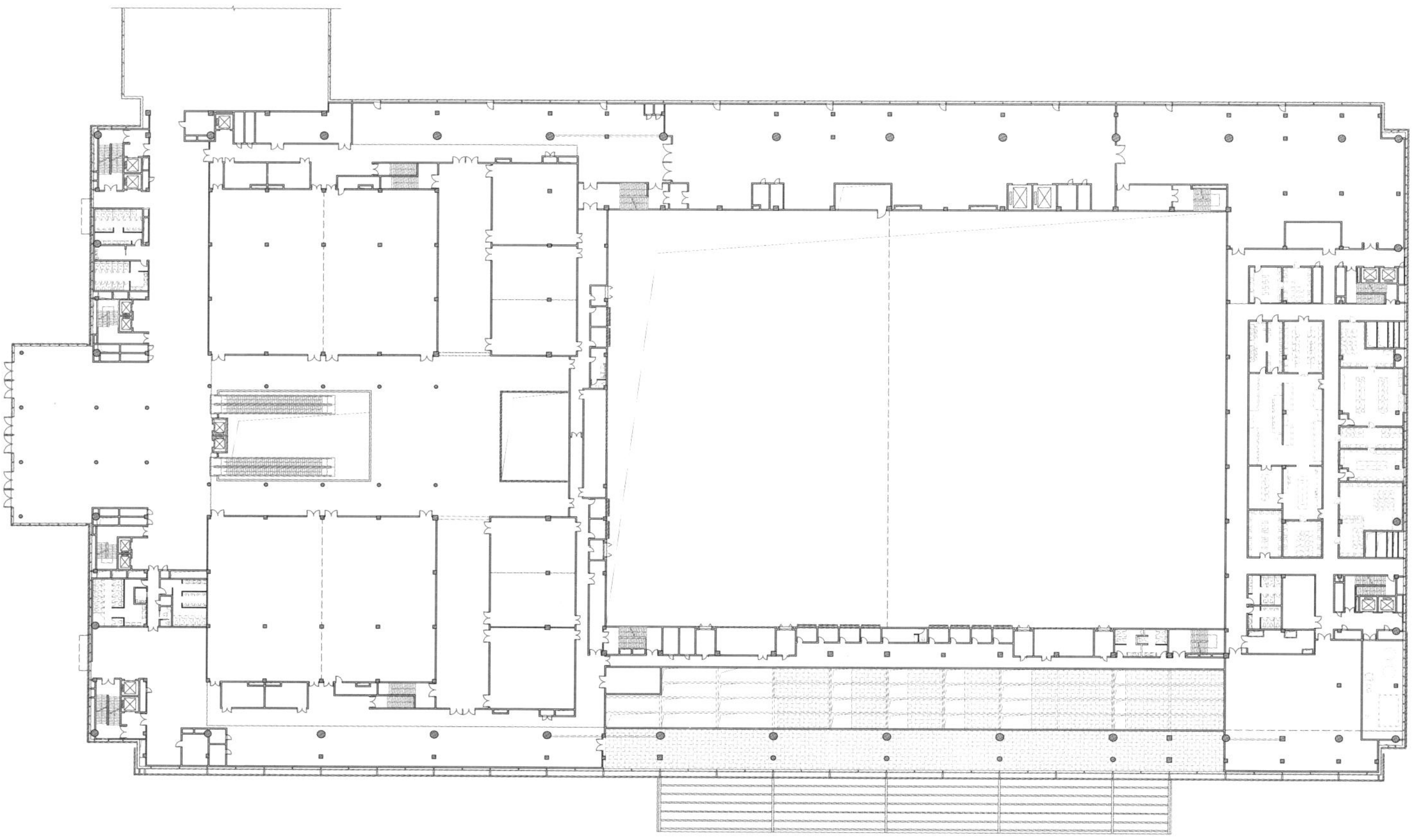

图 3-98 三层平面图
Figure 3-98 Floor Plan - Level 3

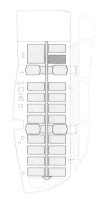

图 3-99-1　18 号展厅索引图
Picture 3-99-1　Hall 18 Key Plan

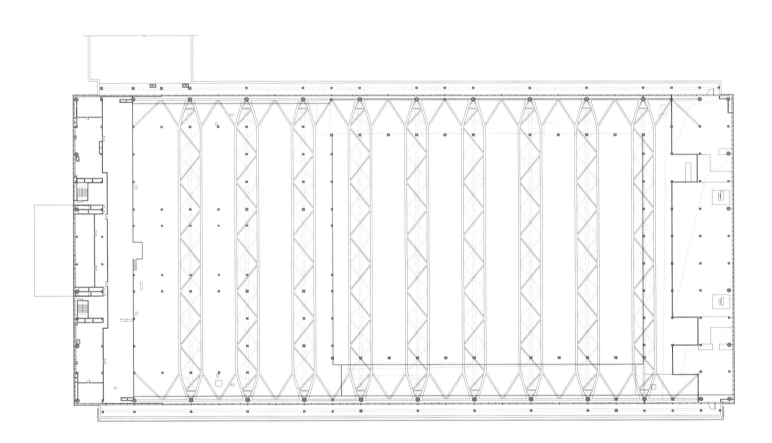

图 3-99　四层平面图
Figure 3-99　Floor Plan - Level 4

0　　10　　20　　　　50　　　　　　　100

20 号展厅：多功能展厅 – 活动中心
No.20 Exhibition Hall: Multi Exhibition Hall-Activity Center - Plan

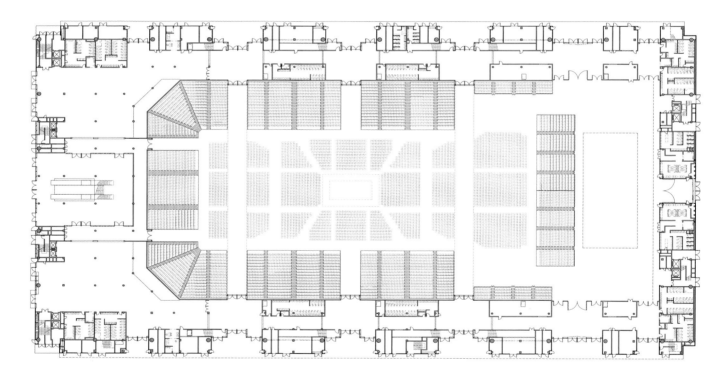

图 3-100 一层平面图
Figure 3-100 Floor Plan - Level 1

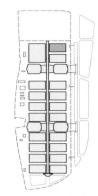

图 3-101-1　20 号展厅索引图
Picture 3-101-1　Hall 20 Key Plan

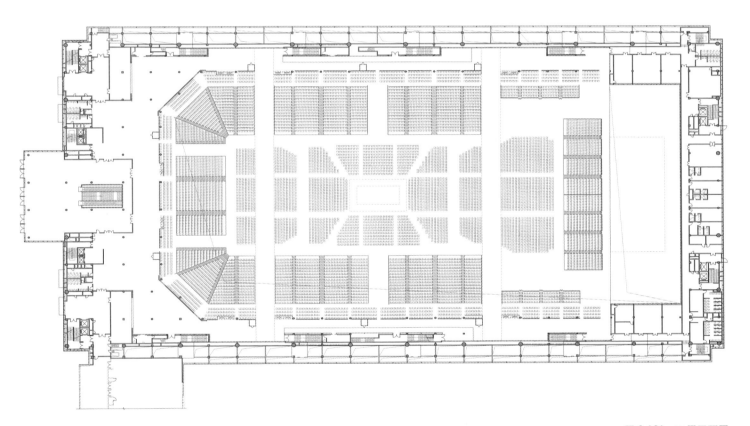

图 3-101　二层平面图
Figure 3-101　Floor Plan - Level 2

0　　10　　20　　　　　50　　　　　　　　　　100

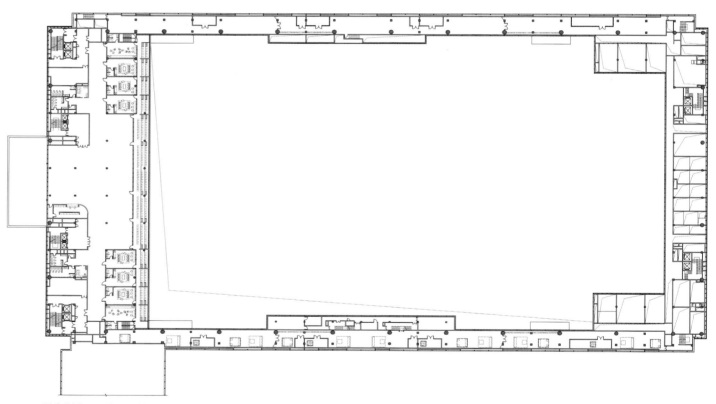

图 3-102　三层平面图
Figure 3-102　Floor Plan - Level 3

图 3-103-1　20 号展厅索引图
Picture 3-103-1　Hall 20 Key Plan

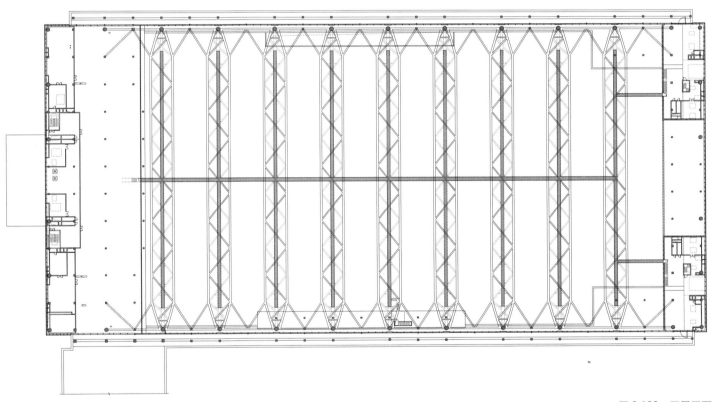

图 3-103　四层平面图
Figure 3-103　Floor Plan - Level 4

0　　10　　20　　　　　50　　　　　　　　　100

会展配套
The Supporting Facilities

展厅是会展的核心空间，是举办展览活动的基础，而会展配套的规划设置水平是体现一流会展标准的关键因素。周边的城市配套资源以及会展内的餐饮，会议，办公，仓储，后勤，地下停车等配套是必不可少的核心功能。

深圳国际会展中心以地域特征为出发点，结合国内外先进经验，全面打造国际一流的会展配套设施。

The exhibition hall is the core space of an exhibition and convention center and also the foundation of exhibition activities. The exhibition supporting facilities are the key factor to reflect the first-class standard of an exhibition and convention center. Surrounding supporting resources and the catering, conference, office, storage, logistics, underground parking and other supporting facilities in the exhibition and convention center are all essential functions.

Based on the geographical characteristics, the Shenzhen World has been built based on the advanced experience of both home and abroad, to create international first-class supporting facilities.

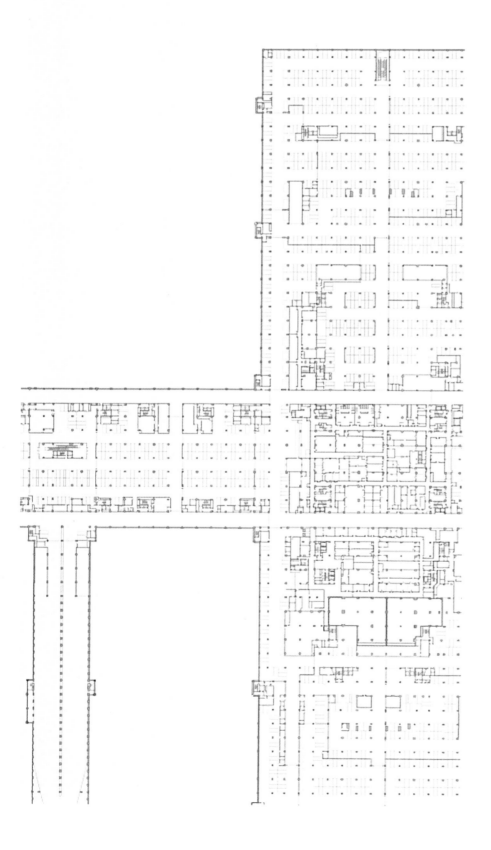

图 3-104 报告厅 VIP 室效果图
Picture 3-104 VIP Room Lecture Hall Rendering

图 3-105 多功能厅效果图
Figure 3-105 Multifunctional Hall Rendering

图 3-106 报告厅效果图
Figure 3-106 Lecture Hall Rendering

图 3-107 会见厅效果图
Figure 3-107 Meeting Hall Rendering

会议配套：会议与展览的多元共生
Conference Supporting: the multiple symbiosis between conference and exhibition

深圳国际会展中心内的 3.4 万 m² 的会议空间，连通项目北侧的国际会议中心，将共同打造一个会展产业的会议聚落群，规模从 100m²、500m²、2000m²、3000m² 到甚至 2 万 m² 的会议产业链；会展自身展览面积约 40 万 m²、餐饮面积约 4.5 万 m²，深圳国际会展中心将真正成为集会展、会议、综合配套为一体的会展湾综合体。

基于此，深圳国际会展中心可打破常规展馆运营模式，非展览期间也可以单独开放会议功能，供会展湾、空港新城甚至为整个粤港澳大湾区提供会议服务，争创会议场馆租借率到达 70% 以上。

会展业作为现代服务业的重要支柱之一，其全球市场规模正在逐步扩大。会展设施建设也已经突破传统的单一展览馆建设，正在向会展综合体、会展集聚区发展；会展内容也日渐丰富多样，现代意义上的"会展"或者"展会"，并不是孤立的"展"或者"会"，而是展中带会，会中有展，"展＋会"、"会＋展"的融合发展。

The 34,000 square meters of conference space within the Shenzhen World, along with the convention and exhibition center on the north side of the Shenzhen World, would jointly build a conference community, with different scale of conference rooms ranging from 100 square meters, 500 square meters, 2,000 square meters, 3,000 square meters to even 20,000 square meters. The Shenzhen World itself has an exhibition area of approximately 400,000 square meters and a catering area of about 45,000 square meters, which making it a real exhibition bay complex blending exhibitions, conferences, and comprehensive supporting facilities altogether.

Therefore, the Shenzhen World is expected to break the conventional operation mode of convention and exhibition center. Even during the non-exhibition period, it could also open and be used for conference, providing services for the exhibition bay, the Airport New Town, and even the whole GBA, striving to achieve an occupancy rate of more than 70%.

Exhibition industry is one of the most important pillar industries of modern service industry, and its global marketing size is expanding steadily. The construction of exhibition facilities has already broke through the traditional and single-used way, and moving towards the development features of exhibition complex, and exhibition cluster, with more diversed contents and services. Modern exhibition is not just exhibiting or meeting, but exhibitions with meeting, or meeting with exhibitions, it is a blend of harmonious "exhibition+meeting" or "meeting+exhibition".

深圳国际会展中心总计设有 150 多个各种规模、类型的会议室，共计 3.4 万 m²；配备类型多样，功能丰富，设备齐全的会议设施，满足各种类型会议需求（图 3-108）。

The Shenzhen world has a total of more than 150 conference rooms of various sizes and types, total 47,000 square meters area, equipped with a wide range of features, various functions, fully equipped conference facilities to meet all types of meeting needs (Figure 3-108).

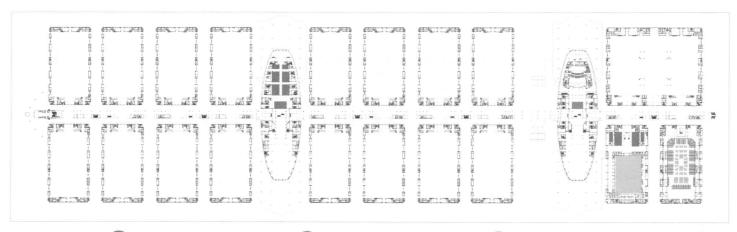

1F 图例：
Legend

● 会议室、媒体发布中心
Conference Room, Press Center

1F 18个会议室
1F 18 meeting rooms

2F 媒体发布中心2个
2F Media Publishing Center 2

● VIP室
VIP Room

1F VIP室1个、VVIP室1个
1F VIP Room, 1 VVIP Room

2F VIP室6个、媒体发布中心1个
2 F VIP Rooms, 1 Press Center

● 北多功能厅
North multi-function hall

6200座会议厅、2700座宴会厅
6,200 Conference Rooms and 2,700 ballrooms

展览、会议、宴会
Exhibition, Conference, Banquet

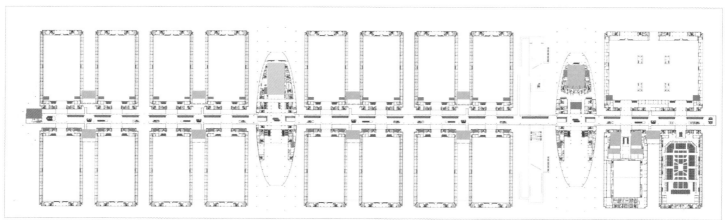

2F 图例：
Legend

● 会议中心
Conference Rooms

1F 12个会议室
1F 12 Conference Rooms

2F 13个会议室
2F 13 Conference Rooms

3F 4个VIP、1个行政酒廊
3F 4 VIPs, 1 Executive Lounge

● VIP室
VIP Room

南入口1个VIP室
South Entrance 1 VIP Room

● 展厅会议室 760 ㎡
Exhibition Room 760m²

展厅会议室(9个)可一分为三
Exhibition Room (9, can be divided into three)

● 南登多功能厅 3600 ㎡
Nandeng Multi - function Room (3600 m²)

2400座会议厅、1100座宴会厅。
2,400 Conference Halls and 1,100 Ballrooms

● 北登国际报告厅 2200 ㎡
Beideng International Auditorium（2200 m²）

舞台区+观众区 （地座+楼座+VIP升降坐席）
Stage Area + Audience Area（Floor + Building + VIP Lifting Seat）

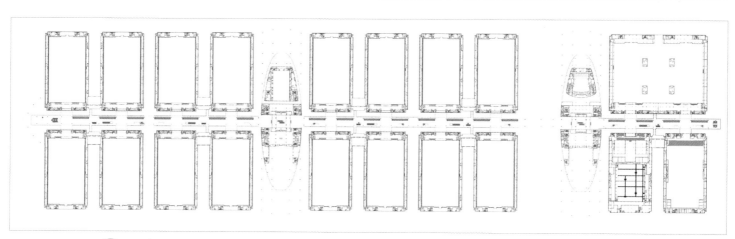

3F 图例：
Legend

● VIP室
VIP Room

8个VIP包间、1个行政酒廊
8 VIP Rooms, 1 Executive Lounge

图 3-108 会展会议与展览多元共生图
Figure 3-108 Exhibition and Convention Multi-symbiosis Diagram

餐饮配套：高品质的供餐服务
Catering: High Quality Catering Services

高品质的供餐服务，是国际一流会展中心的必备条件。会展运营方必须有条件在最短时间内为顾客提供便捷优质的餐饮服务。深圳国际会展中心根据平面规划特征（图 3-109、图 3-110），结合展厅和登录大厅，采用集中布置和分散布置结合的方式，提供了 4.5 万 m² 的就餐空间，可满足约 15 万人 / 天的就餐需求。

High quality catering service is essential for a world-class international exhibition and convention center. The operator of the exhibition and convention center must be capable of providing convenient and high-quality catering services for the customers within the shortest time period. According to the features of the plan (Figure 3-109, 3-110), the Shenzhen World applied the arrangement of integrate both the concentrated and separated pattern, proving a catering space of 45,000 square meters, which can provide enough food and drinks for 150,000 people per day.

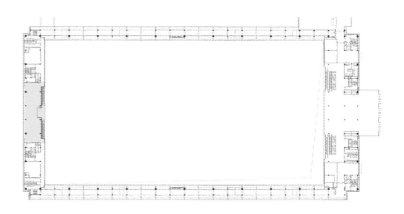

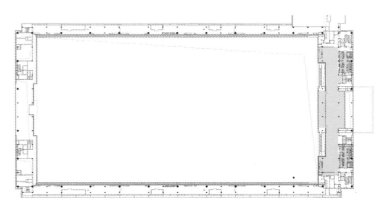

图 3-109　餐饮配套布局图
Figure 3-109　Supporting Catering Layout Diagram

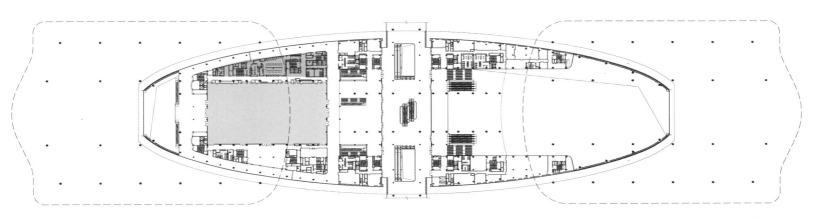

南登录大厅二层平面图
Floor Plan of South Arrival Hall - Level 2

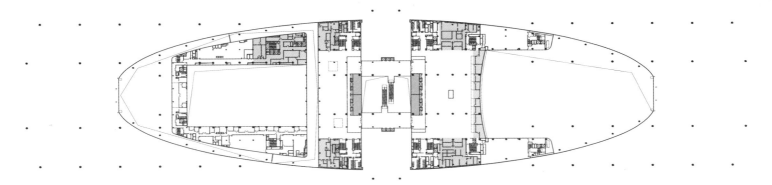

南登录大厅三层平面图
Floor Plan of South Arrival Hall - Level 3

图例
Legend

  VIP包房
VIP Room
 厨房
Kitchen
 餐厅/宴会厅
Restaurant/Banquet Room
  外卖餐厅
Takeaway Restaurant

图 3-110　餐饮配套布局图
Figure 3-110　Supporting Catering Layout Diagram

超大展厅餐饮布局

17 号展厅（超大展厅）靠近中央廊道一侧的三层设两处美食档餐饮，单个餐厅面积约 1900m²，餐厅附近配置热炒厨房，单个厨房面积约 560m²；17 号展厅二层远离中央廊道一侧设两处盒饭外卖餐厅，单个餐厅面积约 1442m²，食物由南登录大厅地下一层中央热灶厨房提供。

The Catering Layout of No. 17 Super Large Exhibition Hall

The No.17 exhibition hall (super large exhibition hall) has two food stalls on the third floor near the central concourse. Each restaurant is about 1900 square meters , with a kitchen about 560 square meters nearby. On the second floor, the side away from the central concourse, there are two takeout restaurants, each has area around 1442 square meters, where the food is provided by the central hot stove kitchen on the first floor of the south arrival hall.

图 3-111　登录大厅三层清真餐厅室内效果图
Figure 3-111　Halal Restaurant on the Third Floor of the Arrival Hall - Rendering

图 3-112　员工餐厅室内效果图
Figure 3-112　Employee Restaurant Rendering

图 3-113　登录大厅三层中餐厅效果图
Figure 3-113　The Chinese Restaurant on the Third Floor of the Arrival Hall - Rendering

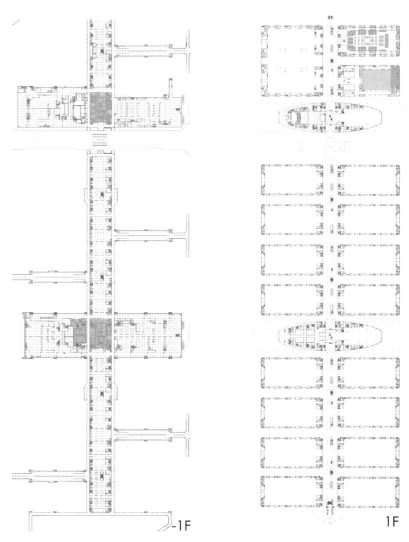

图 3-114　餐饮配套布局图
Figure 3-114　Layout of Supporting Catering Facilities

登录大厅餐饮布局

南、北登录大厅三层靠近中央廊道两侧分别设置中餐厅、清真餐厅、VIP 餐厅和美食广场，单个登录厅餐饮面积共计约 4100 m²，各个餐厅附近分别配置热炒厨房，登录厅三层厨房面积共计约 1400 m²；南登录大厅二层设置 3400m² 左右多功能宴会餐厅，餐厅附近二层、三层分别配置厨房面积共计约 1200 平方米（图 3-114）。

The Catering Layout of the Arrival Hall

There are Chinese restaurant, halal restaurant, VIP restaurant and food court on both sides of the central concourse on the third floor of the arrival hall in the south and north. Each catering area is about 4100 square meters, each restaurant has its own kitchen, and the total area of the kitchen on the third floor of each arrival hall is about 4100 square meters. On the second floor of the south arrival hall, there is a multi-functional banquet restaurant with an area about 3400 square meters. The kitchen area on the second and third floors near the restaurant is about 1200 square meters (Figure 3-114).

标准展厅餐饮布局

南区 16 个标准展厅，靠近中央廊道一侧三层平面设美食档餐饮，单个餐厅面积约 1020m²，餐厅附近配置热炒厨房，单个厨房面积约 230m²；展厅二层远离中央廊道一侧设盒饭外卖餐厅，单个餐厅面积约 660m²，食物由位于南登录大厅地下一层约 4000 m² 的中央热灶厨房提供（图 3-115）。

The Catering Layout of the Standard Exhibition Hall

There are 16 standard exhibition halls in the south of the Shenzhen World. There are food stalls and restaurants on the third floor near the central concourse. Each restaurant is about 1020 square meters, with a single kitchen about 230 square meters.

On the second floor of the exhibition hall, away from the central concourse, there are takeout restaurants, which is about 660 square meters each, the food is provided by the central hot stove kitchen on the underground floor one of the south arrival hall, and it is about 4000 square meters (Figure 3-115).

图 3-116　18 号展厅宴会厅室内效果图
Figure 3-116　effect diagram of banquet hall in hall 18

图 3-117　登录大厅三层美食广场室内效果图
Figure 3-117　the Food Court on the third Floor of the Arrival Hall - Rendering

图 3-118　标准展厅餐厅室内效果图
Figure 3-118　Standard Exhibition Hall Restaurant - Rendering

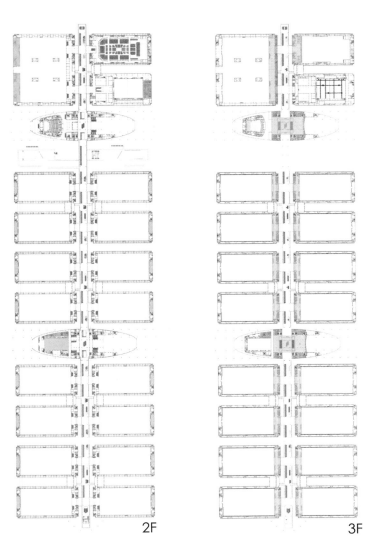

2F　　3F

图 3-115　餐饮配套布局图
Figure 3-115　Layout of Supporting Catering Facilities

18 号展厅餐饮布局

18 号展厅一层设置了 6800m² 左右的大宴会餐厅，餐厅附近一层、二层分别配置厨房面积共计约 2320m²；南北登录大厅地下一层分别设有约 6200m² 的大型中央加工厨房为地上各热炒厨房提供食物半成品。

The Catering Layout of the No. 18 Exhibition Hall

There is a 6800 square meters banquet restaurant on the first floor of the No.18 exhibition hall, its kitchen is on the first and the second floor near the restaurant, with a total area around 2320 square meters. Two large central processing kitchen with an total area of 6200 square meters is set respectively on both the north and the south side of the ground floor arrival hall, providing semi-finished food for each hot cooking kitchen above ground.

后勤配套

会展中心是集展览、会议、餐饮、娱乐、办公于一体的超大型会展综合体，其职能已远远超出最初单纯举办展览的范畴。要带动这一庞大工程高效运转，其后台的配套功能及设施缺一不可，因此需设置相应的配套服务功能。展厅安保、办公、仓储区、垃圾站、海关国检等配套功能及其规模配比规划，都在设计中进行了全面考量，并以分布均匀、就近布局和分级配置的原则进行规划布局（图3-119）。

Logistics

The Shenzhen World is a super large exhibition complex integrating exhibition, conference, catering, entertainment and office functions. Its functions are much over the original simple exhibitions scope. To guarantee the efficient operation of this huge machine, its supporting functions and backstage facilities are indispensable and important. For example, the exhibition hall security, office, storage area, garbage station, customs and CIQ, other supporting functions, as well as their scale and proportion planning, all have been comprehensively considered in the design process. Moreover, the layout also has been planned according to the principles of uniform distribution, proximity layout and hierarchy configuration (Figure 3-119).

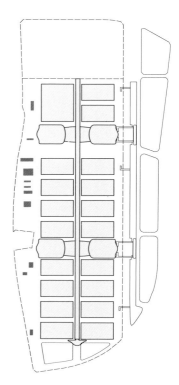

图 3-120　西侧配套索引图
Figure 3-120　West Side Supporting Facilities - Key Plan

安保用房
Security Room

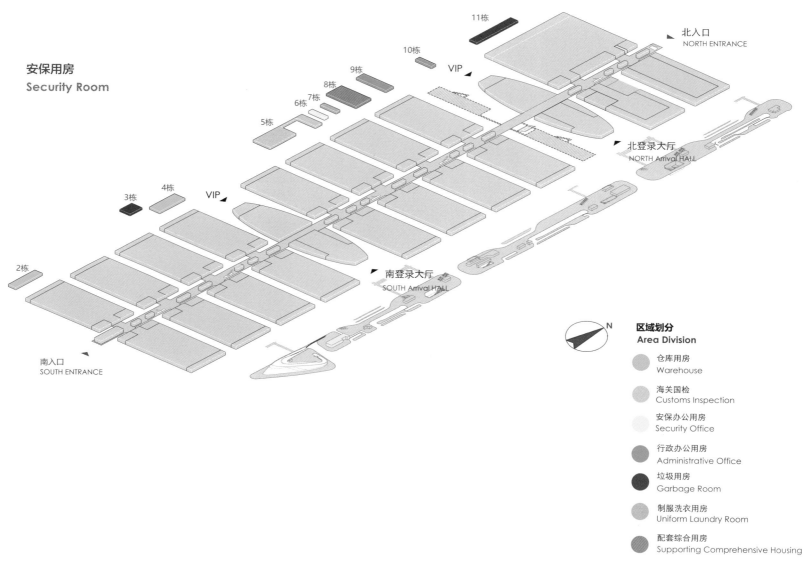

图 3-119　会展中心西侧配套
Figure 3-119　Supporting Facilities on the West Side of the Exhibition and Convention Center

图 3-121　8 栋一层多功能活动中心效果图
Figure 3-121　Building 8 Multi-functional Activity Center on the First Floor - Rendering

图 3-122　8 栋二层前台接待效果图
Figure 3-122　Building 8 Reception on the Second Floor - Rendering

图 3-123　8 栋开放办公效果图
Figure 3-123　Building 8 Open Office - Rendering

垃圾用房
Garbage

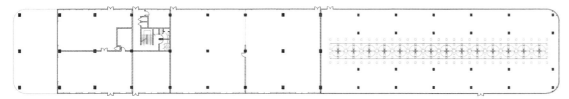

图 3-124　11栋 垃圾站 一层平面图
Figure 3-124　Building 11 Garbage Station Plan - Level 1

图 3-125　11栋 垃圾站 二层平面图
Figure 3-125　Building 11 Garbage Station Plan - Level 2

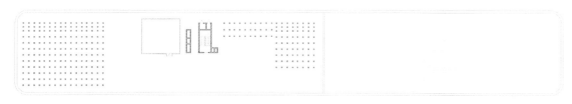

图 3-126　11栋 垃圾站 三层平面图
Figure 3-126　Building 11 Garbage Station Plan - Level 3

海关用房
Customs

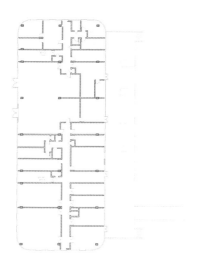

图 3-127　5栋 海关国检办公 一层平面图
Figure 3-127　Building 5 Customs Inspection
Office-Plan - Level 1

图 3-128　5栋 海关国检办公 二层平面图
Figure 3-128　Building 5 Customs Inspection
Office-Plan - Level 2

图 3-129　5栋 海关国检办公 三层平面图
Figure 3-129　Building 5 Customs Inspection
Office-Plan - Level 3

仓储用房
Warehouse

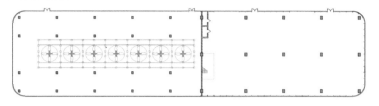

图 3-130　2栋 会展仓库 一层平面图
Figure 3-130　Building 2 Exhibition Warehouse Plan - Level 1

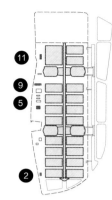

图 3-137-1　西侧配套索引图
Figure 3-137-1　West Side Supporting Facilities Key Plan

图 3-131　2栋 会展仓库 二层平面图
Figure 3-131　Building 2 Exhibition Warehouse Plan - Level 2

办公用房
Office

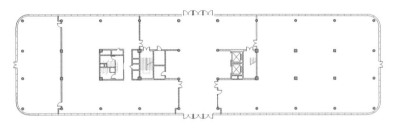

图 3-132　9栋 行政办公 一层平面图
Figure 3-132　Building 9 Administrative Office Plan - Level 1

图 3-133　9栋 行政办公 二层平面图
Figure 3-133　Building 9 Administrative Office Plan - Level 2

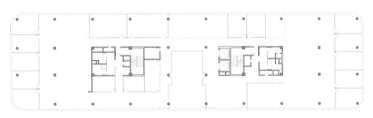

图 3-134　9栋 行政办公 三层平面图
Figure 3-134　Building 9 Administrative Office Plan - Level 3

图 3-135　9栋 行政办公 四层平面图
Figure 3-135　Building 9 Administrative Office Plan - Level 4

图 3-136　9栋 行政办公 五层平面图
Figure 3-136　Building 9 Administrative Office Plan - Level 5

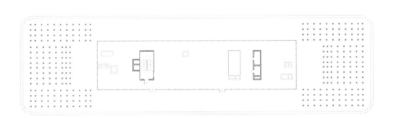

图 3-137　9栋 行政办公 六层平面图
Figure 3-137　Building 9 Administrative Office Plan - Level 6

0　10　20　50　100

交通配套：货运与停车
Transportation Supporting: Freight and Parking

鉴于大型会展特殊的交通需求特点，如果设计照搬常规项目地下车库以停车位效率为首位的思维惯性，会造成高峰时段地下车库拥堵不堪，会有进不去，出不来的现象。因此，深圳国际会展中心对地下车库设计提出了新的评价标准：以人为本，保证将全部地下车辆快进快出的时间控制在 1.5 小时以内。

Regarding the features of special traffic demand of large-scale exhibition, if we continue to use the conventional thinking of parking space efficiency first, it will cause traffic congestion in the underground parking during peak hours, and it would be difficult for cars to get in out get out. Therefore, the Shenzhen World puts forward a new underground parking design standard: people first, and guarantee all the vehicles get in or out of the underground parking within 1.5 hours.

交通设施规模建议

1. 大型会展中心主要靠集约交通组织模式，公交分担率普遍较高，国内公交分担率一般超过 80%，高峰时段更是达到 90%。
2. 国外的会展集散交通中小汽车出行条件相对宽松，配建泊位基本高于 7 个 / 百平米展览面积。
3. 国内配建泊位一般不超过 2.5 个 / 百平米展览面积。

Suggestions on the scale of transportation facilities

1. Large-scale convention and exhibition centers mainly rely on intensive transportation organization models, and the public transport sharing rate is generally high. The domestic public transport sharing rate usually exceeds 80%, and it reaches 90% during peak hours.
2. The conditions for travelling in small and medium-sized automobiles in foreign trade fairs are relatively loose, and the number of berths allocated is basically higher than 7 exhibition spaces per 100 square meters.
3. Domestical allocated berths usually do not exceed 2.5 per 100 square meters of exhibition area.
4. Save resources. There will be no parking spaces for trucks and buses on site, and they will be shared in time.

表 3-2   国内外著名会展中心交通设施规模相关案例一览表
Table 3-2   List of Cases Related to the Scale of Transportation Facilities of Famous Convention and Exhibition Centers at Home and Abroad

	建筑面积 （万平方米） Floor area(hectare)	展览面积 （万平方米） Exhibition area(hectare)	轨道交通 Rail traffic	公共交通分担率 （日常高峰） Public Transport Share (Daily peak)	停车位 Parking lot	停车配建标准 （个 / 百平方米展览面积） Parking configuration standards (Pcs / 100 m² exhibition area)
国家会展中心（上海） National Convention and Exhibition Center (Shanghai)	147	50	3 线 4 站 3 lines / 4 stations	80%（90%）	4000+	2.75
广州琶洲会展中心 Guangzhou Pazhou Exhibition Center	110	33.8	3 线 4 站 3 lines / 4 stations	80%（85%）	7200	2.13
汉诺威会展中心 Hannover Convention Center	100	46.6	3 线 3 站 3 lines / 3 stations	65%	50000	10.73
米兰新国际会展中心 Fiera Milano	100	40.5	2 线 2 站 2 lines / 2 stations		31000	8.99

展会高峰小时客流比例

1. 离去高峰（晚高峰）比到达高峰（早高峰）集中。
2. 专业展晚高峰小时流量占全天 35%~50%；专业展以外地客流为主，一般先至酒店住宿，再由酒店至会展中心。
3. 消费展晚高峰小时流量占全天 25%~30%。消费展以本市客流为主，以中心城区方向为主要方向，外地客流相对较少。

Peak Hour Passenger Flow Ratio

1. The departure peak (late peak) is usually more concentrated than the arrival peak (early peak).The peak hour traffic of professional exhibition evenings accounts for 35-50% of the total. For professional exhibitions, the main visitor source is from outside the city, they usually arrived at the hotel first and then go from the hotel to the exhibition center.
2. For the peak hours of consumer type exhibition the evening traffic accounted for 25%-30% of the total. The consumer exhibition is mainly based on the passenger flow in this city, with the direction from the central city area to the exhibition area as the main direction, and the passenger flow from other places is relatively small.

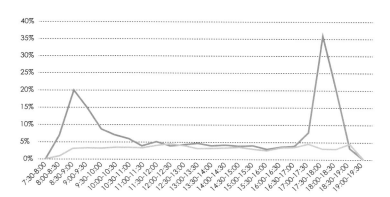

图表 3-6   会展高峰小时客流比例表
Chart 3-6   Passenger Flow Ratio in Peak Hours of Exhibition

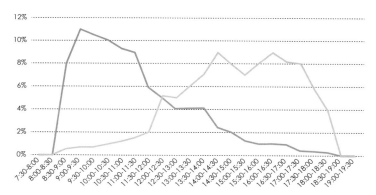

图表 3-7   会展高峰小时客流比例表
Chart 3-7   Passenger Flow Ratio in Peak Hours of Exhibition

会展的交通特征与停车指标

会展中心的交通特征主要有两点：客运交通需求量大，高峰时段需求集中。经过交研对周边市政交通和相似规模会展的数据分析和评估，预测在 40 万平方米展览面积全部开展的情况下，高峰期会展客运交通需求为 24 万人次 / 日，高峰小时 7.2 万人次 / 小时。高峰日出现在展览开幕式当天，高峰小时出现在晚高峰。

从大型会展中心使用经验数据来看，展馆同期出租率不超过 80%。由于是开幕和闭幕时间的固定规律性决定了车库中车辆通行无阻、快速进出的首要需求，因而停车效率都要退而求其次，以满足使用者的心理和生理需求。

深圳国际会展中心客运交通由地铁、公交巴士、出租车和社会车辆等交通方式构成。其中会展中心地下车库的客量是由周边市政道路承载能力决定，并最终于会展中心中廊下二层地下车库设计了 9000 多辆社会停车位，并与东侧休闲带 4000 多辆地下车库相连通，以满足会展峰值时段客运交通需求。

The Transport Characteristics and Parking Index

There are two main transportation characteristics of exhibition center : one is the massive demand for passenger transport, the other is the concentrated demand during the peak period. Based on the data analysis and evaluation of the surrounding municipal transport and other exhibitions of similar scale, when the whole 400,000 square meters exhibition area is fully developed and used, the demand of passenger transport during peak period would be 240,000 person/day, and 72,000 person/hour during the peak hour. Usually, the peak day is on the day of the opening ceremony, and the peak hour appears in the evening of the same day.

From the experience and data of other large exhibition centers, the occupancy rate of exhibition halls in the same period is usually less than 80%. Due to the regular timing of the opening and the closing of an exhibition, it's much more important to guarantee the fast, accessible and free passage, and the efficiency of the parking is of secondary importance, so as allowed to satisfy both the psychological and physiological needs of users.

The passenger transport for the Shenzhen World includes subway, bus, taxi and social vehicles. The capacity of the underground parking is determined by the passerger load of the surrounding municipal roads. There eventually designed with more than 9000 social parking spaces at the second floor of the underground parking under the central concourse, which is connected with the underground parking space for 4000 vehicles under the leisure belt on the east side so as to meet the demand of passenger transport during the peak period.

会展中心的交通特点

1. 会展交通需求大，展会高峰时段需求集中；
2. 交通构成复杂、交通方式多样；
3. 具有季节性和波动性特征，日常需求与高峰需求差异大。

Transportation characteristics of the Convention and Exhibition Center

1. There is a large demand for exhibition transportation, and demand is concentrated during the peak hours of the exhibition;
2. The traffic composition is complex and the traffic modes are diverse;
3. With seasonal and volatile characteristics, the daily demand and the peak demand are greatly different.

会展交通设施构成

1. 货车：需考虑货车轮候区、货车停车区；
2. 专线大巴；
3. 公共交通：含公交车与轨道交通；
4. 商务巴士：观展或参展商团队用车，可利用场内停车位；
5. 出租车；
6. VIP 车队；
7. 私家车：含观展车辆与参展商车辆；
8. 其他场内用车：含穿梭巴士（大型场馆）、工程叉车、检修车辆等。

Composition of exhibition transportation facilities

1. Trucks: The truck waiting area and truck parking area need to be considered;
2. Green bus
3. Public transportation: including buses and rail transit;
4. Business bus: car for exhibition or exhibitor team, can use on-site parking space;
5. taxi
6. VIP fleet
7. Private cars: including visitors 'vehicles and exhibitors' vehicles;
8. Other on-site vehicles: including shuttle buses (large venues), engineering forklifts, maintenance vehicles. etc .

交通设施组织原则

1. 会展交通应与城市交通尽量分离（管道化组织及高速公路出入口分工使用），小汽车交通快进快出；
2. 结合会展交通需求和周边交通条件，确定各类交通的主要集散方向；
3. 公交化、枢纽化、立体化组织交通；
4. 以人为本就近方便；
5. 人车分流、客货分流；
6. 会展内部货车流线成环路，尽量不交叉不回头，多展同期互不干扰；
7. 节约资源，场地内不分设货车和大客车停车位，分时共享。

Organization of Transportation Facilities

1. Convention and exhibition traffic should be separated from urban traffic as much as possible (pipelined organization and division of labor between highway entrances and exits), and car traffic should enter and exit quickly
2. Determine the main distribution directions of various types of transportation based on the needs of exhibition traffic and surrounding traffic conditions;
3. Transitization, hub, and three-dimensional organization of transportation;
4. People-oriented and convenient nearby;
5. Passenger and car diversion, passenger and cargo diversion;
6. The trucks in the exhibition are streamlined in a loop, try not to cross or turn back, and do not interfere with each other during multiple exhibitions;
7. Save resources. There will be no parking spaces for trucks and buses in the site, and they will be shared in time.

交通疏解策略

借鉴国内外经验，结合会展周边交通规划条件，可采用如下疏解策略：

1. 日常配套与临时应急措施相结合，日常配套应对日常高峰客流，临时应急措施应对极端高峰客流；
2. 以大运量交通方式（轨道、公交、巴士）为主疏解客运交通；
3. 固定线路公交与定制公交、团体巴士相结合，提高公交服务水平；
4. 兼顾会展运营和路网容量约束，满足合理的小汽车交通需求；
5. 货运与客运时空分离，布撤展货车分时段调度、分级轮候。

Traffic resolution strategy

Drawing on domestic and foreign experience and combining the traffic planning conditions around the exhibition, the following resolution strategies can be adopted:

1. The combination of daily support and temporary emergency measures, daily support to deal with daily peak passenger flow, and temporary emergency measures to deal with extreme peak passenger flow;
2. Relieve passenger transportation mainly by mass transportation mode (rail, bus, bus);
3. Combine fixed-line buses with custom buses and group buses to improve bus service levels;
4. Take into account the convention and exhibition operation and road network capacity constraints to meet reasonable car transportation needs;
5. Freight transportation and passenger transportation are separated in time and space, and trucks are deployed and dispatched in different time periods, and waiting in stages.

地下车库出入口设计

如图3-138，地块共设置9组（个）地下车库（小汽车停车场）出入口，其中西侧于海滨大道设置3组，通过定向匝道与海滨大道衔接（4进/5出）；东侧依托海汇路辅路设置3组，均为单向进出，进出分离；南侧依托海云路辅路设置1组，均为单向进出；结合凤塘大道下穿市政路设置2个地下车库出入口（图3-138）。

地下车库出入口设计考虑会展交通的潮汐性特征，为避免设施浪费及占用较大地下空间，在西侧出入口处设置了潮汐车道；在入口闸机前设置了掉头车道，供无证或未预约的车辆掉头至出口侧车道驶离。并将部分出入口坡道直接连通了地下二层车库，减少绕行提高进出效率。

The Design of the Entrance and Exit of the Underground Parking

There are 9 (group) entrances and exits of underground parking (car parking lot), among which 3 are set on the west side of Haibin avenue, connecting with Haibin avenue by directional ramp (4 in /5 out). Six entrances (3 groups) are set on the east side of Haihui road, all of which are one-way access, with separate entrance and exit. Two entrances (1 group) are set on the south side of Haiyun roadside road, all one-way access. 2 entrances and exits are set under the municipal road connecting the Fengtang road (Figure 3-138).

The design of the entrance and exit of the underground parking has considered the tidal characteristics of exhibition traffic, and to avoid wasting the use of facilities and large underground space, and sets up tidal lanes at the entrance and exit of the west side. Provide a turnaround lane in front of the entrance gate for unlicensed or unreserved vehicles to turn around and leave. In addition, some entrance and exit ramps are directly connected to the second floor underground parking to reduce the detour and improve the efficiency.

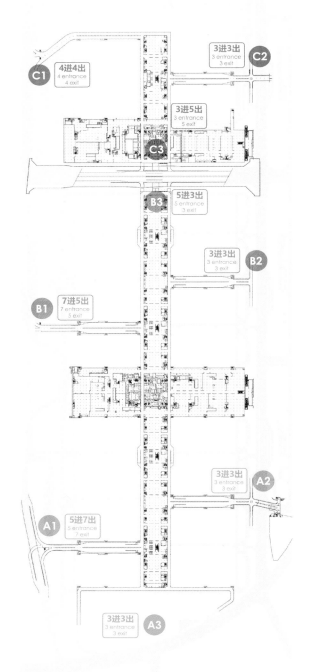

入口闸机数量：36　出口闸机数量：36

图 3-138　地下车库出入口
Figure 3-138　Entrance and Exit of Underground Parking

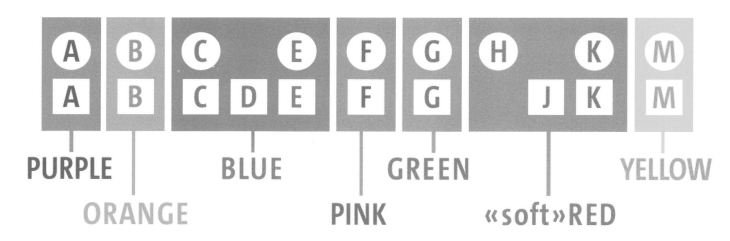

图 3-139　地下车库标识图
Figure 3-139　Underground Parking Instruction Signs

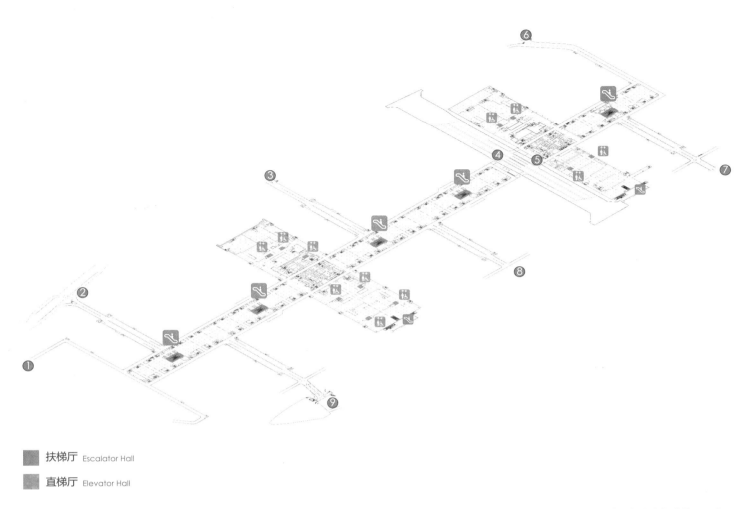

■ 扶梯厅 Escalator Hall

■ 直梯厅 Elevator Hall

图 3-140 地下车库直扶梯位置示意图
Figure 3-140 Underground Parking Escalator Position Diagram

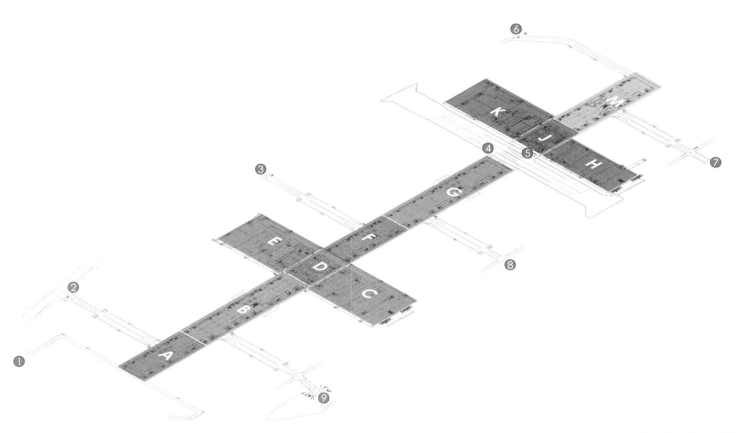

图 3-141 地下停车分区示意图
Figure 3-141 Underground Parking Division Diagram

地下车库三级车道设计

为了构建与分为主、次、支三级通道体系（图 3-142）的城市道路系统相对应的路网系统。主车道采用单向三车道，宽 9.25m 逆时针单向外环，不承担停车功能。次通道采用单向逆时针双车道，宽 7m，外接主通道、内连支通道，不承担停车功能。次通道距出入口的间距大于 25m，次通道之间的间距在 50~120m 之间。支通道在停车分区内部，用于停车，且是双向交通组织，尽端连通，与主通道分离，车道宽 5.5m。

Three Driveway Design for the Underground Parking

To build the road network system corresponding to the urban road system, which is divided into three levels of channel system: primary, secondary and the subchannel (Figure 3-142). The main lane is three counterclockwise one-way outer ring, with a width of 9.25m, without parking function. The secondary channel adopts one-way counterclockwise dual lane, with a width of 7m. Externally, it connected to the main channel, and internally, it connect the subchannel without parking function. The distance between the secondary channel and the entrance is more than 25m, and the distance between each secondary channel is between 50~120m. The subchannel is inside the parking zone, and it is used for parking purpose. It is two-way traffic organization, connecting the ends and separated from the main channel with a width of 5.5m.

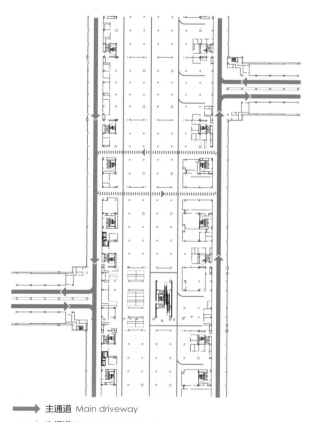

➡ 主通道 Main driveway
⋯⋯ 次通道 Secondary driveway
—— 支通道 Branch driveway

图 3-142　主、次、支三级通道体系图
Figure 3-142　Main, Secondary and Branch Channel System Diagram

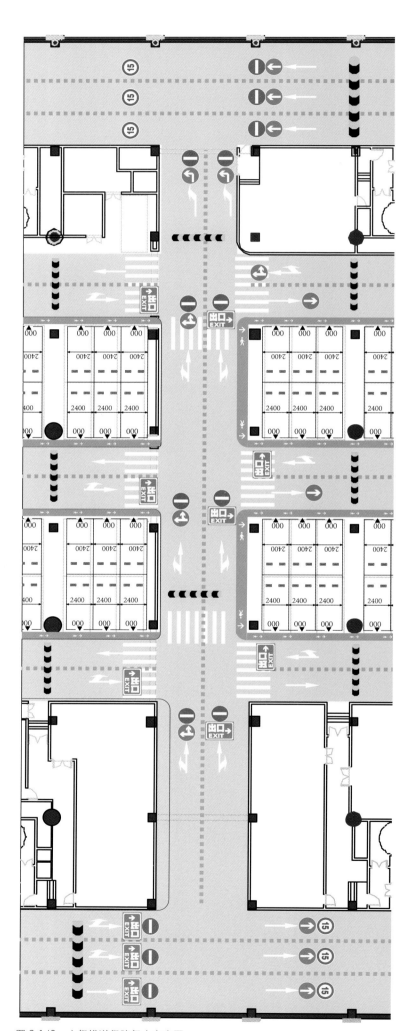

图 3-143　人行横道保障行人安全图
Figure 3-143　Pedestrian Safety Guarantee Chart

会展休闲带综合配套
—— 特殊会展配套
Special Supporting Facility – The Leisure Zone of the
Exhibition City

深圳国际会展中心休闲带（以下简称休闲带）长约 1700m，宽 60m，平行于会展主体建筑。位于会展岛中轴线，东西向连接会展与城市住宅、酒店、商业、办公、商贸等地块，南北向连接海上田园与西部湾区景观带。

The Shenzhen World leisure belt (leisure belt for short) is about 1700m long, and 60m wide, paralleled to the main exhibition & convention building. Located on the central axis of the exhibition island, it connects the exhibition center and urban residential, hotel, commercial, office, commercial and other plots from east to west, and connects the sea garden and the landscape belt of the western bay area from south to north.

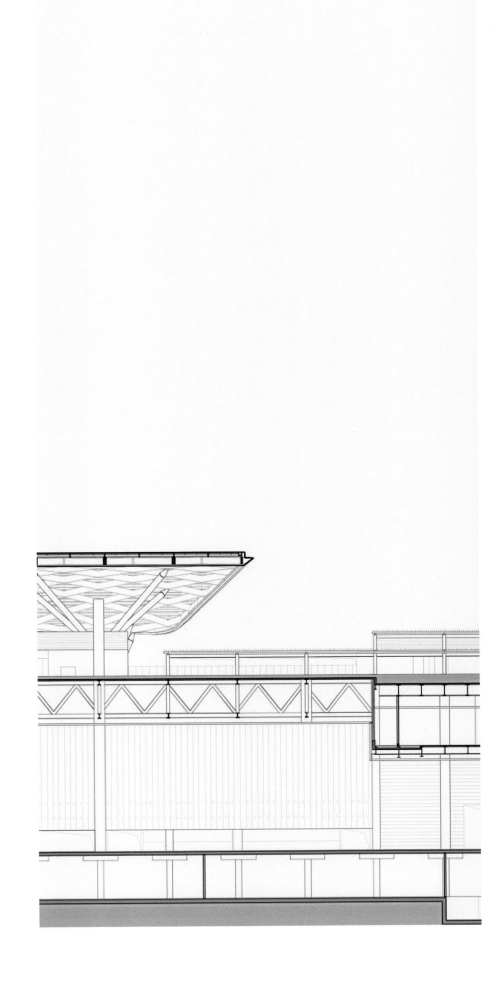

延伸至会议
中心、海上田园
Extends to the conference center,
maritime countryside

酒店
Hotel

展贸
Exhibition
& trading

公寓
Apartment

购物
中心
Shopping
mall

会展
Exhibition

以会展中心为核心
Centered on the
exhibition center

会展核
Exhibition core

连廊/地下空间
Corridor/under ground space

旅游核
Tourism core

以旅游主题
商业为核心
Centered on
tourism business

社区
商业
Community
commerce

休闲带
Leisure belt

产业带
Industry belt

消费带
Consumming belt

办公
Office

图 3-144   二核两带空间布局图
Figure 3-144   Spatial Layout of Three Cores and Two Belts

缝合带的作用与利用
The Function and Utilization of the Suture Zone

空间过渡带

会展中心与城市之间的缝隙之间嵌入了一条缝合带作为功能补充及空间过渡（图 3-144），其意图在于消解因会展中心的运营需求导致的与城市环境之间的割裂，从而让会展占用的场地重回城市的公共生活。

Spatial Transition Belt

The gap between the exhibition center and the city is embedded with a Leisure zone as a function supplement and a space transition (Figure 3-144), which is intended to eliminate the separation between the exhibition center and the urban environment caused by the operational needs of the exhibition center, so as to return the space occupied by the exhibition public life of the city.

功能互补带

休闲带以双线复合公园的形式，将城市绿色公共开放空间与城市公共交通等配套服务功能规划在同一地块，并最大化利用占地面积，为城市高度集约使用土地提供解决方案（图3-145，图3-146，图3-147）。

Complementary Function Belt

With a form of a double-line compound park, the leisure belt plan combined the urban green public open space and the supporting service functions such as urban public transportation within the same plot to maximizes the land use. and provides a solution for intensive use of land in the city.(Figure 3-145, 3-146, 3-147)

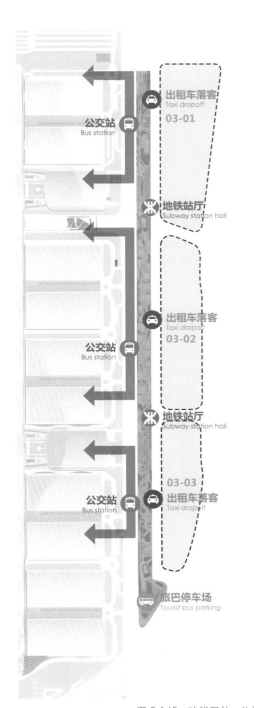

图 3-145　功能互补 - 公共交通图
Figure 3-145　Complementary Functions - Public Transportation Map

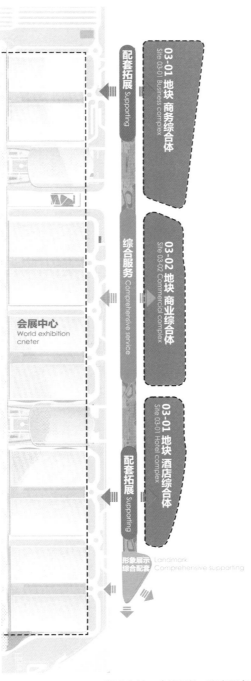

图 3-146　功能互补 - 配套服务图
Figure 3-146　Functional Complementation - Supporting Services

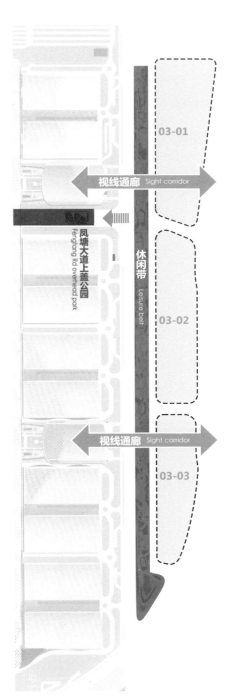

图 3-147　功能互补 - 公园
Figure 3-147　Functions Complementation - Park

多元生活带

在休闲带公园上，AUBE 欧博设计从公共活动、跨领域业态植入、生态系统等多维度构建多元生活的载体，使会展片区获得商业贸易之外的丰富多样的城市生活功能（图 3-149）。

Diverse Life Belt

In the leisure belt park, we construct multiple life carriers from multiple Perspectives, such as public activities, cross-field implantation and ecological system, so that the function of the exhibition area can get richer and acquire diverse urban life beyond just commercial functions (Figure 3-149).

图 3-148　多元生活形成因素
Figure 3-148　Multiple Life Formative Factors

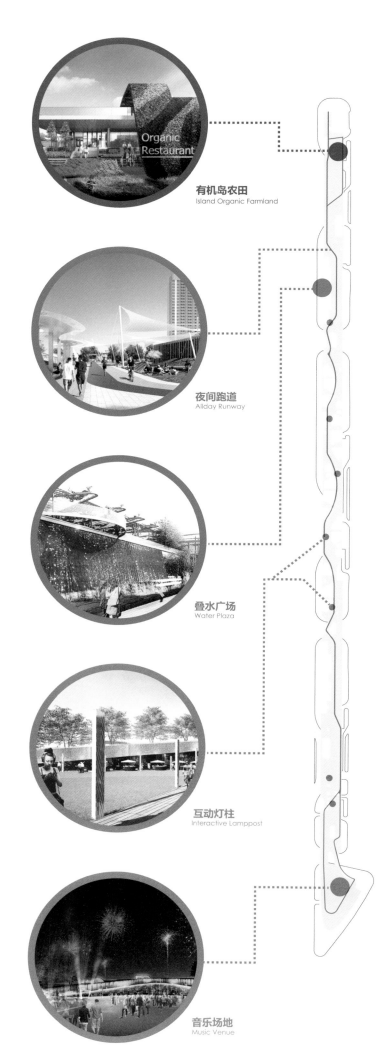

有机岛农田
Island Organic Farmland

夜间跑道
Allday Runway

叠水广场
Water Plaza

互动灯柱
Interactive Lamppost

音乐场地
Music Venue

图 3-149　多元生活带分析图
Figure 3-149　Life Belt Analysis Diagram

微气候改良带

在 8m 高线上构建公园,除基础绿色系统构建以外,面临两项特殊挑战。一方面,深圳在亚热带海洋性季风气候下,全年相对湿度较高,日照时间长且强度大,公园需具备应对突发暴雨与湿热暴晒天气的能力。另一方面,会展地块分时序开发,公园也需具备应对未来城市功能升级的改造弹性(图 3-150)。

Microclimate Improvement Zone

Building the park on the 8m high line presents two specific challenges in addition to the basic green system construction. On the one hand, the subtropical Marine monsoon climate of the city indicates a relative high humidity throughout the year, and the daylight duration is long and intense. Therefore, the park should be able to cope with unexpected rainstorm, humidity and hot weather. On the other hand, the exhibition plot is developed in time-sharing sequence, and the park also needs to be resilient to the urban functions upgrading in the future.(Figure 3-150)

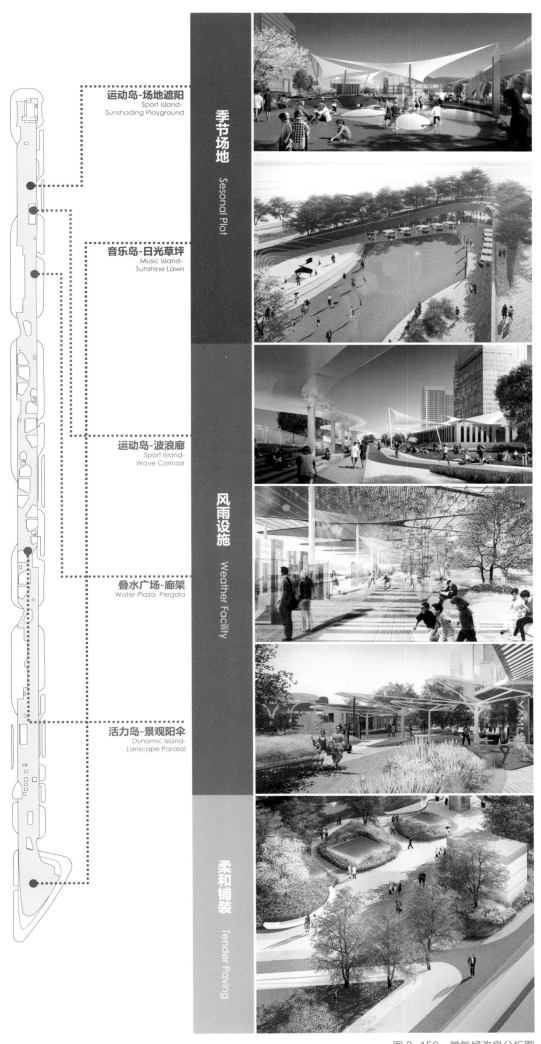

运动岛-场地遮阳
Sport Island-
Sunshading Playground

音乐岛-日光草坪
Music Island-
Sunshine Lawn

运动岛-波浪廊
Sport Island-
Wave Corridor

叠水广场-廊架
Water Plaza- Pergola

活力岛-景观阳伞
Dynamic Island-
Lanscape Parasol

季节场地 Sesonal Plot

风雨设施 Weather Facility

柔和铺装 Tender Paving

图 3-150 微气候改良分析图
Figure 3-150 Microclimate Improvement Analysis Diagram

交通疏导带

会展中心可能会在人流疏散呈现瞬时集中时出现爆发的特点。在会展人流的疏散高峰期，休闲带可供一部分人停留、休憩、等候的空间，休闲带紧邻两条地铁线路，其地下空间也与地铁展厅空间产生了紧密的连接（图 3-151）。

Traffic Dispersion Belt

The Shenzhen World is the largest convention and exhibition center around the world at present, therefore it could bring instantaneous concentration outbreak during the crowd evacuation.

During the evacuation peak period of the crowd, the leisure belt can provide a temperary space for people to stay, rest and wait. The leisure belt is adjacent to two subway lines, and the underground space is also closely connected to the subway space.(Figure 3-151)

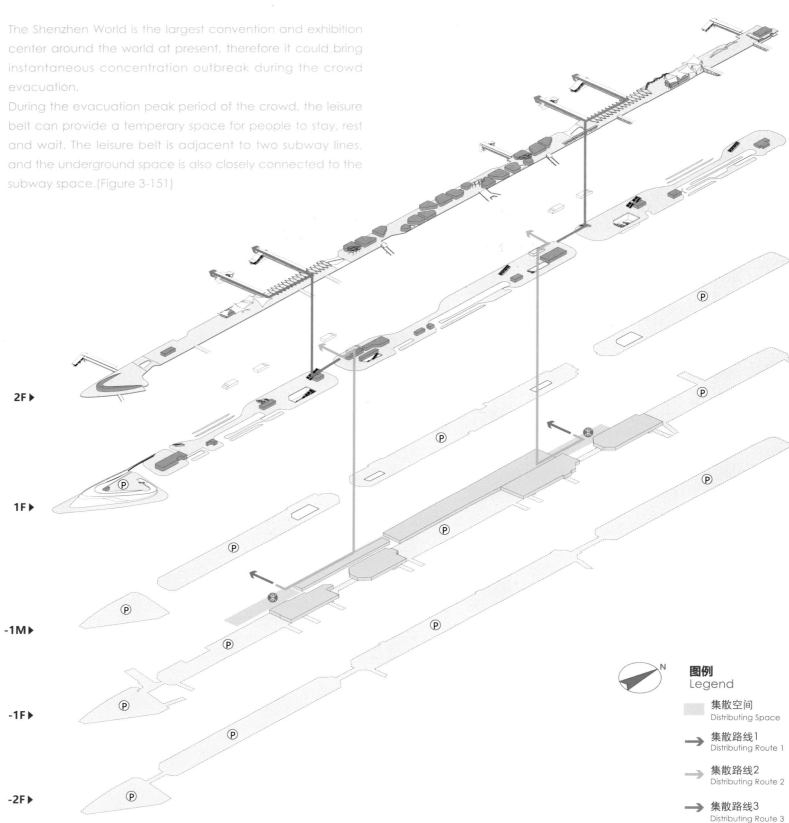

2F ▶

1F ▶

-1M ▶

-1F ▶

-2F ▶

N

图例
Legend

集散空间
Distributing Space

集散路线1
Distributing Route 1

集散路线2
Distributing Route 2

集散路线3
Distributing Route 3

图 3-151　会展休闲带交通疏导分析图
Figure 3-151　Traffic Distribution Analysis of the Exhibition Leisure Zone

共享停车带

会展中心有着庞大的停车泊位需求，而基于展厅的特殊荷载要求，会展中心场地内不宜建设大面积地下车库。休闲带的介入弥补了这一功能空缺，展时可作为会展中心辅助停车位，非展时亦可作为周边配套商业设施停车位，集约利用土地资源，提高土地使用效率（图3-152）。

Co-parking Spaces

There is a huge demand for parking in the exhibition and convention center area due to the traffic load requirements of the exhibition hall and it is not suitable to build a large underground parking space under the exhibition and convention center. However the leisure belt fixes this proplem, it can be used as the supporting parking space during the exhibition seasons, and can be used by the supporting commercial facilities nearby during non-exhibition seasons. It improves the efficiency of the land use. (Figure 3-152)

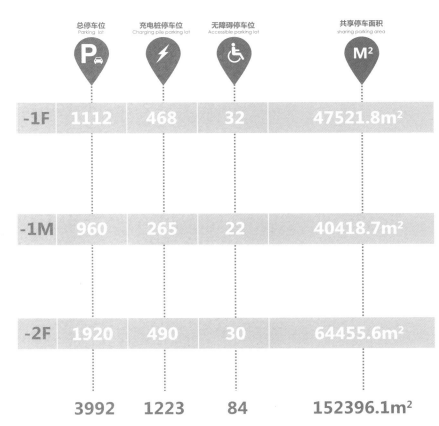

	总停车位 Parking lot	充电桩停车位 Charging pile parking lot	无障碍停车位 Accessible parking lot	共享停车面积 sharing parking area
-1F	1112	468	32	47521.8m²
-1M	960	265	22	40418.7m²
-2F	1920	490	30	64455.6m²
	3992	1223	84	152396.1m²

图表 3-8　会展休闲带地下停车指标表
Chart 3-8　Indicators of the Underground Parking in the Exhibition Leisure Zone

−1 层夹层停车分布
-1m F Parking Distribution

−1 层停车分布
-1 F Parking Distribution

−2 层停车分布
-2F Parking Distribution

图 3-152　会展休闲带地下停车分布图
Figure 3-152　Underground Parking Distribution of the Exhibition Leisure Zone

交通组织
The Organization of Transportation

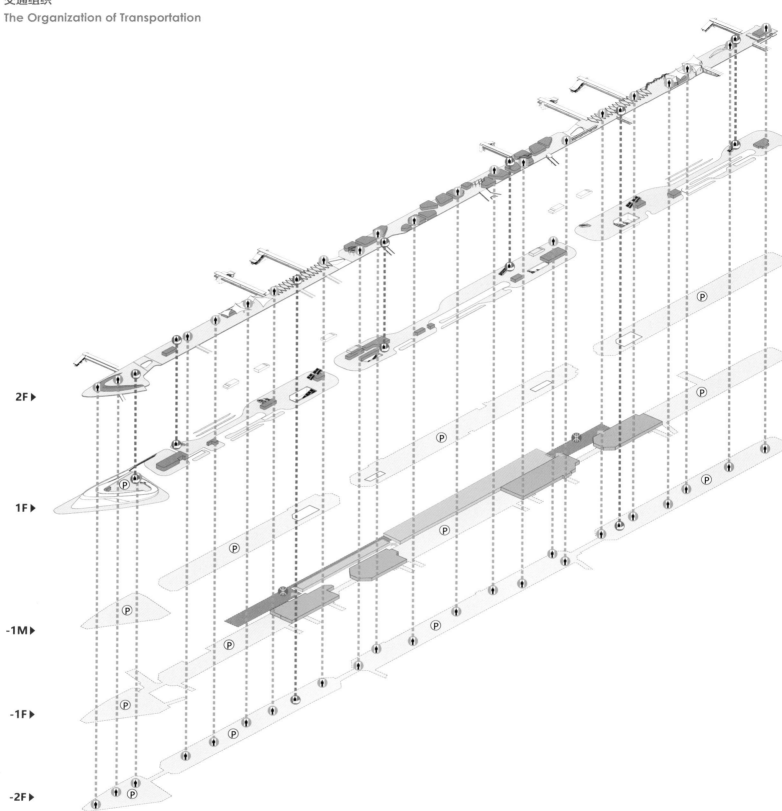

2F▶

1F▶

-1M▶

-1F▶

-2F▶

图 3-153　会展休闲带交通组织分析图
Figure 3-153　Traffic Organization Analysis Diagram of the Exhibition Leisure Zone

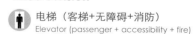

区域划分
Area Division

电梯（客梯+无障碍+消防）
Elevator (passenger + accessibility + fire)

扶梯+电梯
Escalator + Elevator

停车场
Parking Lot

配套商业
Supporting Commercial Facilities

配套功能
Supporting Functions

地下联系
Underground connection

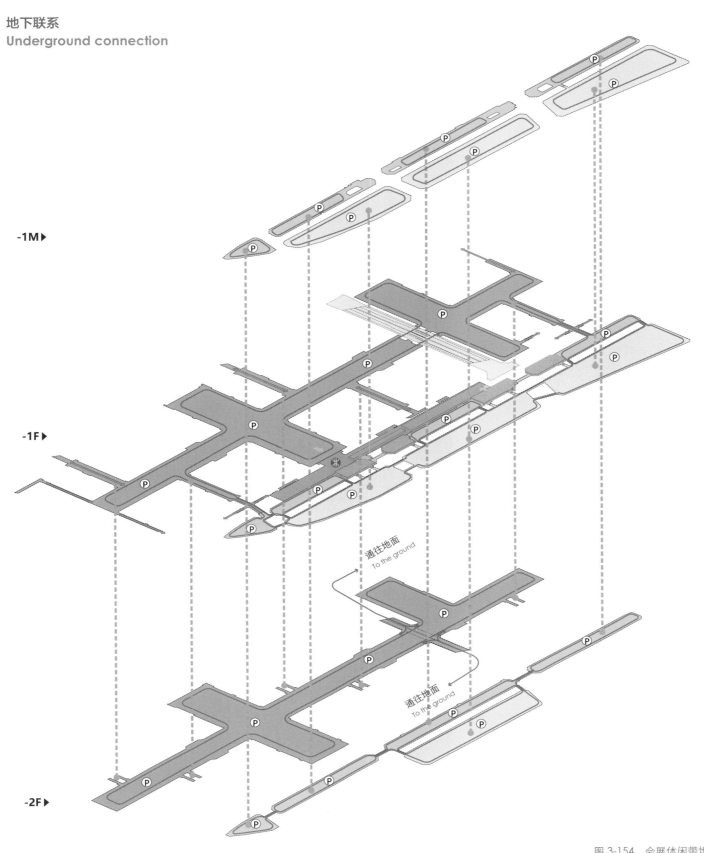

-1M▶

-1F▶

-2F▶

通往地面
To the ground

通往地面
To the ground

图 3-154　会展休闲带地下交通联系分析图
Figure 3-154　Underground Transportation of the Exhibition Leisure Zone

N

区域划分
Area Division

● 会展地下停车
　Exhibition Underground
　Parking

○ 03地块地下停车
　03 Plot Underground
　Parking

● 地铁站
　Subway Station

● 休闲带地下停车
　Leisure Belt Underground
　Parking

● 配套商业
　Supporting Commercial

➤ 地铁站台方向
　Subway Platform Orientation

▭ 地下交通流线
　Underground Traffic
　Circulation

功能组织
The Organization of Function

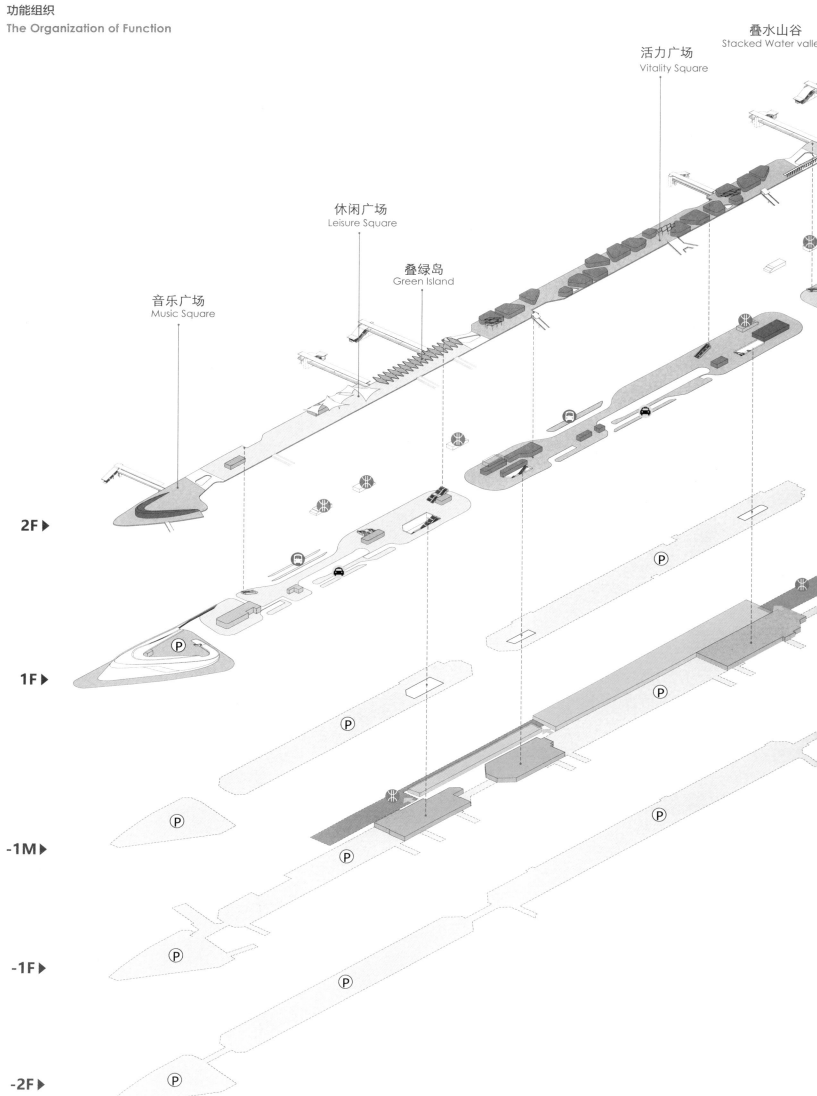

叠水山谷
Stacked Water valle

活力广场
Vitality Square

休闲广场
Leisure Square

叠绿岛
Green Island

音乐广场
Music Square

2F▶

1F▶

-1M▶

-1F▶

-2F▶

动广场
rts Square

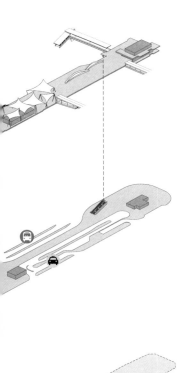

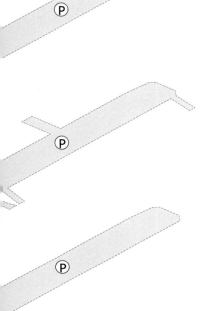

图 3-155　会展休闲带功能分析图
Figure 3-155　Leisure Belt Function

区域划分
Area Division

● 音乐广场 MusicSquare	○ 休闲广场 Leisure Square
● 音乐餐厅 Music Restaurant	● 配套商业 Supporting Commercial
● 地铁站 Subway Station	● 派出所 Police Station
○ 活力广场 Vitality Square	○ 运动广场 Sports Square
● 商业功能 Business	● 有机餐厅 Organic Restaurant
● 综合管廊监控中心 Utility Tunnel Monitoring Center	○ 配套功能 Supporting Function

Ⓟ 停车场 Parking Lot　　地铁出入口 Subway Entrance

公交站 Bus Stop　　出租车站台 Taxi Platform

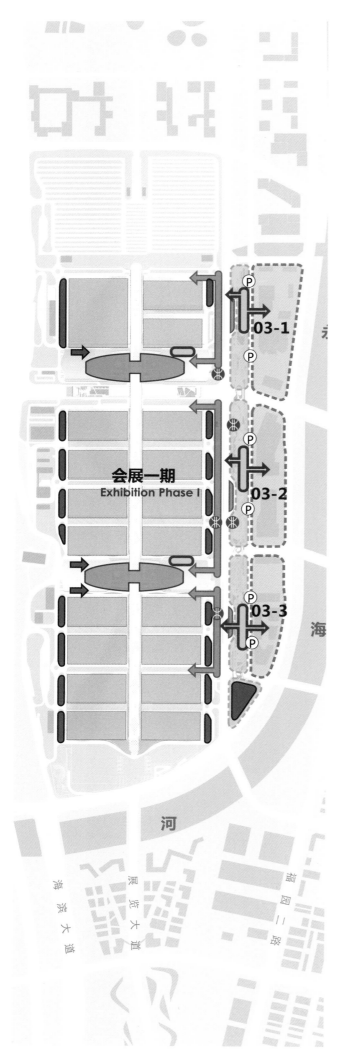

地面交通联系
The Organization of above Ground Transportation

位于休闲带 8m 大板之下的灰空间，可往西直接到达会展登录大厅前广场与周边商业体块。该灰空间（即架空层）紧邻繁忙的市政道路，面临扬尘、噪音等问题，需通过绿色环境的营造降低相应影响。此外，地面层也复合大面积公共交通枢纽，会导致人流的瞬间聚集，因此也需具备快速疏导并喷人流的能力（图 3-156）。

There is a gray space (empty space) under the 8m large board in the leisure zone, which to the west can directly reach the front square of the arrival hall and the surrounding commercial land. This gray space is adjacent to the busy municipal road with problems of dust, noise and others, therefore it needs green belt to help reduce the relative impact. In addition, the ground floor is also an area of public transportation hub, which could lead to the instantaneous aggregation of people, so it has to be able to quickly bypass large crowds.(Figure 3-156)

二层联系
The Connection of the Second Floor

休闲带的二层景观平台在一片绿林环绕的开放公园中，人们可以通过人行天桥前往会展登录大厅前广场。多种路径长短有别，在竖向上分层设置，各有特点，为到达会展的人流提供多样化的行进体验，使整体疏散能力、可容纳人群数量得到了有效的提升。
休闲带东向连接城市公寓、酒店、商业、办公、展贸馆等开发地块，南北向连接海上田园与西部湾区景观带，所有功能构成均围绕着服务会展的目标设置，担负着重要的二层衔接作用（图 3-157）。

In a green open park, the landscape platform on the second floor of the leisure zone is connected to the front square of the arrival hall through the pedestrian bridge. A variety of paths with different lengths and characteristics are set vertically, not only providing different walking experience for people, but also effectively improving the capacity of overall evacuation and load of people. The leisure zone to the east connects urban apartments, hotels, businesses, offices, and exhibition and trade centers, while from the south to the north it connects the coastal countryside and the western bay area landscape belt. All the functions are based on the goal to serve the exhibition center, playing an important role as the cohesion on the second floor.(Figure 3-157)

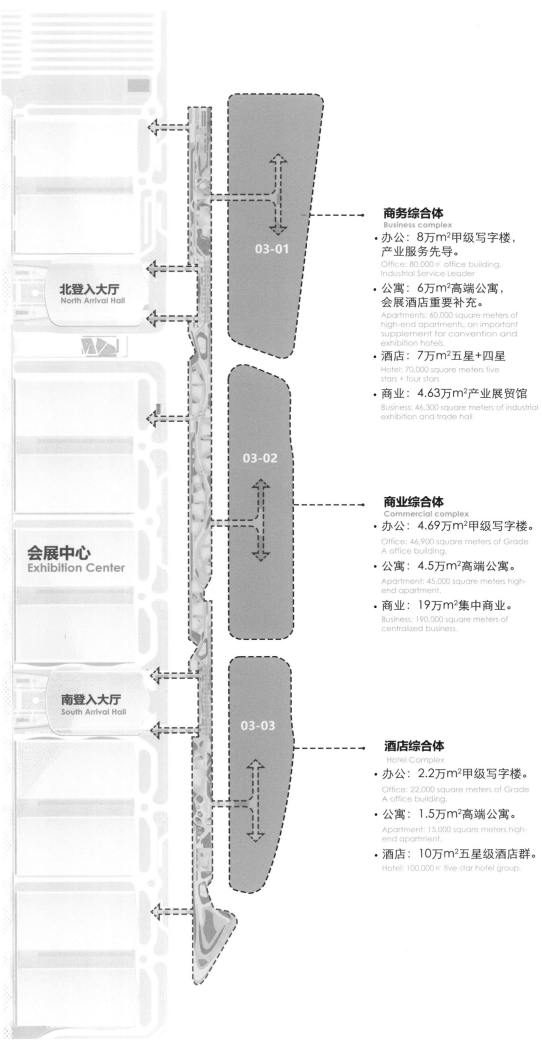

商务综合体
Business complex

- 办公：8万m²甲级写字楼，
  产业服务先导。
  Office: 80,000 ㎡ office building.
  Industrial Service Leader

- 公寓：6万m²高端公寓，
  会展酒店重要补充。
  Apartments: 60,000 square meters of
  high-end apartments, an important
  supplement for convention and
  exhibition hotels.

- 酒店：7万m²五星+四星
  Hotel: 70,000 square meters five
  stars + four stars

- 商业：4.63万m²产业展贸馆
  Business: 46,300 square meters of industrial
  exhibition and trade hall

商业综合体
Commercial complex

- 办公：4.69万m²甲级写字楼。
  Office: 46,900 square meters of Grade
  A office building.

- 公寓：4.5万m²高端公寓。
  Apartment: 45,000 square meters high-
  end apartment.

- 商业：19万m²集中商业。
  Business: 190,000 square meters of
  centralized business.

酒店综合体
Hotel Complex

- 办公：2.2万m²甲级写字楼。
  Office: 22,000 square meters of Grade
  A office building.

- 公寓：1.5万m²高端公寓。
  Apartment: 15,000 square meters high-
  end apartment.

- 酒店：10万m²五星级酒店群。
  Hotel: 100,000 ㎡ five-star hotel group.

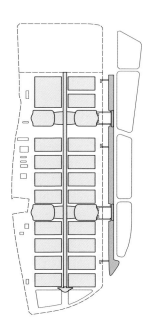

Figure 3-158　Exhibition Leisure Belt
Connection Key Plan

图 3-157　会展休闲带二层联系分析图
Figure 3-157　Exhibition Leisure Zone on the second floor - Connection Analysis Diagram

运动岛
Sport Island

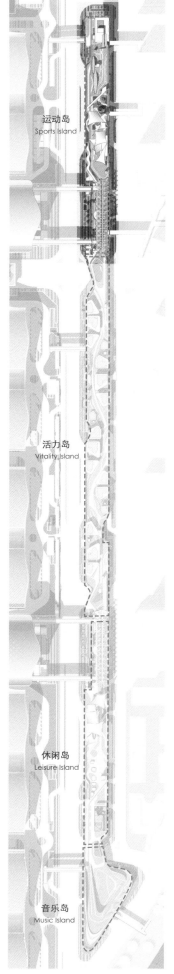

图 3-159　运动岛
Figure 3-159　Sports Island Location

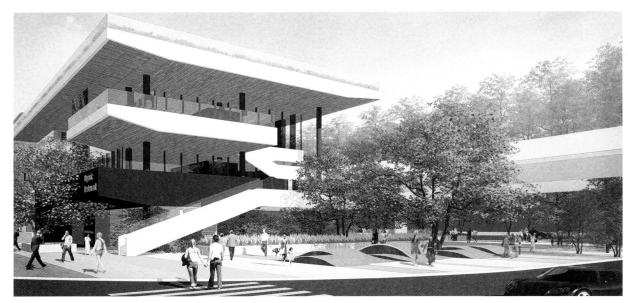

图 3-160　有机餐厅效果图
Figure 3-160　Organic Restaurant Rendering

图 3-161　运动广场效果图
Figure 3-161　Sports Square Rendering

图 3-162　音乐广场效果图
Figure 3-162　Music Plaza Rendering

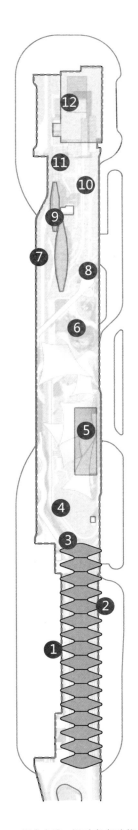

图 3-163 运动岛岛功能分析图
Figure 3-163 Sports Island Function Analysis

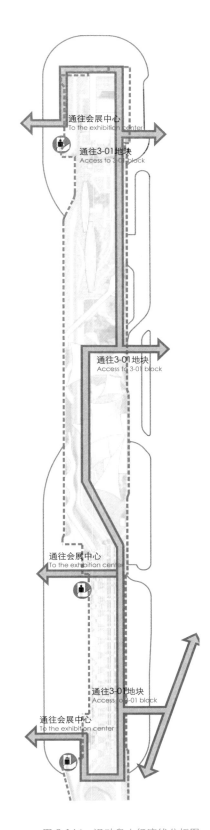

图 3-164 运动岛人行流线分析图
Figure 3-164 Sports Island Pedestrian Circulation

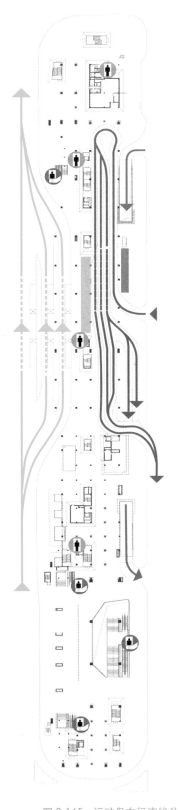

图 3-165 运动岛车行流线分析图
Figure 3-165 Sports Island Vihicle Circulation

图例
Legend

1 水瀑 Waterfall

2 花影长廊 Flower Shadow Gallery

3 艺术廊 Art Gallery

4 活动绿地 Sports Green Land

5 云顶餐厅 Cloud Top Restaurant

6 运动园 Sports Park

7 夜光跑道 Luminous Runway

8 玩乐廊 Playground

9 麦浪廊 Wheat Lang Gallery

10 有机农场 Organic Farm

11 运动园 Sports Park

12 有机餐厅 Organic Restaurant

人行流线 Pedestrain Circulation

公交车流线 Bus Streamline

出租车流线 Taxi Circulation

地下车库流线 Underground Garage Circulation

出租车落客区 Taxi Drop off Area

出租车等候区 Taxi Waiting Area

电梯（客梯+无障碍+消防）Elevator (passenger + accessibility + fire)

扶梯+直梯 Escalator + Elevator

活力岛
Vitality Island

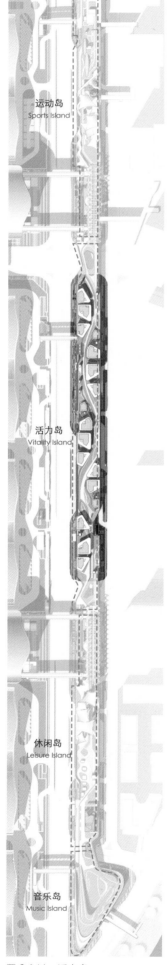

图 3-166 活力岛
Figure 3-166 Vitality Island Location

图 3-167 叠水广场效果图
Figure 3-167 Dieshui Water Plaza Rendering

图 3-168 叠水瀑布效果图
Figure 3-168 Stacked Water Waterfall Rendering

图 3-169 夜光跑道效果图
Figure 3-169 Luminous Runway Rendering

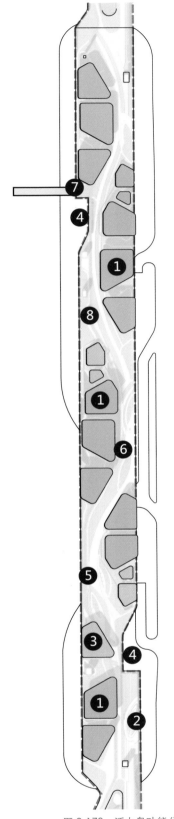

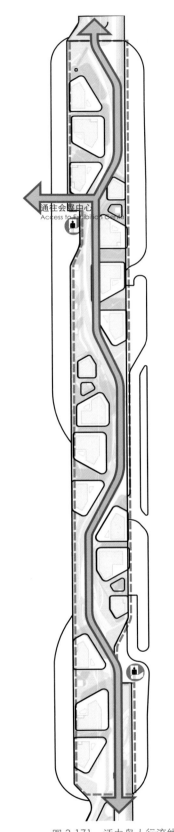

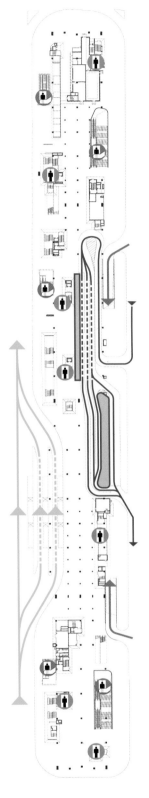

图 3-170　活力岛功能分析图
Figure 3-170　Vitality Island Functional Analysis

图 3-171　活力岛人行流线分析图
Figure 3-171　Vitality Island Pedestrain Circulation

图 3-172　活力岛车行流线分析图
Figure 3-172　Vitality Island Vehicle Circulation

图例
Legend

① 活力商业
Dynamic Business

② 曲水景
Waterscape

③ 餐厅
Restaurant

④ 垂直交通
Vertical Circulation

⑤ 休闲广场
Leisure Plaza

⑥ 流云廊道
Pedestrian Corridor

⑦ 过街天桥
Overpass

⑧ 林下休闲
Forest Recreation

人行流线
Pedestrian Circulation

公交车流线
Bus Circulation

出租车流线
Taxi Circulation

地下车库流线
Underground Garage
Circulation

出租车落客区
Taxi Drop-off Area

 电梯（客梯+无障碍+消防）
Elevator (passenger + accessibility + fire)

 扶梯+直梯
Escalator + Elevator

休闲岛
Leisure Island

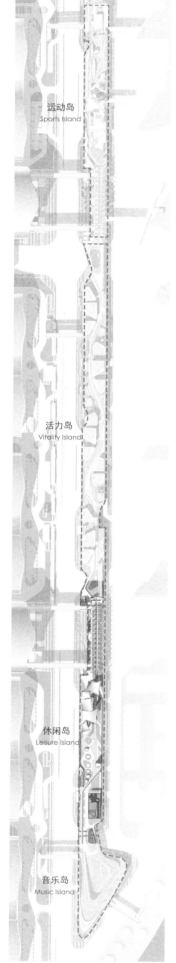

图 3-173 休闲岛
Figure 3-173 Leisure Island Location

图 3-174 休闲岛效果图
Figure 3-174 Leisure Island Rendering

图 3-175 叠水瀑布效果图
Figure 3-175 Stack Water Waterfall Rendering

图 3-176 叠绿广场效果图
Figure 3-176 Stack Green Square Rendering

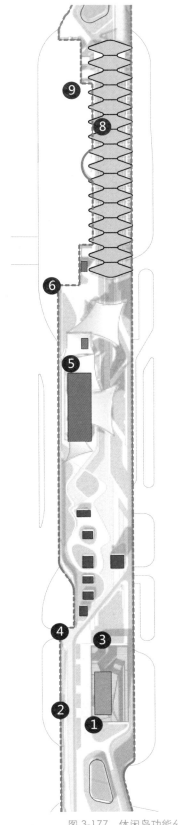

图 3-177　休闲岛功能分析图
Figure 3-177　Leisure Island Functional Analysis

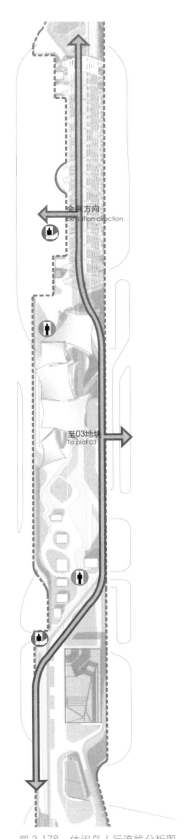

图 3-178　休闲岛人行流线分析图
Figure 3-178　Leisure Island Pedestrian Circulation

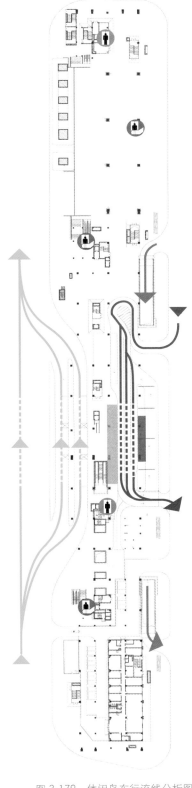

图 3-179　休闲岛车行流线分析图
Figure 3-179　Leisure Island Vehicle Circulation

图例
Legend

① 多功能厅 Muttifuction Room
② 夜光跑道 Luminous Runway
③ 阳光草坪 Sunshine Lawn
④ 垂直交通 Vertical Circulation
⑤ 运动餐厅 Sports Restaurant
⑥ 漫花岛 Manhua Island
⑦ 过街天桥 Overpass
⑧ 花影长廊 Flower Gallery
⑨ 留影台 Photo Deck

 人行流线 Circulation

 电梯（客梯+无障碍+消防） Elevator (passenger + accessibility + fire)

 扶梯+直梯 Escalator + Elevator

 公交车流线 Bus Circulation

出租车流线 Taxi Circulation

地下车库流线 Underground Garage Circulation

音乐岛
Music Island

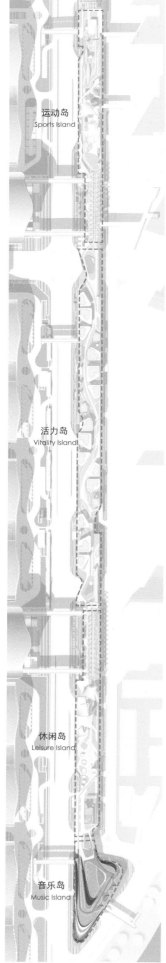

运动岛
Sports Island

活力岛
Vitality Island

休闲岛
Leisure Island

音乐岛
Music Island

图 3-180 音乐岛
Figure 3-180 Music Island Location

图 3-181 活力广场效果图
Figure 3-181 Vitality Square Rendering

图 3-182 音乐岛效果图
Figure 3-182 Music Island Rendering

图 3-183 音乐餐厅效果图
Figure 3-183 Music Restaurant Rendering

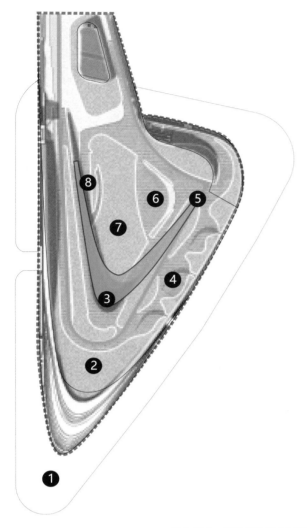

图 3-184　音乐岛功能分析图
Figure 3-184　Music Island Functional Analysis

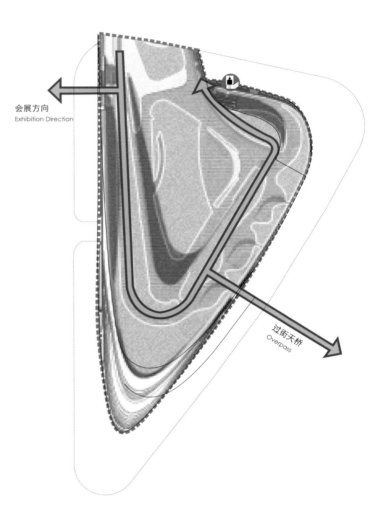

会展方向
Exhibition Direction

过街天桥
Overpass

图 3-185　音乐岛人行流线分析图
Figure 3-185　Music Island Pedestrian Circulation

图例
Legend

❶ 音乐喷泉广场
Music Fountain Square

❷ 甲板观景平台
Observation Deck

❸ 扶手栏杆
Handrail

❹ 天空酒吧
Sky Bar

❺ 夜光跑道
Luminous Runway

❻ 剧场舞台
Theater Stage

❼ 阳光草坪
Sunshine Lawn

❽ 观赏台阶
Ornamental Steps

▶ 商务大巴车库出入口
Business Bus Garage Entrance

🛗 电梯（客梯+无障碍+消防）
Elevator (passenger + accessibility + fire)

🛗 扶梯+电梯
Escalator + Elevator

▭▭▭ 地下车库出入口
Underground Garage Entrance

▬ ▬ ▬ 商务大巴车库流线
Commercial Bus Garage Circulation

▭ 人行流线
Pedestrian Circulation

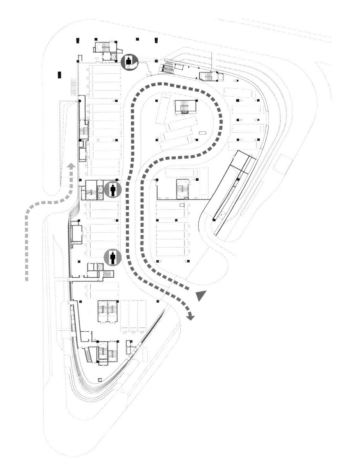

图 3-186　音乐岛车行流线分析图
Figure 3-186　Music Island Vehicle Circulation

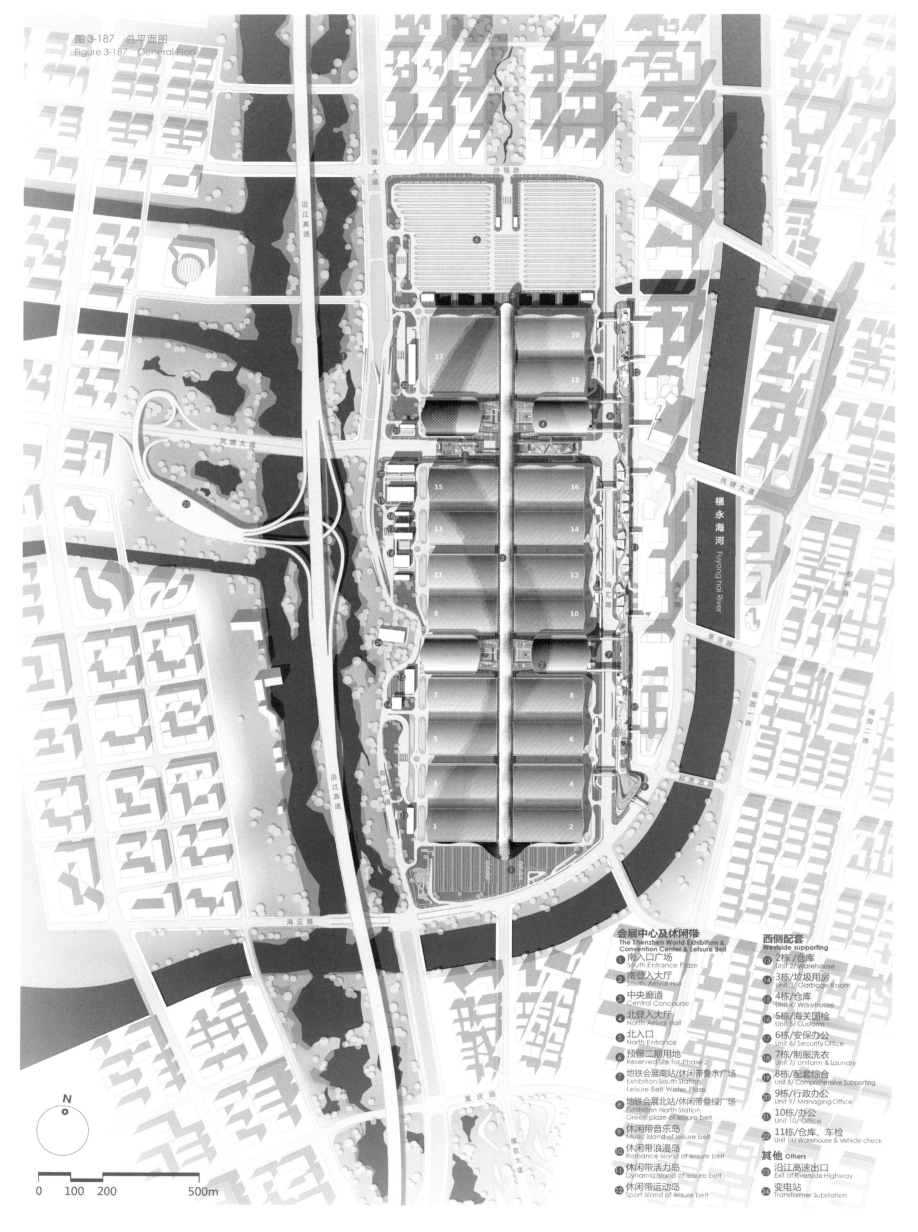

图 3-187 总平面图
Figure 3-187 General Plan

N

0   100   200   500m

会展中心及休闲带
The Shenzhen World Exhibition &
Convention Center & Lefsure Bell

① 南入口广场
   South Entrance Plaza
② 南登入大厅
   South Arrival Hall
③ 中央廊道
   Central Concourse
④ 北登入大厅
   North Arrival Hall
⑤ 北入口
   North Entrance
⑥ 预留二期用地
   Reserved Site for Phase 2
⑦ 地铁会展南站/休闲带叠水广场
   Exhibition South Station
   Leisure Belt Water Plaza
⑧ 地铁会展北站/休闲带绿广场
   Exhibition North Station
   Green plaze of leisure belt
⑨ 休闲带音乐岛
   Music Island of leisure belt
⑩ 休闲带浪漫岛
   Romance island of leisure belt
⑪ 休闲带活力岛
   Dynamic island of leisure belt
⑫ 休闲带运动岛
   Sport island of leisure belt

西侧配套
Westside supporting

⑬ 2栋/仓库
   Unit 2/ Warehouse
⑭ 3栋/垃圾用房
   Unit 3/ Garbage Room
⑮ 4栋/仓库
   Unit 4/ Warehouse
⑯ 5栋/海关国检
   Unit 5/ Customs
⑰ 6栋/安保办公
   Unit 6/ Security Office
⑱ 7栋/制服洗衣
   Unit 7/ Uniform & Laundry
⑲ 8栋/配套综合
   Unit 8/ Comprehensive Supporting
⑳ 9栋/行政办公
   Unit 9/ Managing Office
㉑ 10栋/办公
   Unit 10/ Office
㉒ 11栋/仓库、车检
   Unit 11/ Warehouse & Vehicle check

其他 Others

㉓ 沿江高速出口
   Exit of Riverside Highway
㉔ 变电站
   Transformer Substation

福永海河
Fuyong hai River

经济技术指标表
The Economic and Technical Indicators

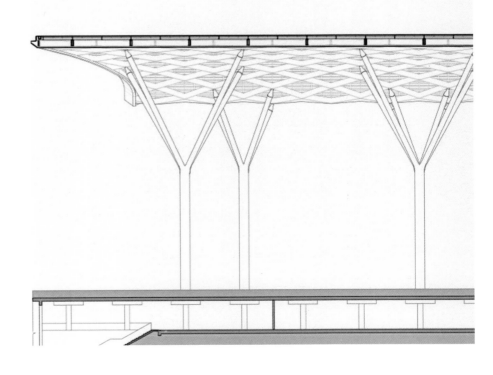

技术经济指标表
Economic and Technical indicator

项目概况
Project Overview

项目名称： project name:	深圳国际会展中心 The Shenzhen World Exhibition and Convention Center	用地单位： Land unit:	深圳市商务局 Commerce Bureau of Shenzhen Municipality
宗地号： Parcel No:	A222-0280	用地位置： Land Location:	宝安区福永机场以北，空港新城南部 North of Fuyong Airport, Baoan District, South of Airport New Town

主要技术经济指标
The main technical and economic indicators

建设用地面积 Construction land area	1210667.79㎡	总建筑面积 Total Construction area	1600526.51㎡
计容率建筑面积 Floor area ratio	1016563.26㎡	容积率/规定容积率 Floor area ratio / Specified floor area ratio	0.84/0.78
地上规定建筑面积 Specified floor area ratio	946483.20㎡	不计容积率建筑面积 Excluding floor area ratio	583963.25㎡
地上核减建筑面积 Above ground reduction of building area	—	地下规定建筑面积 Underground Specified construction area	20274.92㎡
地上核增建筑面积 Above-ground Increased building area	70080.06㎡	地下核增建筑面积 Underground increased building area	563688.33㎡
建筑基底面积 Building based area of construction	602235.00㎡	建筑覆盖率（一/二级） Building coverage (first/secondary) ratio	60%
绿地面积/折算绿地面积 Green area / Converted green area	—	绿化覆盖率 Green coverage	—
最高高度 Maximum height	42.80m	最大层数（地上/下） Maximum layer number (ground/lower)	地上5层、地下2层 5 floors above ground, 2 floors below ground
停车位（地上小车/地上货车/地下小车） Parking space (ground car / ground truck / underground car)	10486（487/914/9085辆）	自行车停车位 Bicycle parking space	5840（辆）

本期建筑面积及分配
Construction area and distribution of the current Phase

			建筑功能 Building function	建筑面积/㎡ Building area / ㎡
				规定 Regulation
总建筑面积1600526.51㎡ The total construction area is 1600526.51m²	计容积率建筑面积1016563.26㎡ Metering ratio floor area 1016563.26m²	计规定容积率建筑面积946483.20㎡ The specified floor area ratio is 946,483.20㎡	展厅净展览建筑面积 Exhibition hall net exhibition building area	403882.23㎡
			配套功能（配套办公、会议、餐饮、零售、仓储等） Supporting functions (supporting office, conference, catering, retail, warehousing, etc.)	542600.97㎡
		地上核增建筑面积70080.06㎡ Above-ground increased building area 70080.06㎡	架空公共空间 Overhead public space	70080.06㎡
	不计容积率建筑面积583963.25㎡ Excluding the floor ratio, the building area is 583,963.25㎡	地下规定建筑面积20274.92㎡ The underground Specified construction area is 20274.92㎡	地下厨房、餐厅、安保 Underground kitchen, dining, security	20274.92㎡
		地下核增建筑面积563688.33㎡ Underground increased building area 563688.33㎡	地下停车设备 Underground parking equipment	563688.33㎡ 其中，人防面积49158.42㎡ Among them, the area of civil air defense is 49158.42㎡

本期地上建筑分栋指标
Current above ground floor building indicators

栋号 Building number	高度（m） Height (m)	层数/地上/地下 Layers /above/under ground	规定功能 Specified function	规定面积/㎡ Specified area/㎡	核增功能 Increased Function	核增面积㎡ Area ㎡
1	23.60（混凝土屋面） 23.60 (concrete roof) 42.8（金属装饰屋面） 42.315 (metal decorative roof)	3/2	展厅净展览建筑面积 Exhibition hall net exhibition building area	403882.23	架空公共空间 Overhead public space	70080.06
			配套功能（配套办公、会议、餐饮、零售、仓储等） Supporting functions (supporting office, conference, catering, retail, warehousing, etc.)	513565.64	—	—
			地下停车库及设备用房 Underground parking garage and equipment room	562211.55	—	—
			地下安保、厨房及座椅储藏间等配套 Underground security, kitchen and seating storage, etc	20274.92	—	—
1栋合计 Building1 total				1570014.40㎡		
2	9.00	1	仓库 Warehouse	833.42	—	—
3	9.00	1	垃圾用房 Garbage house	1302.17	—	—
4	9.00	1	仓库 Warehouse	1561.13	—	—
5	10.10	2	海关、国检 Customs, national inspection	6114.60	—	—
6	13.60	3	安保及医疗中心 Security and medical center	1614.97	—	—
7	13.60	3	制服洗衣 Uniform laundry	1614.97	—	—
8	9.00	1	配套综合 Comprehensive supporting	3135.06	—	—
9	23.50	5	行政办公 Administration	9531.15	—	—
10	13.60	3	办公 Office	1614.97	—	—
11	9.00	1	仓库、垃圾用房、车检 Warehouse, garbage room, car inspection	1712.89	—	—
2-11栋合计 2-11 buildings total				30512.11㎡		

表 3-3　经济技术指标表
Table 3-3　Economic and Technical Indicator Table

展馆数据—展览场地
Exhibition Hall CAPACITY CHARTS - EXHIBITION HALLS + SOUTH PLAZA

展厅 Exhibition Hall No.	面积(平方米) Area(m²)	长度与跨度(米) Dimensions(m)	馆内净高(米) Ceiling Net Heights(m)	地面承重(吨/平方米) Floor Loading(Tonnes/m²)	吊点间距 Suspension Point Spacing	吊点承重 Suspension Point Loads	立柱 Column
1	20,000	210 × 108	16	5	9 × 9	2ton	Column Free
2	20,000	210 × 108	16	5	9 × 9	2ton	Column Free
3	20,000	210 × 108	16	5	9 × 9	2ton	Column Free
4	20,000	210 × 108	16	5	9 × 9	2ton	Column Free
5	20,000	210 × 108	16	5	9 × 9	2ton	Column Free
6	20,000	210 × 108	16	5	9 × 9	2ton	Column Free
7	20,000	210 × 108	16	5	9 × 9	2ton	Column Free
8	20,000	210 × 108	16	5	9 × 9	2ton	Column Free
9	20,000	210 × 108	16	5	9 × 9	2ton	Column Free
10	20,000	210 × 108	16	5	9 × 9	2ton	Column Free
11	20,000	210 × 108	16	5	9 × 9	2ton	Column Free
12	20,000	210 × 108	16	5	9 × 9	2ton	Column Free
13	20,000	210 × 108	16	5	9 × 9	2ton	Column Free
14	20,000	210 × 108	16	5	9 × 9	2ton	Column Free
15	20,000	210 × 108	16	5	9 × 9	2ton	Column Free
16	20,000	210 × 108	16	5	9 × 9	2ton	Column Free
18	20,000	210 × 108	16	5	9 × 9	2ton	Column Free
20	20,000	210 × 108	16	5	9 × 9	2ton	Column Free
17	48,000	219 × 249	18	5	9 × 9	2ton	Central Columns
南广场 South Plaza	31,706	—	—	—	Outdoor Exhibition	—	—
主办方会议室 及休息室 Organizer Offce & Lounge	100	—	—	—	每个展厅的二楼 on level 2 of each exhibition hall	—	—

表 3-4　展览场地信息表
Table 3-4　Exhibition Site Information Table

展馆数据—会议场地
CAPACITY CHARTS - MEETING FACILITIES

	会议室 Meeting Facilities	面积(平方米) Size(m²)	室内净高(米) Ceiling Height(m)	吊点间距 Rigging Grid Spacing	吊点承重 Rigging Point Loads	可容纳人数 Capacity				
						剧院式 Theatre	课桌式 Classroom	围桌式 Banquet	董事会式 Boardroom	U形 U Shape
南登录大厅 西侧 一层 West side of South Arrival Hall	LM 101A	157	6.0	3 × 3	250kg	147	87	81	34	29
	LM 101B	150	6.0	3 × 3	250kg	140	83	77	32	27
	LM 101C	157	6.0	3 × 3	250kg	147	87	81	34	29
	LM 101 combined(合并)	465	6.0	3 × 3	250kg	434	257	231	—	—
	LM 102A	157	6.0	3 × 3	250kg	147	87	81	34	29
	LM 102B	150	6.0	3 × 3	250kg	140	83	77	32	27
	LM 102C	157	6.0	3 × 3	250kg	147	87	81	34	29
	LM 102 combined(合并)	465	6.0	3 × 3	250kg	434	257	231	—	—
	LM 103A	157	6.0	3 × 3	250kg	147	87	81	34	29
	LM 103B	150	6.0	3 × 3	250kg	140	83	77	32	27
	LM 103C	157	6.0	3 × 3	250kg	147	87	81	34	29
	LM 103 combined(合并)	465	6.0	3 × 3	250kg	434	257	231	—	—
	LM 104A	157	6.0	3 × 3	250kg	147	87	81	34	29
	LM 104B	150	6.0	3 × 3	250kg	140	83	77	32	27
	LM 104C	157	6.0	3 × 3	250kg	147	87	81	34	29
	LM 104 combined(合并)	465	6.0	3 × 3	250kg	434	257	231	—	—
	LM 105	79	4.5	—	248kg	74	44	40	17	14
	LM 106	79	4.5	—	248kg	74	44	40	17	14
	LM 107A	117	6.0	3 × 3	250kg	109	65	60	25	21
	LM 107B	117	6.0	3 × 3	250kg	109	65	60	25	21
	LM 107 combined(合并)	234	6.0	3 × 3	250kg	218	130	120	50	42
	LM 108A	134	6.0	3 × 3	250kg	125	74	69	29	24
	LM 108B	134	6.0	3 × 3	250kg	125	74	69	29	24
	ML 108 combined(合并)	268	6.0	3 × 3	250kg	250	148	138	58	48
登录大厅 (1) 西侧 Lobby 1 West	南多功能厅 South Ballroom	3446	9.5	4.5 × 9	2ton	3,221	1902	2088	—	—
二层展厅 间会议室 Exhibition Hall Meeting Rooms (level 2)	MA1-203(1-A)	236	5.0	3 × 3	250kg	215	127	118		
	MA1-202(1-B)	314	5.0	3 × 3	250kg	282	167	155		
	MA1-201(1-C)	236	5.0	3 × 3	250kg	215	127	118		
	ABC merger	786	5.0	3 × 3	250kg	712	421	391		
	IA1-204(1-D)	56	5.0	—	—	47	28	26	11	9
	IA1-203(1-E)	64	5.0	—	—	47	28	26	11	9
	IC1-204(2-D)	56	5.0	—	—	47	28	26	11	9
	IC1-203(2-E)	64	5.0	—	—	47	28	26	11	9
	MC1-203(2-A)	236	5.0	3 × 3	250kg	215	127	118	—	—
	MC1-202(2-B)	314	5.0	3 × 3	250kg	282	167	155	—	—
	MC1-201(2-C)	236	5.0	3 × 3	250kg	215	127	118	—	—
	combined (合并)	1026	5.0	3 × 3	250kg	712	421	391	—	—
	IA3-204(5-D)	56	5.0	—	—	47	28	26	11	9
	IA3-203(5-E)	64	5.0	—	—	47	28	26	11	9

表 3-5　会议场地信息表
Table 3-5　Conference Venue Information Table

会议室 Meeting Facilities		面积(平方米) Size(m²)	室内净高(米) Ceiling Height(m)	吊点间距 Rigging Grid Spacing	吊点承重 Rigging Point Loads	可容纳人数 Capacity				
						剧院式 Theatre	课桌式 Classroom	围桌式 Banquet	董事会式 Boardroom	U形 U Shape
	IC3-204(6-D)	56	5.0	—	—	47	28	26	11	9
	IC3-203(6-E)	60	5.0	—	—	47	28	26	11	9
	MA3-203(5-A)	236	5.0	3 × 3	250kg	215	127	118	—	—
	MA3-202(5-B)	314	5.0	3 × 3	250kg	282	167	155	—	—
	MA3-201(5-C)	236	5.0	3 × 3	250kg	215	127	118	—	—
	combined (合并)	1026	5.0	3 × 3	250kg	712	421	391	—	—
	IA4-202(7-D)	56	5.0	—	—	47	28	26	11	9
	IA4-201(7-E)	60	5.0	—	—	47	28	26	11	9
	IC4-202(8-D)	56	5.0	—	—	47	28	26	11	9
	IC4-201(8-E)	60	5.0	—	—	47	28	26	11	9
	MA6-203(9-A)	236	5.0	3 × 3	250kg	215	127	118	—	—
	MA6-202(9-B)	314	5.0	3 × 3	250kg	282	167	155	—	—
	MA6-201(9-C)	236	5.0	3 × 3	250kg	215	127	118	—	—
	combined (合并)	1026	5.0	3 × 3	250kg	712	421	391	—	—
	IA6-204(9-D)	56	5.0	—	—	47	28	26	11	9
	IA6-203(9-E)	60	5.0	—	—	47	28	26	11	9
	IC6-204(10-D)	56	5.0	—	—	47	28	26	11	9
	IC6-203(10-E)	60	5.0	—	—	47	28	26	11	9
	IA7-202(11-D)	56	5.0	3 × 3	250kg	73	43	40	17	14
	IA7-201(11-E)	60	5.0	3 × 3	250kg	79	46	43	18	15
	IC7-202(12-D)	56	5.0	3 × 3	250kg	79	46	43	18	15
	IC7-201(12-E)	60	5.0	3 × 3	250kg	73	43	40	17	14
二层展厅 间会议室 Exhibition Hall Meeting Rooms (level 2)	MA6-203(10-A)	236	5.0	3 × 3	250kg	215	127	118	—	—
	MA6-202(10-B)	314	5.0	3 × 3	250kg	282	167	155	—	—
	MA6-201(10-C)	236	5.0	3 × 3	250kg	215	127	118	—	—
	combined (合并)	1250	5.0	3 × 3	250kg	712	421	391	—	—
	IA8-204(13-D)	56	5.0	—	—	47	28	26	11	9
	IA8-203(13-E)	60	5.0	—	—	47	28	26	11	9
	IC8-204(14-D)	56	5.0	—	—	47	28	26	11	9
	IC8-203(14-E)	60	5.0	—	—	47	28	26	11	9
	MA8-203(13-A)	236	5.0	3 × 3	250kg	215	127	118	—	—
	MA8-202(13-B)	314	5.0	3 × 3	250kg	282	167	155	—	—
	MA8-201(13-C)	236	5.0	3 × 3	250kg	215	127	118	—	—
	combined (合并)	1026	5.0	3 × 3	250kg	712	421	391	—	—
	IA9-202(15-D)	56	5.0	—	—	47	28	26	11	9
	IA9-201(15-E)	60	5.0	—	—	47	28	26	11	9
	IC9-202(16-D)	56	5.0	—	—	47	28	26	11	9
	IC9-201(16-E)	60	5.0	—	—	47	28	26	11	9
	MA8-203(14-A)	236	5.0	3 × 3	250kg	215	127	118	—	—
	MA8-202(14-B)	314	5.0	3 × 3	250kg	282	167	155	—	—
	MA8-201(14-C)	236	5.0	3 × 3	250kg	215	127	118	—	—
	combined (合并)	1026	5.0	3 × 3	250kg	712	421	391	—	—
	IA8-204(13-D)	56	5.0	—	—	47	28	26	11	9
	IA8-203(13-E)	60	5.0	—	—	47	28	26	11	9
	IC8-204(14-D)	56	5.0	—	—	47	28	26	11	9

表 3-5-1 会议场地信息表
Table 3-5-1 Conference Venue Information Table

会议室 Meeting Facilities		面积(平方米) Size(m²)	室内净高(米) Ceiling Height(m)	吊点间距 Rigging Grid Spacing	吊点承重 Rigging Point Loads	可容纳人数 Capacity				
						剧院式 Theatre	课桌式 Classroom	围桌式 Banquet	董事会式 Boardroom	U形 U Shape
二层展厅间会议室 Exhibition Hall Meeting Rooms (level 2)	IC8-203(14-E)	60	5.0	—	—	47	28	26	11	9
	MA8-203(13-A)	236	5.0	3×3	250kg	215	127	118	—	—
	MA8-202(13-B)	314	5.0	3×3	250kg	282	167	155	—	—
	MA8-201(13-C)	236	5.0	3×3	250kg	215	127	118	—	—
	combined(合并)	1026	5.0	3×3	250kg	712	421	391	—	—
	B7-D	50	5.0	—	—	47	28	26	11	9
	B7-E	50	5.0	—	—	47	28	26	11	9
	B8-D	50	5.0	—	—	47	28	26	11	9
	B8-E	50	5.0	—	—	47	28	26	11	9
B9会议中心（一层） Conference Center B9 (Level 1)	北多功能厅 North Ballroom	6684	11	4.5×9	1ton	6247	3690	4051	—	—
	CC 101A	231	6.0	3×3	250kg	216	128	118	—	42
	CC 101B	212	6.0	3×3	250kg	198	117	109	—	39
	CC 101C	212	6.0	3×3	250kg	198	117	109	—	39
	CC 101 combined(合并)	655	6.0	3×3	250kg	612	362	336	—	119
	CC 102	78	4.5	3×3	250kg	73	43	40	17	14
	CC 103	78	4.5	3×3	250kg	73	43	40	17	14
	CC 104	67	4.5	3×3	250kg	63	37	40	14	12
	CC 105A	231	6.0	3×3	250kg	216	128	118	—	42
	CC 105B	212	6.0	3×3	250kg	198	117	109	—	39
	CC 105C	212	6.0	3×3	250kg	198	117	109	—	39
	CC 105 combined(合并)	655	6.0	3×3	250kg	612	362	336	—	119
	CC 106	78	4.5	3×3	250kg	73	43	40	17	14
	CC 107	78	4.5	3×3	250kg	73	43	40	17	14
	CC 108	67	4.5	3×3	250kg	63	37	40	14	12
B9会议中心（二层） Conference Center B9 (Level 2)	CC 201A	486	6.0	3×3	250kg	454	268	249	249	88
	CC 201B	486	6.0	3×3	250kg	454	268	249	—	—
	CC 201 combined(合并)	972	6.0	3×3	250kg	908	537	498	—	—
	CC 202	179	6.0	3×3	250kg	167	99	92	38	33
	CC 203A	119	6.0	3×3	250kg	111	66	61	26	22
	CC 203B	119	6.0	3×3	250kg	111	66	61	26	22
	CC 203 combined(合并)	238	6.0	3×3	250kg	—	—	—	—	—
	CC 204A	119	6.0	3×3	250kg	111	66	61	26	22
	CC 204B	119	6.0	3×3	250kg	111	66	61	26	22
	CC 204 combined(合并)	238	6.0	3×3	250kg	—	—	—	—	—
	CC 205	179	6.0	3×3	250kg	167	99	92	38	33
	CC 206A	486	6.0	3×3	250kg	454	268	249	—	—
	CC 206B	486	6.0	3×3	250kg	454	268	249	—	—
	CC 206 combined(合并)	972	6.0	3×3	250kg	908	537	498	—	—
	CC 207A	230	6.0	3×3	250kg	215	127	118	—	42
	CC 207B	320	6.0	3×3	250kg	299	127	164	—	58
	CC 207C	230	6.0	3×3	250kg	215	127	118	—	42
	CC 207 combined(合并)	780	6.0	3×3	250kg	729	431	400	—	—

表 3-5-2　会议场地信息表
Table 3-5-2　Conference Venue Information Table

会议室 Meeting Facilities		面积(平方米) Size(m²)	室内净高(米) Ceiling Height(m)	吊点间距 Rigging Grid Spacing	吊点承重 Rigging Point Loads	可容纳人数 Capacity				
						剧院式 Theatre	课桌式 Classroom	围桌式 Banquet	董事会式 Boardroom	U形 U Shape
B9会议中心 （三层） Conference Center B9 (Level 3)	VIP Room 301	67	6.0	3 × 3	250kg	—	—	—	14	—
	VIP Room 302	77	6.0	3 × 3	250kg	—	—	—	17	—
	VIP Room 303	67	6.0	3 × 3	250kg	—	—	—	14	—
	VIP Room 304	77	6.0	3 × 3	250kg	—	—	—	17	—
	VIP休息室 VIP Lounge	444	6.0	3 × 3	250kg	56				
北登录大厅 （2）西侧 North Arrival Hall West	国际报告厅 Plenary Hall	2137	1,920 pax							
登录大厅（2） 东侧（一层） North Arrival Hall East (Level 1)	VVIP报告厅 VVIP Plenary Hall	69	6.0	—	—	—	—	—	—	—
	VIP休息室 VIP Reception Hall	410	6.0	—	—	—	—	—	—	—
北登录大厅 （2）东侧 （二层） North Arrival Hall East (Level 2)	多媒体室2A Media Room 2A	144	4.0	3 × 3	250kg	135	79	—	31	26
	多媒体室2B Media Room 2B	158	4.0	3 × 3	250kg	148	87	—	34	29
	多媒体室2C Media Room 2C	157	4.0	3 × 3	250kg	147	87	—	34	29
	媒体室 Media Room combined(合并)	459	—	—	—	429	253	—	99	83
	VIP全会厅1 VIP Plenary 1	37	3.5	—	—	—	—	—	—	—
	VIP全会厅2 VIP Plenary 2	82	3.5	—	—	—	—	—	—	—
	VIP全会厅3 VIP Plenary 3	85	3.5	—	—	—	—	—	—	—
	VIP全会厅4 VIP Plenary 4	85	3.5	—	—	—	—	—	—	—
	VIP全会厅5 VIP Plenary 5	79	3.5	—	—	—	—	—	—	—
	VIP全会厅6 VIP Plenary 6	69	3.5	—	—	—	—	—	—	—

表 3-5-3　会议场地信息表
Table 3-5-3　Conference Venue Information Table

展馆数据—会议场地
CAPACITY CHARTS - MEETING FACILITIES

	会议室 Conference Room	可容纳人数 Capacity	面积(平方米) Size(m²)	净高(米) Ceiling Height(m)	固定座椅/伸缩式座椅 Fixed Seating/Retractable	剧院式 Theatre
体育场面积 11,000 m² Event Center B10 Arena Area	底层座位 Stale Seating Capacity	12,526 pax	—	—	—	—
	活动坐席 Retractable Seating	—	—	18.0	4352	—
	楼座 Balcony Seating	—	—	18.0	6220	—
	底层座位 Stall Capacity	—	—	—	1954	—
艺人休息室 后勤办公室 （2层） Artist Lounge Support Offce Level 2	艺人休息室1 Artist Lounge 1	—	67	—	—	63
	艺人休息室2 Artist Lounge 2	—	83	—	—	78
	艺人休息室3 Artist Lounge 3	—	58	—	—	54
	后勤办公室1 Logistic Offce 1	—	67	—	—	63
	后勤办公室2 Logistic Offce 2	—	83	—	—	78
	后勤办公室2 Logistic Offce 2	—	58	—	—	54
VIP包厢 （3层） VIP Box Level 3	VIP包厢301 VIP Box 301	—	51	—	14	—
	VIP包厢302 VIP Box 302	—	48	—	14	—
	VIP包厢303 VIP Box 303	—	49	—	14	—
	VIP包厢304 VIP Box 304	—	52	—	14	—
	VIP包厢305 VIP Box 305	—	52	—	14	—
	VIP包厢306 VIP Box 306	—	49	—	14	—
	VIP包厢307 VIP Box 307	—	48	—	14	—
	VIP包厢308 VIP Box 308	—	51	—	14	—
	深圳国际会展中心休息室 the Shenzhen World Lounge	—	796	—	183	—

表 3-5-4　会议场地信息表
Table 3-5-4　Conference Venue Information Table

展馆数据—停车场
CAPACITY CHARTS - PARKING SPACES

停车场 Car Parks	位置 Location	停车位数量 Number of parking spaces	可停车车辆类型 Type
地下车库1 Underground Parking	展馆地下一层、地下二层 B1&B2	9500	小汽车 Cars
地面VIP停车场 VIP Parking	会议中心B9 Conference Center B9	300	贵宾车辆 VIP car
货车停车区1 Exhibitor Logistics Parking	每个展馆东侧和西侧 East+West of Each Halls	1000	卡车 Trucks
货车轮候区 Truck Marshalling Yard	北广场 North Plaza	1500	卡车 Trucks

表 3-6 停车场信息表
Table 3-6 Parking Lot Information Table

展馆数据—餐饮场地
CAPACITY CHARTS - RESTAURANTS + FOOD OUTLETS

餐厅 Food&Beverage	数量 Quantity	面积 Size	位置 Location
连锁餐饮店 Retail F&B Outlets	6	4301	南登录大厅三层 South Login Hall, third floor
连锁餐饮店 Retail F&B Outlets	6	4302	北登录大厅三层 North Login Hall on the third floor
连锁餐饮店 Retail F&B Outlets	10	varies	中央廊道一层 Central corridor first floor
展厅内餐饮区 Food Loft	16	1050 m²/ hall	所有标准展厅三层 All standard showrooms on the third floor
展厅内餐饮区 Food Loft	1	4200	17号展厅三层 Hall 17 on the third floor
17号展厅外卖区 Takeaway area in Hall 17	2	1900	17号展厅二层 Hall 17 on the second floor
外卖区 Grab n GO	16	400	所有标准展厅一端二层 All standard exhibition halls
外卖区 Grab n GO	4	130	活动中心B10一二层 Activity Center B10, Level 1

表 3-7 餐饮场地信息表
Table 3-7 Catering Site Information Table

"巨"
构

建造结构与系统设计
Construction Structure and System Design

232 结构设计
Structure Design

254 设备系统设计
Equipment System Design

284 消防设计
Fire Protection Design

293 构造与节点
Structure and Node

306 防水设计
Waterproof Design

Giant Structure

结构设计
Structure Design

3mm厚铝单板(表面氟碳喷涂)
m thick aluminum veneer
ce fluorocarbon sprayed)

管立柱(Q235B)表面氟碳喷涂
teel pipe column (Q235B)
ce fluorocarbon spraying

钢管50x5
50x5 steel pipe

EPDM自粘防水胶带（满粘）
proof tape (full adhesion)

50mm厚保温岩棉
m thick thermal rock wool

1.5mm厚镀锌钢板
ck galvanized steel sheet

100mm厚防火岩棉
00mm fireproof rock wool

3mm喷涂钢板
3mm sprayed steel plate

方钢管60x6
60x6 square steel tube

全熔透坡口焊
netration groove welding

100x75x6mm加强钢板
nm reinforced steel plate

方钢管100x6（Q345B）
steel tube 100x6 (Q345B)

L-6 安装螺栓
L-6 mounting bolt

10+2.28PVB+10+12A+:
半钢化双银LOW-E中空双
10 + 2.28PVB + 10 + 12/
Semi-tempered doubl
laminated glass (four p

铝合金扣盖,表面氟碳喷涂
Aluminum alloy buckle
fluorocarbon coating

1.5mm厚镀锌钢板
1.5mm thick galvanize

25mm蜂窝铝板
25mm honeycomb alu

水平变截面百页（200mm
Hundred pages of vari
section (200mm-1000n

25x25x2mm，L=40mm
25x25x2mm, L = 40mn
fitting pendant group

4mm单层铝板
4mm single layer alun

2-100X80X6加强钢肋板
2-100X80X6 reinforced

1.5mm镀锌钢板+保温岩
1.5mm galvanized stee

6mmU型折弯钢板
6mmU-shaped bent st

4mm厚单层铝板
4mm thick single layer

30mm拉索支座（Q345B
+氟碳喷涂，耐火极限1h
30mm cable support (
fireproof coating + flu
fire resistance limit 1h

深圳国际会展中心的建筑宛如一条"巨龙"于"祥云"之上翱翔——以19座展厅为祥云，中央廊道为龙身，南北登录大厅为龙爪。建筑师赋予了它灵魂和表皮与结构和骨架使其活灵活现。为了消除一栋建筑缺少生命地存在于城市地块中的冰冷感觉，结构工程需展现强大的实力，它汇聚了建筑师的创意灵魂。建筑结构一体化协调且精细化设计，让结构成就建筑之美，建筑展现结构之精妙，塑造了深圳国际会展中心这座地标性建筑。

The Shenzhen World architecture is like a giant dragon soaring above the auspicious cloud. The 19 exhibition halls emblemed the auspicious cloud, while the central concourse emblemed the body of a dragon, and the north and south arrival halls are the dragon claws. The architect gave it its soul and skin, as well as its framework, making it alive. In order to eliminate the cold feeling of a building without any life-form, the structure engineering needs to embody great power, so as to guard the spirit that the architect create for the GBA.

会展结构关注点
The Structure Focus

场地处理：软基处理

- 软土地区建设大型会展项目
- 承载能力要求高，特别是定位于承接重型机械展的会展项目
- 软土地基难以满足建筑主体的承载能力和沉降限值的要求
- 室外展场、运货通道等也对承载能力有较高的要求，必须重视

Site Treatment: Soft Foundation Treatment

- Construction of large exhibition projects in soft soil areas
- High bearing capacity is required, especially for exhibition projects which have heavy machinery
- The soft foundation is difficult to meet the requirements of the bearing capacity and sedimentation limit of the main building
- Outdoor exhibition areas and freight passages also have higher requirements on the bearing capacity, which must be paid attention to.

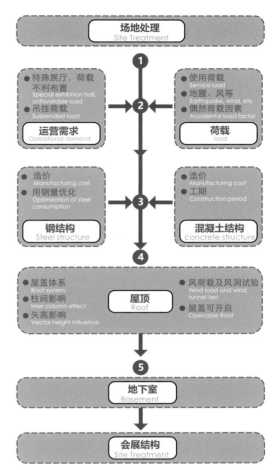

图 4-1 会展结构设计关注点
Figure 4-1 Focus of the Exhibition Structure Design

软基处理设计承载力和变形模量标准见下表
Standard Table for Design Bearing Capacity and Deformation Modulus of Soft Foundation Treatment

范围 Range	荷载 (KN/M2) load	变形模量 (MPa) deformation modulus
展厅 Exhibition hall	100 (50)	≥ 20
室外展场 Outdoor exhibition hall	100	≥ 20
二期范围 Phase 2 Scope	100	≥ 20

沉降控制标准
Settlement control standard

范围 Range	沉降量 (mm) Settlement amount	沉降差 Settlement difference
展厅 Exhibition hall	100	0.002L
室外展场 Outdoor exhibition hall	100	/
二期范围 Phase 2 Scope	100	/

不同地基处理方案的优缺点
Advantages and disadvantages of different foundation treatment schemes

地基处理方案 Foundation treatment plan	优点 Advantage	缺点 shortcoming
排水团结及预压法 Drainage consolidation and preloading method	造价低 Low cost	工期长，堆载土方不易保证 Long duration, not easy to guarantee the stacked earthwork
水泥土搅拌桩 Cement soil mixing pile	造价较低 Lower cost	成桩质量不易保证，养护时间长 The quality of piling is not guaranteed. Long maintenance time
CFG桩复合地基 CFG pile composite foundation	造价较低，环境影响较小 Low cost and little environmental impact	淤泥质土中施工质量难以保证 It is difficult to guarantee the construction quality in muddy soil
预应力管桩复合地基 Prestressed pipe pile composite foundation	施工快，施工质量易于保证 Fast construction, easy to guarantee construction quality	造价较高，打桩对环境影响较大 High cost and great impact of piling on the environment
道路路基（浅层）处理 Treatment of road subgrade (shallow layer)	造价低 Low cost	沉降大，使用期可能需要不定期维修 The settlement is large, and the service life may need irregular maintenance

表 4-1 软基础处理技术
Table 4-1 Soft Foundation Treatment Technology

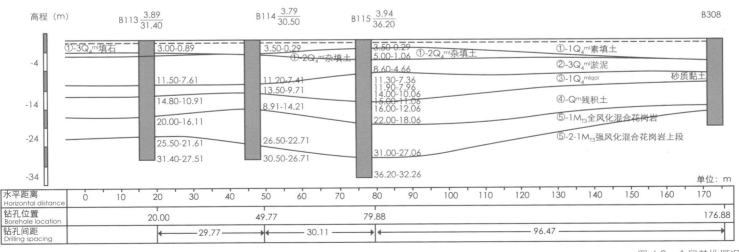

图 4-2 会展基地概况
Figure 4-2 Overview of the Exhibition Site Base

运营需求：设计使用负荷

展厅（地面）承载能力
- 大型综合展览：50kN/m²
- 大型综合展览双层：35kN/m²/ 15kN/m²（首层／二层）
- 超大重载工业型展览：50kN/m² ～ 80kN/m²
- 室外展场：100kN/m²

吊挂荷载
- 标准展厅，常规设计；
- 特殊展厅，后期运营的设计服务；
- 同时兼顾消防需求。

图 4-3　吊挂挂钩示意图
Figure 4-3　Suspension Hook

图 4-4　吊挂示意图
Figure 4-4　On-Site Suspension

Operation Requirements: Designed Service Load

Exhibition hall bearing capacity (above ground)
- Large-scale comprehensive exhibition: 50kN/m²
- Two-story large comprehensive exhibition: 35kN/m²
/ 15kN/m²(first /second)
- Ultra-large heavy-haul industrial exhibition: 50kN/m² ~ 80kN/m²
- Outdoor exhibition space: 100kN/m²

Suspension load
- Standard exhibition hall, conventional design
- Special exhibition hall, design service for later operation
- Meet needs for fire protection at the same time

结构体系研究：钢结构与混凝土结构的选择

国内大型会展屋盖的现状：
- 大多采用空间管桁架结构体系
- 个别采用单层网壳的形式，比如昆明滇池国际会展中心
- 个别采用网架的型式，比如湛江会展中心主展厅
- 个别采用张弦梁的型式，比如深圳会展中心

Structural System Research: selection of steel structure and concrete structure

Current situation of large-scale exhibition roofs in China:
- Mostly space pipe truss structure system
- Single-layer reticulated shell is used in some cases, such as Kunming Dianchi International Convention and Exhibition center
- A few use net racks, for example, the main exhibition hall of Zhanjiang Convention and Exhibition Center

- Some use the beam string structure, for example, the old Shenzhen Convention and Exhibition Center

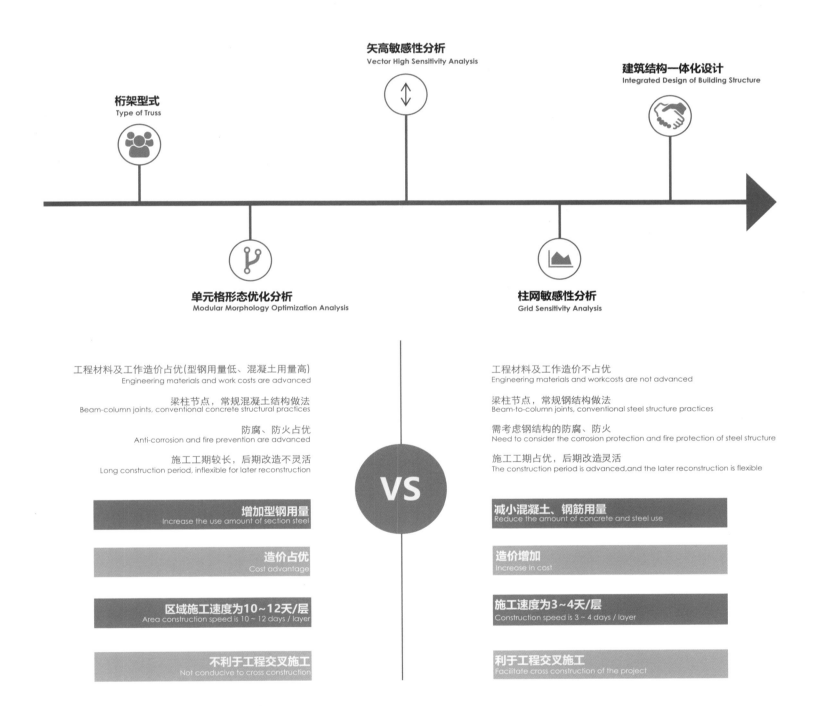

矢高敏感性分析
Vector High Sensitivity Analysis

建筑结构一体化设计
Integrated Design of Building Structure

桁架型式
Type of Truss

单元格形态优化分析
Modular Morphology Optimization Analysis

柱网敏感性分析
Grid Sensitivity Analysis

工程材料及工作造价占优(型钢用量低、混凝土用量高)
Engineering materials and work costs are advanced

工程材料及工作造价不占优
Engineering materials and workcosts are not advanced

梁柱节点，常规混凝土结构做法
Beam-column joints, conventional concrete structural practices

梁柱节点，常规钢结构做法
Beam-to-column joints, conventional steel structure practices

防腐、防火占优
Anti-corrosion and fire prevention are advanced

需考虑钢结构的防腐、防火
Need to consider the corrosion protection and fire protection of steel structure

施工工期较长，后期改造不灵活
Long construction period, inflexible for later reconstruction

施工工期占优，后期改造灵活
The construction period is advanced,and the later reconstruction is flexible

VS

增加型钢用量
Increase the use amount of section steel

减小混凝土、钢筋用量
Reduce the amount of concrete and steel use

造价占优
Cost advantage

造价增加
Increase in cost

区域施工速度为10~12天/层
Area construction speed is 10 ~ 12 days / layer

施工速度为3~4天/层
Construction speed is 3 ~ 4 days / layer

不利于工程交叉施工
Not conducive to cross construction

利于工程交叉施工
Facilitate cross construction of the project

根据项目实际情况选型
Selection According to the Real Situation of the Project

图 4-5 结构体系研究
Figure 4-5 Structural System Research

风荷载及风洞试验

风荷载的特殊性及其与工程设计的进展关系：
- 根据规范先期判断场地粗糙度、风荷载体系系数，预判结构风荷载的参数取值
- 风洞试验及试验配合
- 风洞试验结果的评估及使用

Wind Load and Wind Tunnel Test

The particularity of the wind load and the relationship of engineering design progress with it:
- Anticipate the roughness of the site and the wind load system coefficient, as well as the parameters of the wind load of the structure according to the regulation in advance
- Wind tunnel test and test coordinations
- Assessment and the use of wind tunnel test results

图 4-6　风荷载实验照片
Figure 4-6　Wind Load Test

项次 Items	类别 Category	体型及体型系数μs Shape and Shape Coefficient μS			备注 Remark
1	封闭式落地双坡屋面 Closed touch-down double-slope roof	μ_s　-0.5　α	a	μs	中间值按线性插值法计算 Intermediate values are calculated by linear interpolation
			0°	0.0	
			30°	+0.2	
			≥60°	+0.8	
2	封闭式双坡屋面 Closed double-slope roof	μ_s　-0.5 +0.8　α　-0.5 -0.7 +0.8　-0.5 -0.7	a	μs	中间值按线性插值法计算 Intermediate values are calculated by linear interpolation μs的绝对值不小于0.1 The absolute value of μs is no less than 0.1
			0°	-0.6	
			30°	+0.2	
			≥60°	+0.8	
3	封闭式落地拱形屋面 Closed touch-down arched roof	-0.7 μ_s　f　-0.5 L	f/l	μs	中间值按线性插值法计算 Intermediate values are calculated by linear interpolation
			0.1	+0.1	
			0.2	+0.2	
			0.5	+0.6	

表 4-2　风荷载及风洞试验
Table 4-2　Wind Load and Wind Tunnel Test

可开启屋盖研究

可开启屋面是超大展厅采用的重要消防措施。最大的一块可开启屋面面积达 500 平方米，重达 60 吨，在 9 根导轨上移动。其结构设计需在满足消防报警的情况下，从全闭合状态到全开启状态用时在 60 秒以内（图 4-7）。

Openable Roof Research

The open roof is an important fire protection measure for large exhibition halls. The largest piece of the open roof is nearly 500 square meters, weighs 60 tons and slide on 9 structural rails. In the case of fire alarm, it should take less than 60 seconds to open from full closed to full open.(Figure 4-7)

活动屋盖
Movable Roof

轨道
Track

防撞击装置
Anti-collision Device

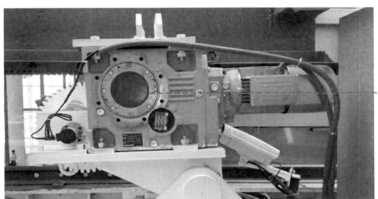

动力单元
Power Unit

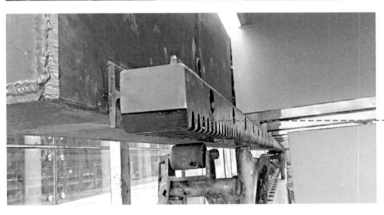

动力组件
Power Components

活动屋盖
Movable Roof

动力单元-传动轴
Power Unit-transmission Shaft

动力单元-齿轮
Power Unit-gear

动力单元-齿条
Power Unit-rack

图 4-7　活动屋盖设备布置形式分析
Figure 4-7　Movable Roof Equipment Layout

地下室结构

在地下室结构设计方面，我们结合建筑功能和正常使用要求，选用了超长无缝钢筋混凝土结构，解决了因分缝引起的建筑防水隐患；同时也依靠设计与构造措施，减少或消除了因超长混凝土结构开裂导致的不利影响；无梁楼盖与梁板式相结合的地下室楼板结构形式使建筑结构轻巧简洁，空间使用率也得到了提高（图4-8）。

Basement Structure

In terms of the structural design of the basement, we chose the super long seamless reinforced concrete structure based on the function of the building and the requirements of normal use to solve the hidden danger of building waterproofing caused by joints split. By means of design and construction measures, the adverse effects caused by cracking of extra-long concrete structures are reduced or eliminated as well. The combination use of the flat floor slab and the beam and slab structure in the basement makes the building structure light and simple, and also improves the utilization rate of the space.(Figure 4-8)

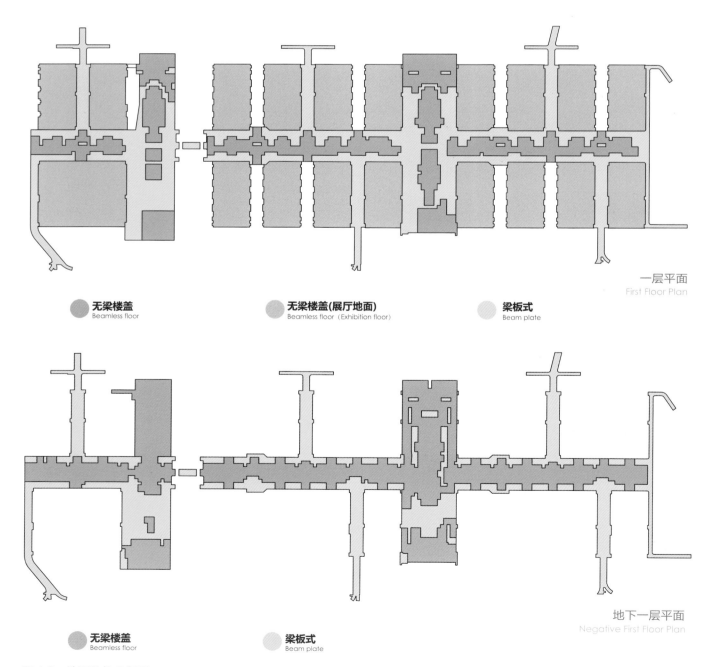

一层平面
First Floor Plan

● 无梁楼盖 Beamless floor　　● 无梁楼盖(展厅地面) Beamless floor（Exhibition floor）　　● 梁板式 Beam plate

地下一层平面
Negative First Floor Plan

● 无梁楼盖 Beamless floor　　● 梁板式 Beam plate

图4-8 地下结构分析图
Figure 4-8 Underground Structure Analysis

重点空间结构设计
Structure Design of Key Space

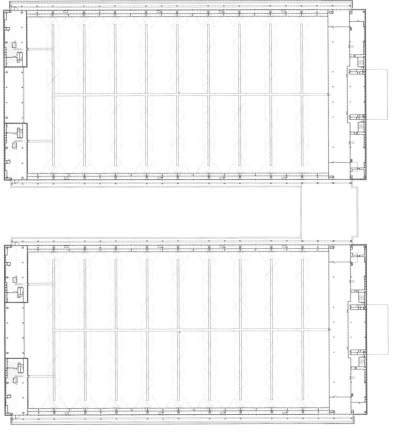

图 4-9　标准展厅 RF 平面图
Figure 4-9　Standard Exhibition Hall RF Plan

标准展厅

标准展厅平面尺度为长 209.5m× 宽 108m，两个展厅之间宽度为 42m，高度为 26m。原中标方案则为沿长向起伏的屋面形态，由于屋顶起拱方向存在严重排水隐患，在不影响整体屋面效果的情况下，在深化设计的过程中将展厅屋面造型进行了 90°旋转以实现建筑排水的要求，并结合展厅大跨度的要求，从而确定了展厅的结构形式（图 4-9）。

结构采用完全拟合建筑外形的单曲立体桁架为结构基本受力单元。结构主体二层、三层及屋面层均采用钢结构框架体系，上部曲面屋盖则利用钢柱支撑屋顶立体桁架，支撑屋面的立体桁架中的 99m 跨度的主桁架跨中矢高 7m，桁架间距 18m，展厅共 9 榀桁架；两个展厅之间的 42m 跨度采用桁架连续、框架连续的结构布置，形成了连续、清晰的结构体系，满足了抗连续倒塌的概念（图 4-10）。

The Standard Exhibition Hall

The standard exhibition hall is 209.5m long and 108m wide. Between two exhibition halls, the width is 42m and the height is 26m. The original bidding scheme had a wavy roof form with ups and downs, which brought hidden trouble for drainage.

Therefore, when came to detailed design, the roof has been rotated 90°without affecting the whole roof shape, which enables it to meet the need of building drainage and build a series of exhibition halls on a large span, eventually led to the ultimate structure.(Figure 4-9)

The structure adopts the single three-dimensional truss which completely fits the shape of the building design as the basic load bearing unit of the structure. The steel structure frame system is adopted in the second, third and roof layers of the structure. The upper curved roof is a three-dimensional truss system supported by steel columns. The span of the main truss that supporting the three-dimensional roof truss is 99m and its rise of arc is 7m tall. The truss spacing is 18m, and there are total 9 trusses in the exhibition hall. The span length of 2 exhibition halls is 42m, which adopted a continuous truss and frame structure, forming a continuous and clear structural system which avoids continuous collapse.(Figure 4-10)

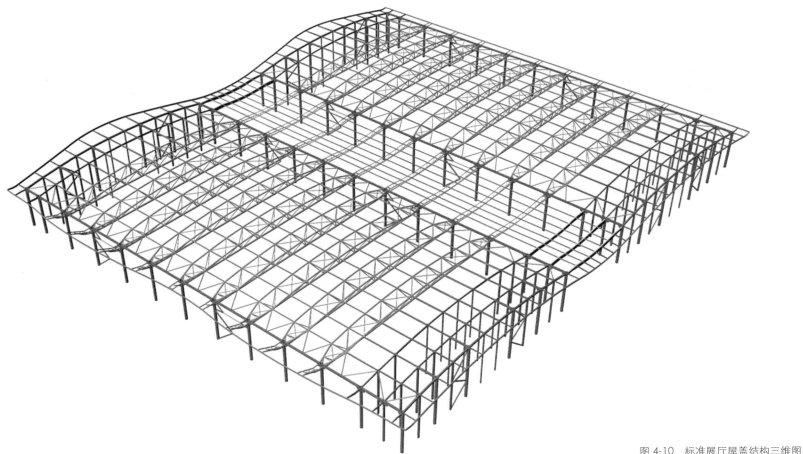

图 4-10　标准展厅屋盖结构三维图
Figure 4-10　Three Dimensional Structure of the Roof of the Standard Exhibition Hall

超大展厅

5 万平方米的 17 号超大特殊展厅作为深圳国际会展中心项目体量最大的一个特殊展厅，承担了举办飞机、游艇、大型机械设备等特殊类型展览的用途，是本项目的标志性场馆。17 号展厅短向跨度达 209.5m，为减小其结构跨度，展厅中央增设了 4×4 个承重柱，并结合超大展厅的消防要求，增设了可开启屋面。

17 号展厅结构主体二层、三层及屋面层均采用钢结构框架体系，上部曲面屋盖则是利用钢柱支撑屋顶立体桁架，主桁架为跨度 209.5m 的 3 跨立体空间桁架，主桁架跨中矢高 9m；次桁架为立体桁架，跨度 81m，桁架跨中矢高 7m，其余 18 榀桁架及立体桁架跨度 60m，桁架跨中矢高 7m。

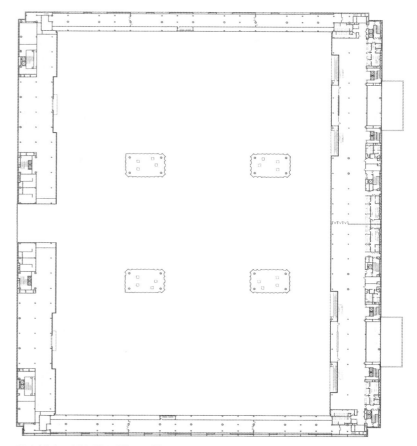

图 4-11-1　A11 展厅 3F 平面图
Figure 4-11-1　Exhibition Hall A11 Plan-Level 3

图 4-11-2　A11 展厅现场实景图
Figure 4-11-2　Exhibition Hall A11

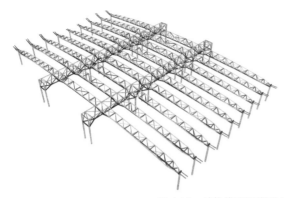

图 4-12　结构模型三维图 1
Figure 4-12　Structure Model 1

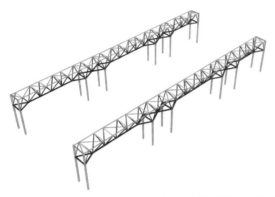

图 4-13　结构模型三维图 2
Figure 4-13　Structure Model 2

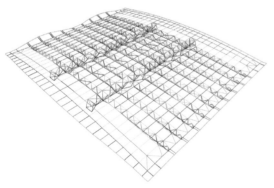

图 4-14　结构模型三维图 3
Figure 4-14　Structure Model 3

Super Large Exhibition Hall

As the largest special exhibition hall of the Shenzhen World, the No.17 hall covers an area of 50,000 square meters and it is the landmark venue of the project, which is able to host exhibitions with large exhibition items such as aircraft, yacht, large-scale mechanical equipment and others. No.17 hall has a short span of 209.5 meters and in order to reduce its structural span and meet the fire protection requirements as a super-large exhibition hall, an open roof is added. Additionally, 4×4 load bearing columns are added in the center of the exhibition hall to meet the Stable requirements of the structure.

The structural system of the No.17 hall consists of steel frame structure system on the second and third floors of the main structure and the roof layer. The upper curved roof adopted a three-dimensional truss supported by steel columns. The span of the main truss is 209.5m, a three-span three-dimensional space truss, the rise of the arc is 9m, the span of the secondary trusses is 81m, and its rise of arc is 5m. There are 18 trusses in total, and the span of the three-dimensional truss is 60m, with a arch rise of 7m.

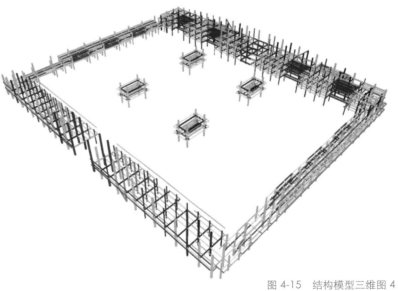

图 4-15　结构模型三维图 4
Figure 4-15　Structure Model 4

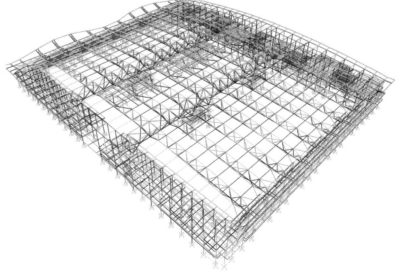

图 4-16　结构模型三维图 5
Figure 4-16　Structure Model 5

可开启屋盖
Openable Roof

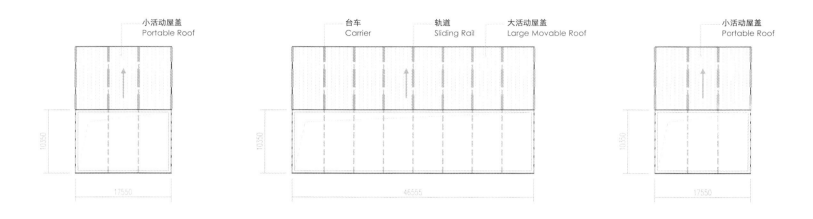

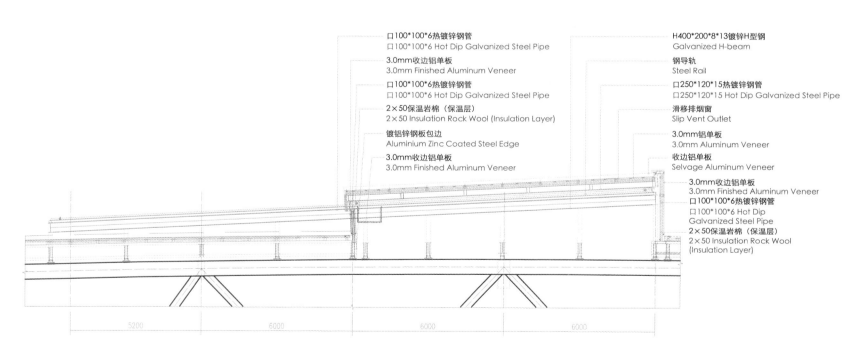

口100*100*6热镀锌钢管
口100*100*6 Hot Dip Galvanized Steel Pipe

3.0mm收边铝单板
3.0mm Finished Aluminum Veneer

口100*100*6热镀锌钢管
口100*100*6 Hot Dip Galvanized Steel Pipe

2×50保温岩棉（保温层）
2×50 Insulation Rock Wool (Insulation Layer)

镀铝锌钢板包边
Aluminium Zinc Coated Steel Edge

3.0mm收边铝单板
3.0mm Finished Aluminum Veneer

H400*200*8*13镀锌H型钢
Galvanized H-beam

钢导轨
Steel Rail

口250*120*15热镀锌钢管
口250*120*15 Hot Dip Galvanized Steel Pipe

滑移排烟窗
Slip Vent Outlet

3.0mm铝单板
3.0mm Aluminum Veneer

收边铝单板
Selvage Aluminum Veneer

3.0mm收边铝单板
3.0mm Finished Aluminum Veneer

口100*100*6热镀锌钢管
口100*100*6 Hot Dip Galvanized Steel Pipe

2×50保温岩棉（保温层）
2×50 Insulation Rock Wool (Insulation Layer)

Scale: 1 : 150

图 4-17   滑动屋盖节点图
Figure 4-17   Joint Detail of the Sliding Roof

登录大厅

登录大厅地上共 3 层，橄榄形平面，东西向长约 360m，南北向宽约 100m，屋顶为绿化上人屋顶（图 4-18）。南北登录大厅的首要功能是为了满足快速疏导大规模观展人流的需要，大厅一层东侧设置了长约 126m、宽约 63m 的登录大厅，大厅上部为结构标高 22.3m 的屋面层，屋面层采用跨度 45m 的平面桁架，营造了宽敞大气的空间感受。

登录大厅屋盖采用单层网壳 + 分叉柱的结构形式，单层网壳最高点距地面约 38m，网壳自然地向登录大厅前广场舒展，将这种大气磅礴的建筑结构风格延续升华。单层网壳由边长约 6.5m 的菱形网格拼接而成，各杆件采用高 900mm、宽 250mm、壁厚 18mm 的方通。支撑此屋面的是最大柱间距达到 45m 的分叉柱，在柱高 18.5m 处分为四枝，各分枝成放射状均匀分布（图 4-19）。

The Arrival Hall

The arrival hall has three floors above ground, with an olive-shaped plan, its east-west length is 360m and south-north width is 100m approximately. The roof is walkable green roof (Figure 4-18). Entering and checking are the primary function of the south and north arrival halls. In order to evacuate the crowd rapidly, a arrival hall about 63m wide and 126m long is set on the east side of the ground floor. The upper part of the hall is a roof layer with a structural elevation height of 22.3m. The roof layer also adopts a plane truss with 45m span to create a spacious atmosphere.

图 4-18 单层网壳整体提升照片
Figure 4-18 The Integral Lifting of the Single Layered Latticed Shell

图 4-19 结构施工照片
Figure 4-19 Construction Process

The roof of the arrival hall adopts the structural form of single-layer reticulated shell + bifurcated column. The highest point of the single-layer reticulated shell is 38m above the ground, and the reticulated shell naturally extends to the square in front of the arrival hall, extending the magnificent architectural structure style. The single-layered reticulated shell is composed of a rhomboid mesh with a side length of about 6.5m, and each member bar adopts a square tube with a height of 900mm, a width of 250mm and a wall thickness of 18mm. The bifurcated column with the maximum column spacing up to 45m are used to support the roof. At the 18.6m Point of this column, it divided into four branches, which are evenly and radially distributed. (Figure 4-19)

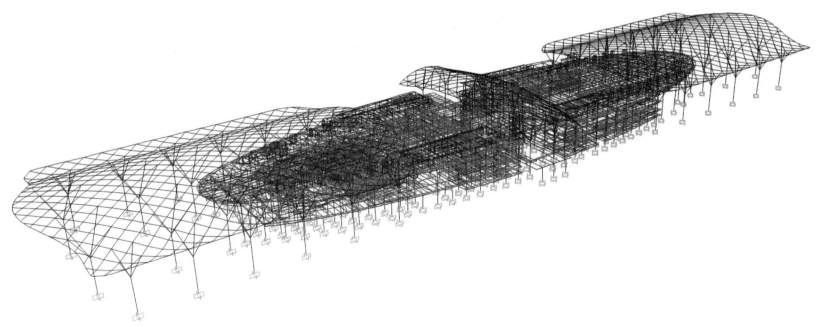

图 4-20 北登陆大厅结构三维图
Figure 4-20 North Arrival Hall Structure Modeling

图 4-21　登录大厅实景图
Figure 4-21　Arrival Hall Reality Image

图 4-22　北登录大厅一层平面
Figure 4-22　The North Arrival Hall First Floor Plan

中央廊道

中央廊道屋盖造型流线起伏，一条长 1750m 的中央廊道汇成"巨龙"之势，承担整个会展中交通枢纽的重要角色。

中央廊道屋盖结构采用钢结构空间网壳结构体系，网壳结构兼有杆件结构和薄壳结构的主要特性，受力合理，可以实现较大的跨度。网壳结构是典型的空间结构，合理的曲面可以使结构力流均匀，结构具有较大的刚度，结构变形小，稳定性高，且节省材料。网壳结构与建筑造型完美的融合，既能表现静态美，又能通过平面和立面的切割以及网格、支撑与杆件的变化表现动态美。

下部结构为正常的钢梁柱结构，结合建筑功能，利用结构分缝及滑动支座的特性，使得中央廊道与各大展厅及登陆大厅进行有效的连接，从而达到枢纽的作用（图 4-23）。

Central Concourse

The roof of the central Concourse has an undulant shape with a total length of 1750m embodies a momentum of dragon, playing an important role as a transportation hub for the whole building.

The roof structure of the central concourse adopts the steel space reticulated shell structure system, which is the combination of both the main characteristics of the truss structure and the thin shell structure, which makes, large span achievable. Reticulated shell structure is a typical spatial structure, the reasonable surface can guarantee an even force flow, and the structure is more rigid with less deformation and higher stability, and could saves materials at the same time. The perfect integration of reticulated shell structure and its architectural form not only shows a static beauty, but also show the dynamic beauty through the building sections and facade as well as the changes of grid, support and pole.

In addition, the substructure is the normal steel beam-column structure. Combined with the architectural functions, the central concourse can effectively connect with the exhibition halls and arrival halls by taking advantage of the characteristics of its structural split and sliding support, thus to achieve its function as transportation hub.(Figure 4-22)

图 4-23 中央廊道实景图
Figure 4-23 Central Concourse Reality Image

地下室

深圳国际会展中心拥有地下二层，局部一层。凤塘大道从深圳国际会展中心中部穿过，将地下室分隔为南北两部分，南侧地下室与北侧地下室通过设于凤塘大道下方的下穿通道相连，下穿通道于南、北地下室交界处设伸缩缝断开。北侧地下室东西向最大宽度 540m，南北向最大长度 415m；南侧地下室东西向最大宽度 516m，南北向最大长度 1250m。南北向不分缝最大长度为 1250m，为目前世界上最长的地下室。

Basement

The basement of the Shenzhen World includes the second-floor underground and the first floor underground in part. Fengtang avenue runs through the middle of the Shenzhen World, dividing the basement into north and south parts. The south and the north are connected by the underpass under Fengtang avenue, which is disconnected by the expansion joint set in the junction of the south and north basements. In the north basement, the maximum east-west width is 540 meters while the south-north length is 415 meters. In the south basement, the east-west width is 516m, and the south-north length is 1250m, which is so far the longest basement in the world.

地下室顶板设计构造措施

各展厅短边与地下室相邻的第一跨皆设置了结构沟以减小面内刚度，每隔约 150m 设置一条后浇带，每 300m 左右留置一条后浇带待工程主体完工后再浇筑。

地下室顶板按施工活荷载 50kN/m² 进行设计。

南登录大厅横贯于中廊中部，其地下室顶板与中廊地下室顶板标高有 1.2m 的高差。

The Structural Measures for the Basement Roof Design

In order to reduce the structural thickness in plane, a structural trench is set in the first span of the basement adjacent to the short side of each exhibition hall.

A post-cast belt is set every 150 meters, and also leave one to cast every 300 meter after the major construction is done.

Basement roof is designed according to the construction load of 50kN/m²

The south arrival hall runs through the middle of the central concourse, and the height difference between the basement roof of the arrival hall and the basement roof of the central concourse is 1.2 meters.

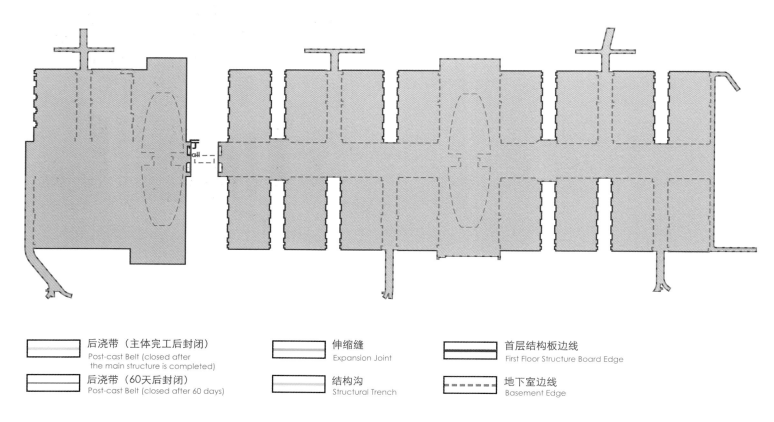

后浇带（主体完工后封闭） Post-cast Belt (closed after the main structure is completed)	伸缩缝 Expansion Joint	首层结构板边线 First Floor Structure Board Edge
后浇带（60天后封闭） Post-cast Belt (closed after 60 days)	结构沟 Structural Trench	地下室边线 Basement Edge

图 4-24　地下室顶板构造措施布置分析图
Figure 4-24　Structural Measures for Basement Roof

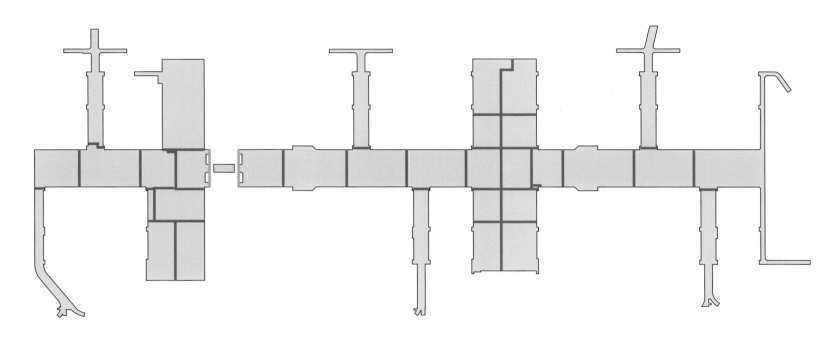

图 4-25　底板结构沟 + 后浇带布置图
Figure 4-25　Bottom Slab Structural Trench + Post Cast Belt

图例：
结构沟+后浇带
Structural Trench + Post-cast Belt

伸缩缝
Expansion Joint

地下室结构

为减少结构的水平刚度并降低竖向构件对楼盖的约束，竖向受力构件采用了钢筋混凝土及钢管混凝土柱。

地下室楼盖尽可能的采用无梁楼盖，而对于部分楼板、电梯洞口、坡道、设备管井等无法布置无梁楼盖的区域，则将采用梁板式楼盖，地下室的结构形式则为带柱帽的板柱与梁板式框架结构。

地下室底板最大长度 1250m，为削弱底板的刚度，每隔150m 左右设置一道 2.5m 宽 ×1.5m 深的结构沟，结构沟结合后浇带设置，结构沟侧壁厚度取 300mm，以充分释放面内刚度。

Basement structure

In order to reduce the horizontal rigidity of the structure and reduce the constraint of the vertical component to the floor slab, the vertical load component is made of concrete and concrete-filled steel tube columns. The basement tried to adopt the flat floor slab as many as possible. As for floor, elevator entrance, ramp, equipment pipe well and other areas where the flat floor slab cannot be arranged, the slab and beam structure is adopted. The structure of the basement is the column with column cap and the beam-plate structure.

The maximum length of the basement floor is 1250 meters. In order to weaken the rigidity of basement floor, a 2.5m wide × 1.5m deep structural trench is set every 150 meters. Along with the post-cast trench, which in. plane structural thickness is 300mm, which can fully release the rigidity.

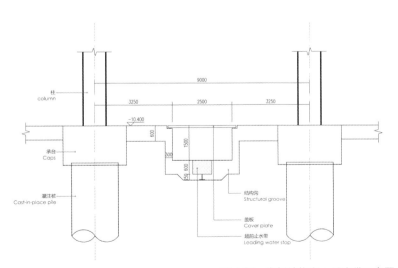

图 4-26　底板结构沟 + 后浇带示意图
Figure 4-26　Bottom Slab Structure Trench + Post. Cast Belt Detail Drawing

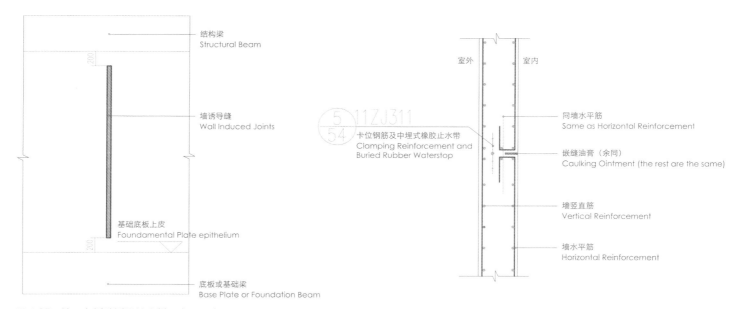

图 4-27　地下室侧壁诱导缝大样及立面示意
Figure 4-27　Induced Joint of the Basement Side Wall and Its Elevation

地下室外墙设计构造措施

根据本工程的具体情况并参考相关工程经验，在地下室侧壁设置了诱导缝，诱导缝沿外墙按间距约每 30m 布置一道，地下室外墙在各楼层后浇带相对应的位置也留置后浇带。

The Structural Measures for the Basement's Exterior Wall Design

According to the specific situation of this project and reference to relevant engineering experience, an inducing joint was set in the side wall of the basement, and arranged every 30 meters along the exterior wall. The external wall of the basement also keeps the post-cast belt in the corresponding position under the post-cast belt of each floor.

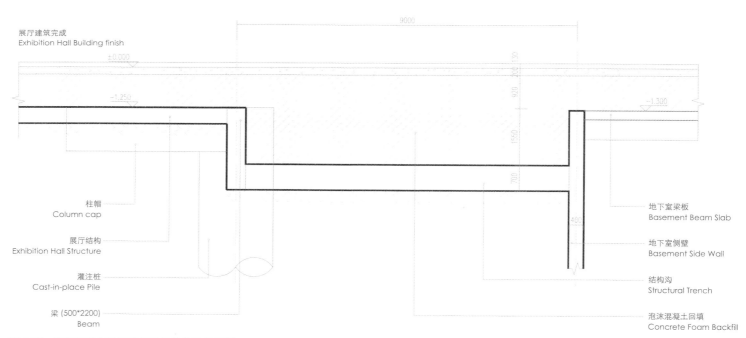

图 4-28　展厅短边与地下室相连处结构沟示意图
Figure 4-28　Schematic Diagram of Structural Ditch at the Connection Between the Short Side of Exhibition Hall and Basement

施工方式

按照跳仓法施工顺序、并考虑后浇带等因素，将进行全过程施工模拟分析。假定地下室底板层、地下一层、顶板层施工周期各为一个月，各楼层跳仓施工间隔为 10 天，后浇带封闭时间均按两侧楼层施工后的 60 天。

The construction Method

The simulation analysis of the whole coustruction process is carried out according to the sequence method, and taking factors such as post-cast belt into consideration. It is assumed that it would take a month each for the construction of basement floor, basement and its roof, 10 days of the sequence method for each floor, and 60 days of closure time for post-cast belt after constructions of floors on both sides.

图 4-29　施工现场实景图
Figure 4-29　Construction Site

图 4-30　施工现场实景图
Figure 4-30　Construction Site

图 4-31　施工现场实景图
Figure 4-31　Construction Site

图 4-32　施工现场实景图
Figure 4-32　Construction Site

风工程研究
Wind Engineering Research

刚体测压风洞试验模型

深圳国际会展中心分别于华南理工大学及同济大学进行了风洞试验与风振分析（图 4-33）（图 4-34），主要工作如下：

1、刚体模型群体建筑测压风洞试验。

2、结构风振响应分析。

考虑到深圳国际会展中心项目地处深圳市临海新区，故采用 A 类地貌风场。

风洞试验基本情况：

- 几何缩尺比 1/150
- 36 个风向角
- 试验转盘直径 6.8m

根据风洞试验结果在负风压较大的区域加密屋面檩条、支座、加强屋面抗风揭能力。本工程因第五立面屋面造型的需要在金属屋面之上增加了一层外装饰层，其对风致响应的影响及极值风压的影响还需进一步研究（图 4-35，图 4-36）。

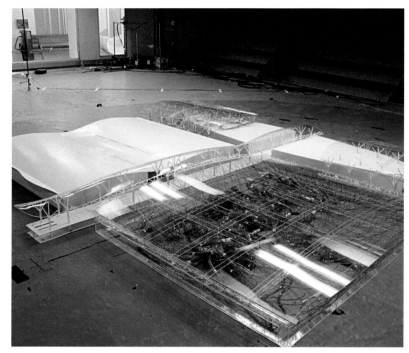

图 4-33　风洞测压模型照片 1
Figure 4-33　Wind Tunnel Pressure Model

Wind Tunnel Test Model for Pressure Measurement of Rigid Body

The wind tunnel test and wind vibration analysis of the Shenzhen World were conducted separately in South China University of Technology and Tongji University, the main work is listed as follows:

1. Pressure measuring wind tunnel test of rigid body group building model.

2. Wind vibration response analysis of the structure.

Considering that Shenzhen World is located in new area near the sea, a geomorphic wind field is adopted.

Basic information of wind tunnel test:

- geometric scale ratio 1/150
- 36 wind directions
- the diameter of the test turntable is 6.8m

According to the wind tunnel test results, the roof purlins and supports are added in the area with high negative wind pressure to strengthen the roof's ability of wind resistance. An external decorative layer was added to the metal roof due to the design need of the fifth facade roof, which needs to be further experimented on to test its impacts on the wind-induced responses and extreme wind pressure.(Figure 4-35, 4-36)

图 4-34　风洞测压模型照片 2
Figure 4-34　Wind Tunnel Pressure Model

风洞试验结果
Results of Wind Tunnel Test

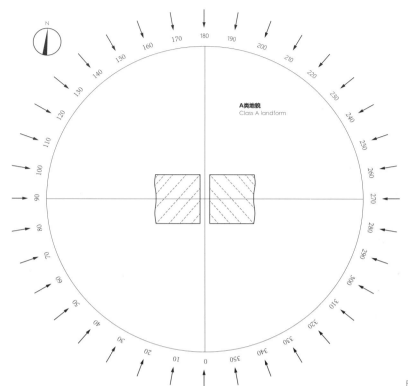

图 4-35　风压试验图
Figure 4-35　Wind Pressure Test Diagram

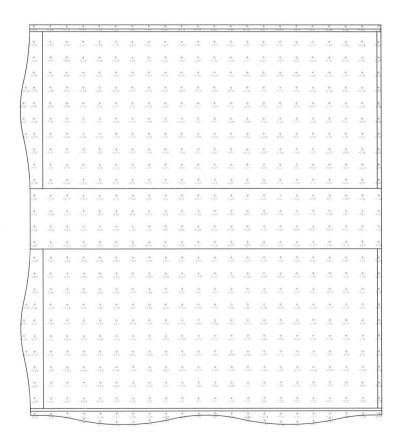

图 4-36　风洞实验结果
Figure 4-36　Wind Tunnel Test Results

结构抗震与节点
Seismic and Joints of The Structure

大震弹塑性时程分析
Elastic-plastic time-history analysis of severe earthquake

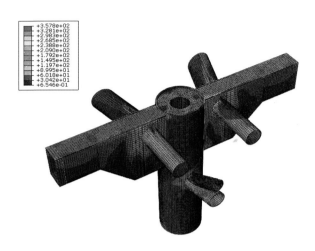

图 4-37 桁架与柱连接节点有限元分析模型
Figure 4-37 Finite Element Analysis Model of Truss Column Joint

图 4-38 管节点有限元分析模型
Figure 4-38 Finite Element Analysis Model of Tubular Joints

图 4-39，图 4-40，图 4-41 施工现场实景图
Figure 4-39，4-40，4-41 Construction Site

	X 向 Direction X		Y 向 Direction Y	
	层间位移角 Interlaminar displacement angle	顶点位移（mm） Vertex displacement (mm)	层间位移角 Interlaminar displacement angle	顶点位移（mm） Vertex displacement (mm)
Ccel波 Ccel wave	1/171	84.9	1/113	126
Ccel波（小震） Ccel wave(earthquake)	1/1015	14.8	1/614	22.9
LveL波 LveL wave	1/143	101.8	1/135	106
人工波 Artificial wave	1/153	95.5	1/114	123

表 4-3 结构抗震与节点分析表
Table 4-3 Aseismic Structure and Node Analysis

钢结构施工情况
Construction of Steel Structure

图 4-42，图 4-43，图 4-44，图 4-45，图 4-46，图 4-47　施工现场实景图
Figure 4-42，4-43，4-44，4-45，4-46，4-47　Construction Site

	X 向 Direction X		Y 向 Direction Y	
	剪力（KN） Shear force（KN）	剪力比 Shear ratio	剪力（KN） Shear force（KN）	剪力比 Shear ratio
Ccel波 Ccel wave	114325	20.7%	107141	19.4%
Ccel波（小震） Ccel wave(earthquake)	18752	3.4%	20141	22.3.6%
LveL波 LveL wave	145370	26.3%	91173	16.5%
人工波 Artificial wave	120118	21.7%	126838	23%

表 4-4　大震弹塑性时程分析表
Table 4-4　Massive Earthquake Elasto Plasticity Timing Analysis

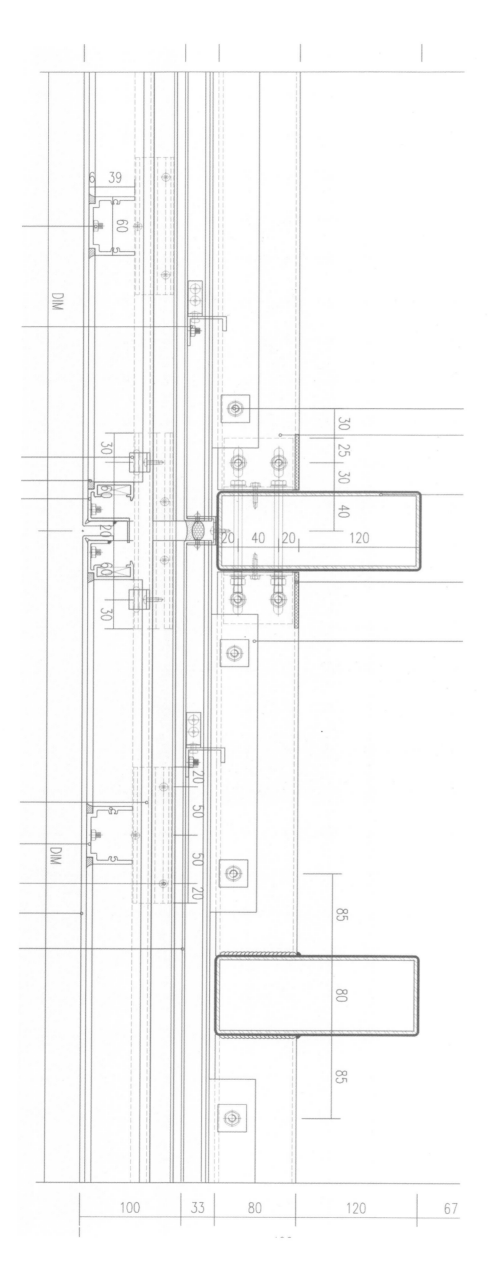

设备系统设计
Equipment System Design

深圳国际会展中心建成后作为全球第一大展馆，绿色会展和智能会展是必然趋势，也是建设目标。机电设计在满足运营功能使用的同时，应用多项绿色建筑技术，最大限度地节约资源（节能、节水、节材）、为场馆方提供最佳的体验并提升运营效率；打造绿色智能化会展。

When the Shenzhen World is completed, it will be the largest exhibition center in the world. As the green and intelligent exhibition is an inevitable trend in the near future, it also became the construction goal of the Shenzhen World.
Therefore, not only the mechanical and electrical design should meet the functional requirements of operation, application of green building techniques is also necessary, so as to save resources (energy, water, materials) provide the best experience, improve the efficiency of daily operation, and furthermore achieve the goal of green exhibition center to the maximum extend.

空调体系
Air-conditioning System

中央冷源的规划要点

本项目主体建筑总建筑面积大，跨度长，对中央冷源的规划带来了较大的难度，通过对本项目的分析并多方论证，最终确定设置 5 个冷站，主要考虑了以下 3 方面因素：

① 根据办展规模的不同，展厅使用的数量及组合存在的较大不确定性，冷站规划将同时使用的场所合并为一个系统。

② 根据展厅的布置，1 至 16 号展厅形成对称布置，因此冷源的选址应位于中心区，以形成天然的平衡环路，降低调试难度，提高运行稳定性。

③ 南登录大厅和北登录大厅地下室均设有大容量消防水池（蓄水容积分别约 7300 立方米和 7800 立方米），为蓄冷空调的使用提供了便利的条件。

根据以上设计原则，共设置了 3 个常规冷站，每个冷站服务半径均为 4 个标准展厅，且位于服务区中心；另设 2 个蓄冷冷站，除服务冷站所在区的登录大厅外，还将附近的展厅也一并纳入了同一中央空调系统（图 4-48）。

Key Points of the Central Cold Source Planning

With a large area and a long span, the planning of the central cold source for this project is quite difficult. Through the analysis and multiple argumentations, 5 cold stations are determined to be set based on the following three factors:
① Due to the different scales of different exhibitions, the number and the combination of exhibition halls are uncertain. Therefore, the cold station planning should be synthesized into one system for all occasions that could happen at the same time.
② According to the layout, the exhibition hall No. 1~16 will be arranged symmetrically. Therefore, the location of the cold source should be in the central area to form a natural balance loop, which can reduce the difficulty of debugging and improve the stability of operation.
③ The basement of the south and the north arrival halls are equipped with large-capacity fire pool (impoundment volumes are 7300 cubic metres and 7800 cubic metres respectively), which provide proper conditions for the cool storage air-conditioning.

According to the design principles above, a total of 3 conventional cold stations are set, each serves 4 standard exhibition halls and is located in the center of the service area. Two other cold storage stations are also set up.
Except for the location of the cold station of the arrival hall, for the nearby exhibition halls is also incorporated into the same central air conditioning system.(Figure 4-48)

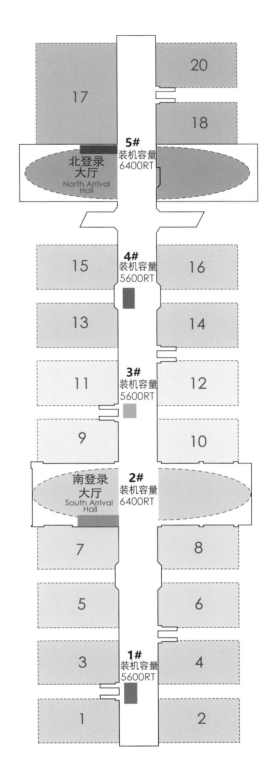

水蓄冷式中央空调冷源
Water-storage central Air-conditioning Cold Source

常规水冷式中央空调冷源
Conventional Water-cooled Central Air-conditioning Cold Source

常规水冷式中央空调冷源
Conventional Water-cooled Central Air-conditioning Cold Source

水蓄冷式中央空调冷源
Water-storage Cooled Central Air-conditioning Cold Source

常规水冷式中央空调冷源
Conventional Water-cooled Central Air-conditioning Cold Source

图 4-48　中央空调冷源 - 冷冻站示意图
Figure 4-48　Schematic Diagram of the Central Air Conditioning Cold Source Refrigeration Station

The Key Points of Air Distribution Design in the Exhibition Hall

The air distribution in the large exhibition hall is complicated. The traditional empirical formula that used to calculate the indoor air distribution is prone to a large deviation, thus the indoor air distribution may not reach the expected level and Lower the comfort level of the indoor environment. Therefore, the height of the side inlet in the exhibition hall of this project was determined by CFD flow organization simulation.(Figure 4-49)

In order to determine the optimal air supply height and verify the indoor comfort under different working conditions, this air distribution simulation studied various relative factors such as different display height, different indoor temperature requirements, thermal resistance of human clothing, different air supply volume and air supply temperature (Figure 4-49). The indoor comfort evaluation indexes were PMV and air age. After multiple tests, the optimal air supply height for the cower and upper inlet was standard exhibition hall (about 180m in length and 97m in width, with air supply outlets arranged symmetrically along the long sides of both sides) finally determined as 13.5m and 10.5m.

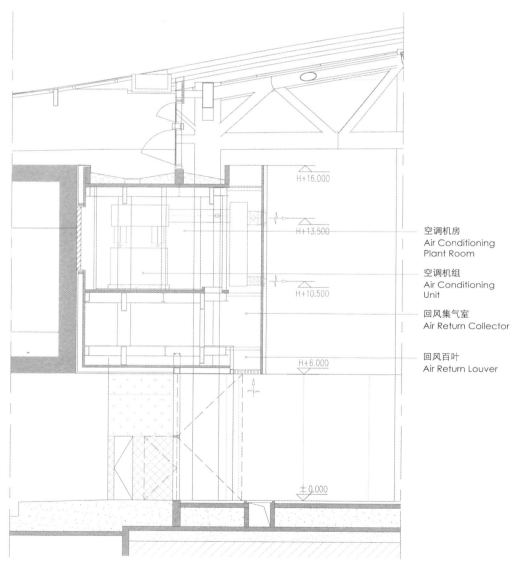

H+16.000

H+13.500 — 空调机房
Air Conditioning
Plant Room

— 空调机组
Air Conditioning
Unit

H+10.500

— 回风集气室
Air Return Collector

H+6.000 — 回风百叶
Air Return Louver

±0.000

图 4-49 空调机房剖面图
Figure 4-49 Air Conditioner Room Section

高大展厅气流组织设计要点

高大展厅的气流组织复杂，若采用传统的经验公式计算室内气流组织容易出现较大偏差导致室内气流组织达不到预期，影响室内环境的舒适度。因此，本项目展厅侧送喷口的高度将通过 CFD 气流组织模拟进行确定（图 4-49）。

为了确定最佳送风高度并验证在不同工况下的室内舒适度，本次气流组织模拟研究了不同展板高度、不同室内温度要求、人体衣物热阻、不同送风量及送风温度等影响因素（表 4-5）。室内舒适度的评价指标选用 PMV 和空气龄。经过反复验证，标准展厅（长度约 180m，宽度约 97m，送风口沿着两侧长边对称布置）最终确定上下两排喷口的最佳送风高度为 13.5m 和 10.5m。

环境温度 Ambient temperature	人体衣物热阻 Body clothing thermal resistance	APMC分布云图1级 APMC Cloud Map Level 1	室内人行活动高度风速云图 Wind speed cloud diagram of indoor pedestrian activity height	AGE空气龄云图 AGE air age cloud map	面积比例 Area ratio
标准 展厅 Standard showroom					
17℃	CLO 1.5				92%
18℃	CLO 1.5	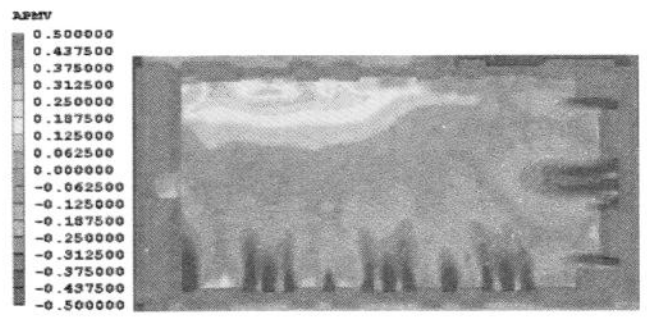	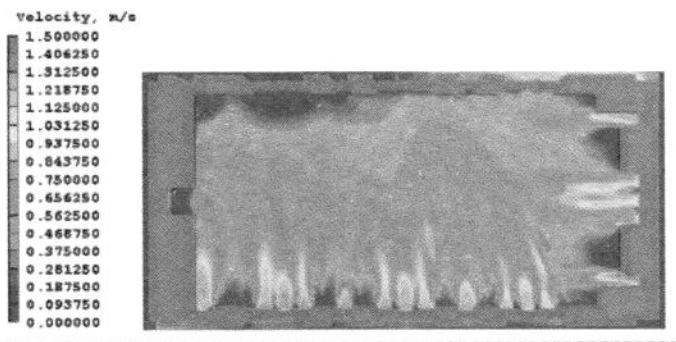	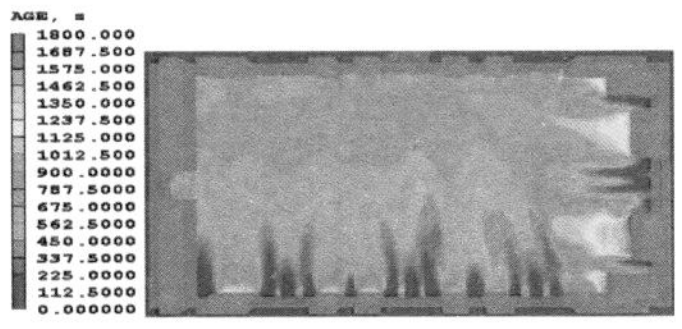	100%
19℃	CLO 1.5	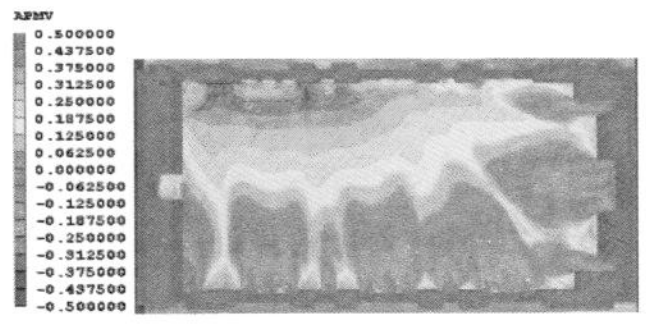	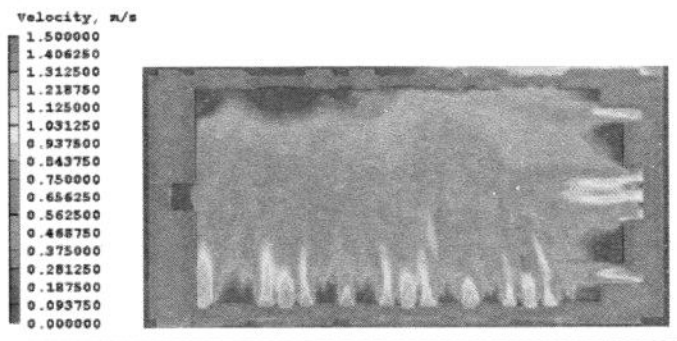	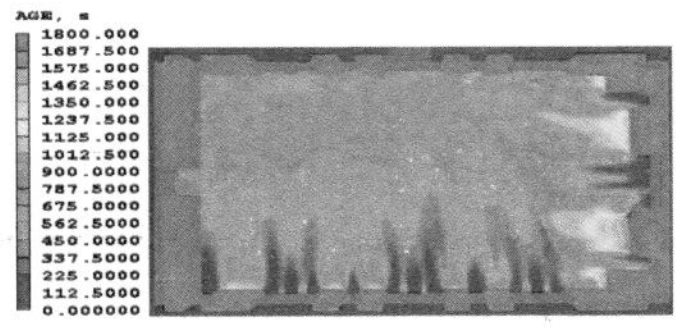	100%
20℃	CLO 1.5	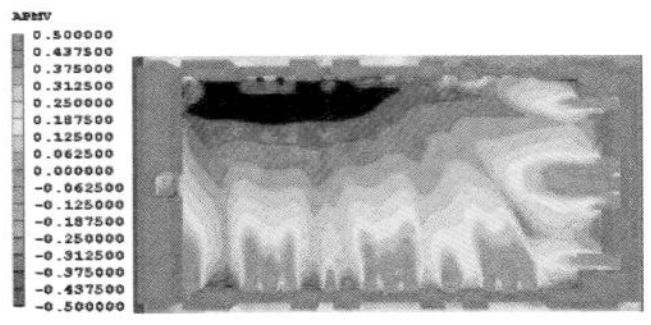		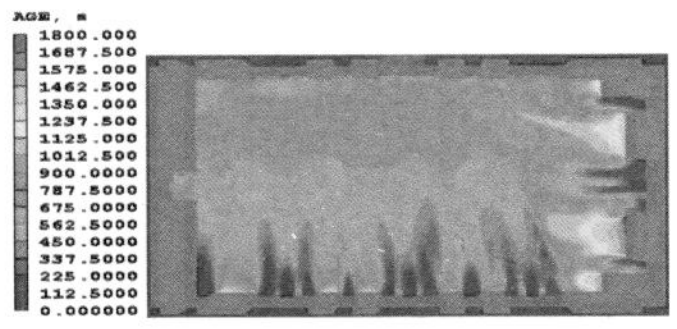	92%
21℃	CLO 1.5	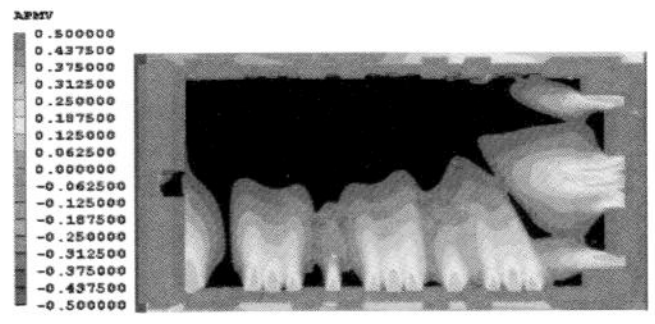	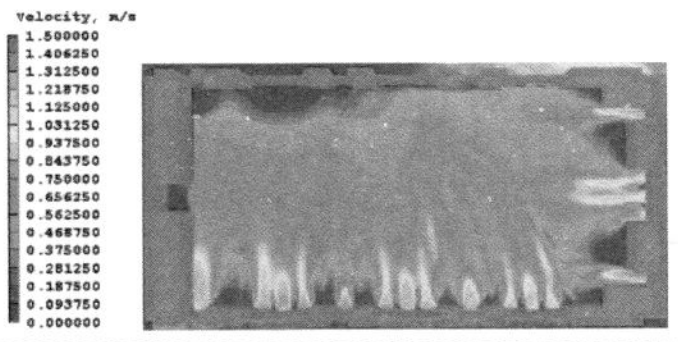	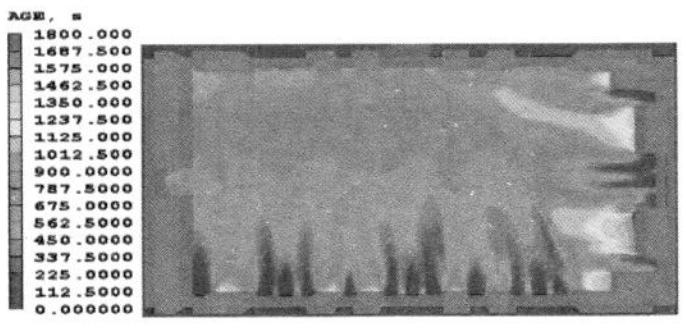	64%
25℃	CLO 1.5	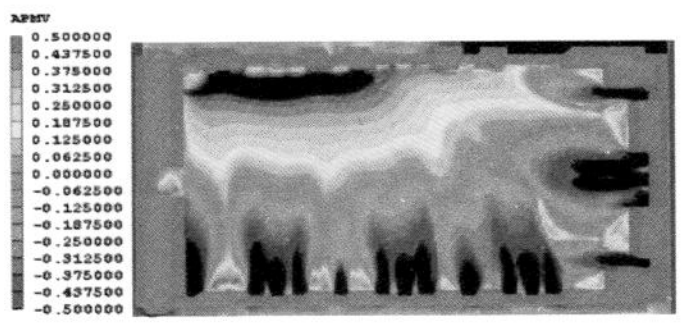	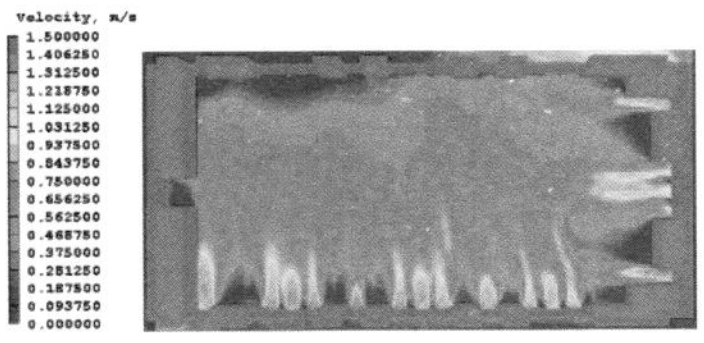	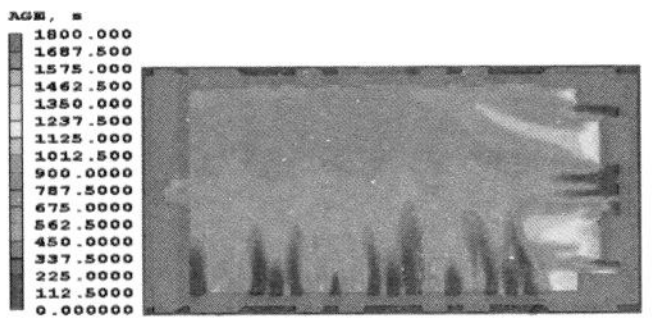	84%
北登陆 大厅 North Arrival Hall					
17℃	CLO 1.5				52%
18℃	CLO 1.5	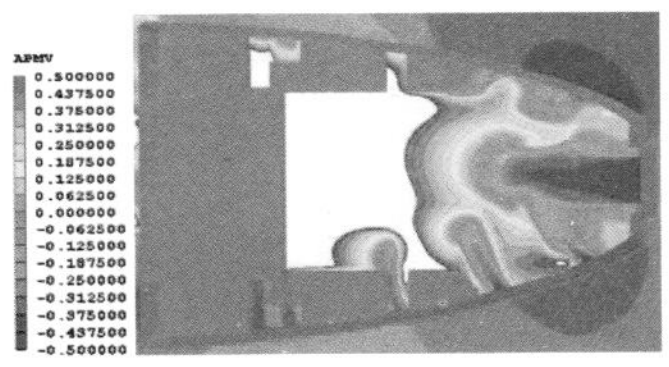		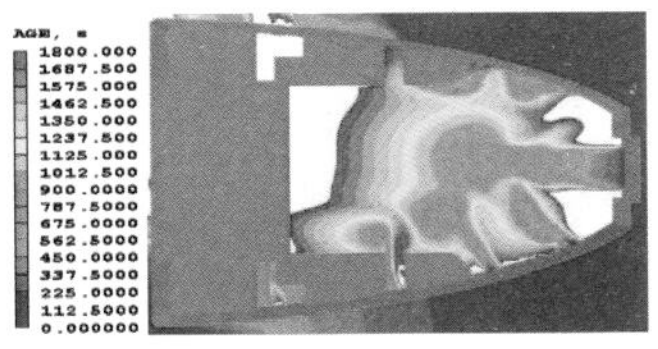	49%
220℃	CLO 1.5	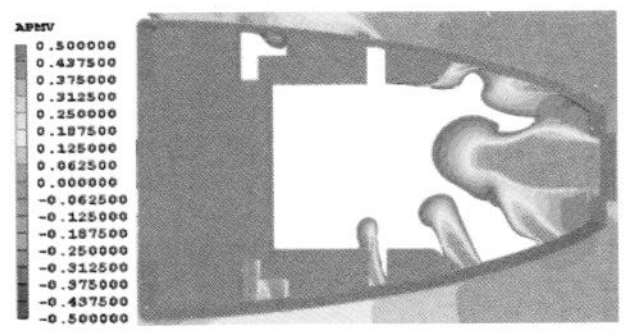	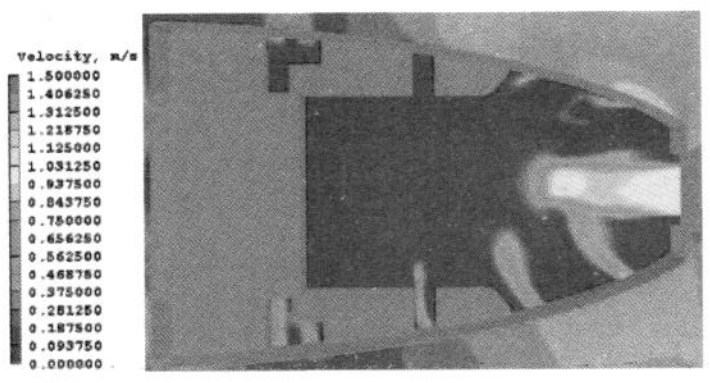	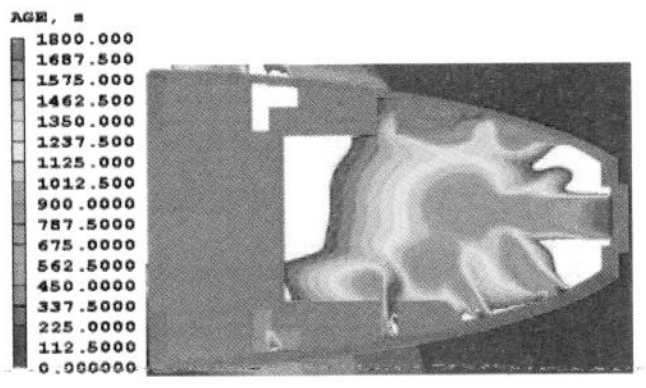	32%
25℃	CLO 1.5			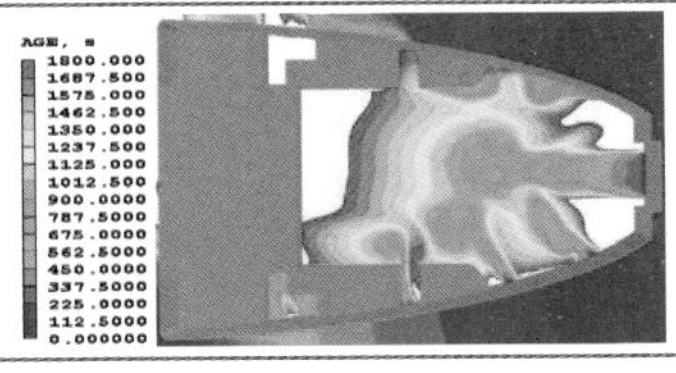	30%

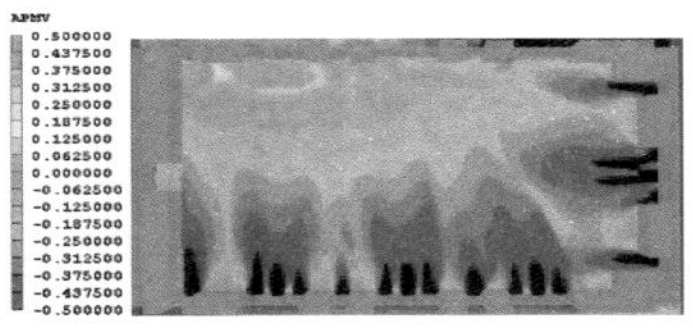

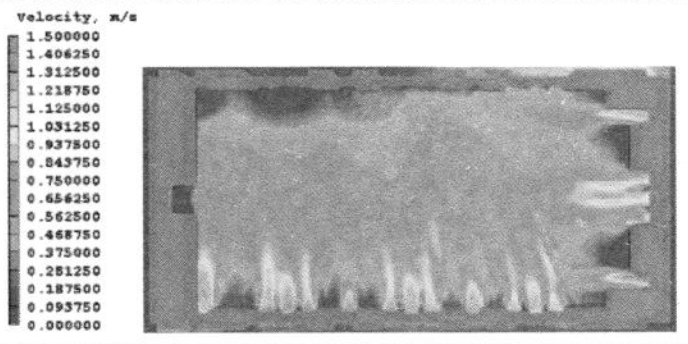

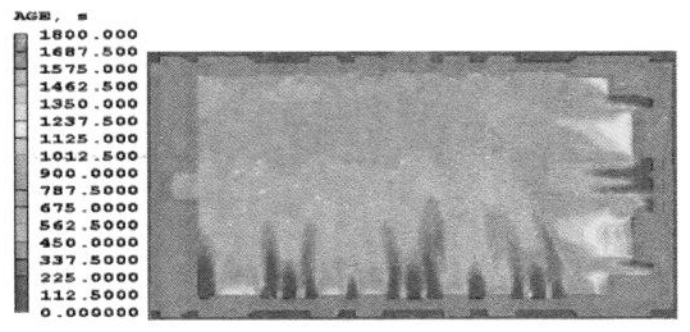

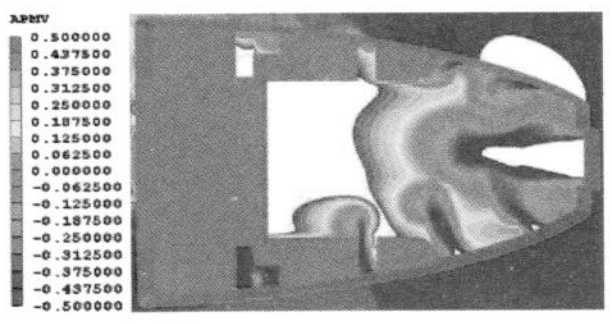

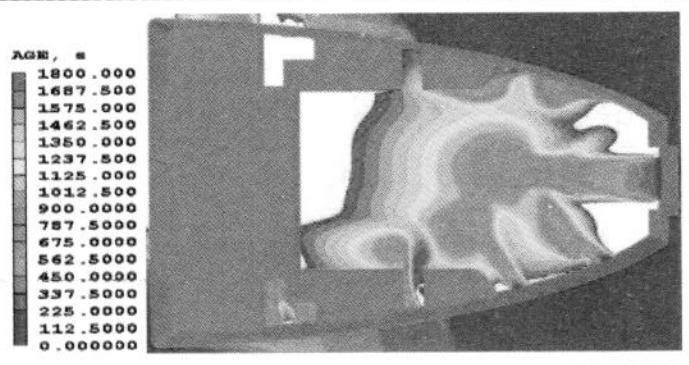

表 4-5 风速分析表
Table 4-5 Wind Speed Analysis

中央空调冷源

根据深圳的气候特点，深圳国际会展中心中央空调系统仅供冷源，且不考虑冬季供热需求（除有特殊要求场合外）。为节省冷冻水系统的输送能耗，采用分区供冷的系统模式，从南到北设置 5 个冷站，以达到规划设计阶段的节能要求。

Central Air Conditioning Cold Source

According to the climate characteristics of Shenzhen, the central air conditioning system of the Shenzhen World would only provide cold sourse without considering the heating demand in winter (except for special occasions).

The north-south span of the Shenzhen World is about 1,750m, and the east-west span is about 470m. In order to save the energy consumption of the cold water system, it adopted the divisional cold supply mode. Five cold stations are set from south to north to meet the energy-saving requirements during the planning and design stage.

中央空调水系统

中央空调冷冻水系统：为降低空调冷冻水泵的输送能耗，各冷站空调水系统均采用供回水温差 9℃的大温差供冷系统。因 1 号，3 号和 4 号冷站位于空调负荷区的中心，且服务的展厅空调负荷特性相似、空调末端均为组合式空调机组并在回水管上设置比例积分调节阀，因此对应的中央空调系统仅在各分支管路设置蝶阀进行水力平衡条件。2 号和 5 号冷站中央空调水系统的系统相对复杂，水力不平衡度较大，因此采取了两级平衡措施：分别在集水器的分支回水管和各楼层设置静态平衡阀，以改善系统的水力平衡。

考虑到办展展厅使用的多样性，以及冷水主机存在故障的可能性，1~3 号冷站将通过连通管进行连通，既可互为备用，又可充分利用 2 号冷站的蓄冷特性为常规冷站供冷，以提升蓄冷站的效益。4~5 号冷站也将采取相同的连通措施。

中央空调冷却水系统：供冷工况下的设计回水温度为 32/37℃；夜间蓄冷工况下的设计供回水温度为 30/35℃。

Central Air Conditioning Water System

Central air conditioning chilled water system: in order to reduce the energy consumption of air conditioning chilled water pumps, all cold stations' air conditioning water system adopt a large temperature difference cooling system with a temperature difference of 9 C between supply and return water. Since No. 1, No. 3 and No. 4 cooling stations are located in the center of the air conditioning loading zone, and the loading characteristics of the air conditioning in the exhibition halls are similar, the air conditioning terminal is a modular air conditioning unit, and a proportional integral control valve is set on the return pipe, the corresponding central air conditioning system only sets butterfly bamper in each branch pipeline for hydraulic balance conditions.

The central air conditioning water system of No. 2 and No. 5 cooling stations is relatively complicated, and lack of hydraulic balance. Therefore, there adopted two balancing measures: several static balancing valves are set in the branch return pipe of the water collector and each floor to improve the balance of the system.

Considering the diverse use of the exhibition halls and the possible failure and breakdown of the main cold water engine, No. 1~3 cold stations are connected by pipes, which not only allow them to back up each another, but also make full use of the cold storage characteristics of No. 2 cold stations to provide cooling for conventional cold stations, so as to improve the benefit of cold storage stations. No. 4~5 cold stations also adopt the same connecting measures.

Cooling water system of central air conditioning: the return water temperature under the cooling operation is 32/37 C, and the same nieasure at night is 30/35 C.

空调通风系统

展厅、登录大厅等高大空间均采用一次回风全空气系统，展厅采用侧送侧下回的气流组织形式，登录大厅采用顶送侧下回的气流组织形式。

Air Conditioning Ventilation System

The exhibition hall, landing hall and other tall spaces all adopt the all-air primary return air system, the exhibition hall adopts an airflow organization form of side supply and side bottom return, and the arrival hall adopts an airflow organization form of top supply and side return.

特殊场合空调系统

中央空调系统存在使用时段集中、低负荷下系统能效低、高温高湿天气下除湿能力弱等缺点，本项目针对特殊场合的自身特点及运营需求，将会采用适用性更强的空调系统（图 4-50）。

① 常年制冷的场所：弱电机房、变配电房等存在常年发热设备的功能房间。其中设置了独立 VRV 多联式空调系统用于辅助制冷，在通风系统无法满足室内的排热需求时打开降温。

② 灵活使用的场所：展厅配套的餐饮、会议室、门厅、走道及卫生间等功能房间和南、北登录大厅 VIP 贵宾室、包间及其配套的疏散通道、安保室、消防控制室等功能房间在非展览期间也需要使用，因此，为了满足非展览期间的日常使用需求，此类功能房间都将设置 VRV+ 新风系统（新风机组自带室外机）。

③ 北登录大厅的国际报告厅、媒体发布中心和南登录大厅的大型会议室：国际报告厅的空调系统需要设置除湿能力强的空调系统以满足全年工况下的室内舒适度需求。设计选用热泵热回收型溶液除湿空调，既达到了深度除湿能力以应对极端天气的除湿需求，又可以对室内排风的冷量进行回收达到节能目的。同时，可结合大温差供水系统实现温湿度独立控制，溶液除潜热控制室内相对湿度，中央空调大温差水系统除湿热控制室内温度，进一步提升了室内温湿度控制精度。

④ 登录大厅的小会议室：设置风机盘管 + 新风系统的温湿度独立控制空调系统。其中，热泵热回收型溶液除湿空调除潜热控制，风机盘管除湿热控制室内温度。

Special Occasion Air Conditioning System

The central air conditioning system has several disadvantages, such as its use period is concentrated, it has low energy efficiency under low usage load, and its dehumidifying ability is weak under high temperature and high humidity, etc. This project adopts the air conditioning system with stronger applicability according to the site's characteristics and operational requirements in special occasions.(Figure 4-50)

① Spots need cooling all year round: weak electricity engine room, transformer room and other functional rooms with perennial heating equipments are equipped with independent VRV multi-connected air conditioning system to assist the cooling, which would be turned on when the ventilation system cannot meet the cooling equiment.

② Places with Flexible use of air conditions: system Spaces like dining areas, conference room, lobby, aisle, restroom, as well as VIP room, private room, evacuation exit, security room, fire control room also adopts individual VRV plus fresh air system, its own outdoor unit.

③ International lecture hall at the north arrival hall, media release center and large conference room at the south arrival hall: the international lecture hall needs to be equipped with an air-conditioning system with strong dehumidifying equipment to guarantee the indoor comfort under all year-round working conditions. The design selects heat pump heat recovery liquid desiccant air conditioning, which not only has the deep dehumidifying capacity to handle dehumidification in extreme weather, but also can recycle the indoor exhaust cooling air to achieve the purpose of energy saving. At the same time, the temperature and humidity can be independently controlled by combining use of the large temperature difference water supply system. The solution can remove the latent heat to control the indoor relative humidity, and the large temperature difference water system of central air conditioning can remove sensible heat to control the indoor temperature, further improving the control accuracy of indoor temperature and humidity.

④ Small meeting room at the arrival hall: set up the air conditioning system with independent temperature and humidity control: the fan coil + fresh air system. Among them, the heat pump heat recovery liquid desiccant air conditioning removes latent heat control, and the fan-coil removes sensible heat to control indoor temperature.

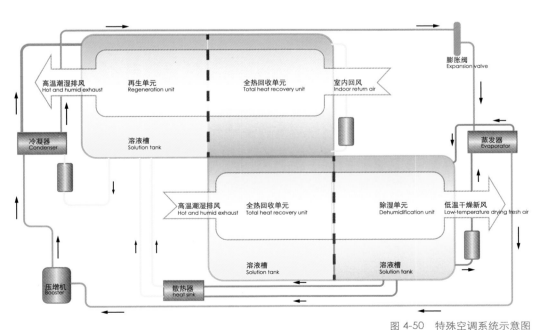

图 4-50　特殊空调系统示意图
Figure 4-50　Schematic Diagram of Special Air Conditioning System

电气体系　Electrical System

10/0.4KV 变配电系统

① 用电负荷等级

各种消防用电设备、消防控制中心、应急照明等消防用电为一级负荷；安防系统，通信基站、数据中心、电子信息设备机房、客梯、生活水泵等用电等为一级负荷，其中安防系统、数据中心、重要的计算机系统、应急响应系统的用电为一级负荷中特别重要负荷，展览用电、展厅照明、通风机、闸口机按二级负荷，其余负荷为三级负荷（图 4-51）。

② 主要用电指标

用电容量：深圳国际会展中心总用电安装容量为 182500kVA，其中高压机组总容量 20220 kVA，变压器总装机容量167700kVA；共采用 2000kVA 变压器 64 台，1600kVA 变压器 12 台，1250kVA 变压器 8 台，800kVA 变压器 8 台，1200kVA 高压机组 9 台，1350kVA 高压机组 6 台，总计变压器或高压机组共 107 台，变压器负荷率约 80%。

③ 变配电所设置

另 13~17 号变配电所设置于地下二层 1~5# 冷站处，18~19 号变配所设置于西侧配套用房处。

④ 供电电源

本工程共设置 7 个公共开关房，具体位置如上图。总共将从 2 个不同的 110kV 变电站中引出 21 路 10kV 电源进入地下一层各高压配电站，现阶段已投入 14 路；同时还设有 14 座柴油发电机作为消防负荷或重要负荷的备用电源。以此从高压、低压两层保障用电可靠性。在数据中心、通信基站、应急指挥中心、消防控制中心、安防控制中心、消防分控室、安防分控制室、各智能化控制室、网络机房、其他重要计算机系统、演艺厅的灯光音响控制室等的设备用电中，将再加 UPS 作为备用电源。

⑤ 低压供电系统接线型式

低压采用单母线分段接线方式，正常情况下各段母线分段运行，当其中一台变压器因故停运时，将切除部分非重要负荷，并手动投入母联开关，只对消防负荷及重要负荷进行供电；同时，变压器出线主开关将与母联开关实行电气联锁，防止变压器并列运行。

市电与柴油发电机电源间设机械和电气联锁切换装置，对各消防设备及重要负荷采用双回路末端切换。当两台变压器同时失电后，发电机将启动（图 4-52）。

⑥ 计量

本工程采用高压总计量，在 1~12 号变配电所 10kV 进线电源处分别设有高压专用计量柜。

各展位用电在展位配电箱出线回路上设有带液晶屏显示和通讯功能的断路器，以便运营内部核算参展商申请用电的电费成本。

⑦ 功率因数补偿

本工程在低压侧设集中电容补偿，由于展览建筑配有大量电子整流器、LED 灯、空调变频器和各种电子信息设备，因此采取抑制谐波电流和降低多次谐波（特别是三次谐波）是非常必要的。

⑧ 主要变配电设备选型

变压器选用新型节能环保的 SCB13 干式变压器（带 IP30 防护等级外壳）。

高压配电选用中置柜，高、低压电缆均采用下进下出，柜下设电缆沟。柴油发电机应为应急自启动型，而应急启动装置及相关成套设备则将由厂家成套供货。

10/0.4kV Transforming and Distributing System

① Electrical Load Level

All kinds of fire control equipments, fire control center, emergency lighting are categorized as the first order load; Security systems, communication base station, data center, electronic information equipment room, passenger elevator, water pump are first order load, among them, the security systems, data center, important computer system, and emergency response system are of extra significance; exhibition, exhibition hall lighting, ventilator, gate machine are the secondary order load, and the rest equipments are the third order load.(Figure 4-51)

② Main Electricity Consumption Indicators

Power capacity: the total installed capacity of the Shenzhen World is 182,500kVA, among which the total installed capacity of high-voltage units is 20,220kVA and the transformers is 167,700kVA. There are 64 2000kVA transformers, 12 1600kVA transformers, 8 1250kVA transformers, 8 800kVA transformers, 9 1200kVA high voltage units, and 6 1350kVA high voltage units. There is a total of 107 transformers or high voltage units , with the transformer load rate is about 80%.

③ Power Substation

In addition, No. 13-17 power substations are set at the No. 1~5 cold stations on the second floor underground, and No. 18-19 power substations are set at the supporting houses on the west side.

④ Power Supply

This project sets 7 public switch rooms in total. From two different 110kV transformer substations, there prepare to draw 21 lines of 10kV power sources to transfer the power to various high-voltage distribution stations on the underground first-floor, among which 14 lines were put into operation at the present stage. Moreover, there set 14 diesel generators as fire load or backup power supply for important load. These two levels of high and low voltage guarantee the reliability of electricity.

Additionally in the data center, communication base station, emergency command center, fire control center, security control center, fire control room, security control room, intelligent control room, network room, other important computer systems, lighting and audio control room of performance hall, etc., there would also equip with power plus UPS as backup power.

⑤ Wiring Type of Low-voltage Power Supply System

The low voltage adopted the single busbar with subsection connections. In normal conditions, each subsection of the busbar operates indipendently. When one of the transformers shut down, part of the non-important load would be cut off, then the busbar switch will be turn on manually, so as to supply power only to the fire load and important load. The outlet main switch and busbar switch of the transformer are interlocked together to prevent parallel operation of the transformer.

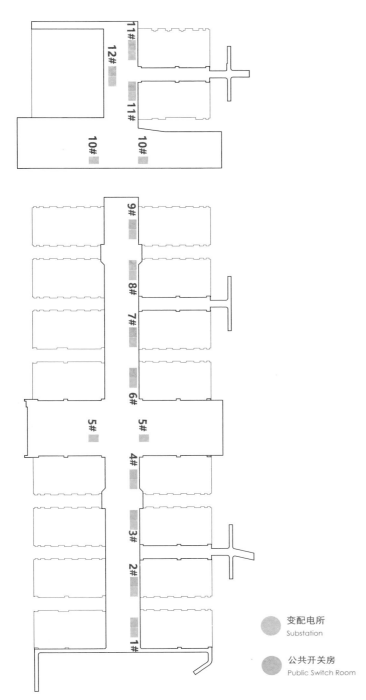

图 4-51 地下一层自备发电机示意图
Figure 4-51 Underground Floor One-Self-Sufficient Generator

变配电所
Substation

公共开关房
Public Switch Room

There are mechanical and electrical interlock switching devices between the mains power and the diesel generator power supply. The double-circuit terminal switching devices are used for every fire control equipment and important loads. When both transformers lose power, the generator will be switched on.(Figure 4-52)

⑥ Measurement
This project adopts the high voltage general measurement and sets special measurement cabinet for the high voltage at the 10kV inlet power source of No. 1-12 power substation. For each booth, the circuit interrupter with LCD display and communication function is set on the outlet loop of the booth distribution box, so that the operator can calculate the electricity cost of the exhibitor's application.

⑦ Power Factor Compensation
In this project, a centralized capacitance compensation is set at the low-voltage side. Because the exhibition building is equipped with a large number of electronic rectifiers, LED lights, air conditioning inverters and various electronic information equipments, it is necessary to suppress the harmonic currents and reduce the multiple harmonics (especially the third harmonics)

⑧ Main Transformer and Distribution Equipment Selection
There equipped the energy-efficient SCB13 dry-type transformer (with IP30 protection).
The high voltage power distribution chooses Medium cabinet, while both the high and low voltage cables are set at the bottom with the cable trench is under the cabinet. The diesel generator is emergency auto-start, the device and related complete sets of equipment are supplied by the manufacturer.

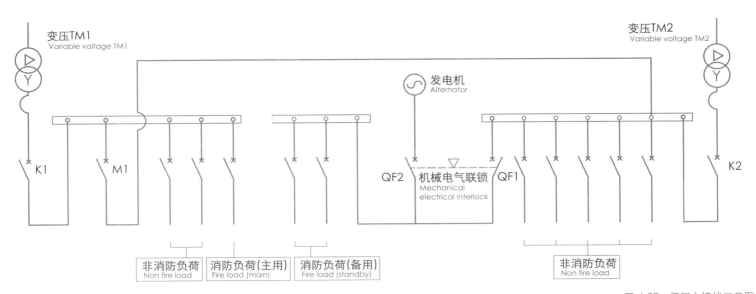

图 4-52 低压主接线示意图
Figure 4-52 Low Voltage Main Wiring Diagram

动力配电系统

① 本工程的消防负荷，如消火栓泵、喷淋泵、消防电梯、应急指挥中心、消防控制中心、消防分控制室等的配电均采用由低压配电室直接双回路至设备房电源箱的配电方式；防排烟风机、排水泵、稳压泵等的配电则将在适当位置设一级配电总箱，再由一级配电总箱再放射至设备控制箱，设备控制箱则设双电源互投（图 4-53）。

② 本工程的重要负荷，如客梯、生活水泵、数据中心、通信基站、智能化机房、重要的计算机系统、舞台机械、音响、餐饮中心厨房应急动力等的配电采用由低压配电室直接双路放射至设备房电源箱的配电方式；智能化分控制室等则将在适当地方设一级配电总箱，并由一级配电总箱再放射至设备控制箱 / 配电箱，设备控制箱 / 配电箱设双电源互投。

③ 制冷主机、空调水泵、冷却塔、新风机组等大动力系统将采用低压配电室单回路放射式配电至设备房电源箱的方式；楼层新风机、排风机等小的动力单元则采用单回路分段树干式配电，即每层均设有动力总配电箱，再分配至各设备控制箱。

Power Distribution System

① The power distribution system of the fire load of the project, such as hydrant pump, sprinkler pump, fire elevator, emergency command center, fire control center, fire sub-control room, and etc. would adopts the direct double circuit from the low-voltage distribution room to the power box of the equipment room; Smoke control and exhaust fans, drainage pumps, pressure stabilizing pumps and etc., are provided with a primary distribution box at appropriate location. The primary distribution box is re-radiated to the equipment control box, and the equipment control box is equipped with a dual power supply mutual switching.(Figure 4-53)

② The power for the important load of the project, such as passenger elevator, water pump, data center, communication base station, intelligent machine room, important computer system, stage machinery, acoustics, kitchen emergency power supply of catering center and etc., is transmitted from the low-voltage distribution room to the equipment power box room directly through a two-circuit radial system; primary power distribution box is set in appropriate places for the intelligent distribution control room and others. The primary distribution box is re-radiated to the equipment control box/distribution box, and the equipment control box is equipped with a dual power supply mutual switching.

③ The power for the main refrigeration unit, the water pump of air conditioning, cooling tower, the fresh air unit and other large power demanding function are transmitted from the low-voltage distribution room to the equipment power box room directly through an one-circuit radial system. Fresh air ventilator, exhaust fan and other small power distribution system apply single circuit sectional trunk system, which sets power distribution box on each floor, then redistributes to each equipment control box.

图 4-53 IP67 工业插座
Figure 4-53 IP67 Industrial Socket

明装插座
Surface-mounted Socket

带螺钉接线端子
Screw Terminal

用于水平安装
For Horizontal Installation

带开关
With Switch

带DUO-机械联锁
DUO-mechanical Interlock

插座可配装安全挂锁
Socket can be equipped with safety p

明装插座
Surface-mounted Socket

带螺钉接线端子
Screw Terminal

软接触工艺
SoftContact process

带开关
With switch

带DUO-机械联锁
DUO-mechanical interlock

插座可配装安全挂锁
Socket can be equipped with safety p

展位用电

① 展览用电采用单独变压器，跟其他公共用电完全分离，展厅二层设置综合管廊，便于管线维护管理、避免潮湿、避免设备腐蚀。

② 展厅地面设展沟，沟内由压缩空气、给排水管、强电管线、弱电管线共用。

③ 展位配电采用接驳井，接驳井间距为 6m×9m，标准展厅共 290个，示意如下图：标准展厅接驳井示意。

每个接驳除配备 16A/220V、32A/380V、63A/380V 插座各一个外，另在展厅配电间内预留了一定数量的 125A、200A、250A、400A 工业插座以满足大电流需求。展位用电采用放射式配电，这不仅提高了用电安全的可靠性，且还最大限度地缩小了故障范围，便于管理维护（图 4-54）。

④ 展厅储藏间内常备一定数量的展位活动箱及活动电缆，将根据运营需求及展览类型而进行使用。

⑤ 室外展场部分，按间距 35m×35m 设置综合井，每个井内设置 63A/380V IP67 工业插座一个。

The Electricity usage of Booth

① The electricity usage for the exhibition is provided with a separate transformer, which is completely separated from other public electricity equipment. An untility tunnel is set on the second floor of the exhibition hall to facilitate pipeline maintenance and management to avoid dampness and equipment corrosion.

② Trenches are set within the floor of the exhibition hall, for compressed air tunnel, supply and drainpipes, strong electric pipelines and weak electric pipelines.

③ The power distribution system of the exhibition booth are connecting wells with a spacing of 6m×9m. There are 290 standard exhibition halls in total.
Each connecting device is equipped with 16A/220V, 32A/380V and 63A/380V sockets. In addition, a certain number of 125A, 200A, 250A and 400A industrial sockets are reserved in the power distribution room of the exhibition halls to meet the need for large electric current. The exhibition booth applied radiative power distribution system, which not only improves the safety and reliability of electricity consumption, but also minimizes the fault range to facilitate management and maintenance.

④ There are a certain number of event boxes and event cables for the exhibition booths in the storage room of the exhibition hall, which would be determined according to the operation needs and the exhibition types.

⑤ In outdoor exhibition area, comprehensive wells are set with a spacing of 35m×35m, and each well is equipped with a 63A/380V IP67 industrial socket.

图例 legend

■ 配电间(共10间)
Distribution Room(10 in total)

┅ 管沟 （内铺设电缆桥架等管线）
Pipe Trench (laying cable bridge and other pipelines inside)

■ 接驳井 （设工业插座及信息插座） （6m×9m，共290个）
Connecting Well (with industrial socket and information socket)

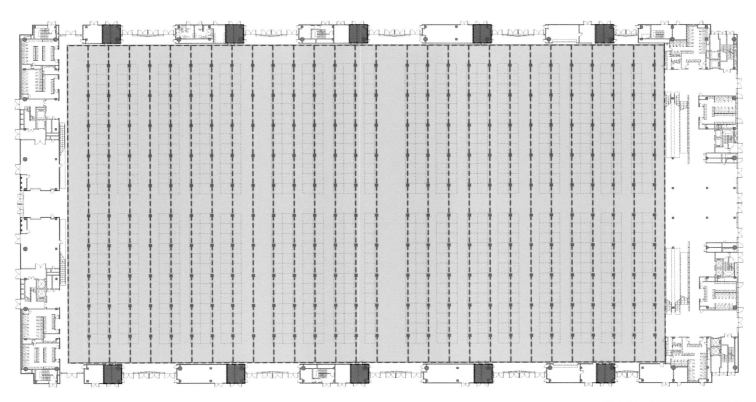

图 4-54 标准展厅接驳井示意图
Figure 4-54 Connecting Well of the Standard Exhibition Hall

电气节能设计

① 合理设计的供配电系统

深圳国际会展中心总设备安装容量为 18.25 万 kVA，项目用电负荷密度为 120VA/m²。合理设计供配电系统，合理确定用电负荷指标，合理确定变压器负荷率，合理设计变配电所位置使其尽量接近负荷中心，节约有色金属，以及减少线路电能损耗是此次电气节能设计的目标。变压器按两台一组的原则设计，每两台变压器中间加联络开关，以便小负荷工况使用一台变压器。

② 光照效率

采用高效 LED 灯具，视觉舒适度高于传统 LED 灯 56%，照明功率密度设计值低于标准值 60% 以上，节能 75%，年节电量可达 220 万度。

③ 灯光控制

最大限度采用节能措施，几乎所有公共区域灯具采用智能照明系统控制。

④ 节能设备

所有电梯、风机、水泵、变压器等选用节能设备，以及空调主机、空调水泵、空压机组、生活水泵等设备将采用变频控制。干式变压器选用 SCB13 型，能效等级满足 GB20052-2013 之不低于 2 级的规定。

电梯采取节能控制措施（电梯群控、扶梯自动启停），柴油发电机组的排放标准符合节能要求。

⑤ 负荷能耗独立计量分析

电力自动监控系统对照明、动力、空调等负荷能耗进行独立分项计量和分析。示意如表 4-6

多功能电表具有测量电压、电流、频率、功率因数、有功、无功、有功电度、无功电度、相位角等的功能，且具备独立计量芯片，与多时段分时计量电能，也可对整时、整日、整月电能数据进行采集和分析。

⑥ 太阳能利用

在 2~11 栋西侧配套设置的屋顶太阳能光伏发电系统，采用低压并网方式运行，经交流计量并网配电柜并入 18 号、19 号变电所（图 4-55）。

图 4-55　太阳能光伏发电系统
Figure 4-55　Solar Photovoltaic Power Generation System

Electrical Energy-saving Design

① Rational Design of Power Supply and Distribution System
The total installed electricity capacity of the Shenzhen World is 182.500 kVA, and the power load density is 120VA/m². Rationally design the power supply and distribution system, determine the electricity load index and the transformer load rate, and design the location of power distribution, which should be as close to the load center as possible to reduce the line power loss is the primary goal of the energy-saving design. According to the design principle which two sets of transformer a group, a connection switch is added in the middle of each two transformers, so that one transformer can be shut down during small load condition.

② Illumination Efficiency
Use highly efficient LED lighting, which is 56% more comfortable than traditional LED to eyes. The value of calculation for lighting power density is 60% lower than the standard value, which saves 75% more energy, and the annual electricity saving can reach 2.2 million kWh.

③ Lighting Control
Use energy saving measures as many as possible. Almost all public lightings are controlled by intelligent lighting system.

④ Energy-Saving Equipment

All elevators, fans, pumps, transformers and other equipment are all energy-saving, and air conditioning main engine, air conditioning pumps, air compressor units, pumps for daily-use and other equipment adopt variable frequency control.
SCB13 is used for dry type transformer, and the energy efficiency grade is no less than grade 2 according to the GB 20052-2013. The elevator adopts energy-saving control measures (group control of elevator, automatic start and stop of escalator), and the emission standards of diesel generator is qualified according to energy-saving requirements.

⑤ Independent Measurement and Analysis of Energy Consumption Load

Automatic power monitoring system can respectively measure and analyze the lighting, power, air conditioning and other energy consumption load .

The multi-function electricity meter can be uses to measure voltage, current, frequency, power factor, active power, reactive power, active power, reactive power, phase angle, etc., and has an independent measuring chip, which can be used to measure electrical energy in different time and periods. It can collect and analyze electrical energy data respectively for the whole hour, the whole day and the whole month.

⑥ Solar Energy Utilization

Solar photovoltaic power generation system is set up on the roof of the 2-11 supporting house on the west side, which is operated by low-voltage and grid-connected mode. The solar energy is gathered through the AC measurement and integrated into the substation No. 18 and No. 19.(Figure 4-55)

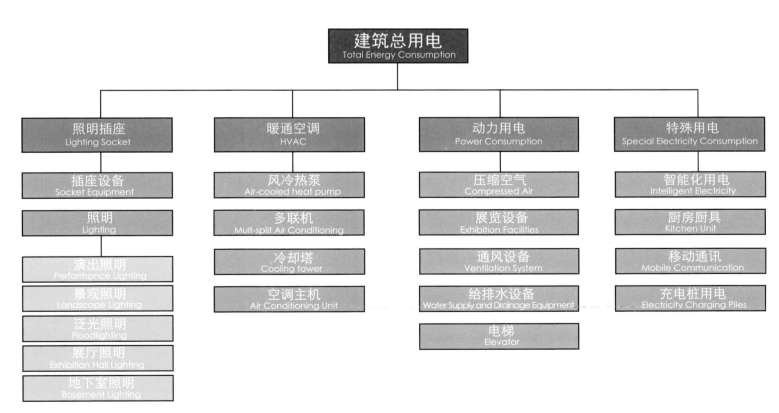

表 4-6 分项计量表
Table 4-6 Itemized Measurement Table

给排水体系
Water Supply and Drainage System

给排水系统

① 生活冷热水系统

项目水源为城市自来水，接入点水压力为 0.28MPa。项目合理利用市政水压，且首层均为市政直供。在南北登陆大厅负一层各设置一座生活加压水泵房，以 A7/A8 展厅分界，分别加压供给南北两区二层及以上楼层的用水点。项目每个展厅、登陆大厅都设有厨房，设计对厨房供给热水，热源采用空气能，伴电辅助加热的方式。示意如图 4-56：热水系统原理图

② 净水系统

南北登陆大厅负一层设有中央加工厨房，由于中央加工厨房设施多、功能复杂，设计将对中央生活用水通过石英石沙钢、活性炭、精密过滤器等过滤工艺处理后再供给。

③ 中水系统

中水将用作对室外绿化补水以及道路冲洗，中水水源为市政中水。

④ 污废水系统

餐饮废水将会在通过隔油池处理后接入室外污水管网；室外污水经过化粪池处理后排入市政污水管网。

⑤ 雨水系统

屋面雨水在通过虹吸雨水系统收集后将接入水外消能井，流速降低后排入室外雨水管网。屋面同时设有溢流管道系统，及溢流口系统。室外场地雨水则将在通过设置在路边的截水沟收集后接入雨水管道。项目将收集部分屋面雨水并经处理后回收用做冷却塔的补水。示意图 4-57：雨水利用系统流程图

Water Supply and Drainage System

① Domestic Cold and Hot Water System

The water source of the project is urban running water, and the water pressure of the access point is 0.28mPa. The project apply the municipal water pressure rationally, and water supply on the first floor are all from municipal supply directly. A pressurized water pump room is set respectively on the underground first-floor of the north and south arrival halls, which is divided by exhibition hall A7/A8 and supply water for the second and above floors of the north and south areas.(Figure 4-56)

Each exhibition hall and arrival hall is equipped with a kitchen. Hot water supply is designed for the kitchen, and the heat source is from air energy accompanied by electric auxiliary heating.

② Water Purification System

Central processing kitchen is set on the underground first-floor of the north and south arrival hall. For it has many facilities and multiple functions, the water supply here is designed to be purified with quartz sand steel, activated carbon, precision filter, and etc. before supply.

③ Recycled Water System

Recycled water is used for outdoor irrigation and road flushing. All tecycled water is from the municipal recycled water system.

④ Sewage and Wastewater System

After being processed by the grease trap, the catering waste water will be drained into the outdoor sewage pipe network. The outdoor sewage is going to be processed by the septic tank and then drained into the municipal sewage pipe network.

⑤ Rainwater System

The roof rainwater is collected by the siphon rainwater system and then drained into to the energy removal well, after the goes velocity is reduced, it goes into the outdoor rainwater network. The roof is also equipped with overflow piping system and overflow port system. Rainwater from outdoor is collected through intercepting ditches set on the roadside, and then gathered into the rainwater pipes. Part of the roof rainwater of this project will be collected and processed to use as replenishing water of the cooling tower (Figure 4-57).

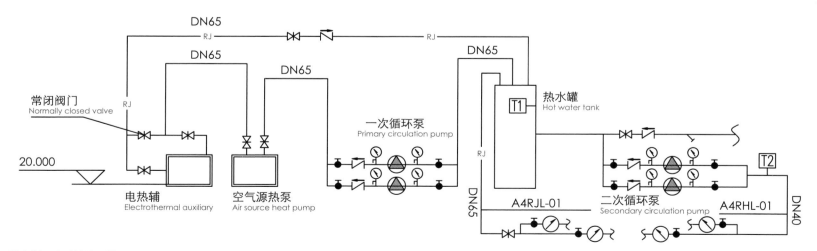

图 4-56　水系统原理图
Figure 4-56　Hot Water System Schematic Diagram

消防给水系统

项目以 7/8 号展厅为界，南北分设置个消防水泵房，分别位于南北登陆大厅负一层。南登消防水池容积 3744 立方米。北登消防水池容积 6768 立方米。稳压设备房分别设置南北登陆大厅屋面。稳压水池容积 36 立方米。

① 室内外消火栓系统
项目室外消火栓系统以凤塘大道为界，南北分成两个系统。分别从周边市政路上接入两个进水点在红线内成环，供室外消防使用。项目除不适用于消火栓系统外，均设置室内消火栓灭火系统，其设计用水量为 40L/S，火灾时延续时间可达 3h。

② 自动喷水灭火系统
除不宜用水扑救的部位及无需设置自动喷水区域外，各区域均设有自动喷水灭火系统（表 4-7）。

③ 自动消防炮灭火系统
项目展厅及登陆大厅大于 12m 大空间区域采用自动跟踪定位射流灭火装置（即自动消防炮）。自动消防炮灭火系统结合 LA100 型火灾安全监控系统对场所进行安全防控（图 4-59）。

④ 雨淋灭火系统
北登录大厅的国际报告厅舞台葡萄架栅顶下部按严重危险等级设计，喷水强度为 16L/min/m²，作用面积为 260m²；设计用水量为 110L/S，火灾时延续时间可达 1h。

⑤ 防火分隔水幕灭火系统
北登录大厅的国际报告厅舞台口部防火幕两侧设置防火冷却水幕，喷水强度为 1(L/S·m)，舞台口长度为 34m；设计用水量为 80L/S，火灾时延续时间可达 3h。
展厅防火隔离带上方设置防火冷却水幕，喷水强度为 2.5(L/S.m)（由国家消防工程技术研究中心消防模拟提供），C12 展厅防火分隔水幕长度为 198m，其它展厅防火分隔水幕长度为 99m；设计用水量为 540L/S，火灾时延续时间为 3h（图 4-60）。

⑥ 气体消防及灭火器系统
展厅按严重危险级 A 类火灾配置灭火器，选用 MF/ABC5 型灭火器，灭火级别 3A、5Kg，每点 2 具，配置要求按照保护距离小于等于 15m 设置；展厅大空间可结合展位情况临时设置灭火器箱。

Fire Water Supply System

Two fire water pump rooms are set in the north and the south divided by No. 7/8 exhibition hall, all located on the underground first-floor of the arrival hall. The fire pool has a capacity of 3744 cubic meters in the south, and 6,768 cubic meters in the north. Voltage stabilization equipment rooms are set on the roof of the arrival halls, with the volume of the tank is 36 cubic meters.

① Indoor and Outdoor Hydrant System
The outdoor fire hydrant system is divided into north and south systems. Two water inlets from surrounding municipal roads form a ring within the red line for outdoor fire protection.The project is protected by indoor fire hydrant extinguishing system, who's water consumption is 40L/S and could last for 3h.

② Automatic Sprinkler System
All areas should be installed with automatic fire-sprinkling system except for the areas that are not suitable to put out fire with water, or the areas that are no need for automatic sprinkler system (Table 4-7).

表 4-7 自动喷水灭火系统设计参数
Table 4-7 Sprinkler System Design parameter

序号 Serial number	区域 region	喷水强度（L/min·m²） Water spray intensity (L / min·m²)	危险等级（m²） Hazard level (m²)	作用面积（m²） Action area (m²)	火灾延续时间 Fire duration	设计水量（L/s） Design water volume (L / s)
1 No.1	车库、商业 Garage, commercial	8	中危级2级 Medium risk level 2	160	1h	40
2 No.2	办公 Office	6	中危级2级 Medium risk level 1	160	1h	30
3 No.3	仓库 Warehouse	16	仓库危级2级 risk level 2 of warehouse	200	2h	80

系统设计用水量取80L/s，最不利点的喷洒头工作压力为0.10MPa
The designed water consumption of the system is 80L / s, and the working pressure of the spray head at the most unfavorable point is 0.10MPa

③ Automatic Fire Extinguishing System

The exhibition halls and arrival halls which are larger than 12 meters are equipped with automatic tracking and positioning stream extinguishing device (i.e. automatic fire water cannon). Combined with the LA100 fire safety monitoring system, the automatic fire extinguishing system can protect the spaces from fire hazard.(Figure 4-59)

④ Deluge System

In the international lecture hall at the north arrival hall, the lower part of the grid on the stage is designed according to the fire risk catagory, with the water spraying intensity of 16L/min/m^2 and its operation area is 260m^2. The amount water consumption is 110L/S and last for 1h.

⑤ Water Curtain for Five Compartment

The international lecture hall is equipped with fire cooling water curtain on both sides of the stage. The water spraying intensity is 1(L/S.m), and the length of the stage is 34m; the amount water consumption is 80L/S last for 3h.

The fire isolation belt of the exhibition hall is also equipped with fire cooling water curtain. The water spraying intensity is 2.5(L/S.m) (provided by the fire protection simulation result of the National Fire Engineering and Technology Research Center). The length of the fire separation water curtain in the exhibition hall C12 is 198m, and in other exhibition halls the length is 99m; the amount water consumption is 540L/S last for 3h.(Figure 4-60)

⑥ Gas Fire and Extinguisher System

The exhibition hall is equipped with MF/ABC5 fire extinguishers according to class A Severe fire dangerous level. The fire level is 3A and 5Kg, 2 extinguishers at each point, with a guard space less than 15m. The large space of the exhibition hall can set temporary fire extinguisher box according to different situations.

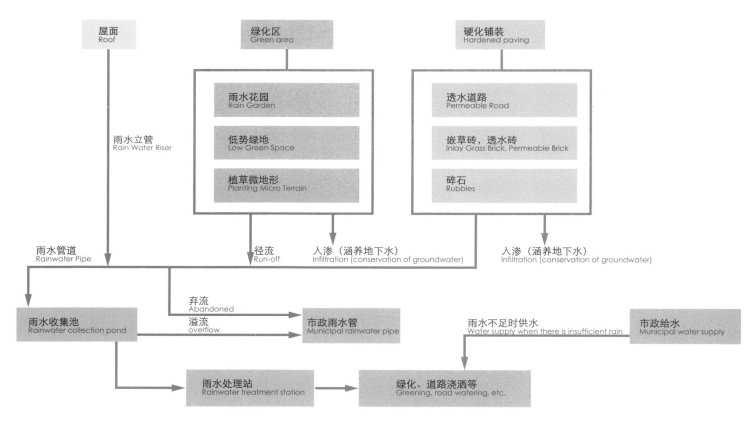

图 4-57 雨水利用系统流程图
Figure 4-57   Flow chart of rainwater utilization system

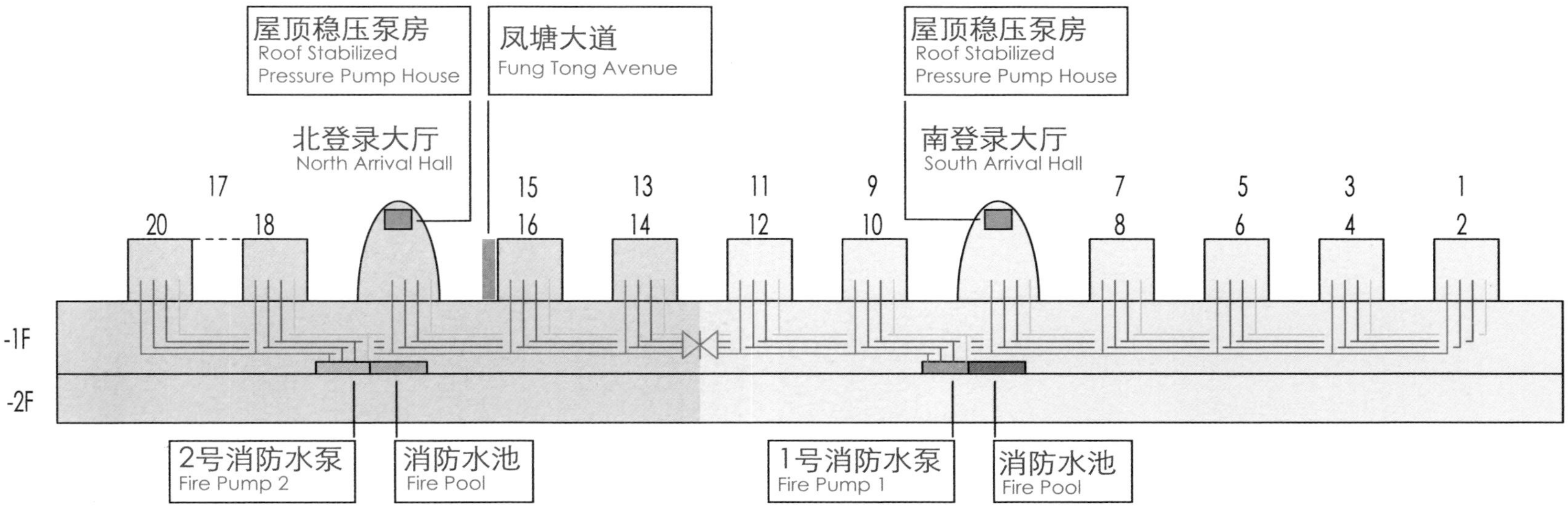

屋顶稳压泵房
Roof Stabilized
Pressure Pump House

凤塘大道
Fung Tong Avenue

屋顶稳压泵房
Roof Stabilized
Pressure Pump House

北登录大厅
North Arrival Hall

南登录大厅
South Arrival Hall

17
20   18

15   13   11   9
16   14   12   10

7   5   3   1
8   6   4   2

-1F

-2F

2号消防水泵
Fire Pump 2

消防水池
Fire Pool

1号消防水泵
Fire Pump 1

消防水池
Fire Pool

图 4-58　消防泵房设置
Figure 4-58　Fire Pump Room Setting

图 4-59　自动消防炮系统测试实景
Figure 4-59　Automatic Fire Water
Cannon System Test

图 4-60　水幕系统测试实景
Figure 4-60　Water Curtain System Test

智能化设计
Intelligent Design

深圳国际会展中心项目智能化设计内容包括：信息设施系统、安全防范系统、建筑设备管理系统、信息化应用系统及智能化集成系统（图 4-61）。

The intelligent design of the Shenzhen World include information facility system, security system, construction equipment management system, information technology application system and intelligent integrated system.(Figure 4-61)

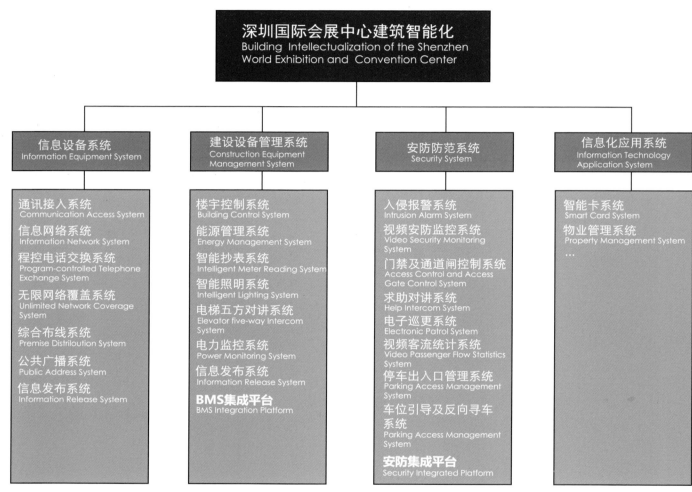

图 4-61　信息设施系统设计
Figure 4-61　Information Facility System Design

信息设施系统设计
Information Facility System Design

信息设施系统由通信接入系统、移动通信室内信号覆盖系统、综合布线系统、信息网络系统、无线网络覆盖系统、用户电话交换系统、集群呼叫对讲系统、背景音乐及应急广播系统、信息发布及引导系统、多媒体会议系统组成。

The information facility system consists of communication access system, mobile communication indoor signal coverage system, premise distribution system, information network system, wireless network coverage system, user telephone switching system, cluster call intercom system, background music and emergency address system, information release and guidance system and multimedia conference system.

通讯接入系统

本系统在南登录大厅地下一层和北登录大厅地下一层预留运营商进线机房及进线路由桥架，通信接入光缆和机房内设备组建由运营商自行设计。

Communication Access System

The space for operator inlet wire machine room and route crane span structure are reserved for this system on the underground first-floor of the south and north arrival halls, communication access cable and equipment in the room are designed seperately by the operator.

移动通信室内信号覆盖系统

室内覆盖分布系统必须涵盖三大电信运营商所使用的移动通讯频率，深圳国际会展中心一期项目，因规模大、移动电话使用密度大，因而局部网络容量不能满足用户的需求，无线信道可能会发生拥塞现象，因此，本工程室内覆盖系统将采用电信、移动和联通各自独立的室内分布系统。

综合布线系统

深圳国际会展中心综合布线采用星型拓扑结构，该结构下的每个分支系统都是相对独立的单元，因而对每个分支系统的改动都不影响其他子系统，只要改变结点连接方式就可使综合布线在星型结构之间进行转换。
深圳国际会展中心综合布线包括公共网、业务网、安防智能化专网、建筑设备专网和信息发布网。每套系统均由工作区子系统、水平布线子系统、管理间子系统、垂直干线子系统、设备间子系统、建筑群子系统组成。

信息网络系统

深圳国际会展中心信息网络系统建立在综合布线系统的基础上，以数据机房中心为核心，以 IP 和 Internet 技术为技术主体，为深圳国际会展中心提供一个整体的数据通信平台，系统可以承载各种类型的语音服务和电子邮件、文件共享、文件传输、信息发布等办公通信服务、各种数据多媒体业务、建筑设备管理等不同的数据应用系统。具体规划如下：

① 公共网络系统，为深圳国际会展中心参展商、参展企业／个人提供有线网络和无线网络。

② 业务网络系统，为深圳国际会展中心物业管理、运营办公提供有线网络和电话通信。

③ 安防智能化专网，为监控、门禁等安保设施提供网络支撑。

④ 建筑设备专网，为建筑设备管理系统提供网络支撑。

⑤ 信息发布系统，为 LCD 屏、IPTV、LED 屏等提供网络支撑。

系统全采用三层星型结构，即核心层、汇聚层和接入层，各套网络根据所承载的数据不同，带宽各不相同。
各个信息网络系统均通过综合布线作为通信介质，为各类物业运维管理系统、公共服务管理系统、公众信息服务系统等提供网络通信平台，满足数字化物业管理与信息服务的需求。

Mobile Communication Indoor Signal Coverage System

Indoor signal coverage distribution system should cover the mobile communication frequency used by three major telecom operators. Due to the huge scale of the project and the potential density of mobile phone use, the local network capacity may not meet the needs of such big amount of users and cause the wireless channel congestion. Therefore, the indoor coverage system adopts independent distribution systems of China Telecom, China Mobile and China Unicom.

Premise Distribution System

The premise distribution system of the Shenzhen World adopts the star topology structure, under which each branch is a relatively independent unit, and the modification of each branch system does not affect other subsystems. The premise distribution can be shifted within the star topology structure just through switching the node connection mode.
The premise distribution system of the Shenzhen World includes public network, business network, security intelligent specialized network, construction equipment specialized network and information distribution network. Each system is composed of working area subsystem, horizontal cabling subsystem, management subsystem, vertical trunk line subsystem, equipment subsystem and building group subsystem.

Information Network System

The information network system of the Shenzhen World is based on premise distribution system. By taking data center as the core, and using IP and Internet technology as the main technology, it provides an integral data communication platform which can carry various types of voice service and email, file sharing, file transfer, information release and other office communications services, various data business, construction equipment management, and other multimedia data applications. Details are as follows:

① Public network system: providing wired network and wireless network for exhibitors, and exhibitor/individuals.
② Business network system: providing wired network and telephone communication for property management and operation.
③ Security intelligent specialized network: providing network support for security monitoring, access control and other security facilities.
④ Construction equipment specialized network: providing network support for construction equipment management system.
⑤ Information release system: providing network support for LCD screen, IPTV, LED screen, etc.

The system adopts three-tier star structure: core layer, convergence layer and access layer. Each network has different bandwidth according to different data.Each information network system uses premise distribution system as communication medium, providing network communication platform for various property operation and maintenance management systems, public service management systems, public information service systems, etc., to satisfy the needs of digital property management and information services.

无线网络覆盖系统

无线网络覆盖系统是信息网络和无线通信技术相结合的产物。无线局域网的基础还是传统有线局域网，且是有线局域网的扩展和替换，是在有线局域网的基础上通过无线集线器、无线访问节点、无线网桥、无线网卡等设备来实现无线通信（图 4-62）的过程。

深圳国际会展中心无线网络具备以下功能：

① 使得展厅、公共活动场所、绿化景观、登录大厅以及地下车库等区域内的客户通过 WiFi 连接到 Internet；

② 通过移动终端（手机、平板电脑等）可以在会展范围内访问会展平台及 APP 终端服务。

无线网络覆盖系统由公共网提供网络支撑，无线 AP 采用 POE 的方式供电以降低布线成本，减少故障点，并提高系统可靠性，使系统可在数据中心机房对前端 AP 进行统一管理和维护。

用户电话交换系统

针对物业管理部门，深圳国际会展中心设置了一套程控数字电话交换系统，以便于满足深圳国际会展中心内外通信、话费结算、管理等运行需求。

图 4-62　连网示意图
Figure 4-62　Networking Diagram

Wireless Network Coverage System

Wireless network coverage system is a combination of information network and wireless communication technology. The base of WLAN is still the traditional wired LAN, and is the extension and replacement of wired LAN. On the basis of the wired LAN, the wireless communication is achieved through wireless hub, wireless access node, wireless bridge, wireless network card and other devices.(Figure 4-62)

The wireless network of the Shenzhen World has the following functions:

① Customers can access WIFI in the exhibition hall, public places, green landscape, arrival hall and the underground parking space.

② Mobile devices (mobile phone, PAD, etc.) can access the the exhibition platform and APP service within the scope of the exhibition.

The wireless network coverage system is supported by the public network. The wireless AP is powered by the POE to reduce the wiring cost and the fault points, while improve the reliability of the system at the same time. The system manages and maintains the front end AP in the data center room.

User Telephone Switching System

For the property management department, a program-controlled digital telephone switching system is set to guarantee the operation of internal and external communication, telephone fee settlement, management and so on.

集群呼叫对讲系统

深圳国际会展中心采用 400MHz 数字集群无线对讲通信系统，覆盖红线内以及建筑物内外的公共区域，为消防、保安、维修、清洁等物业管理部门提供了工作与应急通信工具。

本项目设置了一套 PDT 系统，在南登录大厅监控中心和北登录大厅安防控制室都部署了一套 16 载波基站。在南登录大厅监控中心部署了呼叫控制中心以实现指挥调度管理功能，并通过 IP，实现了监控中心与北登录大厅安防控制室 16 载波基站的互联互通。

Cluster Call Intercom System

The Shenzhen World adopts the 400MHz digital cluster wireless intercom communication system, covering the public areas within the red line, and inside and outside of the building, providing work and emergency communication tools for property management departments such as fire protection, security, maintenance and cleaning.

This project set up a PDT system and deployed a set of 16 carrier wave base stations in the monitoring center of the south arrival hall and the north arrival hall. The call control center is deployed in the monitoring center of the south arrival hall to guarantee the command and dispatch management function. And the monitoring center can intercommunicate with the 16 carrier wave base stations in the security control room of the north arrival hall through the IP.

公共广播系统

深圳国际会展中心公共广播系统为一套兼顾背景音乐、公共广播、火灾应急广播等功能为一体的广播系统。业务广播时可以单独对每一个分区进行广播，也可以编组进行群呼广播；整个系统支持多种不同的输入音源。发生火灾时，可强制转换为火灾应急广播，并向全楼进行广播（图 4-63）。

Public Address System

The public address system is a set of broadcast system that could do background music, public broadcasting, fire emergency broadcasting and others.It can be broadcasted independently in each partition, or marshalling for group calls; the whole system supports a variety of input audio sources. In case of fire, it can be compulsively switch to fire emergency broadcasting and broadcast to the whole building (Figure 4-63).

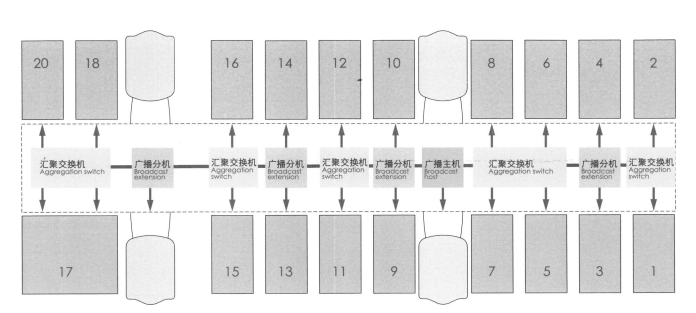

图 4-63　广播示意图
Figure 4-63　Public Address Diagram

信息发布及引导系统

深圳国际会展中心信息发布系统为一套独立网络系统，采用 TCP/IP 传输协议，由中心控制系统和显示终端结合工作，操作人员通过控制服务端进行节目内容采集、编排、发布和管理等功能的操作，节目通过网络传输到各显控终端进行本地存储及信息实时播放。

信息发布系统前端接入屏种类繁多，包括电梯厅 LCD 屏、会议室 LCD 屏、广告屏、室内外 LED 大屏、标识屏等，具备字幕管理、紧急插播、用户管理、日志及报表统计、触摸查询等功能。

Information Release and Guidance System

This system is a set of independent network system, apply the TCP/IP transport protocol, the center control system works with display devices, through which the operation collect, arrange, distribute and manage the program, and transfer through the network to all the display devices to local storage and information real-time playing.

There are many kinds of front-end access screens in the information release system, including LCD screens in elevator, LCD screens in conference rooms, advertising screens, large indoor and outdoor LED screens, sign screens, etc..

All of these screens also has many functions such as subtitle management, emergency interrupting, user management, journal and report statistics, touch query, and etc..

多媒体会议系统

多媒体会议系统是仅针对深圳国际会展中心的会议室、报告厅及多功能厅等设置，系统通过现有的各种电气通讯传输媒体，将人物的静态/动态图像、语音、文字、图片等多种信息分送到各个用户的终端设备上，使得在地理上分散的用户可以共聚一处，通过图形、声音等多种方式交流信息。

Multimedia Conference System

Multimedia conference system is only for conference rooms, lecture halls and multi-function halls. Through the existing various electrical communication transmission media, it can transfer people's static/dynamic images, voice, text, images, and other information to each user's device, bring geographically dispersed users together, and exchange information through various ways such as graphics, sound and etc..

公共安全防范系统设计
Design of Public Security System

本项目共设置 5 间安防控制室，其中，一间安防控制中心位于南登录大厅首层，一间安防控制室位于北登录大厅首层，另外三间安防分控室分别位于中央廊道首层（图 4-64）。

公共安全防范系统由入侵报警系统、视频安防监控系统、出入口控制系统、电子巡更系统、求助呼叫对讲系统、停车场（库）管理系统、车位引导及反向寻车系统等系统所组成。

There are total 5 security control rooms set in the Shenzhen World, one is located on the first floor of the south arrival hall, one is located on the first floor of the north arrival hall, and the other three are located on the first floor of the central concourse.

The public security system include Intruder Alarm System, video security monitoring system, entrance and exit control system, electronic patrol system, mayday call intercom system, parking (garage) management system, parking guidance and reverse vehicle search system, etc..

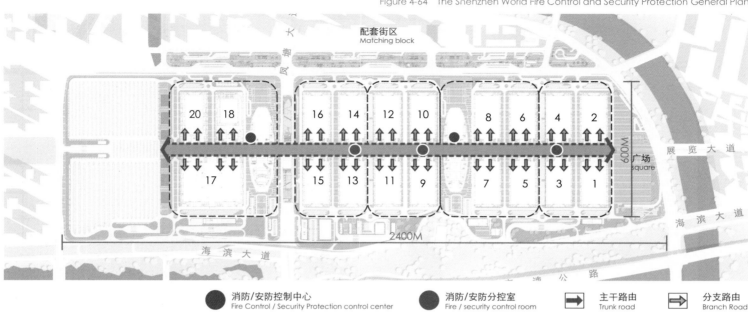

入侵报警系统

深圳国际会展中心入侵报警系统采用总 - 分控架构，在安防控制中心设置入侵报警主机，在北登录大厅首层安防控制室和中央廊道首层的三间安防控制室分别设置入侵报警分机，各入侵报警分机能够独立运行，也能通过安防控制中心入侵报警主机统一管控。如下图所示。

本系统针对园区周边和室内两部分进行设防，室外采用摄像头取代传统周界红外对射，依托于实体围栏，对整个周界进行无死角监控，室内主要在接待台、服务中心、展厅入口等设置传感检测装置，通过防区模块与安防控制中心 / 室报警主机 / 分机通讯，当有入侵时，安防控制中心 / 室发出报警。

Intruder Alarm System

The intrusion alarm system adopts a host-extension controlling structure. Intruder alarm host is set in the security control center, and the intruder alarm extension is set in the security control room in the first floor of the south arrival hall, and other three control rooms in the first floor of the central concourse. Each intruder alarm extension can function independently and is also controlled by the intruder alarm host at the same time.

This system is divided for the surrouding of the park and the indoor space. Camera is used outdoor to replace the traditional perimeter active infrared intrusion detector is set on the fence, it can monitor the entire surrouding without any dead angle; for indoor space, sensor detection device is set up in the reception, service center and exhibition hall entrance, and it would communicate with security control center/room alarm host/extension through protection zones, when invasion happens, the security alarm control center/room will raise the alarm.

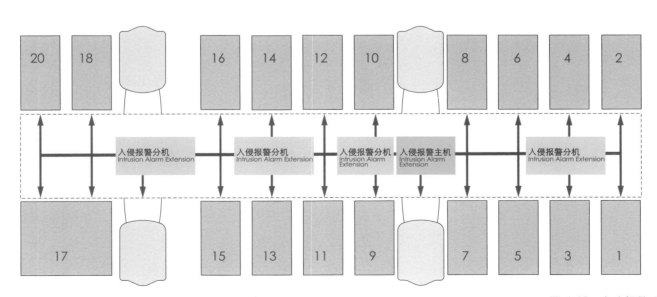

视频安防监控系统

本系统对深圳国际会展中心的展厅内、登录大厅、会议中心、办公以及室外公共区、周边等地方设置高清网络摄像头实施全覆盖。视频监控系统采用总-分架构形式，在南登录大厅首层安防监控中心设置视频监控主机，在北登录大厅首层安防控制室和中央廊道首层的三间安防控制室分别设置视频监控分机，各监控分机能够独立运行，也能通过安防控制中心监控主机统一管控（图 4-66）。

Video Security Monitoring System

This system covers the exhibition hall, arrival hall, conference center, office, outdoor public spaces and surrounding areas of the Shenzhen World with high-definition cameras all over. The video monitoring system adopts a host-extension controlling structure. The video monitoring host is set in the security control center in the first floor of the south landing hall, and the video monitoring extension in the security control room in the first floor of the north arrival hall, and other three control rooms in the first-floor of the central concourse. Each video monitoring extension can function independently and is also controlled by the video monitoring host.(Figure 4-66)

图 4-66 视频安防监控室实景图
Figure 4-66 Video Security Monitoring Room

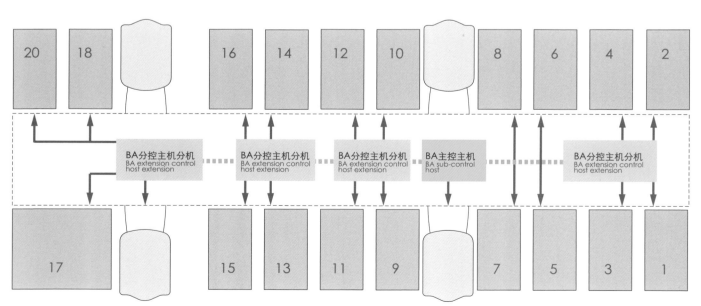

图 4-67 BMS 管理示意图
Figure 4-67 BMS Management Diagram

出入口控制系统

出入口控制系统采用总 - 分架构形式，在南登录大厅首层安防监控中心设置出入口控制主机，在北登录大厅首层安防控制室和中央廊道首层的三间安防控制室设置出入口控制分机，各出入口控制分机能够独立运行，也能通过安防控制中心出入口控制主机统一管控（图 4-68）。

Access Control System

The access control system adopts a host-extension controlling structure. The entrance and exit control host is set in the security control center in the first floor of the south arrival hall, and the entrance and exit control extension is set in the security control room in the first floor of the north arrival hall, and other three control rooms in the first-floor of the central concourse. Each entrance and exit control extension can function independently and is also controlled by the entrance and exit control host.(Figure 4-68)

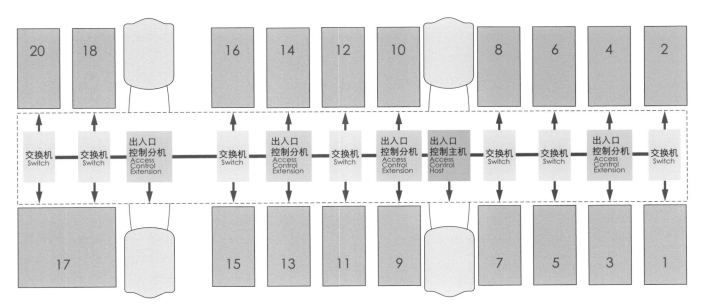

图 4-68　出入口控制系统示意图
Figure 4-68　Access Control System

电子巡查系统

深圳国际会展中心巡更系统采用 RFID 技术，利用 PDT 集群无线对讲系统平台进行数据传输的新型巡更系统。巡更手持机完成巡更点打卡后，打卡信息将实时经由 PDT 集群无线对讲系统回传至巡更服务器，完成打卡信息上报。

Electronic Patrol System

The patrol system adopts the RFID technology and uses the PDT cluster wireless intercom system platform for data transmission. After punching the clock at the patrol spots, the information will be transmitted to the patrol server through the PDT cluster wireless intercom system to report the information.

图 4-69　残疾人卫生间求助对讲系统
Figure 4-69　Help Intercom System of Disabled Toilet

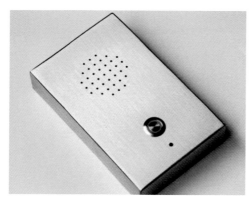

图 4-70　求助对讲系统
Figure 4-70　Help Intercom System

求助对讲系统

深圳国际会展中心求助报警对讲系统主要针对建筑内残疾人卫生间、重要库房、其他人群聚集或存在危险的场所设置求助按钮，以便于需要的人士向安防监控中心发出求助报警呼救（图 4-69，图 4-70）。

Help Intercom System

The system mainly sets help buttons in the toilets for disabled people, important warehouses and other places where crowds are gathering or places with potential hazards, for people to call the security monitoring center.(Figure 4-69, 4-70)

停车场管理系统

深圳国际会展中心的车流大，停车场管理采用车牌识别管理系统，无人值守，同时具有发／收卡机，辅助无牌车和污损车牌车辆进出场。如下图所示：系统采用一体式智能道闸，对车辆进出实现智能化高效管理。一体式智能道闸集合了车牌识别高清摄像机、补光灯、信息指示屏、自动道闸、控制机、停车管理软件等功能，可通过简便智能化的硬件设施与管理软件，并结合相关收费系统，实现出入口的车牌自动识别、自动放行（图 4-71）。

Parking Management System

For the potential large traffic flow of the Shenzhen World, the parking management adopts the license plate recognition management system with no need of supervision. It also has a card issuing/collecting machine to assist unlicensed vehicles and vehicles with uncleaned license plates to enter and exit.

The system adopts the integrated intelligent gate to guarantee the high efficiency of the intelligent vehicle management. This intelligent road gate integrates the equipment of high-definition camera for license plate recognition, supplementary light, information indicating screen, automatic road gate, control machine, parking management software and etc. Through these simple and intelligent hardware facilities and management software, and incorporate with the relevant charging system, the parking management system can achieve the automatic recognition of license plate and automatic release at the gate (Figure 4-71).

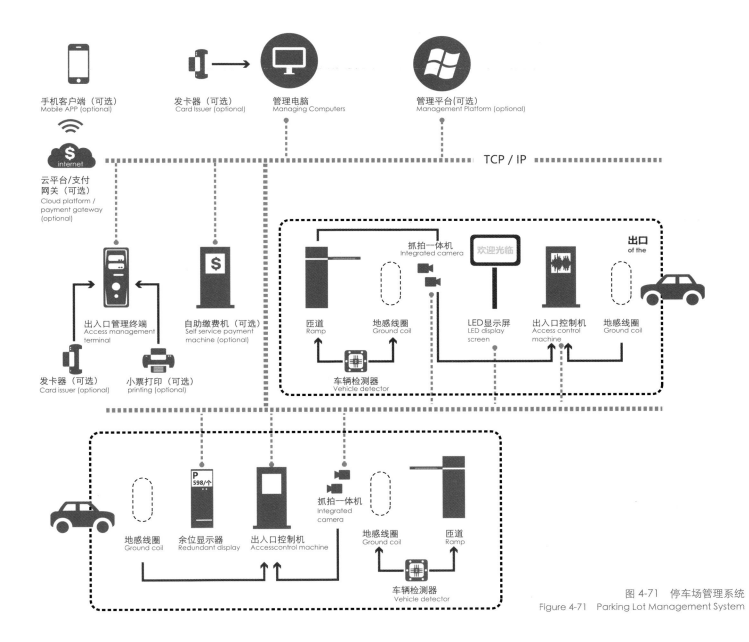

图 4-71　停车场管理系统
Figure 4-71　Parking Lot Management System

车位引导及反向寻车系统

视频车位引导及反向寻车系统利用视频探测器对车牌信息和车位状态进行识别和采集，并将数据传输到后台。设备运行状态、车位占用状态、车牌停放位置等信息均会上传至视频车位引导及反向寻车系统平台。视频车位引导子系统由前端视频车位探测器、局域网络与后台中心组成（图4-72）。

Parking Guidance and Reverse Find-Vehicle System

Video parking guidance and reverse Find-vehicle system adopts video detectors to identify and collect license plate information and parking status, and transmit the data to the system backstage. Equipment running state, parking space occupation state, license plate parking position and other information are uploaded to the system platform.The video parking guidance subsystem is composed of front-end video parking detector, local network and background center. (Figure 4-72)

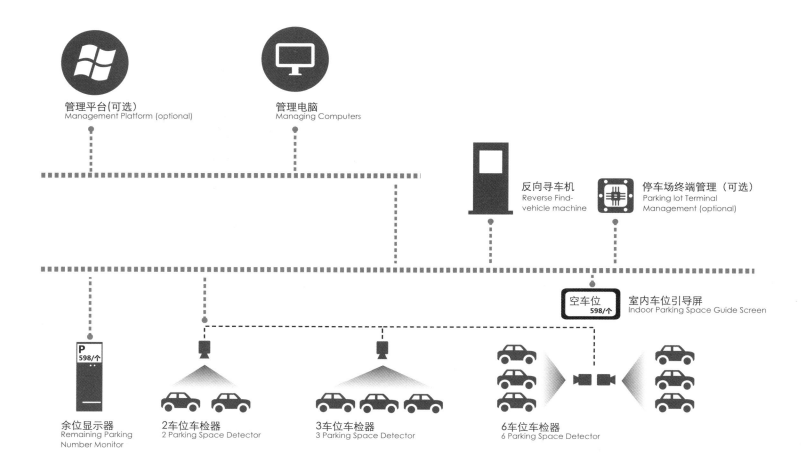

图 4-72　视频车位引导子系统
Figure 4-72　Video Parking Guidance Subsystem

建筑设备管理系统设计
The Design of the Construction Equipment Management System

深圳国际会展中心共设置 5 间楼宇控制室，其中一间为楼宇控制中心，位于南登录大厅首层，另外四间分别位于北登录大厅和中央廊道首层，贴临安防控制室，如下图所示：

There are 5 building control rooms are set, and one of them is the building control center which located on the first floor of the south arrival hall, and the other four are located in the north arrival hall and the first floor of the central concourse, close to the security control room.

建筑设备管理系统由建筑设备监控系统、能耗管理及智能抄表系统、电梯五方对讲系统等组成。

本系统针对深圳国际会展中心一期项目的水、暖、电等建筑机电设备实施自动化控制，提供卫生、健康、舒适的建筑空间环境，通过精确调节控制、降低人力成本、控制建筑能耗来满足建筑物群的节能、绿色、智慧等运行需要。
建筑设备管理系统采用总 - 分架构形式，如下图（图 4-73）所示：

Construction equipment management system is composed of building equipment monitoring system, energy management and intelligent meter reading system, and elevator five-sided intercom system.

This system implements automatic control of water, heating, electricity and other building mechanical and electrical equipment for the first phase of the Shenzhen World project, and provides a hygienic, healthy and comfortable architectural atmosphere, and guarantees the energy-saving, green and smart operation of buildings through precise adjustment and control to reduce labor cost and control building energy consumption.
The construction equipment management system adopts the host-extension structure as shown in the figure below:

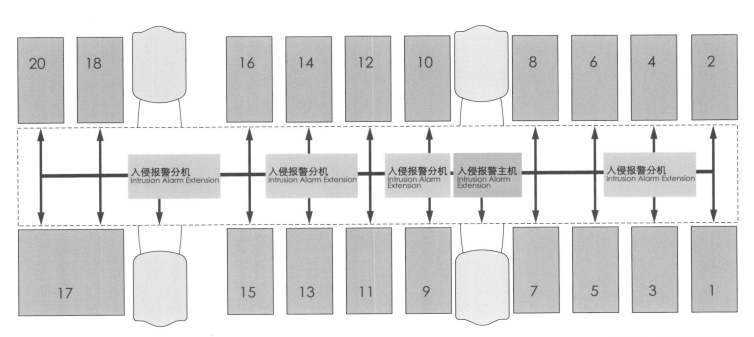

图 4-73　安防报警系统示意图
Figure 4-73　Schematic Diagram of Security Alarm System

建筑设备管理系统（BAS）

深圳国际会展中心 BAS 系统采用数字式 BA 系统，由建筑设备专网支撑网络，系统由管理服务器、工作站、现场直接控制器（DDC）、网关及传感器等设备组成。

系统用于对建筑内机电设备的远程监控，通过标准通讯协议网关接入 BMS 服务器，实现系统的集成监控。

能源管理及智能抄表系统

能源管理系统由管理服务器、工作站、打印机、系统软件组成。用于建筑内机电设备耗能的监测及数据分析。

系统从智能抄表系统、供配电监控系统、楼宇设备控制系统获取建筑能耗信息，实施能效分析与管理。

智能抄表系统由抄表主机、智能采集终端与接口构成。用于对建筑内水表、电表、燃气表的用量进行数据采集和数据分析。

Construction Equipment Management System (BAS)

The BAS system adopts digital BA system, which is supported by the specialized network of construction equipment. The system is composed of management server, workstation, direct digital controller (DDC), gateway, sensor and other equipments.

The system is used for remote monitoring of mechanical and electrical equipment in the building. It is connected to the BMS server through the gateway of standard communication protocol to implement integrated monitoring of the system.

Energy Management and Intelligent Meter Reading System

Energy management system is composed of management server, workstation, printer and system software, they are for monitoring and data analysis of energy consumption of mechanical and electrical equipment in buildings.

The system obtains building energy consumption information from the intelligent meter reading system, power supply and distribution monitoring system and building equipment control system to implement energy efficiency analysis and management.

The intelligent meter reading system consists of a meter reading host, an intelligent acquisition terminal and an interface, which is used for data collection and analysis of water meter, electricity meter and gas meter consumption in the building.

信息化应用系统设计
The Design of the Information Application System

信息化应用系统包括物业管理系统、智能卡应用系统、信息安全管理系统等。

Information application system includes property management system, smart card application system, information security management system, etc.

物业管理系统

物业管理系统包括：人事管理子系统、客户信息管理、租赁管理、租赁合同管理、收费管理、工程设备管理、客户服务管理、保安消防管理等等。

Property Management System

Property management system includes personnel management subsystem, customer information management, lease management, lease contract management, charging management, engineering equipment management, customer service management, security fire management and etc.

智能卡应用系统

系统采用物联网技术，并在安防监控中心设置了智能卡服务器与工作站。系统提供制卡、发卡、授权功能，预留考勤、签到模块，提供智能卡的激活、录入、修改、封闭、发送、挂失、解挂、灭失、销户、充值、结算、查询服务（图4-74，图4-75，图4-76）。

信息网络安全管理系统

信息网络安全是信息网络系统的重要组成部分，包括：物理网络的安全，含网络设备物理层的安全和网络结构的安全；操作系统级的安全；应用系统的安全；最高层是信息内容，或称为数据安全。

本智能化工程网络安全的目标是提供一个安全可靠的网络业务承载平台，既要防止局域网外非法用户的侵入，又要保证局域网内的工作人员分级按权限操作，同时也要保证系统长期稳定地运行。

Smart Card Application System

The system adopts the IoT technology and sets up smart card server and workstation in the security monitoring center. The system provides card making, card issuing, card authorization, reserved attendance, sign-in module, also provides smart card services, such as activation, entry, modification, closure, transmission, loss reporting, unhook, loss, account cancellation, recharge, settlement, query, etc. (Figure 4-74, 4-75, 4-76)

Information Network Security Management System

Information network security is an important part of information network system, which includes physical network security, network equipment physical layer security and network structure security; operating system level security; application system security; and the highest level is information content, or data security.
The goal of the intelligent project network security is to provide a safe and reliable network business platform, not only to prevent the invasion of illegal users from the LAN, but also to guarantee the authority operation according to the staff classification in the LAN, and guarantee the long-term stable operation of the system at the meantime .

图 4-74　停车场交费管理
Figure 4-74　Parking Fee Management

图 4-75　通道闸口管理
Figure 4-75　Access Gate Management

图 4-76　门禁卡
Figure 4-76　Access Card

智能化集成系统
Intelligent Integrated System

智能化集成系统将分散的、相互独立的各个智能化子系统，用相同的环境、相同的软件界面进行集中监视。深圳国际会展中心一期项目智能化集成系统考虑到以后系统扩展的需求，智能化集成系统将由独立于其他各子系统的第三方集成软件平台实现，IBMS的服务器构筑在主干网络之上，将建筑内的所有智能化系统包括信息设施系统、建筑设备管理系统、安全防范系统等系统通过TCP／IP协议集成在IBMS系统上。

IBMS集成系统通过对各智能化子系统的集成来提供更有效地对建筑内的各类事件的全局管理，以提高建筑对突发事件的快速响应能力。管理者通过事先的软件设置和编制联动响应程序以及集成一体化界面的联动状态图形，可达到全局事件的联动控制，实现IBMS系统整体的防灾能力。

The intelligent integrated system monitors intensively each scattered and independent intelligent subsystem within the same background and same software interface. Considering the possible demand of system extension in the future, this system is independent from other subsystems and operates on the third-party integrated software platform, the IBMS server is built outside the backbone network, therefore, all the intelligent building system includes information infrastructure system, building equipment management system, security system and other systems can be integrated into the IBMS system via TCP/IP protocol.

By Mtegrating various intelligent subsystems, the IBMS integrated system can provide more effective overall management of various events in the building and improve the building's emergency response speed. By setting up the software in advance and compiling the linkage response program, and through the linkage state graph of the integrated interface, the manager can control the whole event, and realize the overall disaster prevention of the IBMS system.

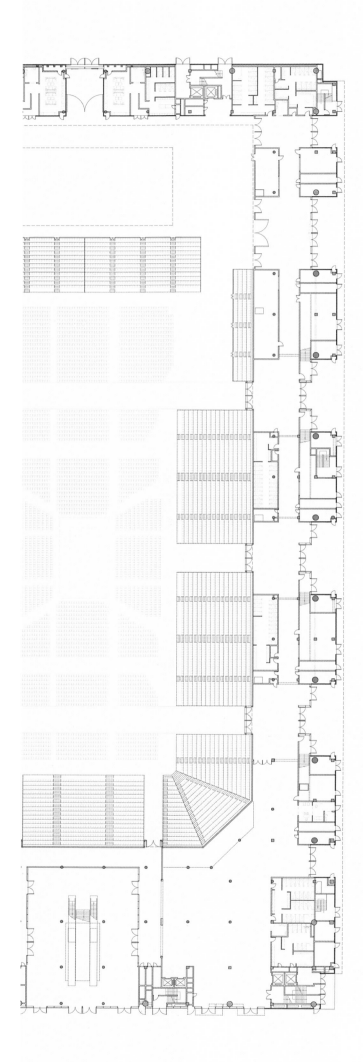

消防设计
Fire Protection Design

由于深圳国际会展中心项目空间尺度庞大，开展时的人员密度集中，因此确保其消防安全是首要课题，同时这也意味着建筑消防设计的新挑战。

Due to the Shenzhen World's large space scale, and the high crowd density during exhibition period, fire safety is the primary mission, which also bring a new challenge for the building fire design.

防火设计难点：防火分区和疏散距离
Difficulties in Fire Design: Fire Partition and Evacuation Distance

以标准展厅为例，因其定位以承接大型展览为主要功能，单个展览空间面积达 2 万 m² （图 4-77，图 4-78）。按照《建筑设计防火规范》要求，展览厅单层建筑空间每个防火分区最大允许建筑面积不应大于 10000m²，疏散距离不应超过 37.5m。因此如何在保证大空间完整及高使用效率的前提下解决防火分区过大、疏散距离超长的消防问题，是项目的重点（图 4-79，图 4-80）。

Take the standard exhibition hall as an example, which is positioned to undertake large-scale exhibition with a space area of 20,000 square meters.(Figure 4-77, 4-78) According to the Fire Regulation in Architectural Design, the maximum area of each fire partition for single-story building should not exceed 10,000 square meters, and the evacuation distance should not exceed 37.5 meters. The key point of the project is how to solve the fire control problem of large fire partition and long evacuation distance on the premise of retaining the large space integrity and high efficiency.(Figure 4-79, 4-80)

图 4-79　超大展厅 17 平面示意图
Figure 4-79　Super Large Exhibition Hall No. 17 Plan

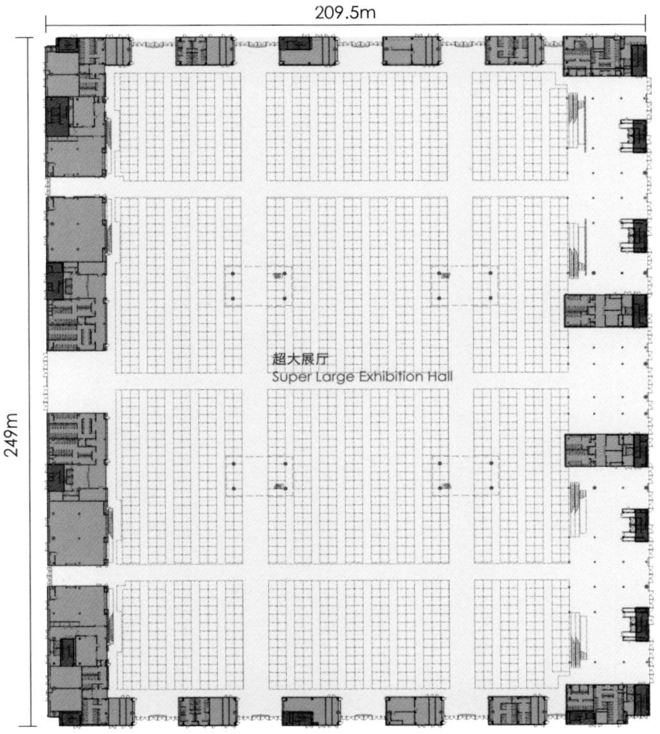

图 4-77　标准展厅平面示意图
Figure 4-77　Standard Exhibition Hall Plan

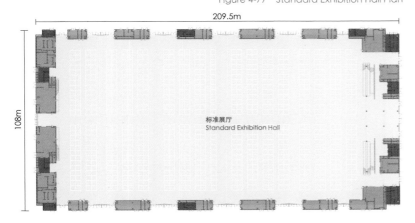

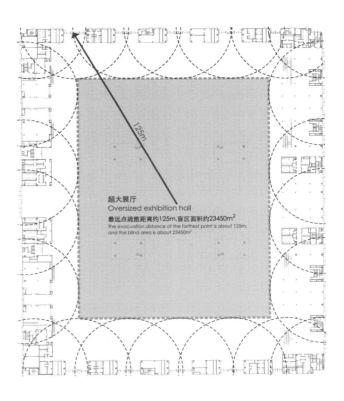

图 4-80　超大展厅 17 疏散距离示意图
Figure 4-80　Evacuation Distance Diagram of Super Large Exhibition Hall No.17

图 4-78　标准展厅疏散距离示意图
Figure 4-78　Evacuation Distance Diagram of Standard Exhibition Hall

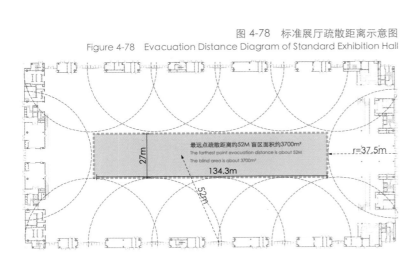

消防难点解决策略
Strategies for Difficulties in Fire Control

超大空间的消防加强措施

针对上述防火设计难点，以及会展建筑自带的人员密度大、展品火灾性复杂等特点，项目须在保证超大空间通透性、灵活性的同时，对超大空间进行必要的消防加强措施，最大限度地降低超大空间的火灾荷载及风险，是消防设计的第一步。

Fire Strengthening Measures for Extra Large Space

According to the design difficulties mentioned above, and the characteristics of exhibition buildings: high crowd density and complicated fire danger; this project must guarantee the permeability and flexibility of the large space, and at the meantime strengthen the necessary measures of fire control to minimize its fire load and risk.

增设具备消防车登高的操作场地

考虑本项目是广东省重点项目，定位较高，且体量超长、超大，势必在展会时的人员密度会较高。因此为提高危急救援的可靠性，在消防总图的设计时，在每栋展厅和登录厅的一个长边均增设了具备消防车登高的操作场地，以确保展厅空间附属多层配套房间的及时救援（图4-81）。

Increase lift-up fire truck operation area

The Shenzhen World is a key project with high positioning in Guangdong Province. Due to its super long and large building volume, the crowd density would be very high during exhibition times. In order to improve the reliability of emergency rescue, each exhibition hall and arrival hall has set up certain Lift-up fire truck operation area to guarantee the rescue of multi-layer supporting rooms affiliated in the exhibition.(Figure 4-81)

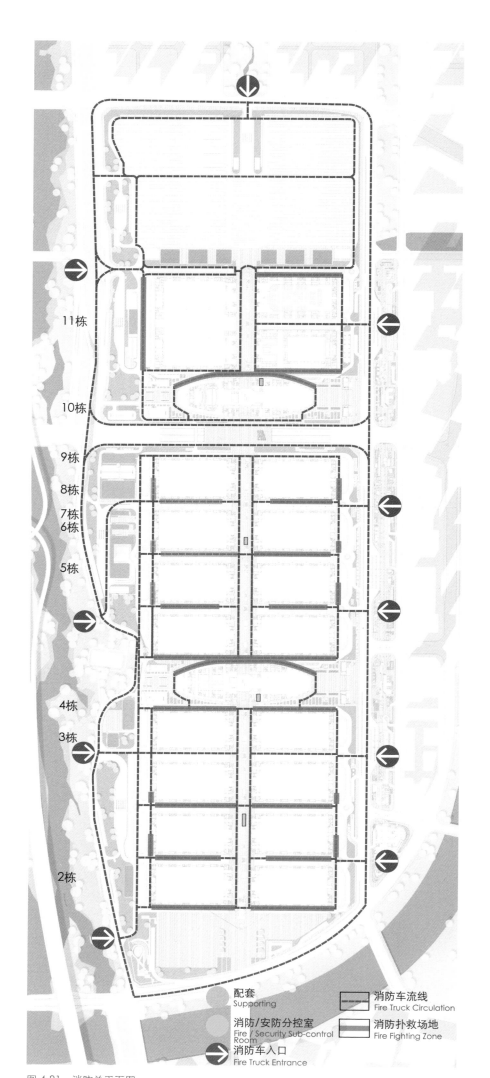

配套
Supporting

消防/安防分控室
Fire / Security Sub-control Room

消防车入口
Fire Truck Entrance

消防车流线
Fire Truck Circulation

消防扑救场地
Fire Fighting Zone

图 4-81 消防总平面图
Figure 4-81 Fire Protection General Plan

加强控制灾害蔓延

从降低火灾风险、控制灾害蔓延方面，项目采取了多项加强措施。首先，各展厅金属围护屋面的保温隔热材料耐火等级均提高到 A 级；其次，各展厅和登录大厅等的防火分区面积超过规范要求的单层空间建筑的楼板耐火极限均提升为 2 小时、同时作为各展厅联系的中央廊道的楼板（8 米标高）耐火极限也提高至 2 小时；再次，各展厅内库房（仅存放丙二类物品）、厨房的房间门的耐火极限均提为甲级防火门，最大限度地降低灾害的蔓延。

Reinfore the Disaster Spreading Control

In terms of reducing fire risk and controlling of disaster spreading, this project has adopted a number of reinforcement measures.
First of all, the fire resistance rating of the thermal insulation material of the metal enclosure roof in each exhibition hall has been raised to grade A; Secondly, the floor fire endurance of all exhibition halls, arrival halls and other fire compartments that beyond code requirement is increased to 2 hours, and the fire endurance of the central concourse floor (8m elevation), which is connected to all exhibition halls, is also increased to 2 hours. Thirdly, the fire endurance of the storeroom inside each exhibition hall (store only class C items),the kitchen door is raised to class A, so as to reduce disaster spreading as far as possible.

提高消防设施配置

项目在消防设施配置上提高了要求。如，项目设置两套独立的消防泵房和水池互为备用；所有净高为 8~12m 的空间均采用快速响应喷头；所有展厅、登录大厅及中央廊道的 8 米平台等净高大于 12 米的室内外空间，设计采用消防水炮，消防炮的布置满足同一平面有两门水炮的水射流同时到达被保护区域的任一部位；单层展厅内设置机械排烟系统（超大展厅 17 除外），排烟量按 60m³/h·m² 计算，排烟量增加 10% 的漏风系数；同时在超规空间的外立面上设置一定量的自然排烟窗作为辅助排烟设施。排烟窗有效面积不小于地面面积的 3%。

Improve the Configuration of Fire Facilities

The configuration of fire facilities of the project has been improved. For example, the project is equipped with two sets of independent fire pump rooms and pools, which are mutual backup; All Spaces with a clear height of 8~12m are equipped with quick response sprinkler head; All exhibition halls, arrival halls, central concourse and other indoor and outdoor spaces with a clear height of more than 12 meters are designed with fire water cannons. The arrangement of fire cannons guarantees that water jets with two water cannons on the same plane can reach any part of the protection area at the same time. A mechanical smoke exhaust system is set up in the single-floor exhibition hall (except the superlarge exhibition hall 17). The smoke exhaust is 60m /h·m², and has increased the air leakage coefficient by 10%. At the same time, a certain amount of natural smoke exhaust windows are set on the facade of the supernormal space as auxiliary smoke exhaust facilities. The effective area of smoke exhaust window is no less than 3% of the ground area.

严格的防火空间分隔设计

在防火空间划分上，项目进行了严格的分隔设计。为确保展厅空间的相对消防独立，其周边设备房、办公、餐饮等配套空间与展厅空间均采用防火墙及防火门进行防火分隔。登录大厅空间作为项目中比较特殊的空间（位于 L1、L2 东侧，为单层空间，面积约 1.5 万平米），被界定为纯交通空间使用，在确保登录大厅无任何火灾荷载的同时，将周围配套用房（如登记、行李寄存等空间）用防火墙、防火卷帘等与之进行完全分隔。

The Strict Design of fireproof space Division

On the division of fireproof space, the project has carried out strict design. In order to guarantee the relative independence of each exhibition hall, the surrounding equipment room, office, catering and other supporting spaces and exhibition hall are separated by firewalls and fire doors. As a special space in the project, the arrival hall (located in the east of L1 and L2, single-story space with an area of 1.5 square meters) is defined as a pure traffic space, which should be guaranteed with no fire load, and at the meantime, the surrounding supporting space (such as registration, luggage space) should use firewall and fire shutter to separate from it completely.

图 4-82　登录大厅防火分区平面示意图
Figure 4-82　Fire Compartment Plan of Arrival Hall

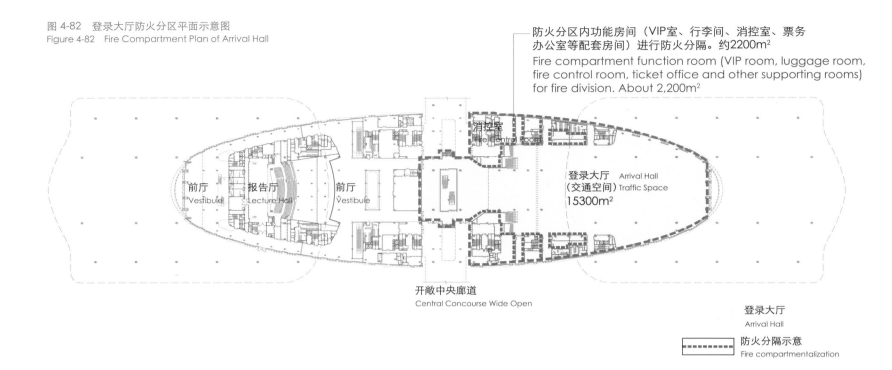

防火分区内功能房间（VIP室、行李间、消控室、票务办公室等配套房间）进行防火分隔。约2200m²
Fire compartment function room (VIP room, luggage room, fire control room, ticket office and other supporting rooms) for fire division. About 2,200m²

消控室
Fire Control Room

前厅
Vestibule

报告厅
Lecture Hall

前厅
Vestibule

登录大厅
（交通空间）Traffic Space
15300m²
Arrival Hall

VIP

开敞中央廊道
Central Concourse Wide Open

登录大厅
Arrival Hall

防火分隔示意
Fire compartmentalization

防火分隔、疏散距离的解决方案

在上述一系列消防加强措施保障下，超大空间才拥有细化消防设计的基础。结合不同空间的使用要求、尺度大小，寻求合理的防火分隔手段，既解决各空间消防难题，又能保全空间完整和使用效率，是本项目的最大挑战。

Spatial Transition Belt

The gap between the exhibition center and the city is embedded with a suture zone as a function supplement as well as a space transition, which is intended to eliminate the separation between the exhibition center and the urban environment caused by the operational needs of the exhibition center, so as to make the space occupied by the exhibition return to the public life of the city.

防火分隔水幕与疏散通道的应用

首先，为确保展厅的快速疏散，在各展厅内设置 6 米宽的主消防疏散通道，此通道仅作为人员通行使用，不得放置任何可燃物；同时在此消防通道上方设置 6 米宽的防火分隔水幕（防火分隔水幕的设计参数按照实体试验研究结果设计），将整个展厅划分为不大于10000 平米的防火单元，从而满足消防规范的要求（图 4-83）。
项目在各展厅内两个对向的安全出口之间增设宽度不小于 6 米的仅供人员通行的通道，在外墙内侧增设宽度不小于 3 米的仅供人员通行的通道，更好地保证人员疏散通道的均匀性（图 4-84）。

The Application of water curtain for fire compartment and evacuation exit

First of all, in order to guarantee the rapid evacuation of the exhibition hall, the main fire evacuation exit with can width of 6 meters is set up in each exhibition hall, which can only be used as personnel passage, and should not allowe any combustible goods within. At the same time, a water curtain for fire compartment with a width of 6 meters was set above the fire evacuation exit (the parameters of the water curtain were designed according to the actual and physical test results), and the whole exhibition hall was divided into fire prevention units no larger than 10,000 square meters, so as to meet the requirements of the fire protection code.(Figure 4-83)
A passage with a width not less than 6 meters is added between the two safety exits in each exhibition hall, and a passage with a width not less than 3 meters is added inside the outer wall, so as to better guarantee the uniformity of the evacuation passage.(Figure 4-84)

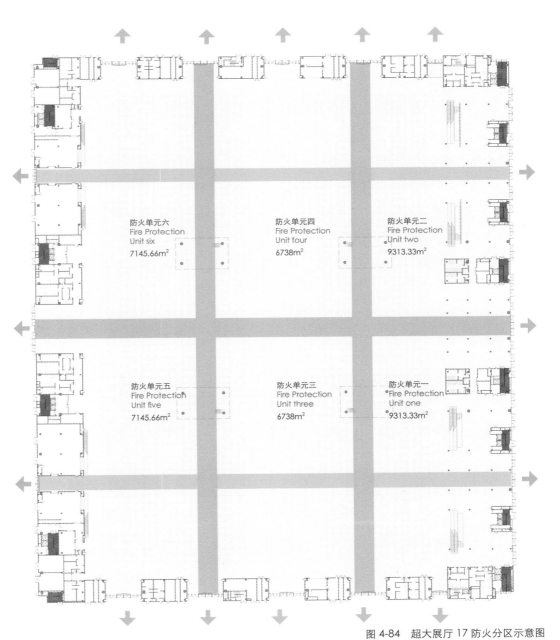

防火单元二 9973m²
Fire Protection Unit two

防火单元一 9869m²
Fire Protection Unit one

图 4-83 标准展厅防火分区示意图
Figure 4-83 Fire Compartment Diagram of Standard Exhibition Hall

防火单元六
Fire Protection
Unit six
7145.66m²

防火单元四
Fire Protection
Unit four
6738m²

防火单元二
Fire Protection
Unit two
9313.33m²

防火单元五
Fire Protection
Unit five
7145.66m²

防火单元三
Fire Protection
Unit three
6738m²

防火单元一
Fire Protection
Unit one
9313.33m²

图 4-84 超大展厅 17 防火分区示意图
Figure 4-84 Fire Compartment Diagram of Super Large Exhibition Hall No. 17

9米消防疏散通道
9m Fire Escape Passway

6米防火水幕
6m Fireproof Water Curtain

6米消防疏散通道
6m Fire Escape Passage

楼梯间
Stairwells

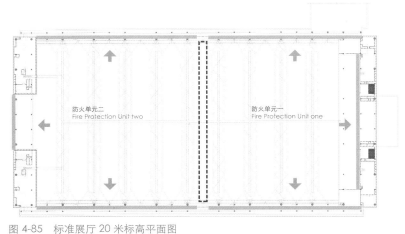

图 4-85   标准展厅 20 米标高平面图
Figure 4-85   The Standard Exhibition Hall 20 Meter Elevation Plan

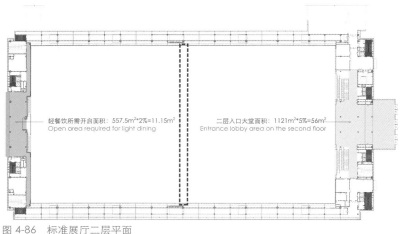

图 4-86   标准展厅二层平面
Figure 4-86   The Standard Exhibition Hall Second Floor Plan

图 4-87   标准展厅三层平面
Figure 4-87   The Standard Exhibition Hall Third Floor Plan

屋顶重力式开启天窗及气动排烟窗的应用

超大展厅 17 号的面积和体量是标准展厅的两倍多，其单层展厅的面积接近防火分区面积规范限值的五倍。防火分隔水幕的解决方式受水量和经济压力的影响已经不再适用。需要引入新的消防策略来解决 17 号展厅的安全设计。

17 号展厅的防火设计策略最终确定为设置三条不同方向的 9 米宽主疏散通道，并将其上方屋面设置重力式开启天窗或气动开启扇（开启面积不低于通道地面面积 25%）；同时在展厅立面上增设不小于展厅地面面积 5% 的气动排烟窗，将展厅物理划分为 6 个小于 10000 平米的防火单元；三条消防通道上仅供人员疏散使用，且无任何火灾荷载。

Application of Gravity Openable Skylight on Roof and Pneumatic Smoke Exhaust Window

The area and volume of the super-large exhibition hall No.17 are more than twice that of the standard exhibition hall, and the area of the single-storey exhibition hall is nearly five times the limit of the fire compartment. Water curtain for fire protection is not applicable due to its required water volume and cost. A new fire strategy needs to be introduced to solve the fire safety issue of exhibition hall no 17.

The final fire design strategy was determined to set three main evacuation exits with a width of 9 meters in different directions, and equipped the roof above them with a gravity openable skylight or a pneumatic fan (the opening area is no less than 25% of the ground area of the exit). Moreover pneumatic smoke exhaust windows with its area no less than 5% of the floor area of the exhibition hall were added on the exhibition hall facade. The exhibition hall was physically divided into 6 fire prevention units less than 10,000 square meters. The three fire passageways are only used for personnel evacuation without any fire load. After experts'review, this area is considered as an outdoor safety area, which meets solves the design requirement that evacuation distance should less than 37.5 meters.

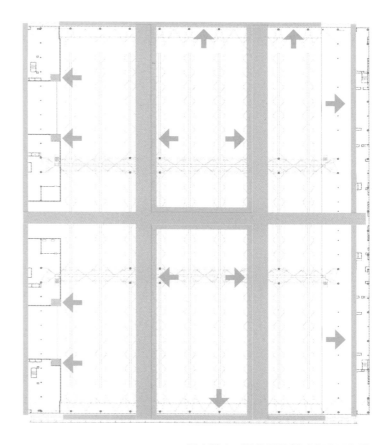

自然排烟方向
Natural Smoke Exhaust Direction

自然排烟风扇（70°下垂或屋面开启）
Natural exhaust fan (70° droop or Roof Open)

消防疏散通道（开启屋面）
Fire Evacuation Passage (Open Roof)

图 4-88-1 17 号展厅 20 米标高平面图
Figure 4-88-1 Exhibition Hall No.17 20 Meter Elevation Plan

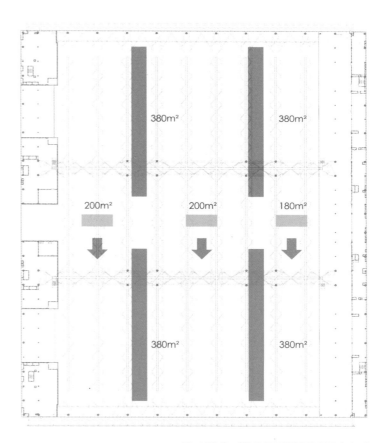

屋盖滑动方向
Roof Sliding Direction

重力滑动屋盖
Gravity Sliding Roof

自动开启天窗
Automatically Open Skylight

图 4-88-2 17 号展厅屋面排烟开启方式
Figure 4-88-2 Roof Smoke Exhaust Opening Mode of Exhibition Hall 17

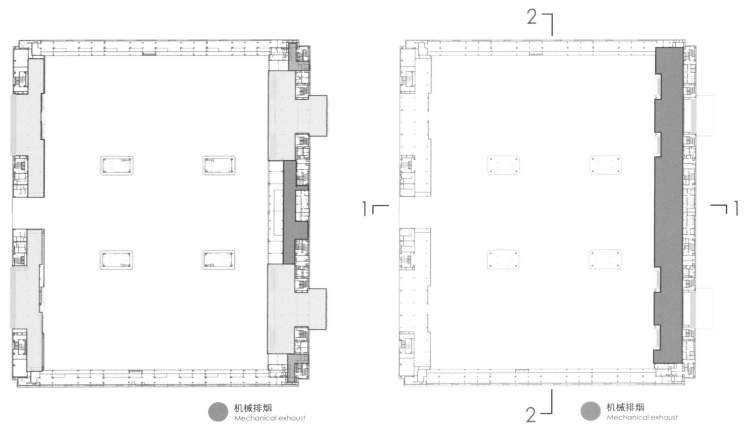

图 4-89-1　17 号展厅二层平面图
Figure 4-89-1　Exhibition Hall No.17 Second Floor Plan

图 4-89-2　17 号展厅三层平面图
Figure 4-89-2　Exhibition Hall No.17 Third Floor Plan

● 机械排烟
Mechanical exhaust

● 机械排烟
Mechanical exhaust

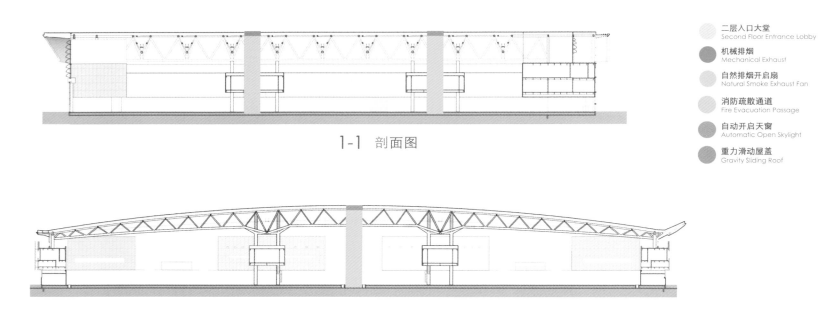

1-1　剖面图

2-2　剖面图

○ 二层入口大堂
Second Floor Entrance Lobby

● 机械排烟
Mechanical Exhaust

○ 自然排烟开启扇
Natural Smoke Exhaust Fan

◐ 消防疏散通道
Fire Evacuation Passage

● 自动开启天窗
Automatic Open Skylight

● 重力滑动屋盖
Gravity Sliding Roof

图 4-89-3　17 号展厅屋面排烟开启方式
Figure 4-89-3　Roof Smoke Exhaust Opening Mode of Exhibition Hall No.17

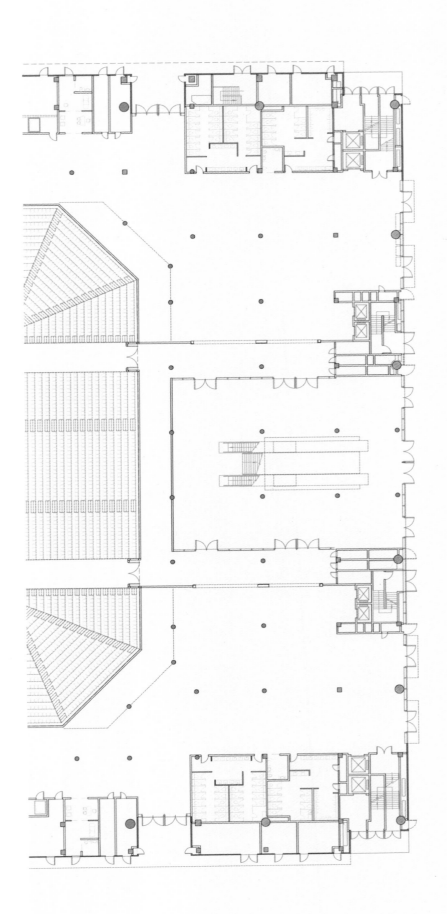

构造与节点
Structure and Node

展厅地面等特殊工艺

展厅地面荷载按 5t/m² 设计，面层做法为合金骨料耐磨地坪，满足机械展、汽车展等重型展览和大型货车运输需求，具体做法如下（图 4-90）：

① 耐磨地坪专用液体硬化剂，合金骨料耐磨面层。

② 130 厚 C30 混凝土随捣随光，配筋并设分隔缝。（兼设备管线层）展位箱、消防坑等洞口四角增设抗裂钢筋。

③ 2 厚聚合物水泥防水涂料。

④ 200 厚 C30 混凝土，双层双向配筋，抹平压光。

⑤ X 厚 6% 水泥石粉渣稳定层，分层压实铺填，压实系数 0.95。

⑥ 钢筋混凝土楼板，板面随捣随抹平。

Special Crafts of the Floor of Exhibition Hall

The floor of the exhibition hall is designed according to a load of 5t/m², and the surface layer is made of alloy aggregate wear-resistant floor, which can meet the requirements of heavy exhibitions such as machinery and automobile exhibition and large truck transportation. The specific methods are as follows(figure 4-90):

① Special liquid hardener is used for wear-resistant floor, alloy aggregate wear-resistant for surface layer.

② 130-thick C30 concrete is rammed and polished, reinforced and separated by joints. (and equipment pipeline layer) booth box, fire pit and other holes are applied with anti-crack reinforcement.

③ Double thick polymer cement waterproof coating.

④ 200 thick C30 concrete, double two-way reinforcement, smoothed and press polished.

⑤ X thick 6% cement powder stable layer, layered compaction and paving, compaction coefficient is 0.95.

⑥ Reinforced concrete floor slab, the surface was rammed and polished.

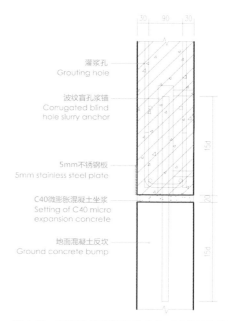

图 4-90　普通内墙底部连接节点（波纹管部位）
Figure 4-90　Connection Node at the Bottom of Common Internal Wall (Corrugeteed Pipe section)

展厅墙体防撞设计

展厅内外围常有货车、叉车穿行应考虑防撞、防剐蹭等的构造做法，本项目对于货车、叉车可能靠近的建筑内、外墙体，采用150mm-160mm厚，3m-6m高的预制混凝土墙体构造做法（图4-91-1，图4-91-2）。

Anti-collision Wall Design of Exhibition Halls

Trucks and forklifts always pass through outside the exhibition hall, so the design for anti-collision and anti-friction should be considered. In this case, precast concrete walls with thickness of 150-160mm and height of 3-6 meters are assembled at the internal and external walls of buildings where trucks and forklifts are more likely to approach.(Figure 4-91-1, 4-91-2)

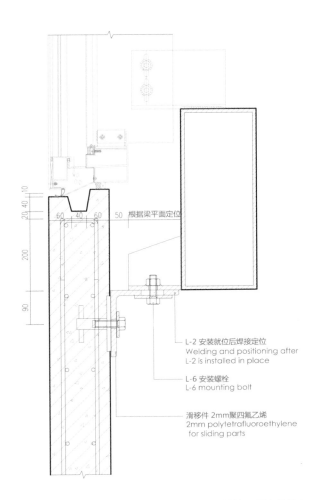

图 4-91-1 外墙顶部连接节点
Figure 4-91-1 Connection Node at the Top of External Wall

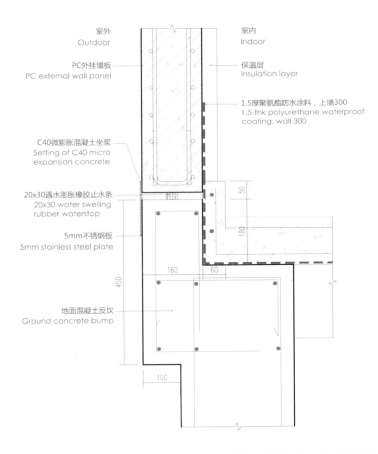

图 4-91-2 外墙底部连接节点（一般部位）
Figure 4-91-2 Connection Node at the Bottom of External Wall (General Part)

展厅大空间隔声吸声构造
Sound Isolation and Absorption Structure of Exhibition Hall's Large Space

① 对产生较大噪声的建筑设备机房墙体进行减振、吸声、隔声设计。

② 针对室外环境的噪声，特别是临近机场带来的航空噪音，在展厅金属屋面、外墙等位置采取了隔声和减噪措施，确保展厅空场时背景噪声的允许噪声级（A声级）不大于55dB且满足运营要求。

③ 展厅室内大空间的内墙面和顶棚采取吸声措施。

① Vibration reduction, sound absorption and sound insulation design shall be carried out on the walls of the equipment rooms which may generate large noise.

② For the noise of outdoor environment, especially the aviation noise brought by the adjacent airport, sound insulation and noise reduction measures were taken at the metal roof and external wall of the exhibition hall to guarantee the permissible noise level (sound level A) of the background noise in the exhibition hall is not more than 55dB.

③ The interior wall and ceiling of the large space in the exhibition hall shall adopt sound absorption measures.

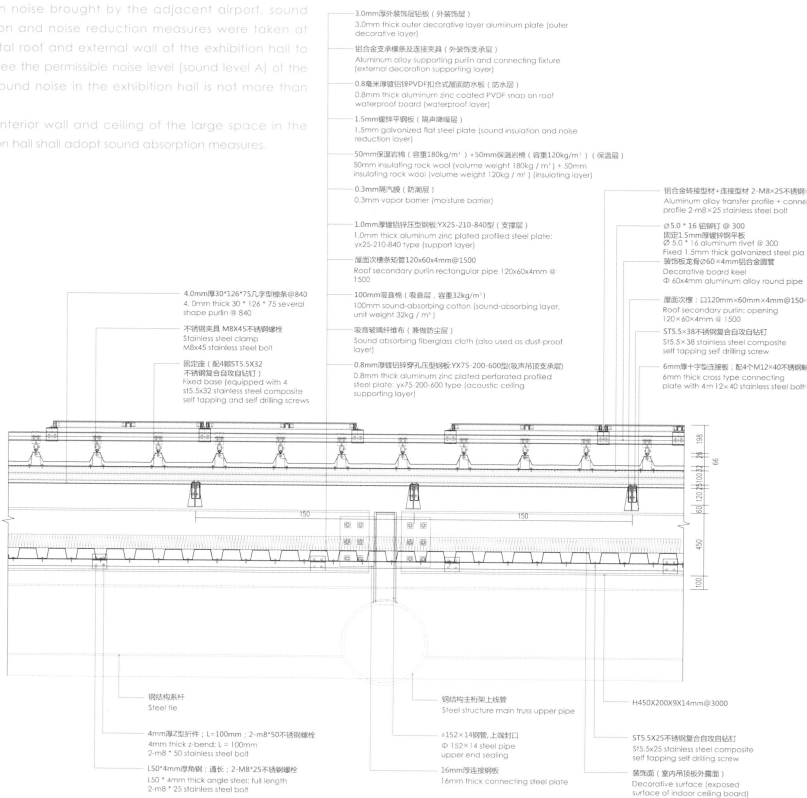

图 4-92 展厅屋面 D 大面区域标准节点详图
Figure 4-92 Typical Node Details of Exhibition Hall Large Area of Roof D

展厅室内大空间的内墙面和顶棚采取吸声措施
Structure Design of Open Roof and Pneumatic Smoke Exhaust Window for the Super Large Exhibition Hall

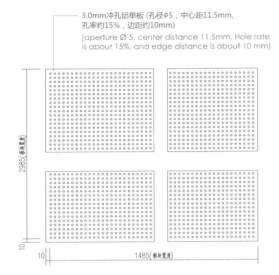

3.0mm冲孔铝单板（孔径ϕ5，中心距11.5mm，孔率约15%，边距约10mm）
(aperture Ø 5, center distance 11.5mm, Hole rate is about 15%, and edge distance is about 10 mm)

图 4-93-1　（东面白色冲孔板，西面白色＋银灰色冲孔板，南北面为银灰色冲孔板）
Figure 4-93-1　(East White Perforated Plate, West White + Silver Gray Perforated Plate, North and South Silver Gray Perforated plate)

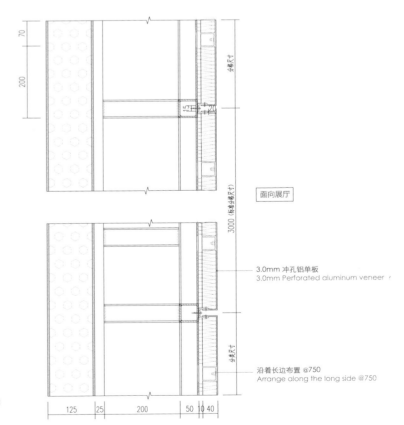

面向展厅

3.0mm 冲孔铝单板
3.0mm Perforated aluminum veneer

沿着长边布置 @750
Arrange along the long side @750

图 4-93-2　（冲孔板后有125厚轻质复合实心墙）
Figure 4-93-2　(125 Thick Light Composite Solid Wall Behind the Perforated Plate)

超大展厅开启屋盖和气动排烟窗构造设计
Structure Design of the Openable Roof and Pneumatic Smoke Exhaust Window for the Super Large Exhibition Hall

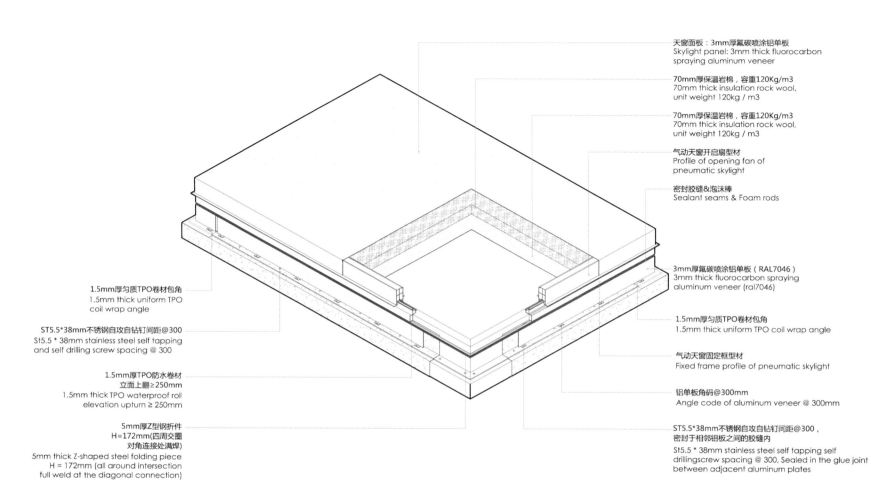

天窗面板：3mm厚氟碳喷涂铝单板
Skylight panel: 3mm thick fluorocarbon spraying aluminum veneer

70mm厚保温岩棉，容重120Kg/m3
70mm thick insulation rock wool, unit weight 120kg / m3

70mm厚保温岩棉，容重120Kg/m3
70mm thick insulation rock wool, unit weight 120kg / m3

气动天窗开启扇型材
Profile of opening fan of pneumatic skylight

密封胶缝&泡沫棒
Sealant seams & Foam rods

3mm厚氟碳喷涂铝单板（RAL7046）
3mm thick fluorocarbon spraying aluminum veneer (ral7046)

1.5mm厚匀质TPO卷材包角
1.5mm thick uniform TPO coil wrap angle

气动天窗固定框型材
Fixed frame profile of pneumatic skylight

铝单板角码@300mm
Angle code of aluminum veneer @ 300mm

ST5.5*38mm不锈钢自攻自钻钉间距@300，密封于相邻铝板之间的胶缝内
St5.5 * 38mm stainless steel self tapping self drillingscrew spacing @ 300, Sealed in the glue joint between adjacent aluminum plates

1.5mm厚质TPO卷材包角
1.5mm thick uniform TPO coil wrap angle

ST5.5*38mm不锈钢自攻自钻钉间距@300
St5.5 * 38mm stainless steel self tapping and self drilling screw spacing @ 300

1.5mm厚TPO防水卷材立面上翻≥250mm
1.5mm thick TPO waterproof roll elevation upturn ≥ 250mm

5mm厚Z型钢折件H=172mm(四周交圈对角连接处满焊)
5mm thick Z-shaped steel folding piece H = 172mm (all around intersection full weld at the diagonal connection)

图 4-94 气动天窗主要构造三维节点图
Figure 4-94 Axonometric of Main Structure Node of Pneumatic Skylight

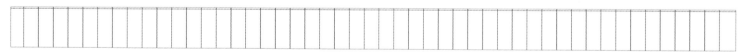

图 4-95-1 A 面视图
Figure 4-95-1 Plane-A View

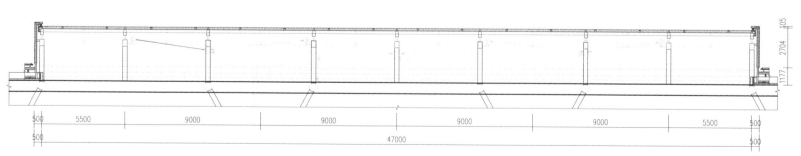

图 4-95-2 A - A
Figure 4-95-2 A-A surface view

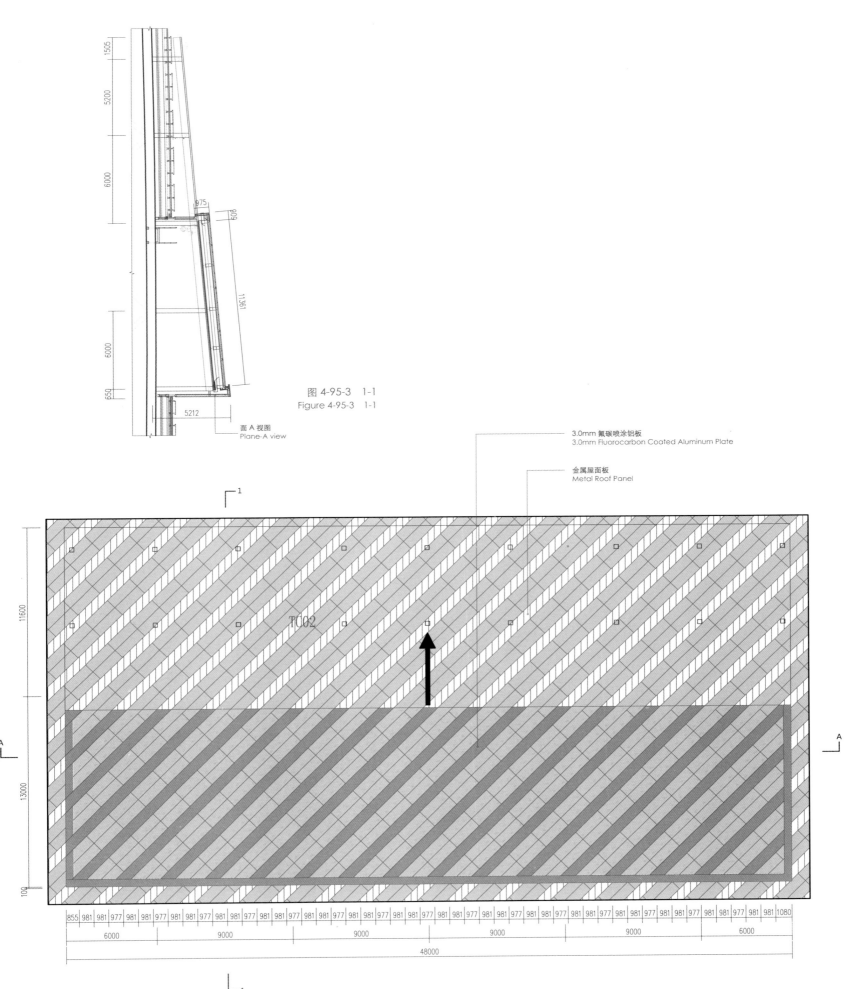

图 4-95-3　1-1
Figure 4-95-3　1-1

面 A 视图
Plane-A view

3.0mm 氟碳喷涂铝板
3.0mm Fluorocarbon Coated Aluminum Plate

金属屋面板
Metal Roof Panel

图 4-95-4　展厅 D 屋面滑动排烟窗大样
Figure 4-95-4　Detail Drawing of Sliding Smoke
Exhaust Window on Roof of Exhibition Hall D

管沟与接驳井

在展厅地面上平行铺设贯穿整个展厅的综合管沟，每条管沟的间距为 6m，并按 6m×9m 间距布置接驳井。主管沟最小截面尺寸为 1000mm×800mm，次管沟最小截面尺寸为 650mm×650mm。为方便开启及更换，管沟上方铺设的可移动盖板采用重量轻，燃烧性能 A 级的复合材料盖板，承载力要求应大于展厅地面设计承载力。

Pipe Ditches and Connection Well

A comprehensive pipe ditch is was laid in parallel through the floor of the whole exhibition hall. The gap of each pipe ditch is 6m, and connection wells were arranged according to a gap of 6x9m. The minimum section size of main pipe ditch is 1000x800, and that of secondary pipe ditch is 650x650. To open and replace conveniently, the movable cover plate laid above the pipe ditch is made of composite material cover plate with light weight and grade A combustion performance, with a load capacity higher than the designed load capacity of the floor of the exhibition hall.

次沟盖板 760*600*80
Secondary Ditch Cover Plate
760 * 600 * 80

2-提手孔
2- handle hole

次沟出线盖板 760*600*80
Cover Plate of Secondary Outlet
760 * 600 * 80

出线块
Outlet

出线块
Outlet

出线口
Outlet

图 4-96-1、2、3、4、5、6、7、8 管沟盖板深化设计图
Figure 4-96-(1-8) Detailed Drawing of Trench Cover Plate

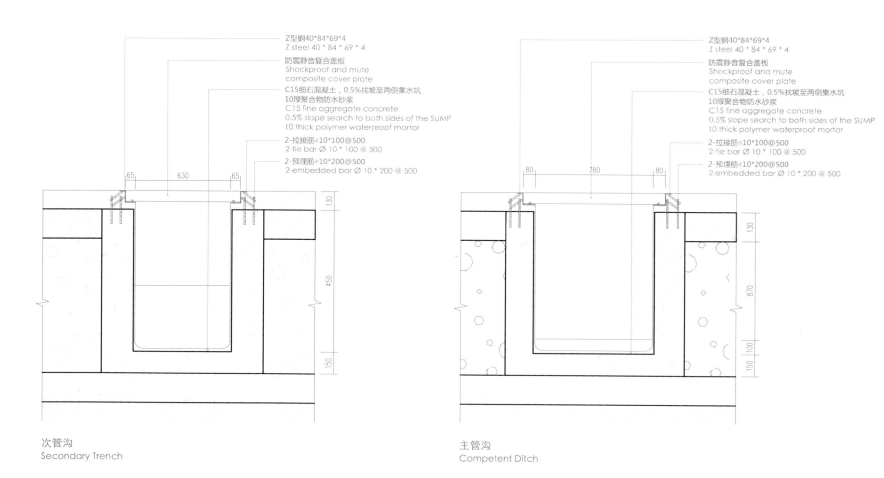

次管沟
Secondary Trench

主管沟
Competent Ditch

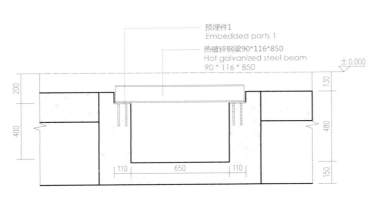

预埋件1
Embedded parts 1

热镀锌钢梁90*116*850
Hot galvanized steel beam
90 * 116 * 850

±0.000

C - C 视图
C - C Section

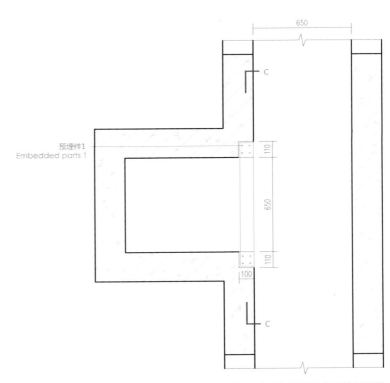

预埋件1
Embedded parts 1

主结构层次沟与接驳井交叉处平面图
Plan of intersection of main structure level
ditch and connecting well

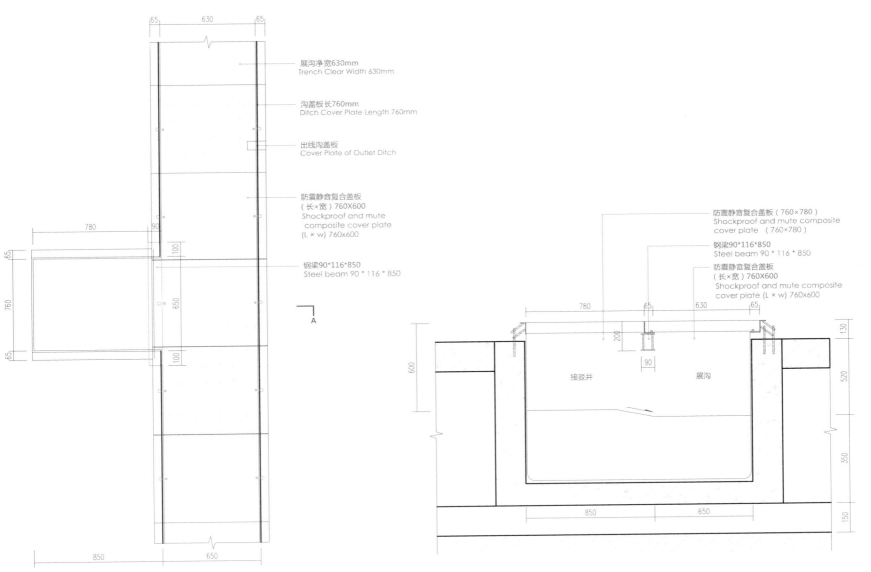

展沟净宽630mm
Trench Clear Width 630mm

沟盖板长760mm
Ditch Cover Plate Length 760mm

出线沟盖板
Cover Plate of Outlet Ditch

防震静音复合盖板
(长×宽) 760X600
Shockproof and mute
composite cover plate
(L × w) 760x600

钢梁90*116*850
Steel beam 90 * 116 * 850

A

防震静音复合盖板 (760×780)
Shockproof and mute composite
cover plate (760×780)

钢梁90*116*850
Steel beam 90 * 116 * 850

防震静音复合盖板
(长×宽) 760X600
Shockproof and mute composite
cover plate (L × w) 760x600

接驳井   展沟

展厅接驳进深化设计图
Exhibition Hall Connection Detailed Drawing

A - A 视图
A - A View

展厅货运门

展厅的货运门需允许大型货车直接驶入展厅并具备消防车驶入及平时人员疏散的功能，因此需要特殊构造设计；设计采用了平开式和滑移式两种做法，大门有 6mx6m，8mx6m，24mx12m 三个尺寸；大门上的疏散小门尺寸为 3mx3m。

Freight Gate of the Exhibition hall

The freight gate of the exhibition hall should allow large trucks or fire trunk to enter directly and should also be able to use as personnel evacuation exit, which requires special structural design; swing gate and sliding gate are used in this case, with different sizes of 6x6m, 8x6m, 24x12m; the evacuation door of the gate is 3x3m.

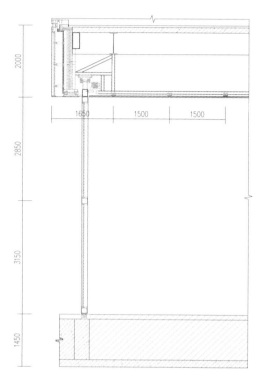

图 4-97-1　8×6 平移门大样图，A-A 剖面
Figure4-97-1　Detail Drawing of 8 × 6 Sliding Door, Section A-A

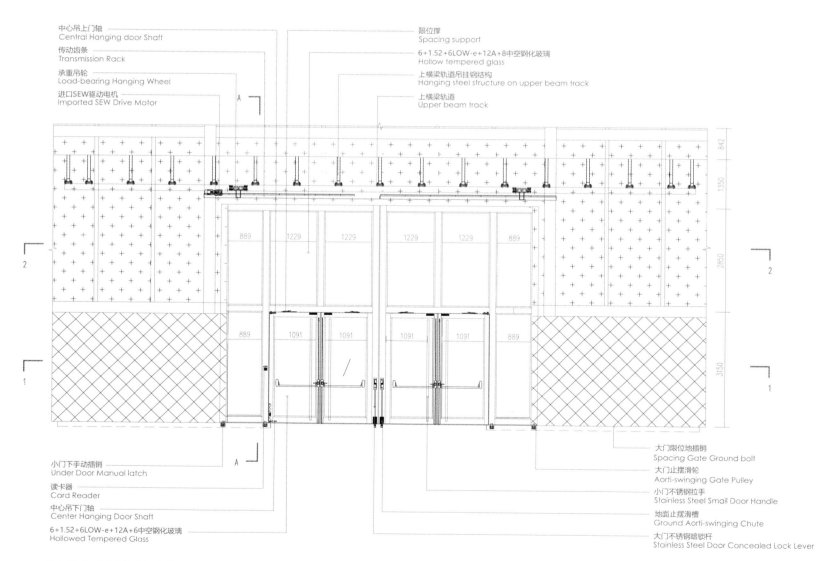

中心吊上门轴　Central Hanging door Shaft
传动齿条　Transmission Rack
承重吊轮　Load-bearing Hanging Wheel
进口SEW驱动电机　Imported SEW Drive Motor

限位撑　Spacing support
6+1.52+6LOW-e+12A+8中空钢化玻璃　Hollow tempered glass
上横梁轨道吊挂钢结构　Hanging steel structure on upper beam track
上横梁轨道　Upper beam track

小门下手动插销　Under Door Manual latch
读卡器　Card Reader
中心吊门轴　Center Hanging Door Shaft
6+1.52+6LOW-e+12A+6中空钢化玻璃　Hollowed Tempered Glass

大门限位地插销　Spacing Gate Ground bolt
大门止摆滑轮　Aorti-swinging Gate Pulley
小门不锈钢拉手　Stainless Steel Small Door Handle
地面止摆滑槽　Ground Aorti-swinging Chute
大门不锈钢暗锁杆　Stainless Steel Door Concealed Lock Lever

图 4-97-2　8 * 6 平移门大样图
Figure 4-97-2　Detail drawing of 8 * 6 sliding door

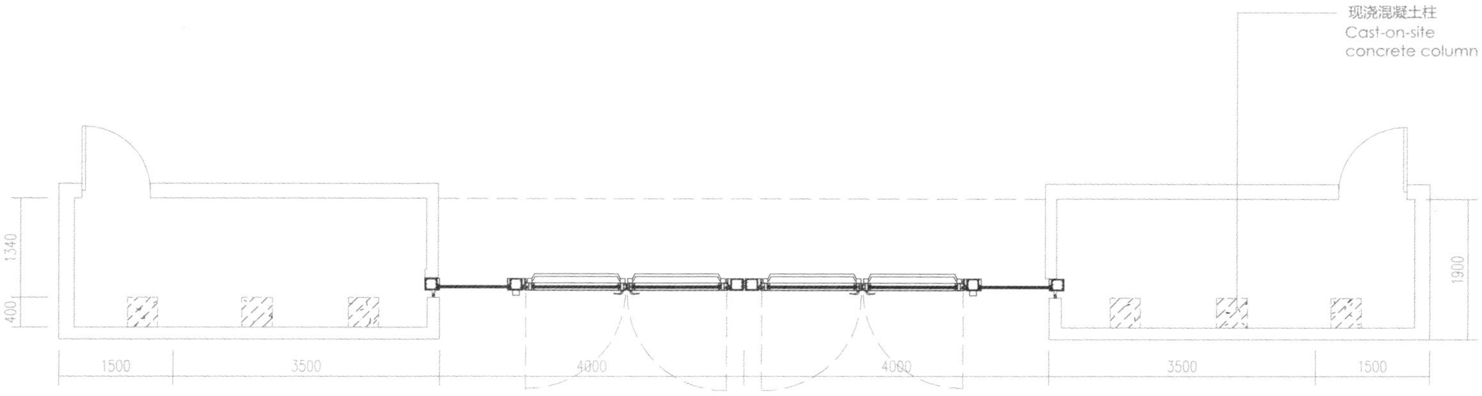

现浇混凝土柱
Cast-on-site
concrete column
1340
400
1900
1500
3500
4000
4000
3500
1500

图 4-97-3 8×6 平移门大样图，1-1 剖面
Figure 4-97-3 Detail Drawing of 8 × 6 Sliding door, Section 1-1

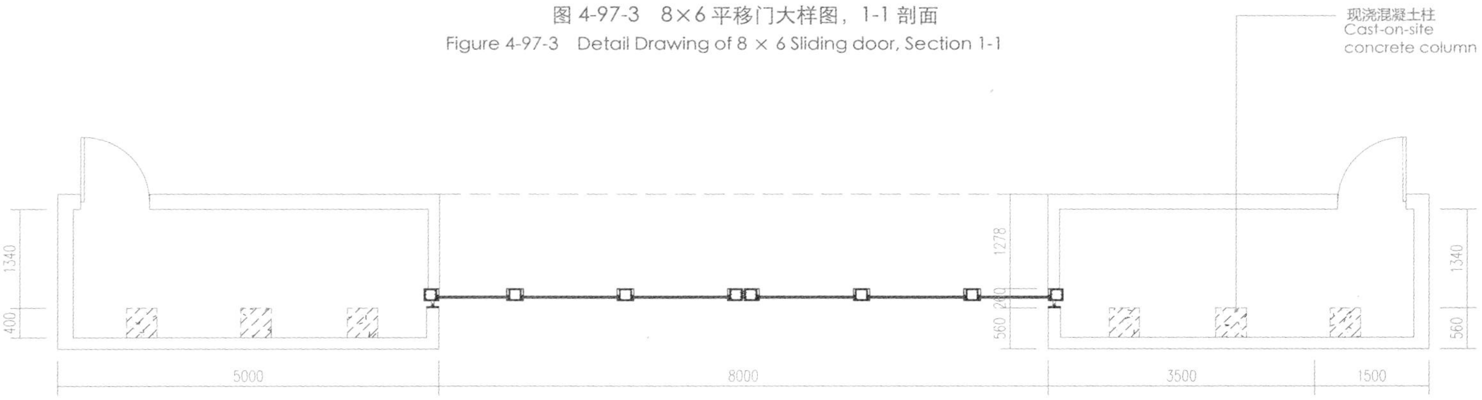

现浇混凝土柱
Cast-on-site
concrete column
1340
400
1278
560
1340
560
5000
8000
3500
1500

图 4-97-4 8×6 平移门大样图，2-2 剖面
Figure 4-97-4 Detail Drawing of 8 × 6 Sliding door, Section 2-2

大门天地插销锁
Gate lock
疏散指示
Evacuation instructions
小门磁力锁
Small door magnetic lock
限位撑（带定位功能）
Spacing Support
大门承重天地轴
Load Bearing Gate Vertical Shaft
供电过线器
Power supply line
大门承重天地轴
Load Bearing Gate
Vertical Shaft
下门轴固定
Lower Door fixed Shaft
中心吊
Center Crane
小门下手动暗插销
Manual concealed
latch under the door
A
1
2
5845
3000
3000
3000
6000
室外立面图

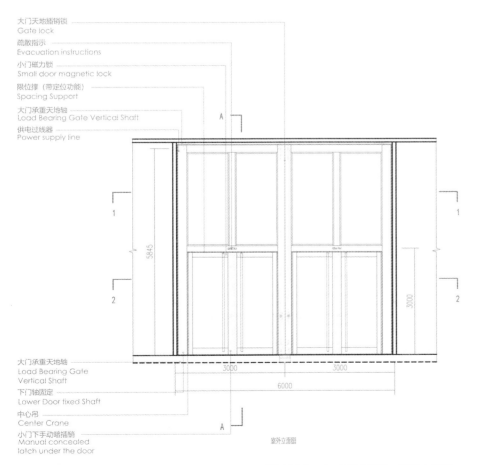

图 4-98-1 6×6 门大样图，室外立面图
Figure 4-98-1 6 × 6 Door Detail Drawing Elevation

1000
500
1000
1500
1570
2200
2370
6000
±0.000

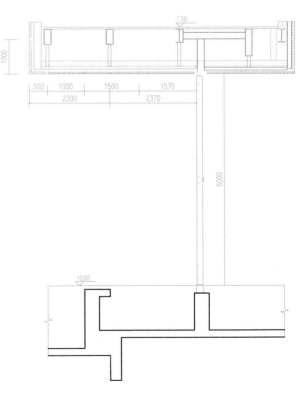

图 4-98-2 6×6 门大样图，A - A 剖面
Figure 4-98-2 6 × 6 Door Detail Drawing, Section A-A

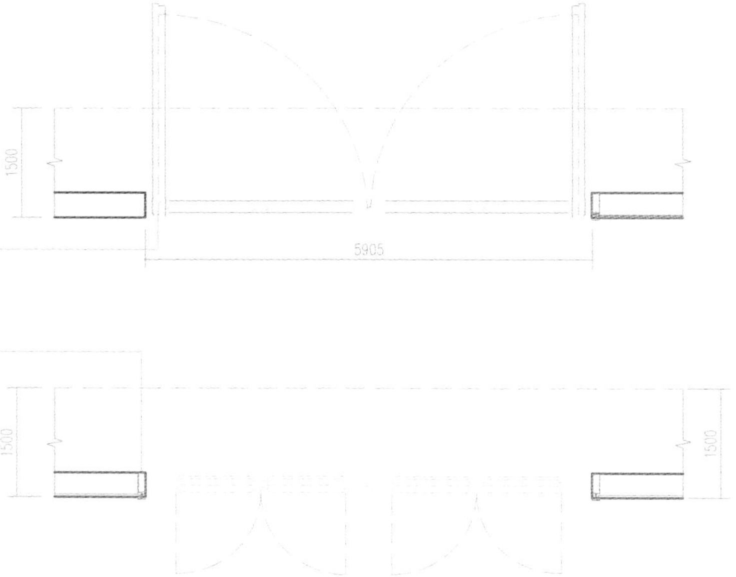

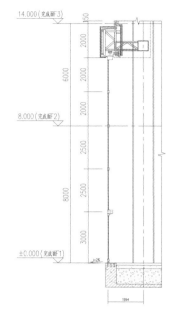

供电边线器
Power supply
line

室内出门按钮
Indoor exit
button

图 4-98-3　6×6 门大样图，1-1&2-2 剖面
Figure 4-98-3　6 × 6 Door Detail Drawing, 1-1 & 2-2 Section

图 4-99-1　超大门大样图，2-2
Figure 4-99-1　Super Gate Detail Drawing, Section 2-2

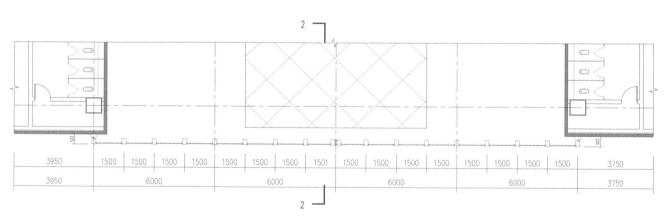

图 4-99-2　超大门大样图，1-1 剖面
Figure 4-99-2　Super Gate Detail Drawing, Section 1-1

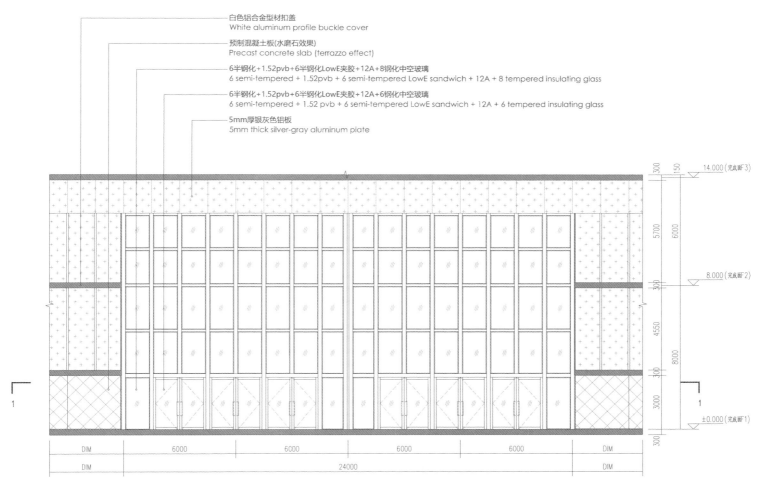

白色铝合金型材扣盖
White aluminum profile buckle cover

预制混凝土板(水磨石效果)
Precast concrete slab (terrazzo effect)

6半钢化+1.52pvb+6半钢化LowE夹胶+12A+8钢化中空玻璃
6 semi-tempered + 1.52pvb + 6 semi-tempered LowE sandwich + 12A + 8 tempered insulating glass

6半钢化+1.52pvb+6半钢化LowE夹胶+12A+6钢化中空玻璃
6 semi-tempered + 1.52 pvb + 6 semi-tempered LowE sandwich + 12A + 6 tempered insulating glass

5mm厚银灰色铝板
5mm thick silver-gray aluminum plate

图 4-99-3　超大门大样图，立面图
Figure 4-99-3　Super Gate Elevation Detail Drawing

地上综合设备管廊

地上设备管廊内有：强弱电桥架、空调冷冻水管、消火栓管、管廊喷淋管。为了方便检修维护，还铺有钢格栅检修。

Above-Ground Utility Turnnel

Inside the above-ground utility tunnel, there are strong and weak electric bridge frame, air conditioning cooling water pipe, hydrant pipe and pipe corridor sprinkler pipe.
In order to facilitate the maintenance, it also paved with steel grid maintenance.

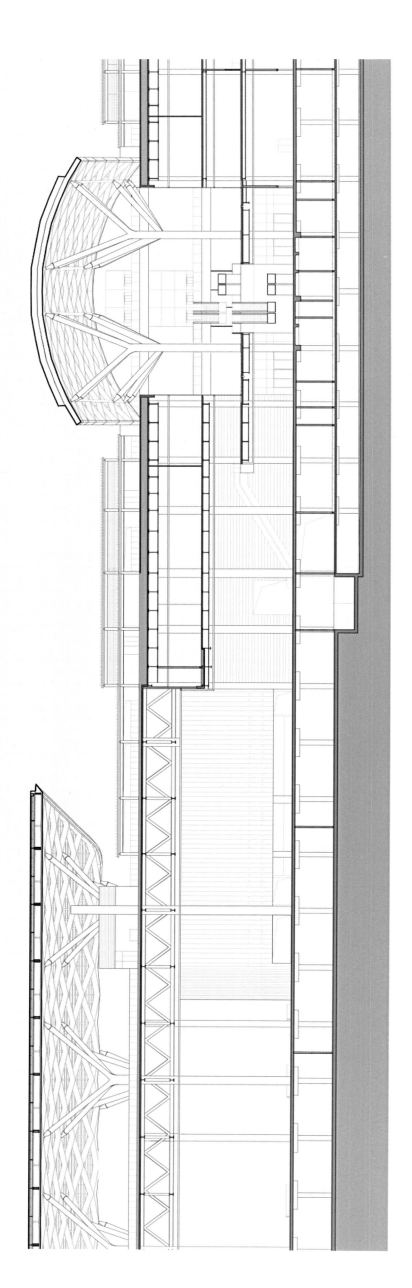

防水设计
Waterproof Design

利用景观设施滞留和净化雨水以达到海绵的作用，是提升雨洪水管理的一个新途径。而坚实可靠的建筑防水，则是确保建筑基本功能得以实现的保障，特别是对于应用海绵城市理念的建筑项目而言更显得重要。

深圳国际会展中心作为一个世界级湾区的"巨无霸"项目，在设计过程中遇到防水设计难点，也多与"超"有所关联，例如"超长"地下室和"超大"屋面的防水设计。

Using landscape facilities to hold and purify rainwater and make it like a sponge is a new way of improving rain and flood management. Solid and reliable building waterproof is to guarantee the basic functions of the building, and more importantly, to realize the sponge city concept for this project.

As a world-class giant project in the bay area, the Shenzhen World encounters difficulties in the waterproof design during the design process, which are mostly related to the "giant" issue, such as the waterproof design of the super long basement and the super large roof.

地下室防水设计的两个难题

在深圳国际会展中心地下室防水设计过程中，存在两个重要难题：

一是单层规模巨大，超长超大地下室的温度效应、收缩、徐变效应等十分显著，易造成混凝土结构的开裂渗漏。

二是场地软弱淤泥层分布深厚，基坑将超挖回填后可能继续产生沉降和不均匀沉降，对防水层造成损害。

这两个难题分别通过优化主体结构设计和选用合适外防水材料的方式，有针对性地进行解决。

Two Major Problems in the Design of Waterproof Basement

During the process of waterproofing design for the basement, there are two important problems:

Firstly, the temperature effect, shrinkage, and creep effect is very severe in this large and super long single-story basement, which would easily cause cracking and leakage of concrete structure.

Secondly, the weak silt layer of the site is deeply distributed, and the foundation pit may continue to settle, and unevenly settlement after over excavation and backfilling, could cause damage to the waterproofing Layers.

These two major problems were solved by optimizing the main structure design and selecting suitable waterproof materials.

图 4-100　金属屋面施工实景图
Figure 4-100　Metal Roof Construction

超长地下室防水设计

深圳国际会展中心项目地下室共有二层，底板绝对标高为 -3.4m，开挖深度约10.4m。地下室防水按一级防水设防。地下水埋深较浅，丰水期高程为 1.45~5.2m，枯水期平均高程为 2.11m。地下室总长度约为1750m，功能为汽车库，设备机房，厨房及安保用房。地下室被凤塘大道分隔为南北两部分，北侧地下室最大长度 570m，南侧最大长度1250m。

Waterproof Design for the Super Long Basement

There are two floors in the basement of the Shenzhen World. The absolute height of the baseplate is -3.4 meters, and the cutting depth is about 10.4 meters.

The waterproof of the basement is graded as class A. The underground water is buried shallowly, and the height is about 1.45 ~ 5.2 meters during the wet season and 2.11 meters during the dry season. The total length of the basement is about 1.75km, and it is used as parking lot, equipment room, kitchen and security room. The basement is divided into north and south parts by Fengtang avenue, with a maximum length of 570 meters on the north and 1,250 meters on the south.

结构自防水设计

地下室围护体系抗渗等级按 P10 设计，但结构防水不仅是抗渗，更重要的是减少裂缝。在设计前期，设计团队通过计算分析，取消伸缩缝，优化后浇带，实现超长无缝地下室，减少因缝引起的渗漏问题。具体措施有：

地下室底板跳仓浇筑，底板上间隔设置用于削弱底部刚度、吸收结构变形量的结构沟，用于代替传统的底板变形缝，并在结构沟内设置后浇带。后浇带采用超前止水及预注浆技术，通过注浆管系统密封接缝区域的缝隙和孔洞。

地下室外墙设置诱导缝，使混凝土可以无缝连续浇筑。

通过优化混凝土材料、外加剂、掺合料、混凝土强度、施工措施、养护要求等，减少混凝土构件裂缝的产生。

Structure Self-waterproof Design

The basement enclosure system is designed according with the impermeability grade P10. However, the structural waterproof should not only be impermeable, but should also need to reduce the cracks. In the early stage of the design, the design team eliminated the expansion joint through calculation and analysis, and optimized the post-cast belt, made an ultra-long seamless basement to reduce the leakage caused by the joint. Specific measures taken include:

The basement baseplate applied the sequence easting method, the spacing above is set to weaken the bottom stiffness, and the structural groove absorbing structural deformation is used to replace the traditional deformation seam of the baseplate, and the post-cast belt is set in the structural groove. The post-cast belt adopts the advanced stop water and pre-grouting technology to seal the cracks and holes in the joint area throughout the grouting pipe system.

The exterior wall of the basement is provided with induction joints so that the concrete can be continuously casted seamlessly.

Through optimized concrete materials, admixtures, admixtures, concrete strength, construction measures, maintenance requirements, etc., the cracks in concrete unit can be reduced or avoided.

采用成熟可靠外防水材料和做法

除了对结构体系进行裂缝控制外，合理的防水系统也至关重要。由于地下室基坑超挖近 1m，石粉渣回填后，底板下砼垫层、防水保护层及侧墙保护层有可能一起沉降。

因此在进行地下室外防水设计时，取消了底板防水保护层，采用非沥青基高分子自粘胶膜防水卷材预铺反粘法施工，自粘卷材面层能与后浇筑混凝土结构层粘结为一体，避免沉降导致的防水层剥离，并能有效控制"窜水"。对于地下室侧墙，在基层修补后，采用渗透环氧处理，避免采用找平层的开裂风险。

图 4-101 金属屋面完成实景图
Figure 4-101 The Metal Roof

Mature and Reliable External Waterproof Materials and Practices

Except for crack control of the structural system, reasonable waterproof system is crucial. Since the foundation pit of the basement is over dug for nearly 1 meter, the concrete cushion, waterproof protective layer and side wall protective layer under the baseplate may be settled together after the stone powder slag is backfilled.

Therefore, in the design of the underground outdoor waterproof measures, the waterproof protective layer of the baseplate was canceled, and instead, there applied the pre-laying anti-bonding method of non-asphaltene-based polymer self-adhesive film waterproof rolling material. The self-adhesive rolling material surface layer could be bonded with the post cast concrete structure layer, to avoid the stripping of the waterproof layer caused by settlement, and effectively controlling the water interflow. For the basement side wall, after the base is repaired, the penetration epoxy treatment is adopted to avoid the cracking risk of the screed-coat.

超大金属屋面的防水设计

金属屋面系统属于构造排水系统，常因构造层缺陷、板型不牢固、材料热胀冷缩等原因引起渗漏。在深圳国际会展中心项目中，除了超大体量使得热胀冷缩尤为突出外，巨型尺度的屋面造型对排水组织和抗风性能的影响也不可忽视。

因此在屋面设计中针对性的采取了如下解决措施：

首先，优化屋面曲面的排水坡度，坡度随着汇水面积的增加而加陡，控制平均坡度在 10% 左右。并通过第五立面造型设计，将东西向长边起拱的屋面调整为南北向短边起拱，将位于展厅使用空间上方的内天沟调整为位于展厅使用空间两侧的室外天沟，避免渗漏风险出现在室内空间的正上方。金属屋面直立锁边板相应的采用超长板型，不设纵向搭接。最大长度约为 240m。

其次，屋面尺度大，临海台风多，因此选取了力学性能和抗风揭性能较好的镀铝锌 PVDF 压型钢板屋面系统。并针对檐口等特殊部位增设抗风锁夹。
而屋面构造设计采用的是镀铝锌压型钢板直立锁边屋面板 +TPO 防水卷材的两道防水形式。TPO 防水卷材安装于固定支座的底部，简化了穿透部位防水处理。

最后，为解决超大屋面的热胀冷缩问题，方案采用的是摇摆头构造的不滑动固定支座，以避免构件之间因互相约束伸缩变形造成的损坏漏水。屋面板的伸缩量控制在两个固定支座之间，通过每个跨内板面的轻微变形，将整体伸缩量化整为零。

Waterproof Design of Oversized Metal Roof

The metal roofing system is part of the structural drainage system, which often easily cause leakage because of the defects in the structural layer, the unstable plate type, the thermal expansion and the cold shrinkage of materials. In this project, due to its extra-large size, the expansion and contraction problem is more severe, and in addition, the influence of the surface modeling on drainage and wind resistance should not be ignored either.

Therefore, in the roof design, following measures were taken:

Firstly, the drainage slope of the roof surface is optimized. The slope becomes steeper with the increase of the catchment area, and the average gradient is controlled around 10%. Through the fifth façade design, the roof with a long side arched from east to west was adjusted to a short side arched from south to north, and the inner gutter above the exhibition hall was adjusted to an outdoor gutter on both sides of the exhibition hall so as to avoid the risk of leakage directly above the indoor space. The vertical locking edge plate of metal roof uses corresponding extra-long plate, without any longitudinal lap. The maximum length is ground 240 meters.

Secondly, the roof scale is large, and there are relative more typhoons within the area, therefore it adopted the aluminum zinc PVDF pressure steel plate roof system with advanced mechanical properties and anti-wind peeling performance. There also set up wind resistant lock on the eaves and other special parts.
The roof structure is designed with two forms of waterproof: the aluminum and zinc plate vertical locking edge roof plate and TPO waterproof roll. The TPO waterproof roll is installed at the bottom of the fixed bearing to simplify the waterproof treatment of the penetration part.

Finally, in order to solve the problem of hot expansion and cold contraction of the oversized roof, the non-sliding fixed bearing with a swing head structure is adopted to avoid the damage and water leakage caused by the mutual constraint of expansion deformation between components. The expansion of roof plate is controlled between two fixed bearings, so the overall expansion can be ignored due to the slight deformation of each plate surface.

图 4-102 金属屋面版型
Figure 4-102 Metal Roof Corrogated Plate

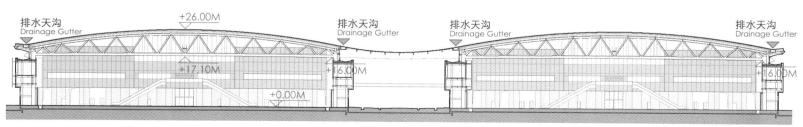

图 4-103 优化后屋面造型方案
Figure 4-103 Optimized Roof Modeling Scheme

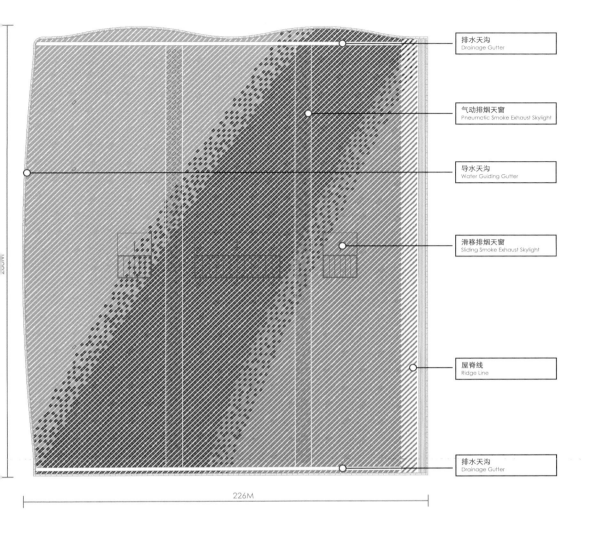

排水天沟
Drainage Gutter

气动排烟天窗
Pneumatic Smoke Exhaust Skylight

导水天沟
Water Guiding Gutter

滑移排烟天窗
Sliding Smoke Exhaust Skylight

屋脊线
Ridge Line

排水天沟
Drainage Gutter

226M

图 4-104 金属屋面完成效果
Figure 4-104 Completed Metal Roof

图 4-105 摇摆支座节点
Figure 4-105 Sway Support Node

3.0mm厚外装饰层铝板（外装饰层）
3.0mm Thick Outer Decorative Layer Aluminum
Plate (outer decorative layer)

0.8mm镀铝锌氟碳辊涂直立锁边钢
屋面板
0.8mm Aluminized Zinc Fluorocarbon Roller
Coated Upright Seam Steel Roof Panel

主结构
Main Structure

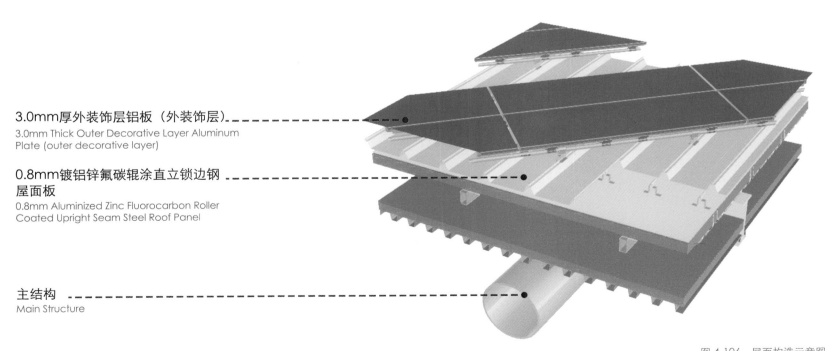

图 4-106 屋面构造示意图
Figure 4-106 Roof Structure Detail

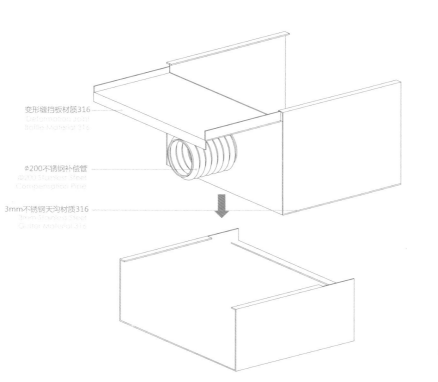

变形缝挡板材质316

φ200不锈钢补偿管

3mm不锈钢天沟材质316

图 4-107　普通内墙底部连接节点（波纹管道）
Figure 4-107　Common Inner Wall Bottom Connection Node (Corrugated Pipe)

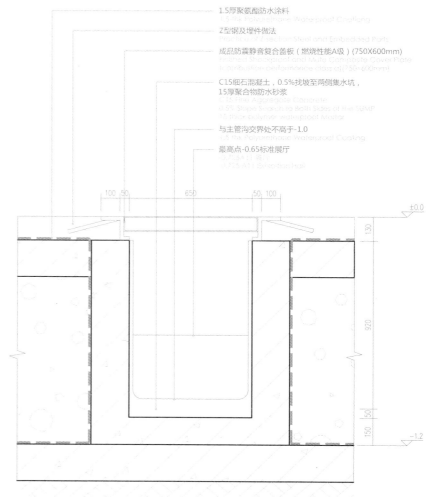

1.5厚聚氨酯防水涂料

Z型钢及埋件做法

成品防震静音复合盖板（燃烧性能A级）(750X600mm)

C15细石混凝土，0.5%找坡至两侧集水坑，15厚聚合物防水砂浆

与主管沟交界处不高于-1.0

最高点-0.65标准展厅

图 4-108　次管沟节点图
Figure 4-108　Secondary Pipe Trench Node

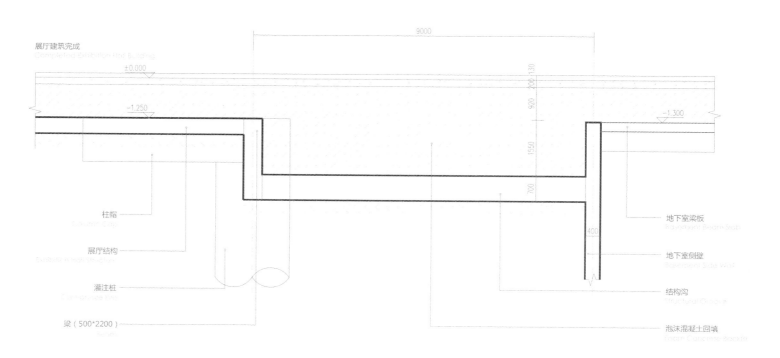

展厅建筑完成

柱帽

展厅结构

灌注桩

梁（500*2200）

地下室梁板

地下室侧壁

结构沟

泡沫混凝土回填

图 4-109　展厅短边与地下室相连处结构沟示意图
Figure 4-109　Structural Ditch at the Connection between the Short Side of Exhibition Hall and Basement Detail

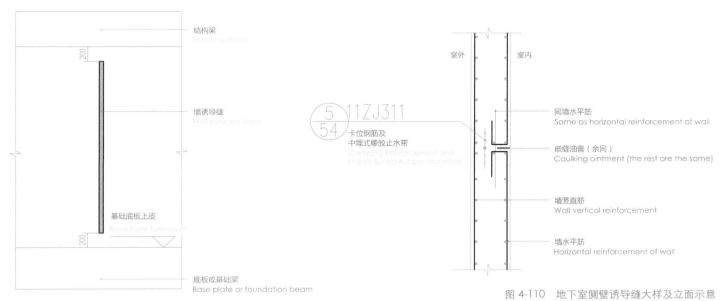

结构梁
Structural beam

墙诱导缝
Wall induced joint

基础底板上皮
Basic Plate Epithelium

底板或基础梁
Base plate or foundation beam

5
54
11ZJ311
卡位钢筋及
中埋式橡胶止水带
Changing Reinforcement and Middle Buried Rubber Waterproof

室外 室内

同墙水平筋
Same as horizontal reinforcement of wall

嵌缝油膏（余同）
Caulking ointment (the rest are the same)

墙竖直筋
Wall vertical reinforcement

墙水平筋
Horizontal reinforcement of wall

图 4-110　地下室侧壁诱导缝大样及立面示意
Figure 4-110　Detailed Drawing and Elevation Sketch of Induced Joint of Side Wall of Basement

0.8毫米厚镀铝锌PVDF扣合式屋面防水板（防水层）
0.8 mm thick aluminum-zinc PVDF buckle
roof waterproof plate (waterproof layer)

图 4-111　屋面板详图
Figure 4-111　Roof Panel Details

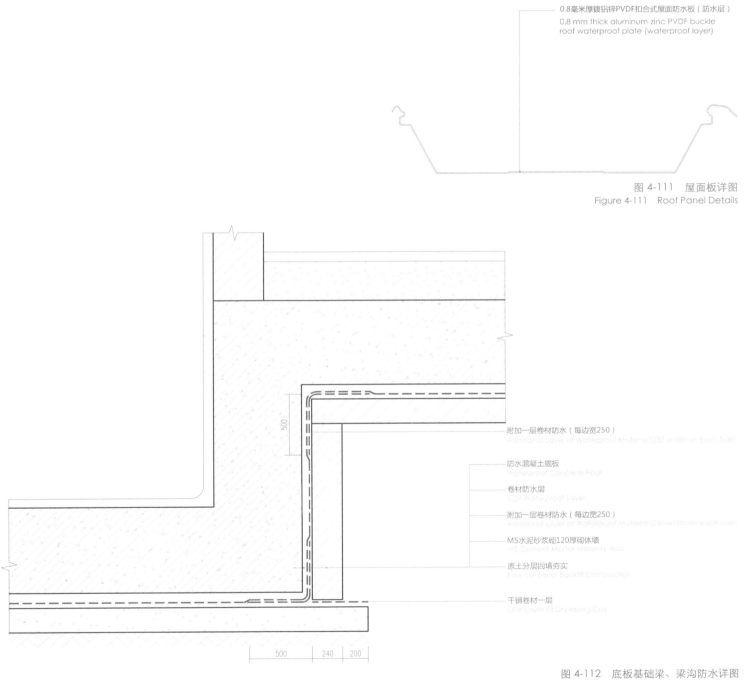

附加一层卷材防水（每边宽250）
Additional Layer of Waterproof Material (250 width on Each Side)

防水混凝土底板
Waterproof Concrete Floor

卷材防水层
Coil Waterproof Layer

附加一层卷材防水（每边宽250）
Additional Layer of Waterproof Material (250 width on each side)

M5水泥砂浆砌120厚砌体墙
Cement Mortar Masonry Wall

原土分层回填夯实
Raw Soil Layered Backfill Compaction

干铺卷材一层
One Layer of Dry Laying Coil

图 4-112　底板基础梁、梁沟防水详图
Figure 4-112　Bottom Slab Foundation Beam and Beam Trench Waterproof Details

图 4-113-1 滑动屋盖
Figure 4-113-1 Sliding Roof

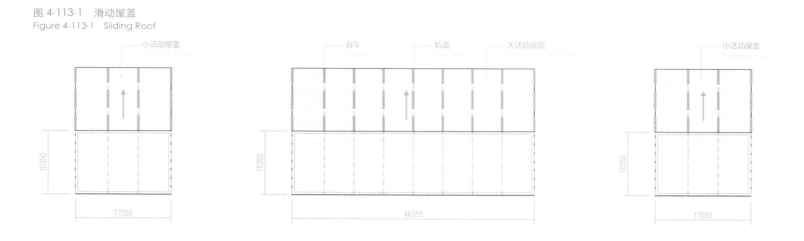

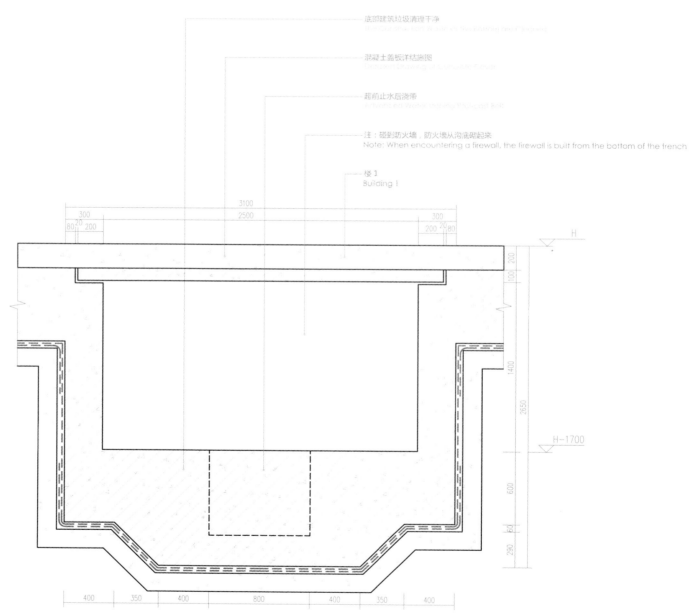

图 4-114 超前止水后浇带地沟 超前止水后浇带地沟
Figure 4-114 Trench with Advanced Water Stop and Post Pouring Belt

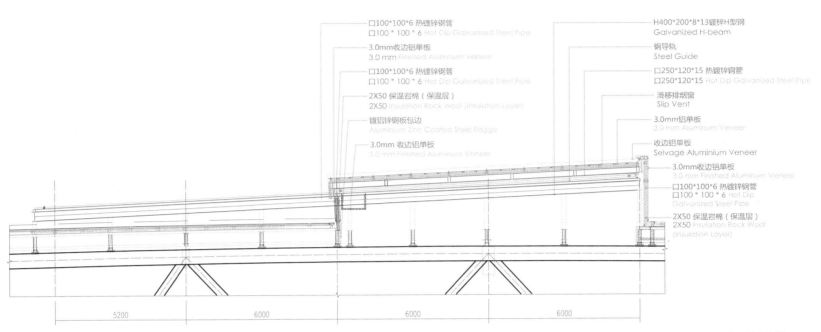

口100*100*6 热镀锌钢管
口100 * 100 * 6 Hot Dip Galvanized Steel Pipe

3.0mm收边铝单板
3.0 mm Finished Aluminum Veneer

口100*100*6 热镀锌钢管
口100 * 100 * 6 Hot Dip Galvanized Steel Pipe

2X50 保温岩棉（保温层）
2X50 Insulation Rock Wool (Intulation Layer)

镀铝锌钢板包边
Aluminium Zinc Coated Steel Eagge

3.0mm 收边铝单板
3.0 mm Finished Aluminum Veneer

H400*200*8*13镀锌H型钢
Galvanized H-beam

钢导轨
Steel Guide

口250*120*15 热镀锌钢管
口250*120*15 Hot Dip Galvanized Steel Pipe

滑移排烟窗
Slip Vent

3.0mm铝单板
3.0 mm Aluminum Veneer

收边铝单板
Selvage Aluminium Veneer

3.0mm收边铝单板
3.0 mm Finished Aluminum Veneer

口100*100*6 热镀锌钢管
口100 * 100 * 6 Hot Dip Galvanized Steel Pipe

2X50 保温岩棉（保温层）
2X50 Insulation Rock Wool (Insulation Layer)

图 4-113-2　滑动屋盖 1:150
Figure 4-113-2　Sliding Roof 1:150

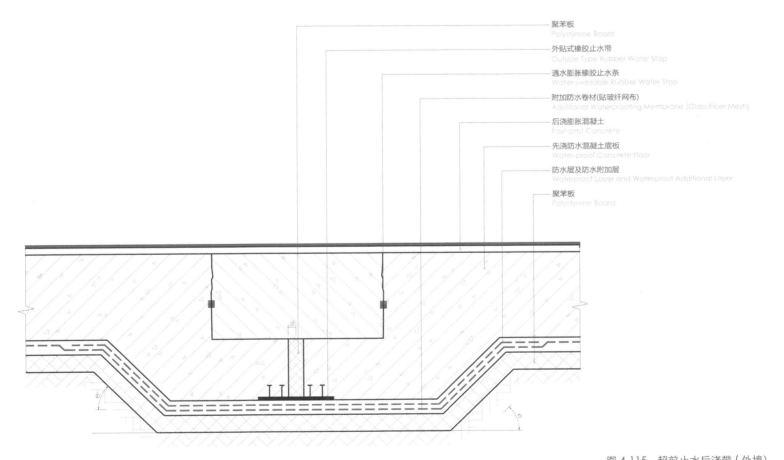

聚苯板
Polystyrene Board

外贴式橡胶止水带
Outside Type Rubber Water Stop

遇水膨胀橡胶止水条
Water-swellable Rubber Water Stop

附加防水卷材(贴玻纤网布)
Additional Waterproofing Membrane (Glass Fiber Mesh)

后浇膨胀混凝土
Post-cast Concrete

先浇防水混凝土底板
Water-proof Concrete Floor

防水层及防水附加层
Waterproof Layer and Waterproof Additional Layer

聚苯板
Polystyrene Board

图 4-115　超前止水后浇带（外墙）
Figure 4-115　Advanced Water Stapping Post-cast Belt (External Wall)

开发模式与设计管理
Construction Mode and Design Management

318 设计总牵头管理
Design and Engineering Procurement Construction

334 会展景观 + 海绵城市
Exhibition Landscape & Sponge City

347 交通设计
Transportation and Circulation Design

378 幕墙设计
Curtain Wall Design

402 标识设计
Signage System Design

411 室内设计
Interior Design

415 垂直交通
Vertical Transportation

422 声学设计
Acoustic Design

435 绿建设计
Green Architecture Design

"重"任

Great Mission

设计总牵头管理
Design and Engineering Procurement Construction

深圳国际会展中心采取的创新 BOD 一体化建设开发及设计总牵头方式
（图 5-1），是立足于促进行业和城市片区可持续发展，围绕会展建设，并
在全面分析深圳会展业定位，综合商业配套开发，场地和周边基建条件，
以及如何带动城市经济等诸多要素后量身定做的成功模式。其不仅为会展
中心工程实施提供了全面的综合技术保障，也为大型公共建筑的设计管理
提供了参考范本。

The Shenzhen World adopted an innovative way of integrated development and construction called BOD, and the design and engineering procurement construction mode (Figure 5-1), which aiming at boosting the sustainable development of industry and urban area. By focus on the construction of the exhibition, the BOD mode gave a comprehensive analysis of exhibition industry positioning in Shenzhen, the comprehensive supporting business development, site conditions and the surrounding infrastructure, urban economy promotion, and many other elements, which made it a successful model that is customized for the Shenzhen World. It not only provides a comprehensive technical support for the construction of the Shenzhen World, but also serves a reference for design and management of large public buildings.

深圳国际会展中心建设开发模式
Construction and Development mode of the Shenzhen World

深圳市政府探索了一条新途径，以会展提升片区经济与活力的城市发展，将场馆建设、后期运营、配套开发综合统筹考虑，提供了一种以会展建设联动片区开发的全新思路。

The Shenzhen municipal government has explored a new way to promote the economic and dynamic urban development of the area through exhibition. The comprehensive consideration of venue construction, operation and supporting facilities development has provided a new way to linked the developments of the surrounding area through exhibition development.

建设新会展的目的

深圳是国内最重要的会展中心城市之一，但目前仅有一座展览面积为 10 万 m^2 的会展中心即深圳会展中心。随着全球会展业发展重心向中国转移，国内会展业进一步大型化发展，现有展馆数量和展览面积不足的问题已成为制约深圳会展业发展的瓶颈。

为满足深圳会展业发展需求，打造推进城市和地区经济发展的重要引擎，深圳市政府按照"一流的设计、一流的建设、一流的运营"三大实施目标，决定在大空港建设一座新的国际会展中心。总展览面积 50 万 m^2，一期建设展览面积 40 万 m^2。

Construction Purpose of the Shenzhen World

Shenzhen is one of the most important convention and exhibition center cities in China. However, there is only one exhibition center with an area of 100,000 square meters. With the development focus of the global exhibition industry shifts to China, and domestic exhibition industry upsizes, the existing exhibition halls and exhibition area are insufficient and will become the bottleneck that restrict the further development of exhibition industry.

In order to meet the development needs of Shenzhen exhibition industry and build an important engine to promote urban and regional economic development, Shenzhen municipal government has decided to build a brand new international convention and exhibition center at the airport area with three implementation goals of "first-class design, first-class construction and first-class operation", the total exhibition area is 500,000 square meters, and the first-phase construction exhibition area is 400,000 square meters.

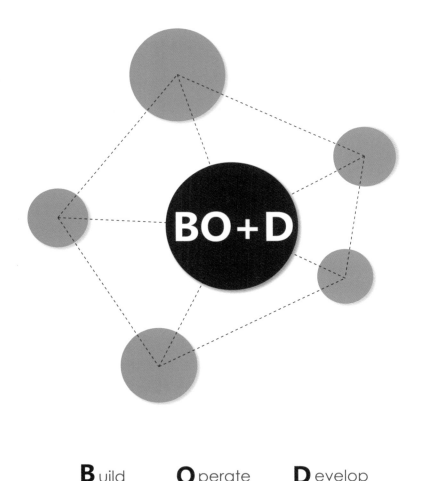

Build   **O**perate   **D**evelop

图 5-1   会展 BO+D
Figure 5-1   Exhibition Bo + D

"建设、运营 + 综合开发" 一体化运作模式

为创建三个 "一流",深圳结合实际情况,引入社会资本同政府合作,创新性采用 "建设、运营 + 综合开发" 的一体化运作模式,即 "BO+D (B-BUILD,O-OPERATE,D-DEVELOP)" 模式。这种模式既有利于发挥社会资本市场化、专业化的运作优势,也有利于确保项目建设和运营符合深圳会展业的发展需求。通过招标方式引入资金实力雄厚、综合开发经验丰富的投资人及国际一流展馆运营机构,负责深圳国际会展中心(一期)建设和二十年运营维护,以及周边配套商业用地的综合开发。

一体化运作的目的是将会展建设、会展运营与会展周边配套商业用地开发通过一个实施主体的整体运作,实现三者目标和利益的统一:实施主体提高建设标准可提升项目会展的运营收益;项目高水平建设和运营又可以提升周边配套商业的价值;周边配套商业的日趋完善又可进一步促进会展的运营收益,三者互为支撑、相互促进,发挥并扩大会展建设开发对城市片区的积极推动作用。

通过一体化运作,一方面可将政府对项目建设、运营的实际要求转化为实施主体的内在激励机制,充分发挥实施主体的主观能动性和创造力,自觉实现项目的高标准建设运营、另一方面、政府通过对配套商业设置预售许可的限制,可建立国际会展中心建设运营与配套商业之间的联动机制,确保实施主体将项目打造成为会展主体和周边配套均是国际一流水准的会展中心。

"Construction, operation + comprehensive development" integrated operation mode

In order to create the three "first-class", Shenzhen, in light of its actual situation, has introduced social capital to cooperate with the government, and innovatively adopted the integrated operation mode of "construction, operation + comprehensive development", namely "BO+D (b-build, o-operate, d-develop)". This mode not only helps to give full play to the operational advantages of social capital marketization and specialization, but also helps to guarantee that the project construction and operation meet the development needs of Shenzhen exhibition industry. Through bidding, we introduced investors with strong financial strength and rich experience in comprehensive development as well as international first-class operating agency of exhibition hall to take charge of the construction, operation and maintenance of the Shenzhen World (phase I) for 20 years, as well as the comprehensive development of surrounding commercial lands.

The purpose of integrated operation is to unify the objectives and interests of exhibition construction, exhibition operation and supporting commercial land development around the exhibition through the overall operation of one implementation body. Raising the standards of subject of implementation can improve revenue from the operation of the project; The high-level construction and operation of the project can also enhance the value of the surrounding supporting businesses; The improvement of surrounding supporting businesses can further promote the operating income of exhibition. These three support and promote each other, which would have positive effect on urban area development.

Through integrated operation, on the one hand, the government's requirements for construction and operation can be transformed into the internal incentive mechanism of the implementation, which encourages initiative and creativity and achieve high-standard construction and operation; On the other hand, by imposing restrictions on the pre-sale license of supporting businesses, the government has established a linkage mechanism between the construction and operation of the Shenzhen World and the supporting businesses, so as to guarantee that the implementer will build an exhibition center with world-class exhibition building and surrounding supporting facilities.

深圳国际会展中心项目建设运营监管机制

深圳市政府在 2015 年 9 月成立了会展建设指挥部，由市领导统筹协调三十多个市、区级政府职能部门通力合作，稳健有序地推进项目实施。

深圳国际会展中心通过国际竞赛于 2016 年 2 月确定了 AUBE 欧博设计 + 法国 VP 联合体的中标设计单位，在引入会展实施主体之前，便已开始深圳国际会展中心片区的城市设计和会展建筑方案设计。会展建设指挥部，规划、国土、土地交易中心等部门，在城市设计的基础上，确定了会展配套商业开发十一宗用地的性质、规模和公共配套设施，为投资主体的引入提供了设计技术条件。AUBE 欧博设计与法国 VP 全程参与了《深圳国际会展中心项目建设运营监管协议书》的拟定，就监管协议中对设计和建造阶段所要达成的目标和要求提供了专业意见。

2016 年 6 月 29 日，深圳市发布土地使用权出让公告，以邀请招标的方式出让了深圳国际会展中心（一期）的建设运营权和深圳国际会展中心（一期）配套商业用地的土地使用权。会展中心（一期）用地面积 125.5 万平方米，土地使用权归政府，由市经贸信息委代表政府持有。会展中心（一期）配套商业用地共计十一宗，土地性质为商业用地，土地使用年期 40 年，用地总面积 52.8 万平方米，总建筑面积 154.3 万平方米，包括为会展中心配套的商业、办公、酒店和商务公寓等分项功能建筑。

本次招标有多家知名企业参与，最终由招商局蛇口工业控股股份有限公司、深圳华侨城股份有限公司、美国 SMG 公司的联合体作为会展实施主体中标，并与政府签署《深圳国际会展中心项目建设运营监管协议书》。

《深圳国际会展中心项目建设运营监管协议书》对中标人的出资方式，金额，建设标准，运营标准，配套开发用地的开发时序和内容及未来转让条件等许多方面，都提出了明确要求，诸如：
① 中标人出资参与会展中心建设，承担除建筑主体外的室内装修和机电设备费用；
② 中标人获得会展建成后的二十年经营权，并要求必须在投标阶段就联合国际知名的会展运营机构；
③ 出让宗地上建设的全部商业用房，且自签订出让合同之日起二十二年内不得转让；
④ 相关宗地内各酒店物业分别限整体转让；
⑤ 十万平方米的商务公寓建成后，将由政府按经审计后的建安成本价收购，作为人才公寓。

The Construction and Operation Supervision Mechanism of the Shenzhen World

In September 2015, Shenzhen municipal government has set up the headquarters for exhibition construction. The municipal leaders coordinated the cooperation of more than 30 municipal and district government departments to promote the implementation of the project steadily and orderly.Through an international competition, AUBE(Shenzhen) and VP(France) are together, as the winning designer team in February 2016, had already started the urban design and architectural design of the Shenzhen World before the subject of implementation was brought in.

Together with the headquarters, the planning department, the ministry of land and resources, the land & real estate exchange center, and other departments, on the basis of urban design, they determined the property, scale and the public supporting facilities of 11 lands for the supporting commercial development, providing design technical conditions to inform the investors. Both the AUBE and VP has participated in the drawing up of the agreement on the supervision of construction and operation of the Shenzhen World, and provided professional advices on the objectives and requirements to be achieved in the design and construction phase.

On June 29, 2016, Shenzhen municipality issued an announcement on land use right granting, offering an invitation for bids of the construction and operation right of the Shenzhen World (phase 1), and the land use right of supporting commercial land of the Shenzhen World(phase1). Shenzhen World (phase 1) has an area of 1,255,000 square meters, which belongs to the Economy, Trade and Information Commission of Shenzhen Municipality, and its land use right is owned by the government. There are totally 11 supporting commercial lands for the Shenzhen World(phase 1), which are commercial lands with 40 years of land use right and a total land area of 528,000 square meters with a total construction area of 1,543,000 square meters, including sub-functional buildings such as commercial, office, hotel and business apartments.

Many established corporations have participated in the bidding, eventually the China Merchants Shekou Industry Holding Co., Ltd, together with Shenzhen Overseas Chinese Town Co., Ltd, and SMG won the bidding and signed the Construction and Operation Supervision Agreement of the Shenzhen World.

The agreement on the supervision of construction and operation of the Shenzhen World puts forward specific requirements on the investment method, amount, construction standards, operation standards, development timing and content of supporting development land, and future transfer conditions on many aspects, such as:
① The winning bidder shall contribute to the construction of the exhibition center and bear the expenses for interior decoration and mechanical and electrical equipments except the main building;
② The winning bidder shall be granted the right of 20 years operation right of the exhibition center after construction, and shall cooperate with international renowned exhibition operation organizations at the bidding stage;
③ All commercial buildings built on the leased land shall not be transferred within 22 years since the date of the leasing contract signing;
④ Each hotel property built on the relevant land is respectively limited to be transferred as a whole;
⑤ After the completion of 100,000 square meters of commercial apartments, the government will purchase them as talent apartments at the audited cost price.

建设开发模式的创新

深圳国际会展中心的建设开发模式，吸取了国内外大型公共建筑的开发经验，在源头上解决了开发运营过程中可能存在的问题。

① 建设与运营的衔接
建设单位绑定国际一流会展运营单位进行联合竞标的方式，保障了会展场馆的后期运营质量。深圳国际会展中心自设计开始至建设完成，运营理念始终贯穿其中，并保持了高度的一致性。避免了场馆建设完成、运营滞后介入等造成的重复建设问题。

② 建设与投资的衔接
政府与建设运营企业联合投资的方式，有利于深圳国际会展中心在社会效益与经济效益之间取得平衡。政府与企业之间的不同诉求需在投资划分中取得相互平衡，以达到投资回报的利益最大化。

③ 政府与企业的责权清晰
《深圳国际会展中心项目建设运营监管协议书》明确了政府与企业对于深圳国际会展中心的各项责权，在赋予建设运营企业权力、激发企业主观能动性的同时，保留了政府对于深圳国际会展中心具备的所有权和有效的监管方式。

深圳国际会展中心的建设、运营+综合开发一体化模式，在前期筹划阶段便未雨绸缪，反复论证调研，结合国际领先的经验和深圳会展业的实际需求与前瞻性，将项目筹备、会展建设和运营、周边配套开发、市政基础设施建设等相关联的诸多内容和实施环节都做到充分预见和考量，引入国际一流设计、咨询、运营团队和国内一流投资建设运营机构，为项目整体顺利实施奠定了坚实基础，为建造国际最高标准的会展中心开创了先河。

The Innovation of Construction and Development Mode

The construction and development mode of the Shenzhen World draws on the experience of large public buildings at home and abroad and solves the problems that may exist in the development and operation process from the beginning.

① Connection Between Construction and Operation
The joint bidding of construction organization and world-class international exhibition operation agency has guaranteed the quality of the operation. The operation idea throughout the whole process from design till the completion has maintained a high degree of consistency, avoiding the repetitive construction problems caused by the lack of consideration of operation after construction.

② Connection Between Construction and Investment
The way of joint investment between the government and construction and operation enterprises is beneficial for balancing the social benefits and economic benefits of the Shenzhen World. Different demands between the government and enterprises should be balanced in the division of investment to maximize the benefit in return.

③ Clear responsibilities of the Government and Enterprises
The agreement on the supervision of construction and operation of the Shenzhen World clearly defines the responsibilities of the government and enterprises for the Shenzhen World, which empowering the construction and operation enterprises to stimulate its subjective initiative while retaining the government's ownership of the Shenzhen World and its effective supervision.

The BO+D mode took precautions in the early planning stage, research repeatedly, combined the international advanced experience and the actual demand of the project to anticipate and consider adequately many contents and implementations such as project preparing, construction and operation, surrounding supporting facilities, municipal infrastructure construction, and many others initial considerations. In addition, it introduced world-class international design, consulting, operation team as well as domestic top investment and construction operators, which laid a solid foundation for the overall project and is a groundbreaking way to establish the highest standards for an international convention and exhibition center in the world.

深圳国际会展中心设计总牵头单位
The EPC (Engineering Procurement Construction) of the Shenzhen World

深圳国际会展中心在投标阶段已明确采用设计总牵头的方式推进各阶段设计工作，AUBE 欧博设计为此组织技术、管理、商务等多部门，在设计和工期要求紧迫的条件下，有序地组织开展各专项顾问分包的甄选和设计统筹工作，针对中标方案的深化，在设计前期充分预判和分析潜在问题，为主体设计和各专项设计顺利推进创造条件。

多专业集成一体化是欧博一直倡导并坚持的设计工作模式，而深圳国际会展中心这一项目又是对该模式的一次集中体现。作为深圳国际会展中心设计联合体牵头单位，为满足业主的各项建设和运营要求，也为使建筑方案创意落地实施，除完成自身承担的建筑主体设计工作外，根据主体设计不同阶段各专业的比选方案，AUBE 欧博设计还统筹顾问分包有针对性地做了大量甄选方案，力求从技术创新、造价控制、快速建造、降低能耗、维护便利等不同角度，寻找多专业综合最优方案，解决各类设计与施工问题。

In the bidding stage, it has already been confirmed to conduct EPC to carry out all the design work in different stages. AUBE has organized the technology, management, business and many other departments under the urgent condition, and orderly selected the consulting agencies for each special project and organized the design work at the same time. Besides, by deepening the bid-winning scheme, we AUBE gave full anticipation and analysis about the potential problems before the design and created precondition for the architectural design and other specialized project designs.
Multi-professional integration is the design method that AUBE consistently advocates and insists on, which is again embodied in the project of the Shenzhen World. In order to satisfy the needs of the owner and the operator and guarantee the implementation of the building scheme. AUBE, the leader of the design cooperation team has not only accomplished its own design work of the building, but also coordinated several solutions for each consulting agency. AUBE tried to find the optimal professional and comprehensive solutions throughs technology innovation, cost control, rapid construction, energy saving, convenient maintenance and many other aspects for all kinds of design and construction problems.

设计总牵头管理的三个阶段

在各类项目中，设计管理的基本目的是相同的，管理方式的呈现也更多样化。每个项目均具有其特殊性，就深圳国际会展中心的设计管理重点，可分为三个主要阶段阐述（图 5-2）。

Three Stages of EPC Management

In all kinds of projects, the basic purpose of the design management is identical, but methods for the design management are various. Each project has its distinctiveness. In terms of the key point of the design management of the Shenzhen World, its design management can be divided into three main stages.(Figure 5-2)

图 5-2　设计管理的三个阶段
Figure 5-2　Three Stages of Design Management

① 设计优化阶段：市政基础条件的协调与落实

深圳国际会展中心选址位于大空港启动区，设计之初各类规划条件与市政设施还不完善。深圳国际会展中心是全球第一个集地铁、市政道路交通、水力河道治理等市政设施为一体同时开发建设的建筑工程，会展设计不但要考虑自身建设实施的时效性和便利性，还要为道路、桥梁、地铁、河道、市政综合地下管廊等市政设施的同步设计、建设创造有利条件。会展设计要考虑的不确定因素，协调工作和复杂程度相比其他建筑工程成倍增长，这对欧博也是一项挑战。各专业工程师要以更宏观的视角提出解决问题的综合有效措施，设计管理也要将所有市政设施的边界条件和复杂因素纳入到设计统筹的大系统之中，由此才能确保会展中心及其配套基础设施同步有序推进。

在此工作过程中，由会展中心指挥部牵头，设计对接了例如：深圳市商务局，市发改委，市规划局，市住建局，市交委等 20 余个政府部门或机构。

在 2016 年初至 2016 年 9 月，组织与政府相关主管部门一起完成了会展片区城市设计以及会展中心建筑初步设计成果的编制，以及主要市政条件的衔接工作，为后续的工程启动创造了坚实的基础。

② 工程启动阶段：设计要求协调与落实

为达到一流设计，一流建造，一流运营的目标，2016 年 9 月深圳市政府宣布招商、华侨城与美国 SMG 公司一同组成中美联合的建设运营单位。德国 JWC 运营顾问同步参与会展建设运营标准的制定。

因此，深圳国际会展中心汲取了德国与美国会展建筑先进经验的同时，也要与国内使用情况相结合，且不影响工程启动实施。在此期间，招华与 SMG 提供了大量国际会展运营标准，并组织国内知名专家就部分关键问题进行了多次征询与讨论。许多重大事项，需要对投资造价、施工工期、设计规范等全方位的评估判断，每一项重大决策，都将对设计方向产生颠覆性的影响。

与此同时，与交通、消防、幕墙、景观、泛光、标识、室内、声学等相关的 30 余家国内外一流的顾问单位先后加入设计总承包团队。所有专项设计要求的衔接与落实，设计进度与质量的把控，难度巨大。

③ 工程全面实施阶段：设计与工程的衔接

2017 年底，深圳国际会展中心进入全面建设阶段，有超过 100 家设计和建设团队成员，每天有超过 2 万建设工人奋战在深圳国际会展中心的施工现场。随着工程建设的推进，市政设施衔接，工程组织，消防设计，屋顶形态，超大金属屋面等各类重大设计与技术课题的不断产生与解决，成为保障项目顺利建设的主要工作。

项目设计单位作为项目建设的源头，任务艰巨，需要与所有参建单位形成工作决策的高度一致性，确保计划与信息的同步是此阶段非常关键的管理内容。

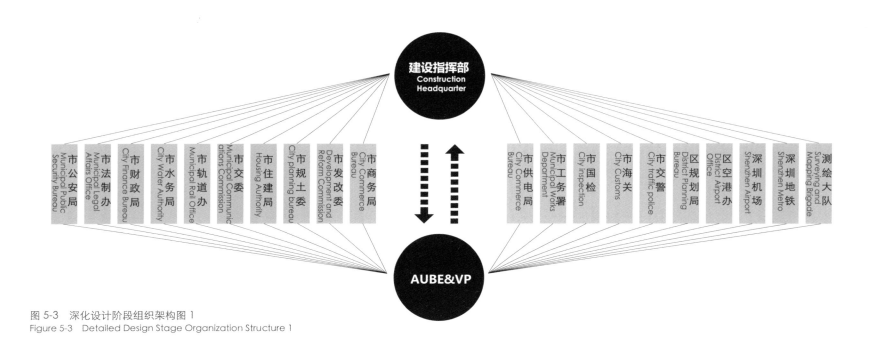

图 5-3　深化设计阶段组织架构图 1
Figure 5-3   Detailed Design Stage Organization Structure 1

1 Design Optimization Stage: Coordination and Implementation of Municipal Basic Conditions

The Shenzhen World is located in the starting area of the Airport New Town, which is lack of various planning conditions and municipal facilities at the beginning of the design. The Shenzhen World is the world's first architectural engineering that integrated the development of the subway, municipal road traffic, hydraulic engineering, and other municipal facilities all together at the same time. Therefore, the design of the Shenzhen World not only should consider the timing and the convenience of its own construction implementation, but also need to create advantage for the synchronous design and constructions of many other municipal facilities such as roads, bridges, subway, river, municipal underground utility tunnel, etc. The uncertainties, coordination and complexity of the design work increased exponentially compared to other construction projects, which is also a challenge for AUBE.

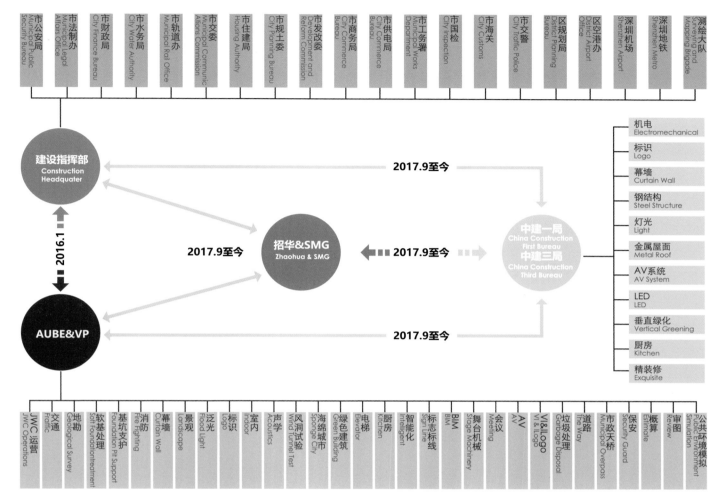

图 5-4 深化设计阶段组织架构图 2
Figure 5-4 Detailed Design Stage Organization Structure 2

All professional engineers should propose comprehensive and effective measures to solve problems from a more macro perspective, and the design management should also integrate the boundary conditions and complex factors of all municipal facilities into the whole system of the design coordination, so as to guarantee the synchronous and sequential constructions of the exhibition center and its supporting infrastructures.

In the process of this work, the headquarters took the lead to connect with more than 20 government departments or agencies, such as the Commerce Bureau of Shenzhen Municipal, the Development and Reform Commission of Shenzhen Municipality, the Municipal Planning Bureau, the Shenzhen Housing and Construction Bureau, and the Municipal Traffic Commission, etc.

② Start-up Stage: Coordination and Implementation of the Design Requirements

In order to achieve the goal of "first-class design, first-class construction and first-class operation", in September 2016, the Shenzhen municipal government announced the establishment of the Sino-American joint construction and operation unit between the China Merchants, Shenzhen Overseas Chinese Town and the SMG. The operation consultant Germany JWC also synchronously participates in developing the standards of the exhibition construction and operation. The Shenzhen World has thus learnt the advanced experience of both German and American exhibition buildings, and also combined the domestic service condition without affecting the implementation. During this period, the China Merchants, Shenzhen Overseas Chinese Town and the SMG has provided a variety of international exhibition operation standards and organized domestic well-known experts to consult and discuss the key issues several times. A lot of important projects needs to be evaluated in terms of the investment cost, construction period, design standards and other comprehensive conditions, with decision of which will have a subversive impact on the design direction. At the same time, more than 30 domestic and foreign first-class consulting units dedicated in such as traffic, fire protection, curtain wall, landscape, lighting, sign design, interior design, acoustics, etc., have successively joined the EPC team. It is extremely difficult to connect and implement every specialized design, as well as to control the process and the quality of the design.

③ Implementation Stage: The Link between Design and Engineering

At the end of 2017, the Shenzhen World started its overall construction with more than 100 design and construction team members and more than 20,000 construction workers working on the site every day. In the progress of engineering construction, we encountered and solved all kinds of big design and technical issues such as the connection with municipal facilities, engineering organization, fire control design, roof form, super-large metal roof, etc., which became the major work to guarantee the construction goes on well.

As the origin of the project, the design is an arduous task. It requires high consistency in decision making with all the precipitants to guarantee the synchronization of plans and information, which is also a critical management in this stage.

项目管理组织架构

深圳国际会展中心设计和施工过程中涉及众多部门和单位，清晰的
管理组织架构便于各参与方根据自身的工作范围和权责开展工作
（图 5-5 ）。

Organization Structure of Project Management

There are many departments and units involved in the
design and construction of the Shenzhen World. The clear
management and organization structure help all participants
to carry out their work according to their work scope and
rights as Well as responsibilities.(Figure 5-5)

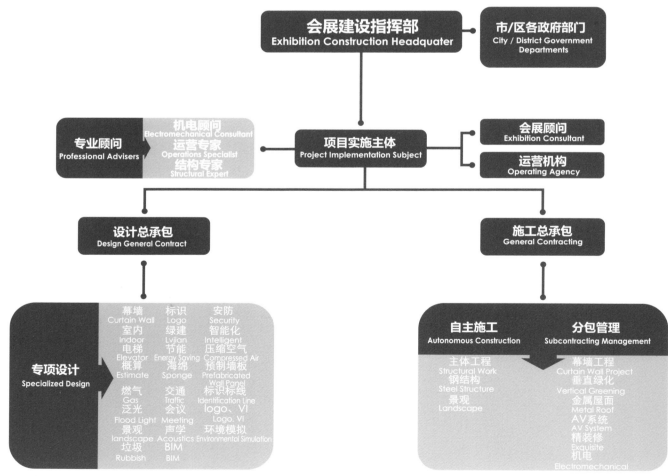

图 5-5   项目管理组织架构图
Figure 5-5   Project Management Organization Structure

深圳国际会展中心设计管理难点与措施

① 合约管理——专业灵活的合约管理，建立国际化专项设计团队

在设计初期（设计优化阶段）完成专项设计的合约管理体系非常重要。合约管理应以满足项目需求为第一原则，搭建优秀专业化的设计团队。所有团队成员需具备满足设计质量与进度的素质。设计合同应充分明确各单位的工作内容、工作界面，提前预判工程可能需要的所有专项设计内容，避免出现设计空白或责权不清的情况。

其中不免会遇到以下难点：众多专项设计，内容复杂，项目环境工期紧张，合约签订往往伴随项目推进同步进行；建设与运营团队、国籍与文化背景多元化，需要对设计要求有准确的把握；合约的签订与最终设计需求的介入存在时间差。

为此，应严格把关各专项设计单位的大型公共建筑设计背景，充分考察各专项设计单位的设计背景与团队成员的一致性。所有专项设计单位均有境外合作经验，这虽然缩小了语言与文化背景差异带来的影响，但沟通成本仍是多方协作付出最多的部分。合约需对服务范围有充分预期并留有余地，明确根据需求合约增补修改的方式，缩短后续合约增补谈判周期。

深圳国际会展中心所具备的国际影响力，吸引了国内外众多的优秀设计团队参与，在设计联合体牵头协调管理下，许多单位倾力的配合与不计成本的付出，为一流的设计做出了卓越贡献。

② 质量管理——协定成果质量标准，成果多向分级审核

建立各个专项设计的成果交付标准与评定方式，并对各家的设计内容进行整合，提高设计的沟通效率与质量管控。设计成果质量把控贯穿整个项目周期，是一个长期而艰巨的工作。根据工程进展不同阶段而重点把控审核区域，是应对超大规模项目的必要方式。

难点一：质量标准的制定

设计文件的质量标准可在合同中参照行业标准进行约定。在社会分工日益细化的今天，部分专项设计（境内境外均有可能）的工作环境、习惯，甚至文化背景的差异，使不同背景的人员对设计文件的深度或内容都有不同的理解，在有限合约谈判时间过程中，无法具体约定准确的成果细节。

因此，设计总承包管理单位需要具备极高的专业性，需要深刻理解相关专项设计内容，以及合作单位提供的服务建议内容。深圳国际会展中心项目在实际操作中，根据具体情况，及时落实了各专项设计成果标准，并与参与方达成广泛一致。

专项设计的成果标准需要考虑造价核算，工程招标，现场施工，现场服务等项目关键环节的地域化因素。

难点二：成果质量审核

鉴于深圳国际会展中心的规模，各专项设计成果的内容篇幅浩瀚，仅欧博主体施工图的图纸就超过10000张，各专项设计图纸完成相互审核，控制设计质量及设计风险的工作量极其巨大。

此时需要重点突破和过程审核。对项目各专项设计存在的技术难点有充分的预期，对重要的设计内容提前进行专项设计评审，利用内外资源提前把控设计方向，控制设计风险（图5-6）。为此，深圳国际会展中心进行的各类专业评审达到上百项，充分保障了各专项顶层设计的先进性。对各单位提供的成果，提前落实各级校对、审核、审定机制，记录校审单，并对不同深度阶段的文件进行审核。设计文件提交后，各连带专项设计均需对设计成果进行评定。对30余项专项设计厘清相互关联，并对所有单位成果对外输出建立接收与审核机制，发动所有相关单位进行过程审核，把控设计成果质量。

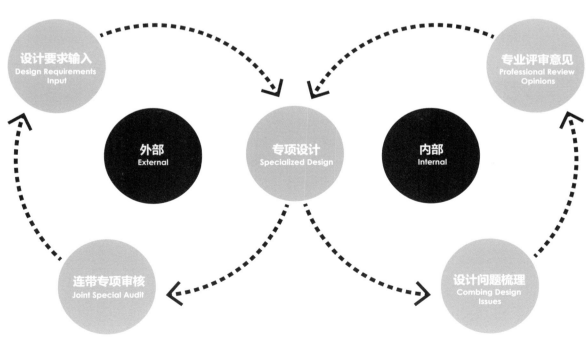

图 5-6　专项设计成果质量控制图
Figure 5-6　Quality Control of Special Design Results

The Difficulties and Measures of the Design and Management of the Shenzhen World

① Contract Management --, Establishing an international specific professional and flexible contract management design team

It is very important to complete the contract management system of special design at the initial design stage (design optimization stage). Contract management should take the project requirements as the first principle and build an excellent and professional design team. All team members must be qualified according to their work quality and timing. The design contract shall clarify the work content and work interface of each unit, and predict in advance all special design contents that may be required for the project, so as to avoid blank design or unclear responsibilities and rights.

It is inevitable to encounter the following difficulties: There are many special designs, and the team has to deal with complex content and tight schedule, and thus the contract usually is signed during the process of the implementation of the project; The construction and operation team has a diverse cultural background and need to grasp the design requirements accurately; There is a time difference between the final contract sign and the intervention of the final design requirements.

Therefore, it is necessary to check strictly each special design unit's design background of large-scale public buildings, and fully examine the consistency of the design background and team members. All special design units have overseas cooperation experience, which equals certain language advantage, yet the communication still cost a lot of efforts due to multi-party collaboration. In the contract, there should be space or possibilities left open for the service scope and make clear the way for addition and modification in the contract according to new requirements, so as to shorten the negotiation time of subsequent contract addition.The international influence of the Shenzhen World has attracted a large number of excellent design teams at home and abroad to participate. Under the leadership and coordination of the design consortium, many units have made outstanding contributions to the first-class design.

② Quality Management – agreed on the quality standard of the outcome, and the outcome shall be audited by multi classification.

Difficulty 1: the Formulation of Quality Standards
The quality standards of design documents shall be written in the contract with referring to the industry standards.

As the increasingly refining of division of labor in society nowadays, the differences of the working environment, habits and even cultural backgrounds of some special designs (both within and outside China are possible) have led to problems of understanding of depth or content of design documents. In the process of limited contract negotiation, it is impossible to specify the accurate details of the outcome.

Therefore, the EPC unit has to be highly professional, and be comprehensive about the relevant special design content, as well as the content of service suggestions provided by the cooperative units. In the actual operation process, according to the specific situation, the EPC unit should guarantee the implementation of the specific design results standards in time and reach a broad agreement with the all participants.

The result standard of special design needs to take account of regional factors in several crucial aspects such as cost accounting, project bidding, site construction and site service, etc.

Difficulty 2: Quality Audit of the Outcome
Due to the scale of the Shenzhen World, the content of each special design outcome is vast. For example, only the construction drawing of the building by AUBE is more than 10,000 pieces. The wordload is tremendous to complete the quality audit of design drawings by each special design unit, as well as to control the design quality and risk. It needs to focus on both breakthrough and process auditing. it is important to fully anticipate the technical difficulties existing in each special design of the project, conduct special design review for important design contents in advance, control design direction and design risks in advance by using internal and external resources. Therefore, there conducted hundreds of professional reviews for the Shenzhen World to fully guarantee the advancement of each special top design.Implement in advance the mechanism of proofreading, examination and approval at all levels for the achievements provided by each special design unit and record the examination list and examine the documents in different depth and stages. After the submission of the design documents, the design results shall be evaluated by each associated special design. More than 30 special designs are related to each other. A receiving and auditing mechanism has been established for the output of all units, and all relevant units have conducted process auditing to control the quality of the design results.

③ 计划管理：合理计划，密切跟踪

大型公共建筑特征，技术难度大、参与单位众多、工期紧张、决策机制复杂。整体设计计划影响因素众多。随着项目的不同阶段，决策者或决策团队在实时变化，决策机制也应适时调整（表5-1，表5-2）。

难点一：计划的适应性
计划赶不上变化，这句谚语虽通俗，但往往是各类大型工程项目的真实写照。深圳国际会展中心是与周边市政设施和商业开发同步进行，各类设计与建设条件相互牵制，错综复杂，部分条件甚至能牵一发而动全身。

此时需要对重大决策事项进行跟踪。与政府以及建设管理单位保持信息高度统一，密切配合，必要进行反向管理和推进是合理制订设计计划的关键。设计计划需根据情况实时调整，然而里程碑式节点需严控，且不应轻易退缩。
深圳国际会展中心实时更新的重大决策事项清单以及定期举行的高层会议，使得计划的可实施性得到了保障。重大决策事项的梳理以及高层决策会议是重大项目推进过程常见的形式，然而关键在于前期的充分研究与准备工作，一事一议，每议必有结论，结论达成也不反复。

计划的适应性与重大事项的处理息息相关，计划的调整与工作的反复会极大的影响团队士气，计划与重大事项处理责任的绑定，能有效地提高计划的适应性与可实施性。

难点二：计划的跟踪
计划的制定需建立在对工作内容与工作对象充分理解的基础之上，并进行实时跟踪。跟踪应避免因某一个环节出现问题，而导致整个体系停滞或相互推诿的情况。
针对该难点，需要建立主要工作信息交互平台。在深圳国际会展中心的设计过程中，各个合作单位建立了每月，每周甚至每日的工作计划交互平台，使得所有合作单位对自己工作在整个设计计划中的位置非常清晰。

3 Plan Management: Reasonable Plan and Close Tracking

Large public buildings are characterized by high technical difficulty, large number of participants, tight schedule and complex decision-making mechanism. The overall design plan is influenced by many factors. As the shift of decision maker or the decision team during different stages of the project, the decision mechanism should be adjusted accordingly.(Table 5-1, 5-2)

Difficulty 1: Adaptability of the Plan
Plans can't keep up with change, especially for such big project. The construction of the Shenzhen World is synchronized with the development of surrounding municipal and commercial facilities. All kinds of design and construction conditions interact with each other, which are very complicated and sometimes extremly relevant to each other.

This is where the big decisions need to be tracked. Keeping highly unified the information with the government and the construction management unit, cooperate with them closely, and sometimes carry out reverse management and promotion is the key to make a reasonable design plan. Design plans need to be adjusted in real time according to the actual situation, while the milestone nodes need to be strictly controlled and should not easily be withdrawn.

The Shenzhen World's updated list of major policy issues and regular high-level meetings guarantee the feasibility of the plan. Sorting out major decision-making matters and high-level decision-making meetings are common forms in the process of promoting major projects. However, the key still lies in the full research and preparation in the early stage of the project.

The adaptability of the plan is closely related to the handling of the major issues. The adjustment of the plan and the repetition of work greatly affects the team morale. The binding between the plan and the handling of major issues can effectively improve the adaptability and implementation of the plan.

Difficulty 2: Plan Tracking
Plans need to be made based on a thorough understanding of what is being done and who is the working subject, and all of these should be tracked in real time. Problems that lead to stagnation or prevarication should be avoid.

In view of this difficulty, it is necessary to establish a main work information interactive platform. In the design process of the Shenzhen World, each cooperation unit has established an interactive platform for monthly, weekly and even daily work plan communication, so that all the cooperation units are very clear about the position of their work within the whole design scheme.

表 5-1 深圳国际会展中心决策事项跟踪表
Table 5-1 The Shen zhen World Project Decision-making Track List

	设计决策事项 Design Decision-making Matters	决策方 Decision maker	决策情况 Decision making	设计责任方 Responsible Party	执行情况 Imple mentation	待完成工作 Work to be done	备注 Remark
建筑 Architecture	1.建筑形态方案 Architectural Form	市政府GOV Municipal Government GOV	展厅特殊立面未确认 Special Exhibition Hall Facade unconfirmed	VP	未完成 undone	特殊立面调整，并汇报 Special elevation adjustment and reporting	市政府意见 City government opinion
			屋面檐口未确认 Roof cornice is not confirmed	VP	未完成 undone	屋面檐口调整，并汇报 Roof cornice adjustment and report	
	2.重点区域室内设计概念方案 Conceptual Plan for Interior Design in the Key Areas	GOV/（CMSK） GOV / (CMSK)	政府未确认 Government not confirmed	VP/ID	未完成 undone	向指挥部汇报 Report to Headquarter	设计深化进行中 Design in progress
	3.方案报建前置条件 Pre-conditions Prior to the Scheme Application	招华（CMSK） CMSK	未完成 undone	VP/AUBE	–	待方案报建 Pending Proposal	市规土委要求政府会议纪要确认形态 Planning Bureau asks minutes of government meetings to confirm form
	4.方案报建相关面积问题 The Area-related Problem Prior to the Application and Construction	招华（CMSK） CMSK	完成 Complete	VP/AUBE	未完成 undone	待方案报建 Pending Proposal	屋面封闭设计，屋顶核增 Roof closed design, roof nuclear increase
消防 Fire Fighting	1.消防设计 Fire Design	省消防厅 Provincial Fire Department	完成 Complete	AUBE	未完成 undone	待方案报建 Pending Proposal	展厅通道按消防专家评审意见执行 The hall passage is implemented according to the review opinions of fire experts
运营 Operation	1.垃圾处理策略 Garbage Disposal Strategy	招华（CMSK）/SMG CMSK / SMG	完成 Complete	AUBE	未完成 undone	施工图落实垃圾处理顾问意见 Construction plan implements garbage disposal consultant's opinion	–
	2.安防策略 Security Strategy	招华（CMSK）/SMG CMSK / SMG	未完成 undone	VP/AUBE	未完成 undone	目前设计暂停 Current Design Suspension	正与奥雅纳联系 Contacting Arup
	3.餐饮策略 Catering Strategy	招华（CMSK）/SMG CMSK / SMG	完成 Complete	AUBE	未完成 undone	深化厨房设计 Detailed Kitchen Design	–
市政接驳 Municipal Connection	1.海滨大道方案 Promenade Plan	市政府GOV Municipal Government GOV	未完成 undone	中冶 MCC	未完成 undone	–	道路方案待市领导确认 Road plan to be confirmed by city leaders
	2.凤塘大道市政管线 Fengtang Avenue Municipal Pipeline	市政府GOV Municipal Government GOV	未完成 undone	中冶 MCC	未完成 undone	–	情况与海滨大道方案相关并相同 The situation is related and identical to the Promenade
	3.海汇路市政管线 Haihui Road Municipal Pipeline	市政府GOV Municipal Government GOV	未完成 undone	中冶 MCC	未完成 undone	–	待会议纪要 Minutes of meeting
	4.登入大厅东侧地铁出入口设计 East side of Arrival Hall Subway Entrance and Exit Design	招华（CMSK） CMSK	完成 Complete	VP/AUBE	未完成 undone	与地铁沟通，VP需要完成出入口深化设计 Communicate with subway, VP needs Complete entrance and exit design	出入口设计与天桥设计需统筹考虑 Entrance and exit design and overpass design need to be considered
	5.海汇路天桥设计 Haihui Road Overpass Design	招华（CMSK） CMSK	完成 Complete	VP/AUBE	未完成 undone	与地铁沟通，VP需要完成出入口深化设计 Communicate with subway, VP needs Complete entrance and exit design	
总图 General Plan	1.货运流线组织 Freight Circulation Organization	招华（CMSK）/SMG CMSK / SMG	未完成 undone	AUBE/VP	未完成 undone	需专题讨论 Need to discuss	交通研究中心已提供概念方案，已翻译并发送给SMG Transportation Research Center has provided the concept plan, translated and sent to SMG
	2.凤塘大道盖板设计 Fengtang Avenue Cover Design	市政府GOV Municipal Government GOV	完成 Complete	AUBE	未完成 undone	深化盖板设计 Detailed Roof Cover Design	凤塘大道条件未稳定 Fengtang Avenue conditions are not stable
	3.场地内大型广告牌 Within Venue Large Billboard	招华（CMSK）/SMG CMSK / SMG	未完成 undone	AUBE/VP	未完成 undone		未确定准确位置与形式，需厂家配合 The exact position and form is not determined, and requires the cooperation of the manufacturer
机电MEP Electromechanical Plan	1.热水与直饮水设计确认 Design confirmation of hot water and direct drinking water	招华（CMSK）/SMG CMSK / SMG	完成 Complete	AUBE	未完成 undone	施工图深化设计 Detailed Design of the Construction Drawings	–
建筑 Architecture	1.标准展厅地面做法 Standard Exhibition Hall Ground Practice	招华（CMSK） CMSK	完成 Complete	AUBE/招华	完成 Complete	–	–
	2.C11，C12安全系统 C11, C12 Safety System	招华（CMSK）/SMG CMSK / SMG	未完成 undone	AUBE/VP	未完成 undone	施工图深化设计 Detailed Design of the Construction Drawings	需要安全网的专业厂家进行配合 Need professional safety net manufacturers to cooperate
	3.C11，C12吊挂系统 C11, C12 Suspension System	招华（CMSK）/SMG CMSK / SMG	完成 Complete	AUBE/VP	未完成 undone	施工图深化设计 Detailed Design of the Construction Drawings	–
	4.立面排烟窗调整至3%（含展厅、南北登、南接） Adjust the Smoke Exhaust Window to 3%	招华（CMSK） CMSK	完成 Complete	AUBE	完成 Complete		调整后的立面还需VP确认 The adjusted facade needs to be confirmed by VP
	5.C12座位优化布置和固定的设计方式 C12 seat optimization And fixed design	招华（CMSK）/SMG CMSK / SMG	未完成 undone	VP/AUBE	未完成 undone	施工图深化设计 Detailed Design of Construction Drawings	需专业厂家配合座椅固定方式 Need professional manufacturers to cooperate with seat fixing method
	6.PC墙详细做法 PC Wall Detailed Practices	招华（CMSK）/SMG CMSK / SMG	完成 Complete	AUBE/VP	未完成 undone	施工图深化设计 Detailed Design of the Construction Drawings	与华艺深化设计配合 Deepening design cooperation with Huayi

深圳国际会展中心决策事项跟踪表(11.25)
The Shenzhen World Decision-Making Track List (11.25)

表 5-2 设计联合周报表
Table 5-2 Design Weekly Report List

深圳国际会展中心联合周报(第五周2017.04.24~2017.04.30)
Shenzhen International Convention and Exhibition Center United Weekly Report (Fifth Week 2017.04.24 ~ 2017.04.30)

单位 unit	主责 Main resp onsibility	已提交历史成果文件 Submitted historical results document	本周完成重要工作实现 (04.24~04.30) Complete important work this week Now (04.24 ~ 04.30)	下周完成重要工作实现 (04.24~04.30) Complete important work next weekNow (04.24 ~ 04.30)	近期将提交业主成果 Owner results will be sub mitted shortly	近期将提交的互提设计成果文件 Design outcome docum ent to be submitted soon	待确定重大事项 Major issues to be determined
AUBE	施工图设计 Construction Design	–	1.2万展厅结构、空调方案确定 1. 20,000 exhibition hall structure and air-conditioning plan determined 2.4万展厅单拱方案确定 2.40,000 exhibition hall single arch plan determined 3.消防评审意见会议纪要 3. Minutes of fire review comments meeting 4.珠海航展楼参观 4.Zhuhai Airshow Building Tour 5.展厅内马道、空气压缩、风管高度已有明确意见,见招商邮件 5. Horseway in the exhibition hall, air compression,There is a clear opinionon the duct height, seeInvestment mail 6.展厅编号已有明确意见,见招商邮件 6. The exhibition hall number has clear opinions.	1.4万展厅消防方案、空调方案确定 1.40,000 exhibition hall fire protection plan and air-conditioning plan determined 2.电梯招标文件审核 2. Review of elevator bidding documents 3.2万展厅消防沟通、消防方案确定 3.20,000 exhibition hall fire commu-nication and fire protection plan determined 4.与海滨大道匝道连接方案 4.Connection scheme with the seaside boulevard ramp 5.混凝土预制墙板材料沟通 5.Communication of precast concrete wall materials 6.桩基础提前开工申请相关文件 6. Relevant documents for the application of pile foundation in advance	1.标准展厅桩基与柱脚钢结构图纸(仅用于备料) 1. Drawing of steel structure of pile foundation and columnfoot of standard exhibition hall (for materialpreparation only) 2.桩基础提前开工申请相关设计文件 2. Design documents related to the application of the pile foundation in advance		1.4万展厅屋面结构与空调方案 1. 40,000 exhibition hallroof structure and air-condition-ing scheme 2.中央廊道结构 2. Central corridor structure 3.2-11栋方案设计 3.2-11 building design 4.登录大厅拉索幕墙方案确定 4.Confirmation of cable curtain wall plan in login hall 5.标准展厅9x9拉索式吊挂可行性分析 5. Feasibility analysis of standard exhibition hall 9x9 cable hanging 6.消防处理措施 6. Fire Fighting Measures 7.垂直绿化顾问单位 7.Vertical greening consulting unit
	景观设计 Landscape Design	–	1.与业主汇报景观概念方案 1. Report the landscape concept plan with the owner 2.设计联合体协商景观概念方案 2.Design the consortium to nego-tiate the concept of landscape	1.深化景观设计 1. Deepen landscape design			1.与运营确认可用于景观绿化的范围 1. Confirm with the oper ation that it can be used for landscaping
	BIM设计 BIM Design	–	1.标准展厅、地下室BIM模型深化 1. Standard exhibition hall, base-ment BIM mold Deepening	1.4万展厅空调方案BIM研究 1. BIM Study of Air Conditioning Solution for Exhibition Hall 40,000			
VP	建筑设计 Architectural Design	–	1.与业主的重要会议 1. Important meeting with the owner 2.协调标识设计出报概算文件 2. Coordinate the logo design and report the budget estimates 3.与华纳幕墙会议 3. Meeting with Warner Curtain Wall 4.景观讨论会 4. Landscape Discussion 5.可开启屋面考察 5. You can open the roof inspection 6.预制混凝土厂家沟通 6. Communication with precast concrete manufacturers 7.钢型材及铝单板实际工程的考察 7. Inspection of the actual engineering of steel profiles and aluminum veneer	1.时间计划的重新组织(重点是哪天提交立面图纸给SUP重新编制展厅幕墙招标图纸) 1. Reorganization of time plan.(em-phasis is to submit façade drawingsto SUP to re-draft exhibition hall curtain wall tender drawings) 2.屋面造型设计的汇报 2.Report on roof design 3.室内概念设计汇报 3.Interior concept design report 4.西侧配套用房设计调整,业确认主 4.The design of supporting houses on the west side was adjusted 5.预制混凝土厂家的考察(可能) 5.Inspection of precast concrete manufacturers (possibly)			1.屋面造型设计 1. Roof shape design 2.西侧配套用房平面功能确认 2. Confirmation of the plane function of the supporting house on the west side
	标识方案设计 Logo Design	–					
	室内概念设计 Interior Concept Design	–					
华纳	幕墙设计 Curtain Wall Design	1.展厅幕墙招标图纸(初版) 1. Tender drawing of exhibition hall curtain wall (first edition) 2.展厅、中央廊道幕墙支座反力 2. Counterforce of the curtain wall support of the exhibition hall and the central corridor 3.登录厅幕墙支座反力修改 3. Modification of the reactionforce of the curtain wall support of the registration hall 4.登入厅入口索网幕墙支座反力 4. Logging curtain wall support at the entrance hall Reaction force 5.金属屋面招标图纸 5. Metal roofing tender drawings 6.金属屋面自重荷载 6. Metal roof self-weight load	1.展厅及中央廊道幕墙招标图纸修改 1. Revision of tender drawings of exhi bition hall and central corridor curtain wall 2.登录厅幕墙招标图纸修改 2. Modification of the bidding drawings of the curtain wall of the registration hall 3.展厅及中央廊道幕墙技术会议 3. Exhibition Hall and Central Corridor Curtain Wall Technology Conference 4.登录厅幕墙技术会议 4. Login to the curtain wall technical meeting 5.南接待大厅索网幕墙支座反力提交 5.Submission of cable net wall support from South Reception Hall	1.展厅及中央廊道幕墙招标图纸修改(包括本次技术会议确定的内容) 1.Modification of the tender drawings of the exhibition hall and the curtain wall of the central corridor (including the content determined by this technical meeting) 2.登录厅幕墙招标图纸修改(包括本次技术会议确定的内容) 2. Modification of the bidding drawings for the curtain wall of theregistration hall (including the content determined by this technical meeting)	1.展厅及中央廊道幕墙招标图纸修改 1. Revision of tender drawings of exhibition hall and centralcorridor curtain wall 2.登录厅幕墙招标图纸修改 2. Modification of the biddingdrawings of the curtain wall of the registration hall	1.中央廊道玻璃盒子幕墙与主体结构关系 1. The relationship between the glass box curtain wall of the central gallery and the main structure 2.登录厅索网幕墙标高0.0米与主体结构连接做法 2. The practice of connecting the 0.0 meter elevation of the curtain wallof the login hall to the main structure	1.展厅入口玻璃盒子主体结构做法 The main structure of the glass box at the entrance of the exhibition hall

④ 信息管理——搭建信息交互平台，集约信息交互节点
信息交互平台。所有设计信息的分类、提交、分发、归档。高效合理的信息交互平台是各单位协作的基础。

在这个设计管理环节，核心的难点在于信息梳理。
目前，互联网技术高度发达，设计信息交互平台的搭建形式丰富。然而，深圳国际会展中心仅各单位参与项目实际管理的人数就达到上千人，参与设计、建设、施工的工作者更是达到几万人，各类项目信息以爆炸的形式在相互传播。如何对信息进行管理筛选，并保持各个设计团队在关键信息中保持高度的同步，是信息管理的难点。

这时需要区分重点，集约信息交互节点，并建立相对唯一的信息出口。在深圳国际会展中心，由设计总承包牵头，各个单位统一建立了唯一的信息出口与交互方式，确保各类信息的一致性。各单位需发出的信息与文件，应经过自身内部审核后，由特殊规定的人或途径对外发送，也根据信息的类型与重要性，约定不同层次的通告方式。

⑤ 协调管理
设计协调管理的核心目的是提高工作效率与工作进度，最大化人力资源利用。明确各单位（包括业主）的决策机制，制定各类事项的会议或决策制度，保障项目顺利推进（图5-7）也是协调管理的重点。

协调管理的难点在于如何正确区分问题的轻重缓急。设计管理单位需要汇总所有内外信息，对各方面临的设计问题进行梳理，综合效果、技术、工程、造价等多方面因素，确定问题的轻重缓急，进而再组织讨论与决策。
此时，需要高度整合的设计资源。在大型复杂的工程建设过程中，各类协调会议的效率以及会议结论的跟踪非常关键。
深圳国际会展中心项目中，AUBE 欧博设计牵头建立了各层级的例会协调制度，以及整个设计周期内的设计会议安排，并根据工程与设计的进度进行了适应性的调整，极大地提高了设计资源的使用效率。

④ Information Management -- Establish Information Interaction Platform and Intensify Information Interaction Nodes

Information exchange platform is for the classification, submission, distribution and filing of all design information. Efficient and reasonable information exchange platform is the basis of a successful cooperation.

In the step of design management, the sorting of information is the core difficulty.

Because of the highly developed internet technology nowadays, the interactive platform for design information is rich in forms. However, for the Shenzhen World project, there are thousands of actual managers and tens of thousands of workers participating in the design, and construction process. All kinds of project information are spread among each other like explosion. The difficulty of information management is how to manage and filter the information and keep each design team highly synchronized with the key information.

At this point, we need to distinguish the key points, intensify the information interaction nodes, and establish a relatively unique singular information outlet. In the Shenzhen World project, led by the EPC unit, each unit has established the unique information exit and interaction way to guarantee the consistency of all kinds of information. The information and documents of each unit shall be sent by the designated person or channel after its own internal examination. According to the type and importance of the information, different levels of notification has been appointed.

⑤ Coordination and Management

The core purpose of the design coordination management is to improve work efficiency and progress, and maximize the use of human resources. Clarify the decision-making mechanism of each unit (including the owner), formulate the meeting or decision-making mechanism of various matters, and ensure the smooth progress of the project (Figure 5-7). is also the key issues.

The difficulty of coordinated management is how to correctly prioritize problems. The design management unit needs to summarize all internal and external information, sort out the design problems of all parties, integrate the effect, technology, engineering, cost and other factors, determine the priority of the problem, and then organize discussion and decision-making process.

At this point, highly integrated design resources are required. In the process of large and complicated engineering construction, it is very important to track the efficiency of various coordination meetings and the conclusions of the meetings.

In the project of the Shenzhen World, AUBE took the lead in establishing the regular meeting coordination system at all levels, as well as the arrangement of design meetings in the whole design period, and made adaptive adjustment according to the progress of engineering and design, which greatly improved the efficiency of design resources.

图 5-7 深圳会展图纸交底会议
Figure 5-7 Drawing Disclosure Meeting

会展景观 + 海绵城市
Exhibition Landscape & Sponge City

会展类景观是城市公共空间设计的特殊领域之一，深圳国际会展中心在超尺度建筑空间形态的探讨中，探索景观如何可以提升场地的高利用率，满足会展功能的专业性，强化生态的可持续性及空间的人性化体验等，以实现多维度综合一体化设计。

Exhibition landscape is one of the special fields of urban public space design. In the landscape design of the Shenzhen World, we have explored how to increase the site use in an ultra-scale building space, not only to meet the needs of the exhibition center, but also to strengthen the ecological sustainability and human experiences, and tealize the multi-dimensional integration design.

会展景观，大尺度下的人性化体验
Exhibition Landscape, Experience of Human Touch on a Large Scale

深圳国际会展中心作为全球最大的会展中心，对城市形象提升有着重要作用（图 5-8）。景观在营造功能合理、空间宜人的物理环境的同时，也会成为深圳历史文脉、地域特征和人文精神的现实载体。主体建筑以南北向鱼骨状分布，而景观则在面相城市重要节点的广场中，采用中轴对称式布局，与建筑相协调，突出主广场的仪式感，同时强调空间的公共性、开放性和实用性。

As the largest convention and exhibition center in the world, the Shenzhen World plays an important role in the promoting of the image of the city (Figure 5-8). While creating a physical environment with reasonable functions and pleasant space, landscape will also become a realistic carrier of Shenzhen's historical context, regional characteristics and humanistic spirit. The main building of the Shenzhen World is distributed in the shape of fish bones from the south to the north, while the landscape in the important node squares facing the city adopts the axisymmetrical layout, which is coordinated with the buildings to highlight the sense of ceremony of the main square, and emphasize the publicity, openness and practicality of the space.

图 5-8　南登录大厅广场效果图
Figure 5-8　Rendering of South Arrival Hall Square

会展景观，大尺度下的人性化体验

① 高效的交通组织

深圳国际会展中心室内展览面积高达 40 万平方米，满展期间，将同时承载最高多达 30 万人流，客流的高强度、瞬时性，显得尤为突出。景观需满足开展期间高效、便捷的人流组织。同时兼顾布展期间，各种车流及人流的需求。

登录大厅前广场

登录大厅是会展中心对外形象展示的重要界面之一，也是观展客流最为密集之地。深圳国际会展中心采用集中安检模式，主要从六个入口进入，其中 A、南接待大厅 B、南登录大厅 C、北登录大厅，三个口为主要登录口（图 5-9），D、北入口，E、F(VIP 入口) 人流量较小。

根据相关顾问数据，合理布局各功能区块在广场位置，分为安检区、等候区以及退场通道区。充分满足人流高峰时段，高效、有序地组织各股人流。广场剩余部分根据空间需求，营造入口景观空间氛围，强化入口形象，同时，根据不同展会需求，适时调整广场安检及人流组织方式（图 5-10），突显其空间的时效性和高效性，并为后续场地运营提供更多可能性。

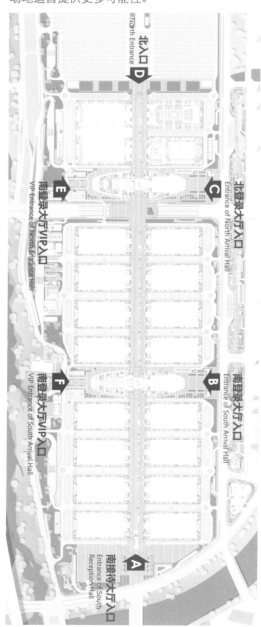

图 5-9　会展主要出入口图
Figure 5-9　The Main Entrance of the Exhibition Hall

中央廊道

作为从登录大厅进入各展厅的主要通道，总长度 1750m，分为两层。其中二层廊道以快速通过为主，依次设置步行道，快速传输带，电瓶车道，让人流快速进入展厅，减少停留；一层廊道以休憩和少量商业配套为主，通过线性铺装肌理，引导人流走向。在展厅入口处，则通过肌理渐变的方式与室内铺地材质相协调，在提升室内外空间整体性的同时，提醒参观者已到达展厅入口（图 5-11）。

Professional and Efficient Site Design

① Efficient Traffic Arrangement

The indoor exhibition area of the Shenzhen World is up to 400,000 square meters. During the full exhibition period, it would carry up to 300,000 people at the same time. The landscape should meet the needs of efficient and convenient organization of people circulation during the exhibition period, as well as the demand of various traffic and people circulation at the same time.

-Front Square of the Arrival Hall

The arrival hall is one of the important interfaces for the Shenzhen World to display its external image and is also the most dense place with visitors. The Shenzhen World adopts the centralized security check with six main entrances, A, south Arrival Hall B, south arrival hall C, north arrival hall are main entrances (Figure 5-9); D, north entrance, E, F(VIP entrance) have less people.

According to relevant consultant data, each function block should be reasonably arranged in the square, divided into security check area, waiting area and exit channel area, to fully guarantee efficient and well-organized crowd at the peak hours. The rest of the square, according to the space needs, creates a space atmosphere of entrance landscape and strengthens the image of the entrance. Meanwhile, according to the different exhibition needs, the square adjusts the security check and crowd organization mode in time to highlight the timeliness and high efficiency of its space and provides more possibilities for the subsequent site operation. (Figure 5-10)

-Central Concourse

As the main passageway from the arrival hall to the exhibition halls, the central concourse has a total length of 1.7km and is divided into two floors. The second floor is mainly for quick passage, with the pedestrian path, fast transmission belt and battery lane to allow people to enter the exhibition hall quickly and reduce the stay. The first floor is mainly for rest and a small number of commercial supporting facilities, with the linear pavement texture guiding the flow of people. At the entrance of the exhibition hall, the gradual texture is coordinated with the material of the interior floor, so as to promote the integrity of indoor and outdoor space and remind visitors that they have arrived at the entrance of the exhibition hall. (Figure 5-11)

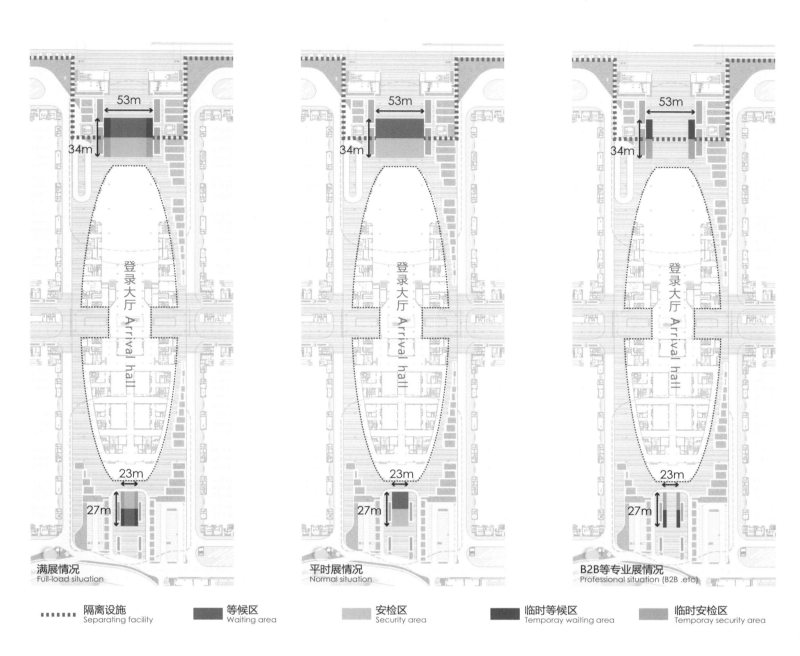

满展情况
Full-load situation

平时展情况
Normal situation

B2B等专业展情况
Professional situation (B2B .etc)

· · · · · · · 隔离设施
Separating facility

等候区
Waiting area

安检区
Security area

临时等候区
Temporay waiting area

临时安检区
Temporay security area

图 5-10 登录大厅安检流线图
Figure 5-10 Arrival Hall Security Check Circulation

图 5-11 登录大厅广场水景效果图
Figure 5-11 Arrival Hall Water Square Rendering

② 弹性的场地模式

在场地设计中，加入"时间"维度，随时间的变化，使空间产生不同的使用需求和方式，设计的弹性预留给后期场地运营留有更多的可能性，以适应未来展览空间的变化与发展（图 5-12）。

场景转换式

主入口广场区域利用"弹性"的策略，在不同场景模式下，功能随之而变。南接待大厅广场作为会展主体建筑的南部起端，成为深圳主城区方向到达会展的第一界面，也是会展与城市空间相互融合的切入点，兼顾城市形象和功能需求（图 5-13）。即可在开幕式时，容纳万人，作为庆典广场使用。同时满足在开展期间，为安保提供需求，成为安检区、等候区。而当需要室外布展时，则成为室外展场，如大型器械展等，从而把场地利用率发挥到极致。

灵活划分式

不同展期的空间分隔，采用灵活划分式，利用移动铁马、伸缩门、轻质隔断等形式划分展厅单元空间。根据需求变化调整单元的大小和形式，便于展厅使用规模及组合方式进行管控，提高场地使用效率和灵活性（图 5-14）。

图 5-12　南接待大厅广场林荫休憩区效果图
Figure 5-12　South Arrival Hall Square Tree-lined Recreation Area Rendering

② Flexible Site

In site design, time is important. Because the space has different needs and ways to use with the change of time. Therefore the flexibility of the design is reserved for more possibilities for site operation, so as to adapt to the change and development of the future exhibition space.(Figure 5-12)

- Scene Conversion

Adopting a flexible strategy,the functions of the square area of the main entrance can change to different scenes. As the southern starting point of the main exhibition building, the south arrival hall square is the first interface when arrives at the exhibition from the main urban area, and it is also the starting point for the integration of exhibition and urban space, which consider both the city image and the functional requirements. (Figure 5-13) Therefore, it can be used as a celebration square for opening ceremony accommodating ten thousand people as a celebration square. It can also serve as security area and waiting area during exhibition period. In addition, it also can serve as an outdoor exhibition area when necessary, for example a large instrument exhibition.

- Flexible partition

The exhibition hall space can be divided flexibly for different event by moving iron horse, telescopic door and light partition. Adjust the size and form according to the change of demands, so as to control scale and combination of the exhibition hall and improve the efficiency and flexibility of the site use.(Figure 5-14)

图 5-13　登录大厅西侧 VIP 入口效果图
Figure 5-13　VIP Entrance on the West Side of the Arrival Hall Rendering

图 5-14　中央廊道一层效果图
Figure 5-14　The First Floor of the Central Concourse Rendering

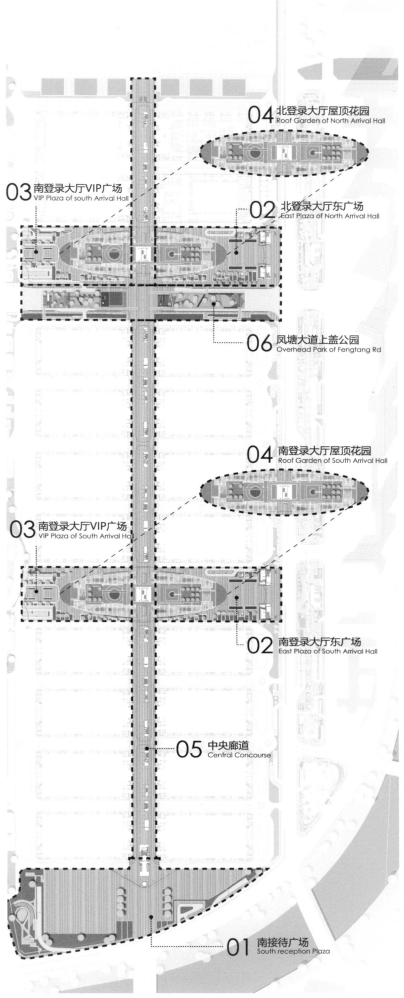

04 北登录大厅屋顶花园
Roof Garden of North Arrival Hall

03 南登录大厅VIP广场
VIP Plaza of south Arrival Hall

02 北登录大厅东广场
East Plaza of North Arrival Hall

06 凤塘大道上盖公园
Overhead Park of Fengtang Rd

04 南登录大厅屋顶花园
Roof Garden of South Arrival Hall

03 南登录大厅VIP广场
VIP Plaza of South Arrival Hall

02 南登录大厅东广场
East Plaza of South Arrival Hall

05 中央廊道
Central Concourse

01 南接待广场
South reception Plaza

图 5-15　景观设计重点区域图
Figure 5-15　Landscape Design Focus Area

人性舒适的空间体验

在会展类公建中，建筑的超大尺度是对人性化体验的最大挑战，设计组织各种集为人所用，为人所享要素的公共空间，使到访者通过各种行为活动，可以获得亲切、舒适、轻松、愉悦、安全、自由的心理感受。一是尺度，如何在超尺度的空间下，使人感到亲切，舒适。二是功能，在人的众多使用诉求中，如何在满足功能的前提下变得更加实用，美观（图 5-15）。

Comfortable Space with Human Touch

In the construction of public building as exhibition hall, the super-large scale of architecture is the biggest challenge for human experience. It needs to design and organize various elements for people to use and enjoy, and provide kind, comfortable, relaxed, pleasant, safe and free psychological feelings through all kinds of activities.

Firstly, how to make people feel comfortable in such a large space, Secondly, how to make it both practical and beautiful for different people (Figure 5-15).

① 人性的尺度
人性的空间尺度，直接关系到人的体验、空间的品质和氛围的形成，如何做到人性尺度是大型会展类建筑中至关重要的一环。

空间的节奏控制
中央廊道一层作为人参与及停留最多的公共空间，总长 1750m，线性的空间体验过于单一。在设计时，采取以四个展厅为 1 个单元的方式进行氛围的营造及功能排布，共 5 个单元，且在保证每个单元均好性的同时，让空间体验更加舒适宜人。

空间的围合
空间尺度过大是会展类建筑的共性。在满足运营需求的前提下，根据建筑功能的排布，利用组团绿化及景观小品进行软围合，有效减少空间尺度。

绿化柔化边界
超长尺度的中央廊道及大体量的两侧展厅，对人的压迫感较强。景观将通过介于建筑与人二者体量之间的植物进行三维空间上的过渡与柔化，从而虚弱建筑给人直接带来的压迫感，让空间变得更加友好。

① The Scale of Human Touch
The space scale of human touch is directly related to human experience, the quality of space and the formation of atmosphere. How to achieve the scale of human touch is a crucial part of large-scale exhibition architecture.

- Rhythm Control of Space
The first floor of the central concourse is the public space that most people participate and stay in, but linear spatial experience is too simple with its total 1.7 kilometers length. Therefore, in the design, four exhibition halls were put into one unit to create atmosphere and arrange supporting function. There are 5 units in total, which makes the special experience more comfortable and pleasant.

- Enclosure of the Space
Large-scale space is common among all the exhibition buildings. On the premise of operation needs, according to the layout of the building functions, the soft enclosure is made of group greening and landscape sketches so as to effectively reducing the space scale.

- Greening and Softening of the Boundary
The ultra-long central concourse and the large exhibition halls on both sides create a strong sense of oppression. Through the transition and softening of the plants in the three-dimensional space between the building and the human, the landscape makes the space more friendly by reducing the sense of oppression caused by the building.

图 5-16　登录大厅屋顶花园效果图
Figure 5-16　Arrival Hall Roof Garden Renderings

图 5-17　登录大厅屋顶花园鸟瞰图
Figure 5-17　Aerial View Renderings of the Roof Garden of the Arrival Hall

② 人性的功能

在重功能，强效率，场地有限的会展空间中，人性化的功能配套显得尤为稀缺。如何在满足展览功能需求的前提下，合理利用空间，把更多的功能空间还给参观者，让他们有不一样的参展体验尤为重要。

屋顶花园

登录大厅屋顶花园作为会展建筑中唯一可上人屋面，让原本功能单一的空间体验变得更加多元。屋面分为两块区域，其一为有建筑屋盖覆盖区域，形成半室外空间，考虑到未来的商业运营模式，此块区域将作为与室内功能相适应的多功能平台，可拓展为咖啡外摆、举办小型活动等（图5-16）。而无建筑屋盖覆盖区域，则作为室外展陈区，可根据需求来更换雕塑展品，让空间更具生长性（图5-17）。

凤塘大道上盖会展公园

凤塘大道上盖公园是会展周边配套中最大的一块公共绿地，全年为市民开放，且在开展期间通过运营管理，为参展商、参观者提供休憩及相关展陈配套功能（图5-18）。上盖区域延续会展建筑对称的设计原则，以中央廊道为中心，呈东西向布局。东侧，提供户外剧场、阳光草坪、创意地形等。西侧，则更偏向于休闲和艺术，依次为跌瀑水景、艺术雕塑、林荫休憩、咖啡吧等。此处将会成为集公众参与性、体验性、功能性、休闲性于一体的上盖公园。

② Functions for People

In the exhibition space with heavy functions, high efficiency and limited space, functional supporting facilities with human touch are particularly scarce. On the premise of meeting the needs of exhibitions, it's also important to make reasonable use of the space to save more functional spaces for visitors, so that they can have different and better exhibition experiences.

- Roof Garden

As the only available roof for visitors among all exhibition buildings, the roof garden of the Arrival hall makes the experience of the original single space more diverse. The roof is divided into two areas, one of which is covered by a building roof to form a semi-outdoor space. Considering the business operation in the future, this area is regarded as a multi-functional platform suitable for indoor functions, which can be expanded to provide outdoor coffee and small activities. (Figure 5-16) The uncovered area of the building roof are used as outdoor exhibition areas, which can change display sculptures according to different needs to make the space more vibrant.(Figure 5-17)

- Upper Park over Fengtang avenue

The park over Fengtang avenue is the largest piece of public green space in the surrounding area of the exhibition center, which is open to the public all year round. Besides, it provides exhibitors and visitors with rest spots and related supporting functions during the exhibition period. (Figure 5-18) The upper area continues the symmetrical design principle of the exhibition building, adopts an east-west layout with the central concourse as the axis. On the east side, there are outdoor theater, sunshine lawn, creative terrain, etc. On the west side, it is more inclined to function of leisure and art with falling waterfall waterscape, artistic sculpture, tree-lined rest, coffee bar and so on. It's an upper park for public with good experience, multi-function and leisure.

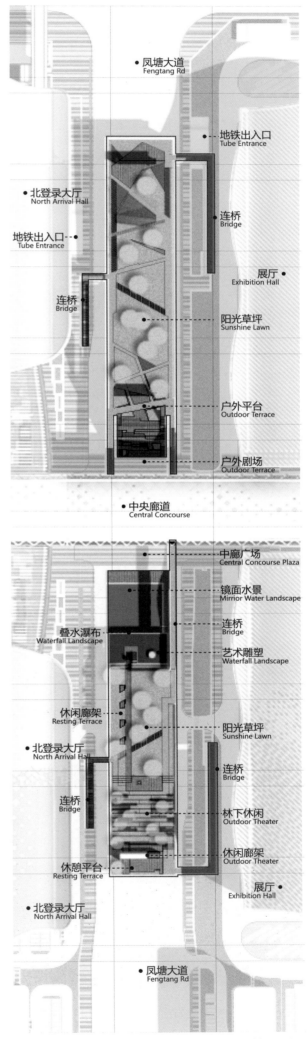

图 5-18　凤塘大道上盖公园总图
Figure 5-18　General Plan of the Fengtang Avenue Upper Park

生态自然的绿化举措

① 雨水花园系统

展馆周边绿地结合高低起伏地形形成下凹绿地，利用地被植物所覆盖，将雨水滞留下渗，来降低暴雨地表径流的洪峰。同时通过吸附、降解、离子交换和挥发等过程减少污染。蓄积的雨水经过净化处理后，为植物进行灌溉，成为可持续发展的生态系统，并与周边截流河形成城市的大海绵效应（图 5-19）。

南接待广场西侧的公共绿地，结合室外展场打造具有可观，可赏、可停留的雨水花园，为人们提供更丰富的参展体验，同时起到科普教育的作用。

② 植物设计

会展以冠大荫浓的常绿本土乔木为基调，保证了园区常年的绿量及遮荫效果，为植物景观形成良好的绿色背景。以观花特色鲜明的植物片植，与高低起伏的地形结合，形成震撼的花林景观。并巧妙地把植物的最佳赏期与大型活动的主要举办时间结合在一起，形成四季有花可观（图 5-20）的效果。

其中，主广场植物利用高大乔木作为主基调，并以对称式排列，地被则以修剪型常绿开花为主，形成强烈的仪式感；西侧配套区域与海滨大道及沿江高速相邻，打造了一个热情开放的滨海花林。在树种选择上以开花乔木为主，视野通透性较好，且不遮挡车行视角从沿江高速观看会展建筑；东侧区域位于海汇路上，并与会展休闲带相邻，植物配置以高大挺拔的常绿乔木朴树列植，形成绿色统一的线性景观。结合会展休闲带以樱花木棉为主的绿岛，打造成大空港区的会展大道（图 5-21）。

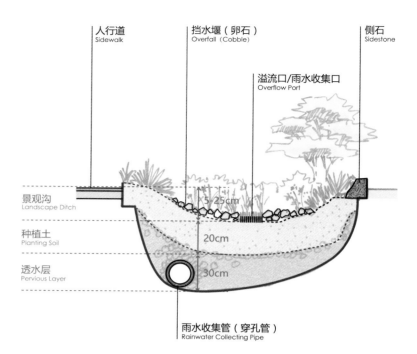

图 5-19　雨水花园结构图
Figure 5-19　Rain Garden Structure

图 5-20　西侧配套区域植物效果图
Figure 5-20　West Side Supporting Regional Plants on the Rendering

图 5-21　东侧海汇路植物效果图
Figure 5-21　East Side Haihui Road Plants Rendering

Ecological and Natural Greening Measures

① Rainwater Garden System

The surrounding green space of the exhibition halls together with the undulating terrain has formed a concave green space, which uses cover plants to stop rainwater infiltration and reduce the flood peak of storm runoff. It also reduces pollution through adsorption, degradation, ion exchange and volatilization and other processes. The collected rainwater is purified and irrigated for the plants, making it a sustainable ecosystem and forming a large urban sponge effect with the surrounding intercepting rivers.(Figure 5-19)

The public green space on the west side of the south reception square is combined with the outdoor exhibition area to create a rain garden that can be enjoyed and stayed, providing people a better and richer experience and playing the role of science education.

② Plant Design

The keynote of the exhibition is the local evergreen trees with thick canopy and shade, which ensures the perennial green quantity and shade effect of the park and forms a good green background for the plant landscape. With the plants of distinctive flower, and the ups and downs of the terrain, it forms an eye-catching flower forest landscape. The best time of flowering is arranged to fit the time of main events and important occasions, which guarantee the flowering views for Four Seasons(Figure 5-20).

Among them, main square uses tall trees as its keynote, which are arranged in a symmetrical pattern, while the ground is mainly pruned evergreen flowers, forming a strong sense of ceremony; The supporting area on the west side is adjacent to the promenade and the high-speed along the river, which creates a warm and open coastal flower forest. In terms of tree species selection, it is dominated by flowering trees with a good visibility which would not block the view from the high-speed road along the river. The east side is located on Haihui road, and is adjacent to the exhibition and leisure zone, where tall and straight evergreen plumbago trees are arranged to form a unified green linear landscape. And the exhibition avenue in the big airport area is built with the combination of the green island features Sakura kapok in the exhibition leisure zone.(Figure 5-21)

图 5-22　凤塘大道上盖休憩空间效果图
Figure 5-22　Open Space Above Fengtang Avenue Rendering

图 5-23　凤塘大道上盖公园西侧效果图
Figure 5-23　West side of Fengtang Avenue Park rendering

图 5-24　凤塘大道上盖公园东侧效果图
Figure 5-24　East side of Fengtang Avenue Park rendering

海绵城市，水和建筑的共生
Sponge City, the Symbiosis of Water and Architecture

为了使城市在适应环境变化和应对雨水带来的自然灾害等方面具有良好的"弹性"，深圳国际会展中心项目在处理水与建筑的关系时，强调让水在城市中的迁移活动更加"有序"和"自然"，并进而建立和完善城市"海绵体"。设计结合会展运营功能和巨尺度的建筑规模，运用防排结合原则，防水、排水、吸水、蓄水、渗水、净水，需要时将蓄存的水"释放"并加以利用，实现自然积存、自然渗透、自然净化的城市发展方式。

In order to make the city more adaptive to the environmental changes and potential natural disasters caused by the rain, the project put an emphasizes on a more order and natural water migration activity among the city when dealing with the relations between water and architecture. According to the exhibition operation functions and huge scale of the buildings, the project builds and perfects a city sponge with a principle of combining waterproof and drainage for waterproofing, water draining, water absorption, water storage, water sewage, water purification, then releases and reuses the reserved water when necessary, in which way it achieves an urban development of natural accumulation, natural penetration and natural purification.

海绵城市与排雨水设计

深圳国际会展中心所在地块属于海绵功能提升区，西部雨型分区，西部珠江三角洲雨型。短时降雨强度大，最大日降雨量 257.3mm. 最大每小时雨量可接近 100mm。现状地下水平均埋深小于 2m，经填海处理后地面高程将大于 6.5m。目前的滩涂底层分布着饱和软土，入渗性较差。雨水落入滩涂直流入海，汇流通畅无内涝。周边现状水系发达，主要为茅洲河支流。场地以东河涌承接了场地东部建成区部分生活污水，现状水质差，水生环境脆弱，生态修复压力大（图 5-25）。

通常人们借助市政排水工程解决城市雨涝问题，铺设大型排水管道或者安装更大功率的水泵，这种方法虽然简单，但往往有生态、经济和环境协调方面的问题。由此，我们将海绵城市概念引入项目的景观设计和排雨水设计中，和谐水与建筑、与景观的关系，化解建筑排水的功能需求与运营使用、空间体验之间的矛盾。

Sponge City and Rainwater System Design

The area where the Shenzhen World stands is a sponge function improvement area. The maximum daily rainfall is 257.3mm and the maximum hourly rainfall is close to 100mm. The average buried depth of the current groundwater is less than 2 meters, and the surface elevation will be more than 6.5m after reclamation. At present, the bottom of tidal flat is distributed with saturated soft soil, which has poor permeability. The rainwater falls into the tidal flat and flows directly into the sea without any waterlogging. The surrounding current water resource is rich, mainly the tributary from Maozhou river. In the east side, the rivers received part of the domestic sewage from the built-up area. With poor water quality and fragile aquatic environment, there is great pressure for ecological restoration.(Figure 5-25)

Usually, municipal drainage is used to solve urban waterlogging problems by laying large drainage pipes or installing more powerful pumps. Although this method is simple, it often has ecological, economic and environmental problems. Therefore, we introduce the concept of sponge city into the landscape design and rainwater system design for this project to deal with the relationship between water, architecture and landscape in a harmony way, and resolve the contradiction between functional demands of the building drainage, operation and use, and spatial experience.

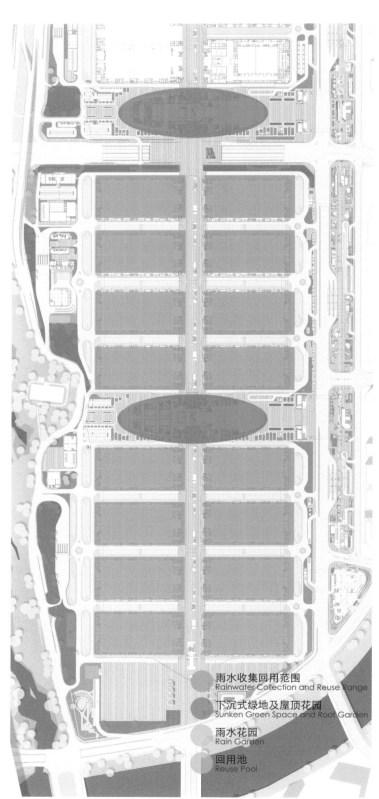

雨水收集回用范围
Rainwater Collection and Reuse Range

下沉式绿地及屋顶花园
Sunken Green Space and Roof Garden

雨水花园
Rain Garden

回用池
Reuse Pool

图 5-25　海绵设施类型及分布图
Figure 5-25　Sponge Facilities Types and Distribution

排水难点

在深圳国际会展中心项目中，排水主要存在以下三个难点：
一是断续不连贯的外围景观带，绿地面积小且分布不均，主要集中在西侧和外围，屋顶面积占比大但多为无承重功能的轻质金属屋面或光伏屋面，且绿化困难；
二是广场道路有重载车辆高频率行驶，地面铺装必须采用传统铺装，雨水下渗空间有限；
三是如何将海绵设施与会展项目难以避免的大尺度公共空间整合设计（图5-26）。

Drainage Difficulties

There are three major difficulties of drainage in this project:
Firstly, the intermittent and incoherent peripheral landscape belt, where the green space area is small and unevenly distributed, mainly concentrated in the west and the periphery area, with a large roof area made of light metal or photovoltaic materials without load-bearing function and hard for greening;
Secondly, there are heavy vehicles passing through with high frequency on the square road, so the ground pavement must use the traditional pavement due to the limited space for rainwater infiltration;
Thirdly, how to integrate the sponge facilities with the design of unavoidable large-scale public space (Figure 5-26).

图 5-26　雨水花园效果图
Figure 5-26　Rain Garden Rendering

设施解决方案

根据《深圳市海绵城市建设专项规划及实施方案》推荐的典型设施类型，结合本项目建设开发条件和建筑、景观需求，选择绿色屋顶、下沉式绿地、雨水花园、雨水调蓄池（回用池）为本项目主要应用的海绵设施类型。

为解决该问题，主要采取措施如下：
一方面，以因地制宜为原则，在南、北两登录大厅及西侧配套建筑、凤塘大道上盖公园等位置设置绿色屋顶，阳光草坪、创意地形等元素，让原本单一的围护结构，具有了展陈、休闲和生态调蓄等多元功能。在绿色屋顶下方设置蓄水层，在提升雨水滞蓄能力的同时为植物生长提供用水；展厅的屋面则以功能适用为原则，在确定虹吸排水能力时考虑一定的安全系数（10年的管道系统，100年的溢流管道系统，200年的溢流口系统），雨水径流经初雨弃流和雨水回用池调蓄净化后，或回用于空调冷却补充水、绿地浇灌，或错峰排入市政排水检查井。

另一方面，结合运营和功能需求及周边截流河等大生态系统，灵活组织排布地块中的下沉绿地，同时利用室外展场的边角空间打造具有可观，可赏、可停留的雨水花园（图5-27），将硬质不透水地面的雨水径流组织到海绵设施内，经净化、下渗后，超标雨水由溢流设施排入场地内雨水排水管网，最终排入市政管网系统，达到净化与削峰的目的。通过这些绿化组团柔化边界，有效地减少了大尺度公共空间带来的压迫感，给参观者带来更加友好和人性化的体验。

通过对项目所在区域自然基础条件、管网建设及建筑、景观功能需求的分析，优化了场地设计和景观布局，合理采用海绵设施，实现净、用、滞、蓄的建设海绵城市目标，最终使年径流总量有效控制在55%，降低了雨水内涝风险，实现雨水的自然积存、自然渗透、自然净化。

Measures and Solutions

According to the typical types of facilities recommended in the Special Planning and Implementation for Shenzhen Sponge City Construction, along with conditions and requirements of architecture and landscape, green roof, sunken green space, rainwater garden and rainwater storage pond (reuse pond) are selected as the main sponge facilities in this project.

To solve the problem, the main measures are taken as follows:
On the one hand, adjusting measures to accommodate local conditions, Green roofs, sunshine lawns, creative terrain and other elements are set in the south and north arrival halls, supporting buildings on the west side, and upper park over the Fengtang avenue, making the building envelope has more functions such as exhibition, recreation and ecological regulation and storage. An aquifer is placed under the green roof to improve rainwater retention and provide water for plant growth. The roof of the exhibition hall can adapt to different functions, take account of certain safety factor (10 years of pipeline system, 100 years of the overflow pipe system, 200 of the overflow mouth system), when determined the siphon drainage ability, after the purification within the rainwater recycling pool, the stormwater runoff is either used as supplement for air conditioning cooling water, landscape irrigation, or flows into the municipal drainage inspection wells avoiding the peak hour.

On the other hand, according to the operational and functional requirements and the surrounding intercepting river ecosystems, the project arranged the sinking green space flexibly, and used the space at the outside corner of the exhibition halls to build a beautiful rain garden for people to stay and enjoy. (Figure 5-27)When the rainwater runoff from the rigid waterproof ground flows into the sponge facilities, it will be purified and infiltrated, then flow into the rainwater drainage pipe network, and finally to the municipal pipe network system, so as to achieve the goal of purifying and peak clipping. Softening the boundaries by these green clusters, it effectively reduces the pressure caused by large-scale public space and provides visitors a more friendly and human experience.

Through the analysis of natural basic condition, pipe network construction and functional requirements of building and landscape, the project optimized the site design and landscape layout, adopted sponge facilities in a reasonable way, achieving the goal of building a sponge city for water purification, water reuse, water retention and water storage, and finally controlled the total annual runoff within 55%, reducing the risk of rain waterlogging, and help saving, infiltrating and purifying the rainwater naturally.

图 5-27 雨水花园效果图
Figure 5-27 Rain Garden Rendering

交通设计
Transportation and Circulation Design

超大型会展中心交通需求具有极强的短时聚集性和方式复杂多样性，目前，国内已投入运营的超大型会展中心数量不多，且展会期会展区域大都存在一定的交通拥堵问题。能高效快捷地疏散会展客流，既是保障会展成功运营的关键，也是会展中心综合运营实力的一种体现。

The traffic of Super-large exhibition center is intense and complex in short time Period.
At present, there are few ultra-large exhibition centers in China. During the exhibition, traffic congestion usually happens, therefore, how to quickly guiding the passenger flow is the key to guarantee the exhibition center operation, and also shows the comprehensive strength of the operator.

整体构思
Overall Gonception

主要挑战

① 从零构建：建筑主体及周边配套交通从零构建，时间进度紧张

深圳国际会展中心与周边配套交通系统均从零构建（除现状沿江高速），计划仅用 4 年时间建成投入使用，时间进度非常紧张，要求设计建设同步推进，必须做好上下统筹，前后协调，减少失误与返工。

② 超大尺度：场地面积达到深圳老展馆的 10 倍，交通组织难度大

交通出行品质是决定会展中心吸引力的重要因素，深圳国际会展中心场地面积达到深圳老展馆的 10 倍，展览面积达到 4 倍，在如此大的空间尺度上组织会展复杂的各类交通方式成为一大挑战，尤其是步行交通，可承受的距离非常有限。

③ 交通集聚：交通需求总量大、高峰集中，交通疏解压力大

会展中心具有交通需求总量大、高峰集中的特征，单位展示面积客流吸引力达到常规商业开发的 2～3 倍，早晚高峰出行率达到 25%～40%，货运交通需求也达到了 1 台车 /100 平方米的净展览面积，如此集聚的交通采用传统增加交通设施供给的方式难以应对，周边城市交通也面临着严重冲击的风险。

Main Challenges

① Construction from Nothing: the building and surrounding supporting traffic is built from zero with a tight schedule.

The Shenzhen World and the surrounding supporting transport system are built from nothing (except the existing high-speed highway along the river), and was planned to finish within 4 years, which meant that the schedule is very tight. Therefore, design and construction should be pushed forward simultaneously, and coordination work was crucial to reduce mistakes and reworks.

② Ultra-large Scale: the site area is 10 times the size of the old exhibition center, which makes traffic organization difficult.

Traffic quality is an important factor for the attractions of exhibitions. The site area of the Shenzhen World is 10 times the size of the old one, and the exhibiting area is 4 times. It is a great challenge to organize various and complicated transportion on such space scale, especially for the pedestrian traffic due to the limited bearable distance.

③ Traffic agglomeration: the total traffic demand is large, the peak is concentrated, and the evacuation is under high pressure.

Exhibition centers has characteristics of large amount of traffic and peak concentration, the traffic attraction is usually 2-3 times more compared to conventional commercial development. And the traveling rate of morning and evening peak is up to 25-40%, while the freight transportation demand reached 1 pallet/displaying area of 100 square meters. Therefore, it's impossible to deal with the traffic demand by simply increasing traffic facilities supply, and the surrounding urban traffic is also facing the risk of severe impact.

展馆名称 Name of exhibition hall	会展名称 Exhibition name	会展类型 Convention and Exhibition types	展示面积（WM²） Exhibition area (WM²)	参展客流（万人／日） Visitors (10000 person times / day)		客流吸引率（人／M²） Passenger attraction rate (person time / M²)	吸引率范围（人／M²） Attraction rate range (person / M²)
上海新国际博览中心 Shanghai New International Expo Center	2011年上海车展 2011 Shanghai Auto Showr	消费展 Consumption Exhibition	17	工作日 Working day	10	0.6	0.4-1.1
				休息日 Rest Day	18.7	1.1	
	2011年华交会 2011 China Fair	专业展 Professional exhibition	10	4		0.4	
	2011年工博会 2011 Industrial Expo	专业展 Professional exhibition	10	4		0.4	
广州琶洲会展中心 Guangzhou Pazhou Exhibition Center	2010年广印会 2010 Guangzhou press conference	专业展 Professional exhibition	12	5		0.4	
	2010年广交会 2010 Guangzhou Fair	专业展 Professional exhibition	33.8	13		0.4	
深圳会展中心 Guangzhou Convention and Exhibition Center	2016年家纺布艺展 2016 Home Textile Exhibition	综合展 Comprehensive exhibition	10	7		0.7	
	2016年国际家具展 International Furniture Exhibition 2016	综合展 Comprehensive exhibition	10	7.5		0.8	

表 5-3　国内主要大型会展相关客流吸引率统计表
Table 5-3　Statistics of passenger attraction rate related to major domestic large-scale exhibitions

总体思路

① 广泛借鉴：广泛调研国内外类似规模会展中心，充分借鉴设计及疏解经验

实地调研了上海国展中心、广州琶洲国际会展中心、深圳老展馆、香港会议展览中心等国内展馆（表 5-3），调研了意大利米兰会展中心、德国汉诺威会展中心、德国慕尼黑会展中心、科隆国际会展中心等欧洲展馆。国内外会展中心均选址交通设施发达的区域，国内主要依托公共交通疏解客流，但欧洲会展中心兼顾非公共交通出行，小汽车使用率相对较高，同时，欧洲的会展中心布撤展周期远高于国内，展馆使用率相对较低。

② 交通先行：在开展建筑设计前，稳定周边交通规划，明确设计条件

会展中心体量规模巨大，需要在大范围内提供完善的交通体系支撑。若急于开展会展中心建筑设计，周边交通规划难以充分考虑并纳入会展中心的诉求，容易存在对外交通功能缺失或相互衔接不畅等问题，会导致后续的返工调整。因此，必须交通规划先行，充分协调对接会展中心诉求，尽快稳定周边的设计条件。

③ 因地制宜：适应并充分利用周边既有交通条件，减小负面交通影响

深圳国际会展中心紧靠沿江高速、海滨快速，南部 7 公里处为深圳国际机场，在会展中心交通设施整体布局时充分考虑利用这些交通资源，构建完善的对外交通系统。由于会展中心占地面积大，紧邻多条道路，会展交通不可避免与城市交通交织，因此周边道路规划设计时宜妥善应对，采取增加道路疏散能力、立体分离等手段尽量减小对城市交通的影响（图 5-28 会展初期交通设计图）。

④ 个性设计：在全面掌握会展交通特征的基础上展开工作，切实满足会展需求。

除需求总量大、高峰集中外，会展中心还存在潮汐现象明显、客流层次差别大、交通组成复杂、出行品质要求高、安检票检必不可少等诸多特征，与城市交通及其他类型交通存在明显差别。因此，交通设计不能套用常规设施的设计方法，必须在充分掌握这些特征的基础上展开，切实满足会展诉求（图 5-31 不同规模展会数量及时间统计表）。

General Strategy

① Extensive Reference: conducted extensive research on exhibition centers of similar scale at home and abroad, (Table 5-3). and fully learned from the design and distribution experience. Investigated the Shanghai International Exhibition Center, Guangzhou Pazhou International Exhibition Center, Shenzhen Exhibition Center, Hong Kong Convention and Exhibition Center, as well as the Milan Exhibition Center, Germany Hanover Exhibition Center, Munich Exhibition Center, Cologne International Exhibition Center and other European exhibition centers. All the exhibition centers at home and abroad are located in regions with developed transportation facilities. In China, they mainly rely on public transportation to relieve passenger flow. However, in Europe, the exhibition centers are usually taken into account of non-public transportation, and the use of cars is relatively high. In addition, it takes longer time to move in and move out during exhibitions in Europe, and the usage rate is relatively lower than in China.

② Traffic is the Priority: before the design of the building, stabilize surrounding traffic planning and clarify the design conditions. The exhibition center has a large scale and needs to be supported by a comprehensive transport system. It would be difficult to consider and incorporate the demands of the exhibition center into the surrounding traffic planning if the design of the building starts first, which would cause problems such as lack of external traffic function or poor connection, and lead to subsequent rework adjustment. Therefore, traffic planning must be done before other design to fully coordinate the demands of the exhibition center and stabilize the surrounding design conditions as soon as possible.

③ Adjust Measures to Local Conditions: adapt to and make full use of the existing traffic conditions in the surrounding areas to reduce the negative traffic impact. The Shenzhen World is close to the high-speed road along the river and the seaside, and 7km to the Shenzhen airport in the south. In the overall layout of the transport facilities, these transportation resources should be fully taken into account to build a perfect external transport system. Since the Shenzhen World covers a large area and is adjacent to many roads, its traffic will inevitably interweave with urban traffic. Therefore, the planning and design of surrounding roads should be appropriately addressed, and the influence on urban traffic should be minimized by increasing road evacuation capacity and three-dimensional separation.(Figure 5-28)

④ Customized Design: based on a comprehensive understanding of the characteristics of exhibition traffic. Except for large amount of traffic demand and concentrated peak, there are many other characterizes such as obvious tidal phenomenon, great difference in passenger flow levels, complex traffic composition, high demand for travel quality, and indispensable security and ticket inspection, which are obviously different from urban traffic and other types of traffic. Therefore, the traffic design can't apply the design method of conventional facilities but must be carried out on the basis of fully grasping these characteristics, and effectively meet the demands of exhibition.

⑤ 公交优先：践行绿色交通理念，通过公交优先应对会展高峰客流带来的交通冲击展会属于商业活动，尤其是专业展，参展人流整体上对交通出行的品质要求较高，普遍偏向于私家车、出租车出行。会展中心的建设对周边交通系统带来较大挑战。欧洲会展中心普遍选址相对偏远，提供充足的交通设施服务，同时满足公共交通与非公共交通的出行需求。但在国内，尤其是深圳国际会展中心，周边均是密集的城市开发用地，必须依赖大运力的公共交通进行疏导，合理引导小汽车使用，减小会展交通对城市日常交通的冲击。

⑥ 集约共享：通过设施共享等手段集约空间，提高效率，降低运营成本。

展会存在淡旺季，淡季展馆的使用率（展示日占比）可低至 30%，同时，展会的规模差异明显，国内绝大部分大型展馆展会的展示面积小于 70%，一年一般仅有数次超大型展会。展会又分为布撤展与展览期两个阶段，布撤展与展览期分布在不同的时段。在这种情况下，若按照高标准配备设施则必然存在大量浪费，完全可以考虑分时共享场地，集约空间。会展中心周边需布置大量商业配套提供服务，也可充分共享会展的交通设施。

⑦ 以人为本：以参展人员的视角看待设计，注重细节，全面提升交通体验。

深圳国际会展中心空间尺度巨大，各类交通组织极为复杂。在超大空间尺度上，设计人员往往容易采用上帝视角，参展人员相对展馆十分渺小，如果没有便捷的交通设施服务，将产生极差的体验效果。因此，必须以人为本，以参展人员的视角看待设计，注重细节，考虑到各类参展人员使用各种交通方式进出及内部活动的方方面面，需不断检校并完善方案。

⑧ 全程介入：深度介入从规划设计到建设运营的各个方面，确保方案落地

大型会展中心建设长达多年，涉及规划、设计、建设、运营多个阶段，其中交通方面包含交通配套规划、设计与评估、综合交通运营保障等内容，需要专门的团队全程参与，确保各阶段有序衔接、规划意图得到落实。本项目深度介入从规划设计到建设运营的各个阶段，开展了多达 35 个专题研究，实现了全过程咨询服务。其中，规划阶段强调搭建轨道、对外道路、交通配套设施等骨架支撑系统。设计阶段重在落实规划意图，注重设施与会展交通特征和需求的匹配。建设阶段主要是在维持规划设计"不走样"的前提下，局部优化完善，并做好运营前的交通保障。运营阶段主要是制定具体展会运营交通组织方案，协助运营组织调度等。

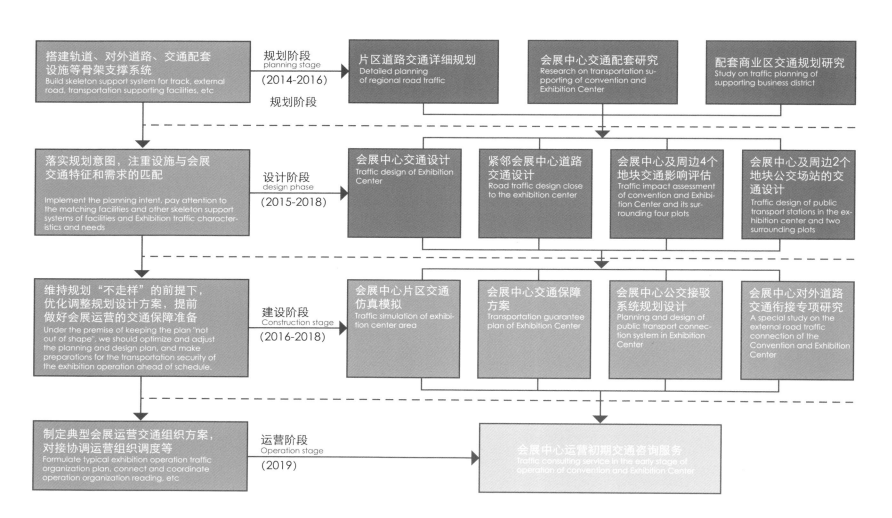

图 5-28　会展初期交通设计图
Figure 5-28　Early stage Traffic Design of Exhibition

展出面积（万平米） Exhibition Area (10000 M²)	展会个数（个） Number of Exhibitions	展出时间（天） Exhibition Time (days)	展示时间比例 Display Time Proportion	总展出面积（万平米） Total exhibition Area (10000 M²)
≥10	10	43	22%	150
5 - 10	16	54	28%	106
≤5	63	98	50%	62
合计 Total	89	195	100%	318

表 5-4　不同规模展会数量及时间统计表
Table 5-4　Statistics of the Number and Time of Exhibitions of Different Scales

⑤ Bus Priority: encouraging green transportation, and deal with the traffic impact caused by peak traffic of exhibition. The exhibition is a commercial activity, especially a professional exhibition. Usually the visitors have higher requirements on the quality of transportation, and generally prefer private cars and taxis. The construction of the exhibition center brings great challenges to the surrounding transport system. Exhibition centers in Europe are generally located in relatively remote areas, providing adequate transportation facilities and services to meet the needs of both public and non-public transportation. However, in China, especially in the Shenzhen World, where is surrounded by dense urban development land, it must rely on mass public transportations to guide the use of cars reasonably and to reduce the impact of exhibition traffic on daily urban traffic.

⑥ Intensive Sharing: intensive utilization of space through facilities sharing, improving efficiency and reducing operating costs. In the low season, the utilization rate of the exhibition hall (the proportion of exhibition days) can be as low as 30%. Meanwhile, the scale of the exhibition varies significantly, the exhibition area of most large domestic exhibition halls is less than 70%, and only holds several super large exhibitions in a year. An exhibition has two different situations: move-in and move out period, exhibiting period. In this case, if the facilities are equipped in accordance with high standards, there will cause a lot of waste. Therefore, time-sharing sites and intensive use of space should be fully considered. A large number of commercial facilities need to be arranged around the exhibition center to provide services, which can also fully share the exhibition transportation facilities.

⑦ Human-oriented: design from the perspective of exhibitors, pay attention to details, and comprehensively improve the transportation experience. The Shenzhen World is 700m wide from east to west, 2.4km long from north to south, with a huge spatial scale and extremely complex traffic organization. On this scale of super large space, designers tend to adopt the perspective of god, in which way the the participants are relatively very small compared with the exhibition hall. Without convenient transportation facilities and services, the experience will be extremely poor. Therefore, it is necessary to see things from the perspective of exhibitors, pay attention to details, take into account of all kinds of exhibitors who could use a variety of transportation to enter and go out as well as the internal activities of all aspects, constantly check and improve the solution.

⑧ Full Involvement: deeply involved in all aspects from planning and design to construction and operation to guarantee the implementation of the plan. The construction of large-scale exhibition center has been going on for many years, involving multiple stages of planning, design, construction and operation, including transportation supporting planning, design and evaluation, and comprehensive transportation operation guarantee. Special teams are required to participate in the whole process to ensure the orderly connection of each stage and the implementation of planning intention. This project is deeply involved in every stage from planning and design to construction and operation. It has carried out as many as 35 special studies and provided whole process consulting services. In the planning stage, we emphasized the construction of rail, external roads, transportation facilities and other framework support system. In the design stage, we focused on the implementation of planning intention and the matching of facilities with the traffic characteristics and needs of the exhibition. In the construction stage, we optimized and improved accordingly, and guaranteed the traffic security before operation. In the operation stage, we formulated specific operation and transportation organization plan and assist the operation and organization.

设计方案
Design Scheme

规划阶段

强调搭建轨道、对外道路、交通配套设施等骨架支撑系统。

① 结合需求及片区条件明确会展主要交通设施配套，强调公共交通服务及集约共享（图 5-29）

2015 年启动会展中心筹建，周边刚完成填海，几乎没有交通设施，需尽快明确交通配套需求并加快实施，开展交通配套研究。

- 充分调研国内外会展中心差异，以大运量交通方式（轨道、公交）作为主要疏解手段，适度限制小汽车交通，以此为原则配置交通设施（表 5-5 ）。
- 日常配套与临时、应急措施相结合，日常配套应对日常高峰客流，临时应急措施应对极端高峰客流，避免设施浪费（表 5-6 ）。
- 优化轨道布局，加强对会展地块覆盖的同时适应会展中心形态，并协调轨道建设时序（图 5-30）。
- 结合会展建筑布局要求，优化调整路网，提出建设改造计划，保障运营基本需求，并避免会展交通对城市交通产生严重影响（表 5-7）。
- 以分级轮候的理念配置货车停车场地；兼顾路网容量及会展运营需求，配建小汽车停车位。（经路网容量分析，车位控制在 9500 个之内）

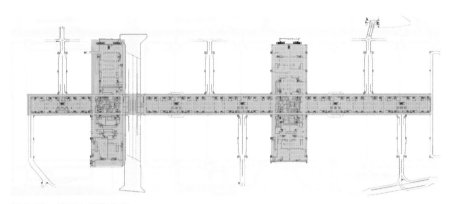

图 5-29　地下车库停车图
Figure 5-29　Underground Parking

远期极端高峰项目离开交通各出行方式分担比例及流量预测表（人/h）							
慢行	团体巴士	常规公交	轨道交通	出租车		小汽车	总计
10%	20%	9%	42%	6%		13%	100%
9600	19200	8640	40000	5760		12480	96000

表 5-5　远期极端高峰项目离开交通各出行方式分担比例及流量预测表
Table 5-5　Share Proportion of Travel Modes and Flow Forecast of Long-term Extreme Peak Projects Leaving Traffic

远期客流数据预测表					
人流属性	会展时间	客流吸引率	日客流量	晚高峰离开比例	晚高峰小时离开人流量（人/h）
专业展观众		0.3	120000	0.4	48000
消费展观众	工作日满展	0.5	200000	0.3	60000
	休息日满展	0.7	280000	0.3	84000
展商人员		0.08	32000	0.3	9600
工作人员		0.02	8000	0.3	2400
专业展总计			160000		60000
消费展总计	工作日满展		240000		72000
	休息日满展		320000		96000

表 5-6　远期客流数据预测表
Table 5-6　Forecast of Long-term Passenger Flow

The Planning Stage

It emphasizes the construction of the framework support system such as the rail, external roads and transportation facilities.

① Clear the main transport facilities according to the needs and conditions of the area, and emphasize public transport services and intensive sharing (Figure 5-29).

The construction was started in 2015, when the surrounding area had just been reclaimed from the sea, and there were almost no transportation facilities. In order to clarify the demand for supporting transportation as soon as possible and accelerate the implementation, research on supporting transportation was carried out.

- Fully investigate the differences between domestic and foreign exhibition centers, use mass transportation (rail and public transportation) as the chief way for traffic evacuation and moderately limit the use of cars, which is the principle to allocate transportation facilities.
- Combining daily supporting with temporary and emergency measures, with daily supporting to deal with daily peak of passenger flow, and temporary emergency measures to deal with extreme peak of passenger flow to avoid waste of facilities (Table 5-5).
- Optimize the rail layout, strengthen the coverage of exhibition land and adapt to the form of exhibition center, and coordinate the timing of rail construction.(Figure 5-30)
- According to the layout requirements of exhibition buildings, optimize and adjust the road network, put forward the construction and transformation plans, guarantee the basic operation needs, and avoid the serious impact of exhibition traffic on urban traffic(Table 5-7).
- Truck parking lots are configured with the concept of graded waiting; considering the capacity of the road network and the operation needs of the exhibition, build up car parking spaces. (according to the capacity analysis of the road network, the number of cars in parking spaces is within 9,500)

方向 direction	道路 Road	车道数 Number of lanes	单向通行能力 One way capacity	过境交通折减 Transit traffic reduction
北向 Northward	沿江高速出入口 Entrance and exit of Yanjiang Expressway	单2 Single 2	2500	1500
	海云路 Hai Yun Road	双6 Double 6	2400	1800
南向 Southward	沿江高速出入口 Entrance and exit of Yanjiang Expressway	单2 Single 2	2500	1500
	海滨大道临时路 Temporary road of Haibin Avenue	双4 Double 4	1800	1400
	展览大道 Exhibition Avenue	双6 Double 6	3000	1800
东向 Eastward	沙福路 Suffolk Road	双6 Double 6	3000	1600
	凤塘大道 Fung Tong Avenue	双6 Double 6	3000	1200
	景芳路 Jing Fang Road	双6 Double 6	2400	1600
	和秀西路 Hexiu West Road	双4 Double 4	1600	1200
通行能力 Traffic capacity				13600
交通分布不均衡折减 Reduction of unbalanced traffic distribution				12240
外围道路限制折减 Peripheral road limit reduction				11016

表 5-7　路网交通容量核算表
Table 5-7　Road Network Traffic Capacity Calculation

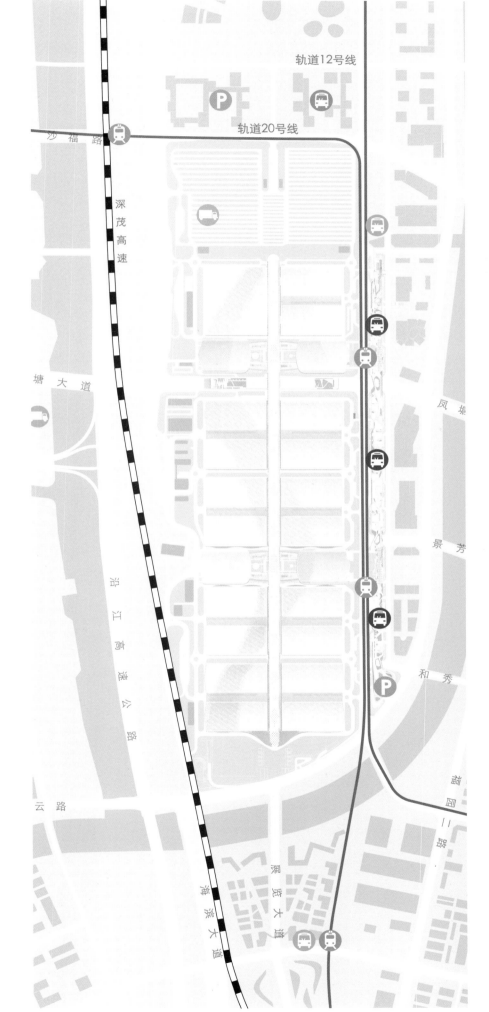

图 5-30　交通设施充分共享图
Figure 5-30　Full Sharing of Transportation Facilities

②片区路网详规充分纳入会展的交通诉求，根据会展交通特征配置或预留充足的疏解通道（图5-32）。

结合片区道路交通详规的编制，将会展中心的交通诉求纳入其中。根据会展的交通特征及需求，配置或预留充足的对外疏解通道，优化调整会展周边的路网结构；合理组织客货运交通，对会展及周边开发实施差异化设施供给（图5-33）。

② The detailed regulations of the regional road network are fully incorporated into the traffic demands of the exhibition, and sufficient channels are configured or reserved according to the traffic characteristics of the exhibition (Figure 5-32).

Combined with the compilation of detailed road traffic regulations in the area, the traffic demands of the exhibition center were included. According to the traffic characteristics and needs of the exhibition, sufficient external channels should be allocated or reserved to optimize and adjust the network structure around the exhibition. Organize the passenger and freight transportation reasonably and supply the exhibition and surrounding development differently (Figure 5-33).

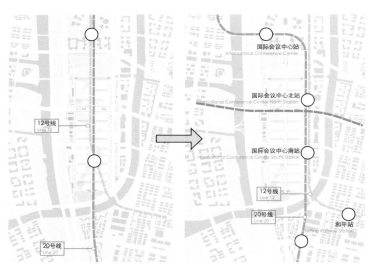

图 5-32　轨道线位调整图
Figure 5-32　Road Alignment Adjustment Diagram

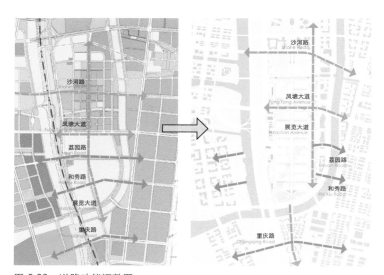

图 5-33　道路功能调整图
Figure 5-33　Road function Adjustment Diagram

图 5-31　配套商业区范围示意图
Figure 5-31　Scope of Supporting Business District

③ 配套商业区充分共享会展交通设施，将相关要求落实至详细蓝图
国际会展中心周边配套开发150万平方米的商业区、办公区、酒店和公寓，为保障良好的内外交通条件与品质，项目结合详细蓝图编制，展开片区交通规划研究（图5-31）。

- 紧邻会展中心，构建立体慢行交通系统，充分共享会展中心轨道、公交等完善的交通配套设施（图5-34）。
- 为避免会展高峰车流与配套区交通的相互冲击，车行交通组织适当分离。
- 充分利用各地块统一开发运营的特点，使地下车库互联互通，加强利用效率，净化地面交通（图5-35）。

③ Exhibition transportation facilities are fully shared in supporting business areas, and relevant requirements are implemented into detailed blueprints

A supporting area with1.5 million square meters of commercial, office, hotel and apartment has been developed around the Shenzhen World. In order to guarantee good internal and external traffic conditions and quality, a detailed blueprint was prepared to carry out a study on the area's traffic planning (Figure 5-31).
- Close to the exhibition center, build the three-dimensional slow traffic system, and fully share the complete supporting transportation facilities such as the rail and the bus (Figure 5-34).
- In order to avoid the impact between the traffic flow peak of exhibition and supporting area traffic, the traffic organization of vehicles should be separated appropriately.
- Make full use of the features of the unified development and operation of various blocks, interconnect underground parking space to enhance utilization efficiency and purify surface traffic (Figure 5-35).

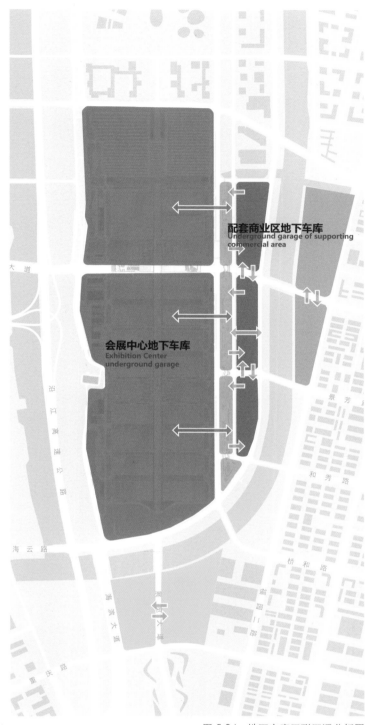

图 5-34 地下车库互联互通分析图
Figure 5-34　The Interconnected Underground Parking

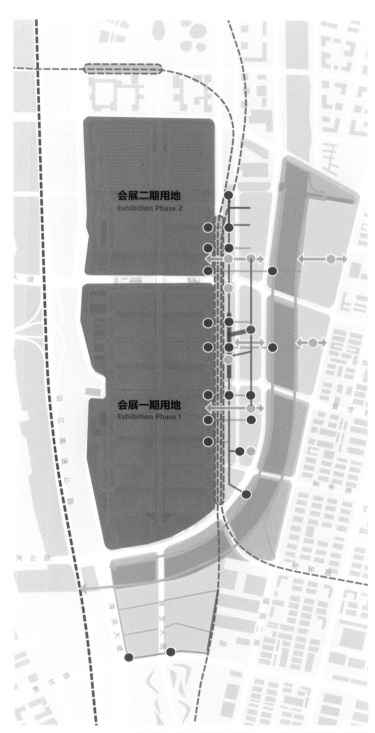

图 5-35 立体慢行交通系统与会展互联互通分析图
Figure 5-35　Three Dimensional Slow Traffic System and Exhibition
Interconnection

图例
Legend

深茂铁路 Shenzhen Maoming Railway	视线通廊 Line of sight corridor	
规划地铁轨道 Planning Subway Track	地铁站点 Subway Stations	
地面散步道 Ground Walk	垂直交通体 Vertical Circulation Body	
空中步行道 Air Walk	地面开放空间节点 Ground Open Space Node	
人行天桥 Footbridge	空中开放空间节点 Air Open Space Node	

图例
Legend

会展与配套商业互联 Exhibition and Supporting Business Interconnection	西岸配套 West Bank Supporting	
配套商业之间互联 Interconnection Inbetween Supporting Businesses	东岸配套 East Coast Supporting	
会展中心 Convention and Exhibition Center	会展休闲带 Exhibition Leisure Belt	
南岸配套 South Bank Supporting		

设计阶段

落实规划意图，注重设施与会展交通特征和需求的匹配

① 全程参与会展建筑设计，以交通专业为主导、多专业合作落实交通规划理念，以交通咨询顾问的身份全过程参与会展中心建筑设计，多专业合作，多部门协调，落实交通规划设计理念（图 5-36）。

- 以客货分离、快慢分离、进出分离、人车分离的设计理念应对会展复杂的交通出行结构，充分考虑会展特殊交通方式的组织需求。
- 货运依托高快速路组织，与客运时空分离，多级轮候；场地内部组织灵活，同时满足大型展会的单向及小型展会的双向组织需求（图 5-39）。
- 人行交通组织充分考虑步行可容忍距离，公共交通设施与展馆出入口布置充分协调，充分布置自动步道、穿梭巴士等代步工具。
- 作为世界最大的地下车库，交通组织强调安全、高效，内部设置三级通道，组织单向交通；出入口设计适应高峰潮汐车流（图 5-38）；满足车辆安检及垃圾车、救护车等特殊车辆的使用需求；构建与车流分离的专用人行通道，人行安检区实现 90 秒步行半径全覆盖（图 5-37）。

Design Stage

Implement of planning intention, and emphasize the matching of facilities with the traffic characteristics and needs of the exhibition

① Participated in the whole process of the architecture design, and implemented the concept of transportation planning through multi-professional cooperation led by transpiration.

Participated in the whole process of the architectural design as a transportation consultant, through multi-professional cooperation, multi-department coordination to implement the transportation planning and design concept (Figure 5-36).

- The design concept is to separate the passenger and cargo, the fast and the slow, the entry and the exit, the people and the vehicle. This is to deal with the complex traffic structure of exhibition, and fully considers the organizational requirements of the special transportation needs of exhibition.
- Relies on high speed road organization, the freight transport is separated from passenger transport with multi-level waiting; the site is flexibly organized to meet the needs of unidirectional organizational requirements of large exhibitions and bidirectional organizational requirements of small exhibitions.(Figure 5-39)
- The tolerable walking distance is fully considered by the pedestrian transport organization, the public transport facilities are fully coordinated with the entrance and exit of the exhibition halls, the automatic footpath, shuttle bus and other transportation tools are fully arranged.
- As the largest underground parking transportation organization in the world, it emphasizes safety and efficiency with three internal channels and one-way traffic; The entrance and exit are designed to adapt to the flow at high peaks (Figure 5-38) ; meet the special needs of vehicle security check, garbage truck and ambulance; build a dedicated pedestrian passage separated from traffic flow; and the pedestrian security zone is fully covered with a 90-second walking radius.(Figure 5-37)

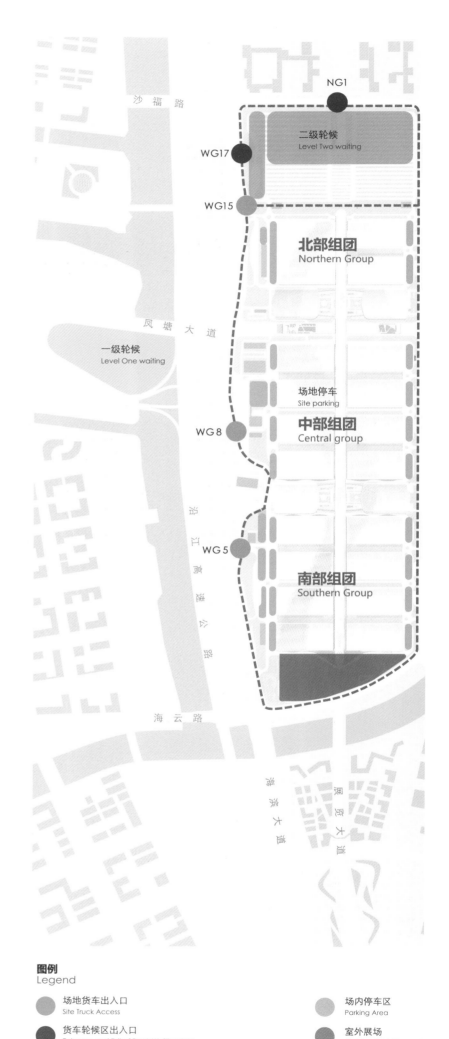

图 5-36　多级轮候区布置图
Figure 5-36　Multi-level Waiting Area Arrangement

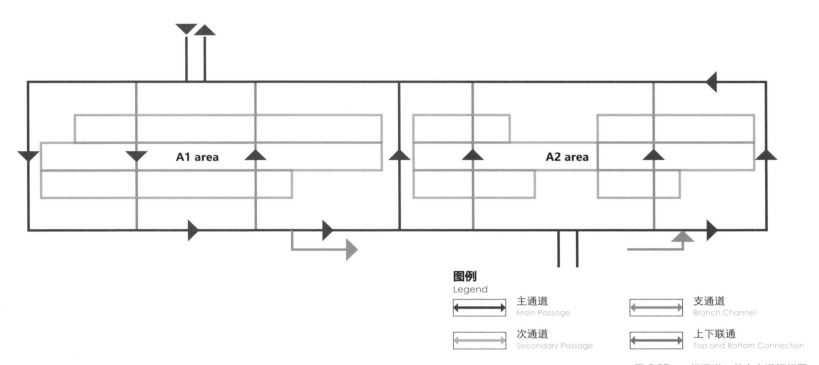

A1 area

A2 area

图 5-37 三级通道、单向交通组织图
Figure 5-37 Three-level access, One-way Traffic Organization

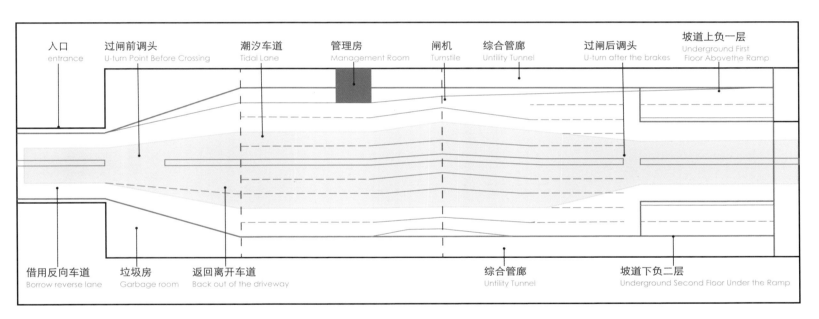

入口 entrance

过闸前调头 U-turn Point Before Crossing

潮汐车道 Tidal Lane

管理房 Management Room

闸机 Turnstile

综合管廊 Untility Tunnel

过闸后调头 U-turn after the brakes

坡道上负一层 Underground First Floor Abovethe Ramp

借用反向车道 Borrow reverse lane

垃圾房 Garbage room

返回离开车道 Back out of the driveway

综合管廊 Untility Tunnel

坡道下负二层 Underground Second Floor Under the Ramp

图 5-38 出入口布置潮汐车道适应高峰车流图
Figure 5-38 Tide Lanes of Entrances and Exits to Accommodate Peak Hour Traffic

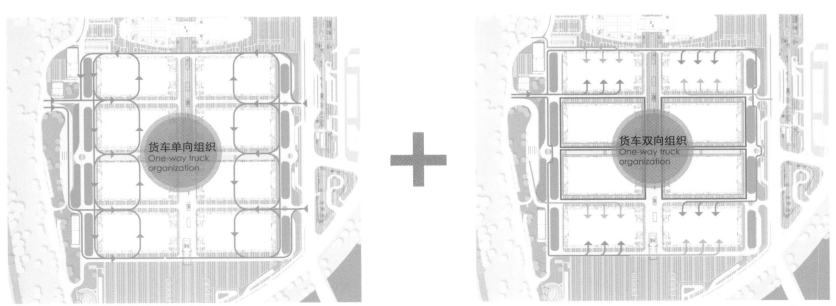

货车单向组织 One-way truck organization

货车双向组织 One-way truck organization

图 5-39 场地同时适应货车单双向组织图
Figure 5-39 The Site is Adapted to Both Unidirectional and Bidirectional Organization of Trucks

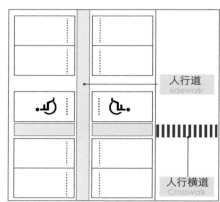

图 5-40　地下车库人行道布置
Figure 5-40　Underground Parking Sidewalk Layout

图例
Legend

服务范围
Service Area

安检区
Security Area

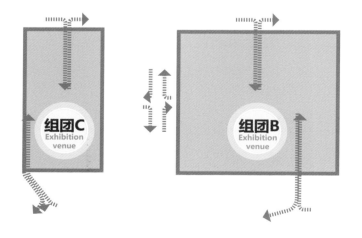

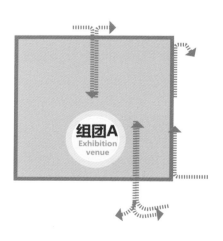

图例
Legend

地下车库对外联系匝道
Underground garage external connection ramp

地下车库组团分区
Underground garage group partition

图 5-41　地下车库与快速路直接衔接
Figure 5-41　Direct Connection Between the Underground Garage and Expressway

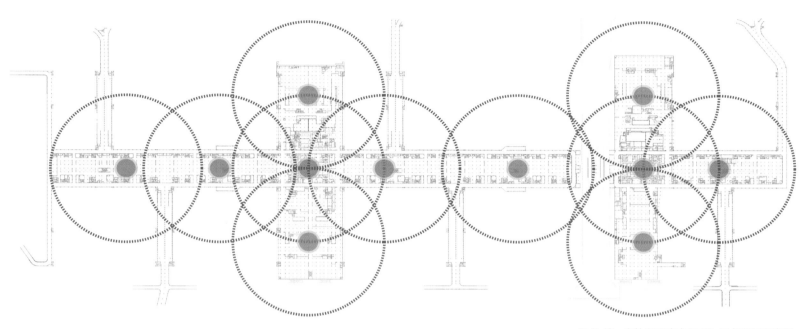

图 5-42　安检及垂直交通 90s 服务范围示意图
Figure 5-42　Schematic Diagram of 90s Service Area for Security Inspection and Vertical Traffic

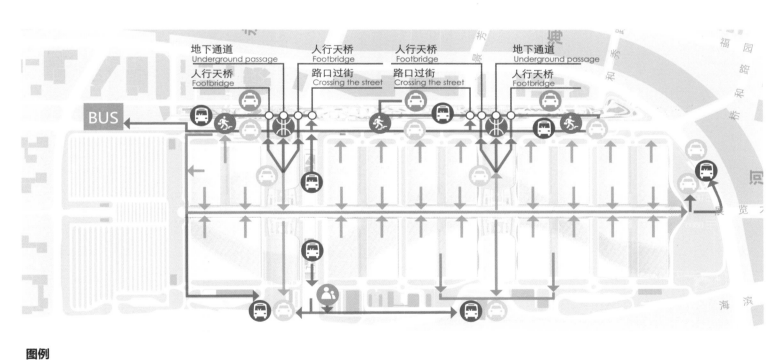

图例
Legend

内部人行流线 Internal Pedestrian Circulation	公交车停靠站 Bus Station	出租车站台 Taxi Platform	地铁出入口 Subway Entrance	员工出入口 Staff Access
外部人行流线 External Pedestrian Circulation	**BUS** 公交车首末站 Bus Terminal	出租车轮候区 Taxi Waiting Area	人行过街天桥 Pedestrian Overpass	

图 5-43　行人离开流线分析图
Figure 5-43　Pedestrian Circulation Analysis in the Exhibition

设计阶段

② 对紧邻会展中心道路开展专项交通设计，加强建筑与周边道路的衔接协调。对紧邻会展中心的海滨大道、凤塘大道、海云路等道路开展专项交通设计，与道路设计单位充分对接，加强建筑与周边道路的衔接协调，在强化周边道路对会展中心服务的同时尽量减小会展交通带来的冲击。

- 快速路重点服务货运交通及地下车库车辆进出，小汽车通过定向匝道快进快出，与地面货运交通立体分离（图5-44）。

- 与高速相连的主干道沿线节点采用立体交叉，会展车流无需等候信号灯，实现快速到达与驶离（图5-45）。

- 对于会展最主要的集散道路，弱化通过性功能，通过主辅分隔带，深港湾公交站等方式增加客运交通的上落客空间，并分离公共交通与非公共交通（图5-46）。

Design Stage

② Carry out special traffic design for the road adjacent to the exhibition center, and strengthen the connection and coordination between the building and the surrounding roads

Special traffic design was carried out for the Haibin avenue, Fengtang avenue, Haiyun road and other roads adjacent to the exhibition center, contacted with the road design unit, strengthening the connection and coordination between the building and the surrounding roads, and minimizing the impact of the exhibition traffic while strengthening the service of the surrounding roads for the exhibition center.

- The expressway is mainly for the in and out of freight traffic and vehicles of the underground parking. Cars enters and exits quickly through directional ramps, separating from freight traffic on the ground.(Figure 5-44)

- The nodes along the main road connected with the high speed express way are vertical crossing, so the vehicles can arrive or leave quickly without waiting for the signal.(Figure 5-45)

- In terms of the main collector-distributor road, the passenger space should be increased by separation of the primary and secondary zones and bus stations, etc., and public transportation and non-public transportation should be separated.(Figure 5-46)

通过立交快速联系高速，
减小会展车流对城市交通的冲击
Quickly contact high speed through interchanges,
Reduce the impact of exhibition traffic on urban traffic

图 5-44　道路与会展场地协调分析图 1
Figure 5-44　Road and Exhibition Site Coordination Analysis

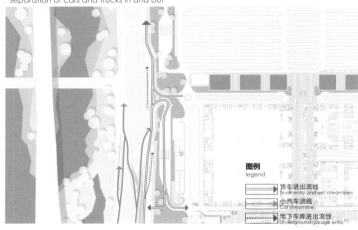

集约利用空间，
小汽车与货车进出立体分离
Intensive use of space, three-dimensional
separation of cars and trucks in and out

图 5-45　道路与会展场地协调分析图 2
Figure 5-45　Road and Exhibition Site Coordination Analysis

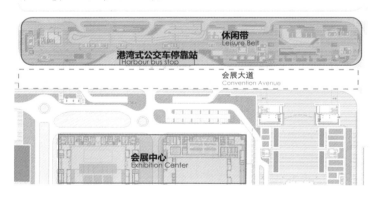

主辅道分隔，分离公共与私有交通
Separation of main and auxiliary roads,
separating public and private transportation

会展大道
Convention Avenue

图 5-46　道路与会展场地协调分析图 3
Figure 5-46　Road and Exhibition Site Coordination Analysis

③ 结合会展交通影响评估全面校验设计方案，通过行政手段敦促设计单位修改完善。项目分别开展了会展中心及周边 4 个商业配套地块交通评估，全面完整校验设计方案，并通过行政手段敦促设计建设单位修改完善方案。

- 会展中心高峰集聚的交通特性对周边的交通影响显著，评估应客观真实，并着重提出改善方案（含管控引导方案），尽量消减其影响。
- 允许会展中心利用周边道路充分布置出入口，规范其具体布置方案。
- 考虑大规模人流、车流安检及票检等排队需求，充分布置缓冲空间。
- 根据评估结果对会展进出交通组织涉及到的主要节点提出平面布置及信号控制方案并在道路设计方案中落实（图 5-47）。
- 在会展中心周边开发地块，合理组织进出交通并与会展车流适度分离，同时充分共享会展的公共交通设施，严格控制其交通配建指标。

③ Assess and verify the design scheme in combination with the traffic impact, urge the design unit to revise and improve through administrative means

Carried out traffic assessment of the exhibition center and four surrounding commercial supporting plots, verified the design scheme fully and completely, and urged the design and construction units to revise and improve the scheme through administrative means.

- The traffic peak concentration will have a significant impact on the surrounding traffic. The evaluation should be objective and real, and focus on proposing improvement plans (including control and guidance plans) to minimize the impact.

- Fully arrange the entrances and exits on the surrounding roads and standardize the specific layout plan.

- Fully arrange the buffer space according to the queuing needs of large crowds, traffic check and ticket check.

- Put forward the plane layout and signal control plan for the main nodes involved in the traffic organization of the exhibition entrance and exit according to the assessment results, and implement them in the road design scheme. (Figure 5-47)

- In terms of the development plots around the exhibition center, reasonably organize traffic and properly separate it from the exhibition traffic flow; meanwhile, fully share the exhibition public transport facilities and strictly control the traffic allocation indexes.

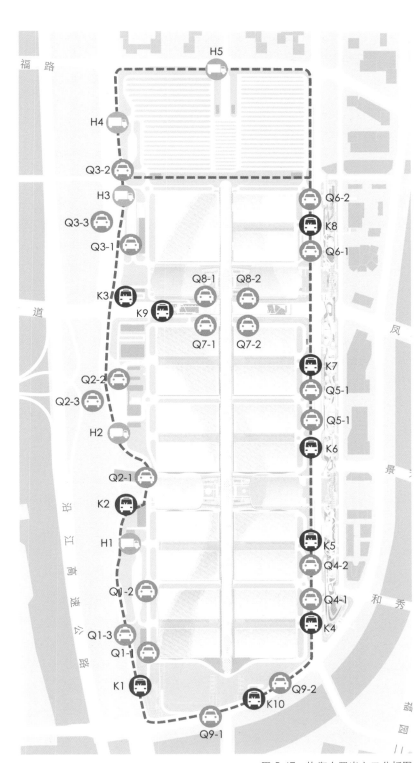

图 5-47 均衡布置出入口分析图
Figure 5-47 Balanced Entrance and Exit Analysis

图例
Legend

 小汽车出入口 Car Entrance
 地面客车出入口 Ground Bus Entrance and Exit
 货车出入口 Truck Access

设计阶段

④ 开展会展及周边地块公交场站交通设计咨询，提出车位及场站配套设施布局方案。

分别开展了会展中心及周边商业配套的两个附建公交首末站交通设计。做好进出口与外部道路的衔接协调，在满足规范的前提下多方案比选以适应上层建筑方案并尽量多布置车位，通过仿真模拟的方法确保所有车位进出安全、顺畅，场站配套设施需满足深圳市的特定要求（表 5-8）。

Design Stage

④ Carry out design consultation of traffic design of exhibition and surrounding bus stations, and propose layout plan for parking spaces and supporting facilities

Design the transport of the first and last station of the exhibition center and the surrounding supporting business. Connect and coordinate the entrances and exits with the external roads.On the premise of the specifications, select the best solution from multiple schemes to adapt the superstructure scheme, and then arrange as many parking spaces as possible, and then, try to guarantee the safety and smooth access of all parking spaces through simulations (Table 5-8).

地块编号 Plot number	停车位测算 Parking space calculation						指标建议 Index suggestion
	酒店 Hotel	办公 Office	商业 Business	公寓 Flats	配套设施 Supporting facilities	合计 Total	
02-01	—	—	—	—	—		1133
02-02	—	—	—	—	—		1350
02-03	—	—	—	—	—		997
02-04	—	—	—	—	—		520
03-01	350	240	217	360	2	1169	1620
03-02	0	149	1520	270	10	1941	1490
03-03	500	66	96	90	1	753	753
04-01	250	0	112	2172	14	2557	2557
04-02	0	0	44	900	4	947	947
05-01	0	128	74	210	0	413	413
05-02	150	203	82	180	2	617	617
合计 Total	1250	785	2145	4182	33	12397	12397

表 5-8　按深标下限 80% 配置配套区停车预测表
Table 5-8　Configure the supporting area parking area according to the 80% of the lower limit of the deep standard

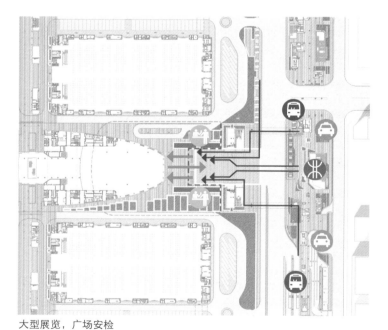

大型展览，广场安检
Large scale exhibition, security inspection

图例
Legend

 离开安检处人行流线
Circulation of People Exit of Security Check

 前往安检处人行流线
Circulation of People towards the Security Check

 通过安检人行流线
Circulation of People Being Checked

 出租车站台
Taxi Platform

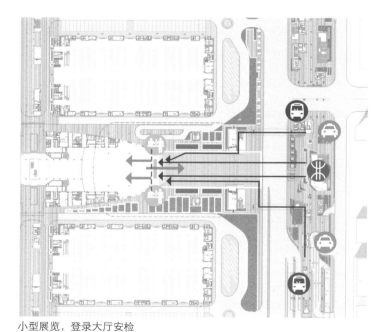

小型展览，登录大厅安检
Small exhibition, Arrival Hall Security Check

 公交车停靠站
Bus Station

 地铁出入口
Subway Entrance

 安检处
Security Inspection Office

 排队缓冲空间
Queuing Buffer Space

图 5-48　预留充足人流排队缓冲空间
Figure 5-48　Reserve Sufficient Buffer Space for Queuing

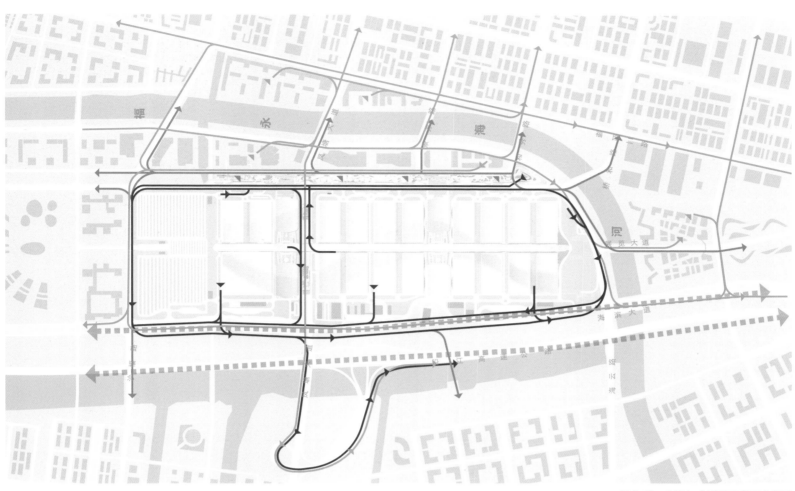

图 5-49　车流与会展适度分离分析图
Figure 5-49　Moderate Separation of Traffic Flow and Exhibition Analysis

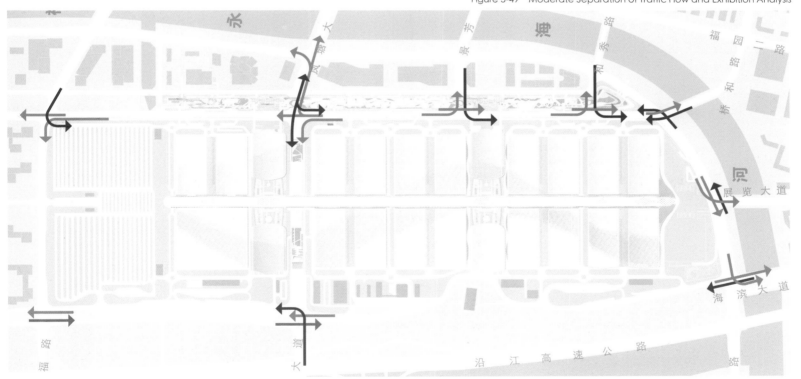

图 5-50　优化瓶颈节点的信控及渠化方案
Figure 5-50　Optimization the Information Control and Channelization Scheme of Bottleneck Nodes

图例
Legend

晚高峰增加绿灯配时
Increase Green Light Timing
During Night Peak Hours

早高峰增加绿灯配时
Increase Green Light Timing
During Moring Peak Hours

高快速路主道
High Speed Main Road

配套商业区车行流线
Vehicle Circulation of Supporting Commercial Area

会展车行主要流线
Main Circulations of Exhibition Vehicle

会展车行次要流线
Secondary Circulation of Exhibition Vehicle

建设阶段

维持规划设计"不走样"的前提下，局部优化完善，并做好运营前
的交通保障。

① 通过对会展中心所在片区开展交通仿真模拟
直面细节问题，会展中心前期已开展了一系列针对交通设施与客流
规模供需不平衡的研究，但缺乏动态、定量的分析手段来支撑片区
精细化设计及交通疏解方案制定（图 5-51）。在会展中心周边道路
即将启动建设时展开片区交通仿真模拟，通过引入交通仿真技术，
采用中微观层面的动态、定量、精细化仿真分析，寻找细节交通问
题，直接指导周边道路施工图的完善，为交通疏解设计方案及后续
保障方案的评估提供支持（图 5-52），指导片区基础设施建设，减
少交通拥堵风险。

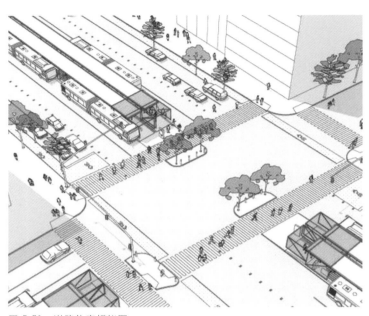

图 5-51　道路仿真模拟图
Figure 5-51　Road Simulation Diagram

Construction Stage

Optimized and improved accordingly and guaranteed the traffic security before operation.

① Face the details by carrying out traffic simulation
In the early stage, a series of studies on the imbalance between the supply and demand of traffic facilities and passenger flow scale has been carried out, but it was lack of dynamic and quantitative analysis to support the refined design of the area and the formulation of traffic distribution plan (Figure 5-51). The traffic simulation was carried out when the road around the exhibition center was about to start constructing. By introducing traffic simulation technology, and through the dynamic, quantitative and refined simulation analysis on a micro-level, we can look for the details of traffic problem, and help to improve the construction drawing of surrounding roads, providing support for the assessment of the traffic evacuation design scheme and subsequent protection programs,guiding the infrastructure construction, and reducing the risk of traffic congestion.(Figure 5-52)

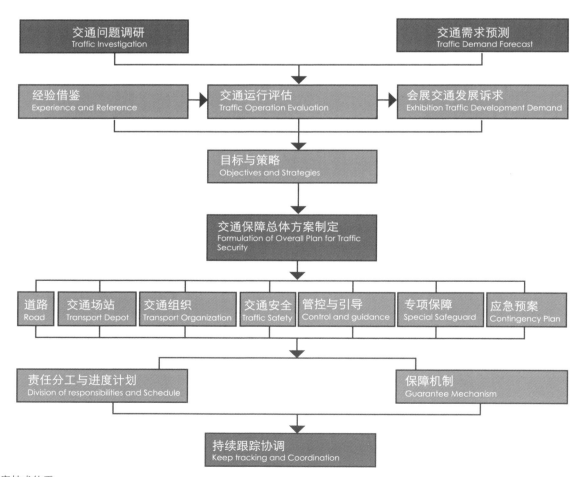

图 5-52　保障方案技术体系
Figure 5-52　Guarantee Scheme Technical System

② 在周边地铁及快速路无法按计划建成的情况下，制定片区交通保障方案，落实会展建设重点任务及运营前的交通保障准备工作

在会展开业前 1 年，由于会展中心周边地铁、快速路等重大交通基础设施无法按计划建成，为确保交通系统的安全、平稳和有序，项目开展了交通保障研究，并制定了片区交通保障方案。以会展开业为截止时间，制定了道路建设、场站设施、交通引导、交通管控、保障措施（含应急预案）等五项总体方案，并成立相应 5 个专门工作组及 1 个统筹协调小组，明确各小组的牵头单位和职责任务，并将重点工作分解成 40 项具体任务，每项任务明确责任单位与完成时限，持续跟踪协调推进。

- 提出最重要、最关键的任务，集中力量加快实施。
- 近期交通设施供给不足，评估交通系统承载能力，通过控制展会规模的方式严格控制交通需求。
- 在路网容量严重不足的情况下，提前安排大量临时公交设施接驳地铁，确保公交有效服务。
- 制定交通引导方案、交通管控方案及应急预案，做好动态交通管控。
- 为应对高速收费口交通拥堵问题，提出高速公路潮汐收费车道方案。
- 任务涉及多个单位，通过持续跟踪协调每项任务的进展及问题，以每月向市领导及相关单位报送简报的方式敦促工作落实。

③ 构建快速智慧公交强化公交接驳提供稳定高效接驳服务

由于近期没有轨道交通直接服务，公交接驳对交通疏解十分关键，而常规公交运力有限，在交通拥堵情况下无法发挥作用，难以满足会展高峰客流期的稳定接驳需求。因此项目提出多专业协同，通过道路渠化调整布置公交专用道、通过智能信号控制系统的构建实现公交专用绿波，同步完善道路品质，打造全国首创车路协同智慧公交系统，实现 15 分钟可达、无缝衔接的一站式高品质接驳体验。由于时间紧迫、方案合理可行，政府部门全力推动，规划方案在提出后直接指导了工可、初设、施工图的同步编制。

② In case the surrounding subways and expressways can't complete as planned, the project formulate the traffic security plan for the area, implement the key tasks of exhibition construction and prepare for traffic security before operation

1 year before the opening of the Shenzhen World, due to the fact that the major transportation infrastructure around the exhibition center, such as the subway and expressways could not complete as planned. In order to guarantee the safety, stability and order of the transportation system, traffic security research was carried out and a regional traffic security plan was formulated. Take the opening as deadline, five overall schemes were formulated, including road construction, terminal facilities, traffic guide, traffic control, safeguard measures (including contingency plans), correspondingly, 5 task forces and 1 overall coordination team were also established, clarifying the leader of each group as well as responsibilities, focusing on 40 specific tasks with clear responsibility and deadline, and continuely tracking of the coordination.

- Identify the most important and critical tasks and focus on accelerating implementation.
- The transportation facilities are in short supply recently, assess the carrying capacity of the transportation system, and strictly control the traffic demand by controlling the scale of the exhibition.
- In case of severe shortage of network capacity, arrange a large number of temporary shuttle buses to connect the MTR in advance to ensure effective bus services.
- Formulate traffic guidance plans, traffic control plans and emergency plans, and take good control of the dynamic traffic.
- Propose a highway tidal toll lane plan to deal with traffic congestion at the expressway toll station.
- Many units were involved. Through continuous tracking and coordination of the progress and problems of each task to urge the implementation of the work by submitting monthly briefings to municipal leaders and relevant units.

③ Construct BRT and enhance bus connection to provide stable and efficient connection service

Because there is no direct rail service recently, shuttle buses are very critical to traffic relief. However, the limited capacity of regular buses cannot play a role in traffic congestion, so it is difficult to meet the stable connection demand during the passenger flow peak period of exhibition. It is proposed that through the multi-professional cooperation, channelization, and adjustment and arrangement of bus lanes, the construction of intelligent signal control system could achieve dedicated green wave for public transportation, which would synchronously improve the road quality, create the national first vehicle and road collaborative smart bus system, and a 15-minute seamless one-stop high-quality connection experience. Due to the tight schedule and the reasonable and feasible plan, the government departments made every effort to promote it. After the proposal of the plan, it directly guided the synchronous compilation of the work plan, preliminary design and construction drawing.

运营阶段

持续提供会展中心运营初期交通咨询服务，做好运营期间的交通组织与调度。

会展中心运营在即，需及时协调解决运营前面临的交通问题，做好详细交通组织部署，与相关政府部门做好交通方面的对接，在运营初期持续完善会展交通保障体系（图 5-53）。主要推动以下工作：

- 针对运营初期具体展会制定详细交通组织方案，明确多级客流响应机制，减小运营浪费。
- 构建运营单位与政府的沟通平台，做好交通管控、公共交通与会展内部的衔接协调。
- 制定公交组织调度方案，协助开展公交接驳实地演练（表 5-9）。
- 及时协调解决运营前及过程中面临的各种交通问题（图 5-54）。

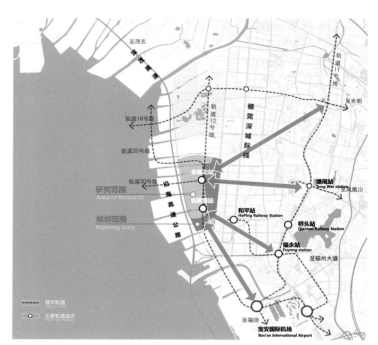

图 5-53　公共交通接驳外围方案图
Figure 5-53　Public Transport Peripheral Connection Scheme

Operation Stage

Continue to provide transportation consulting services at the beginning of the operation of the exhibition center, and organize and dispatch traffic during the operation.

As the operation of the exhibition center is approaching, it is necessary to coordinate and solve the traffic problems before the operation in time, make detailed transportation organization and deployment, coordinate with relevant government departments, and continuously improve the traffic guarantee system at the initial stage of operation (Figure 5-53). Mainly promote the following work:

- Developed detailed traffic organization plans for specific exhibitions at the initial stage of operation, clarified the response mechanism of multi-level passenger flow, and reduced operation waste.
- Establish a communication platform between operating units and the government, and make good connection and coordination between traffic control, public transportation and exhibition.
- Formulated the bus organization and scheduling plan, carrying out the field exercise of bus connection (Table 5-9).
- Coordinate and solve traffic problems before and during operation (Figure 5-54).

开通接驳线路 Open connection line	塘尾站 Tangwei Station	塘尾站，福永站 （和平站） Tangwei Station, Fuyong Station(Peace Station)	塘尾站，福永站 （和平站）桥头站 Tangwei Station, Fuyong Station (Peace Station)Qiaotou Station
客流（万人次/日） Passenger flow (10,000 passengers / day)	< 7	7~12	> 12
晚高峰小时系数 Evening peak hour coefficient	0.3	0.3	0.3
晚高峰小时客流量（万人次/日） Evening rush hour passenger flow (10,000 passengers / day)	< 2.1	2.1~3.6	> 3.6
轨道接驳需求（万人次/日） Rail connection demand (10,000 passengers / day)	< 1	1~1.4	> 1.4

表 5-9　三级公交客流接驳响应机制表格
Table 5-9　Response Mechanism for Third-level Bus Passenger Flow Connection

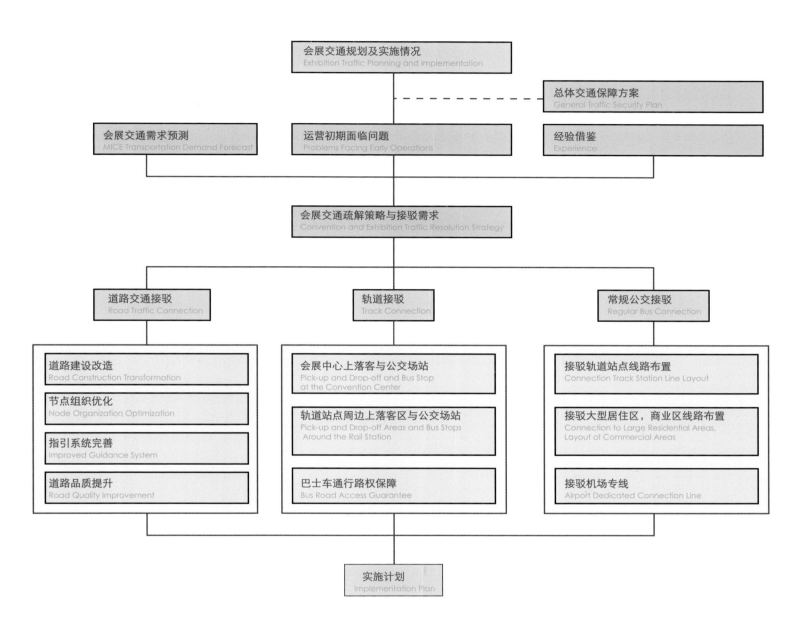

当会展日客流在7万人次/日以下时，仅开通塘尾站接驳线及机场专线；
当会展日客流在7~12万人次/日时，增开通福永站接驳线，兼顾服务和平站客流；
当会展日客流在12万人次/日以上时增开桥头站接驳线，满足高峰疏散需求。

When the number of passengers on the exhibition day is less than 70,000 passengers / h, only Tangwei Station feeder line and airport special line will be opened;
When the number of passengers on the exhibition day is 70,000 to 120,000 passengers / h, the feeder line at Fuyong Station will be opened to take into account the service passenger flow at Heping Station;
When the passenger flow on the exhibition day is more than 70,000 to 120,000 person-times / h, the connection line at Qiaotou Station will be opened to meet the peak evacuation demand.

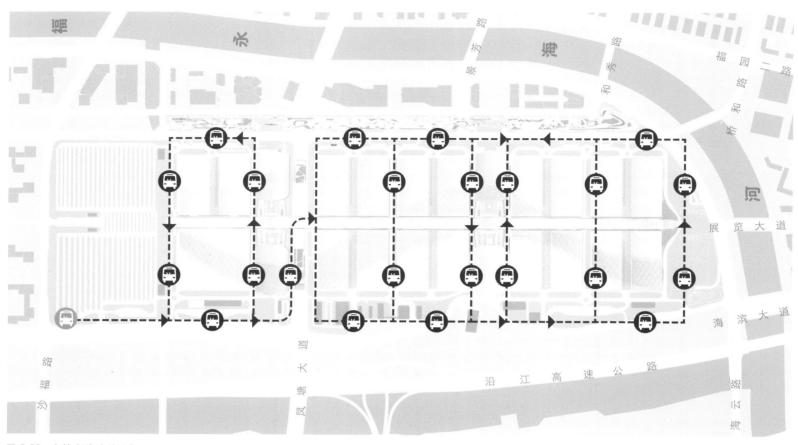

图 5-55　穿梭 BUS 流线分析图
Figure 5-55　Shuttle Bus Circulation

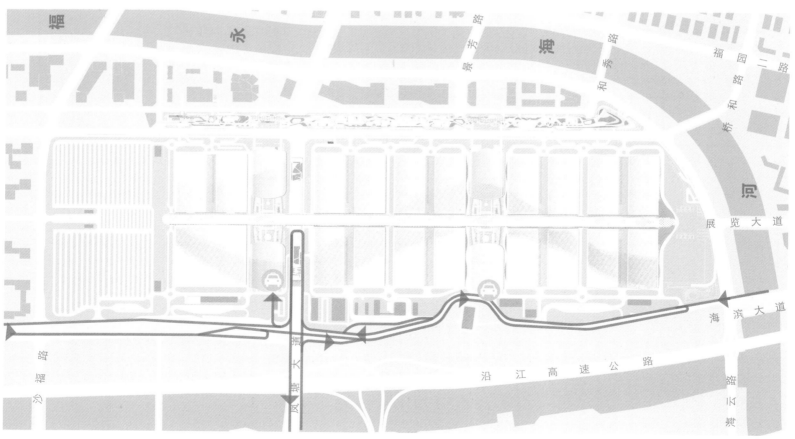

图 5-56　VIP 车行流线分析图
Figure 5-56　VIP Vehicles Circulation

图例
Legend

VIP车辆到达流线
VIP Vehicles Arrival Circulation

穿梭巴士流线
Shuttle Bus Leaving Circulation

穿梭巴士停车靠点
Shuttle Bus Stops

穿梭巴士停车场
Shuttle Bus Parking

VIP车辆停车场
VIP Car Parking

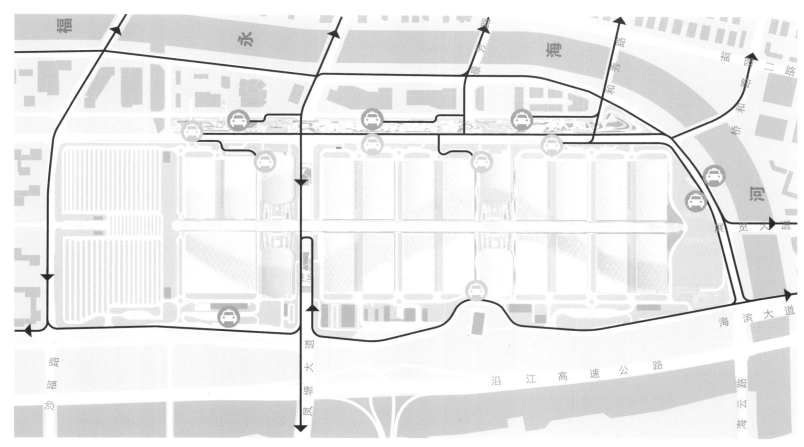

图 5-57 出租车驶出流线分析图
Figure 5-57 Taxi Exit Circulation

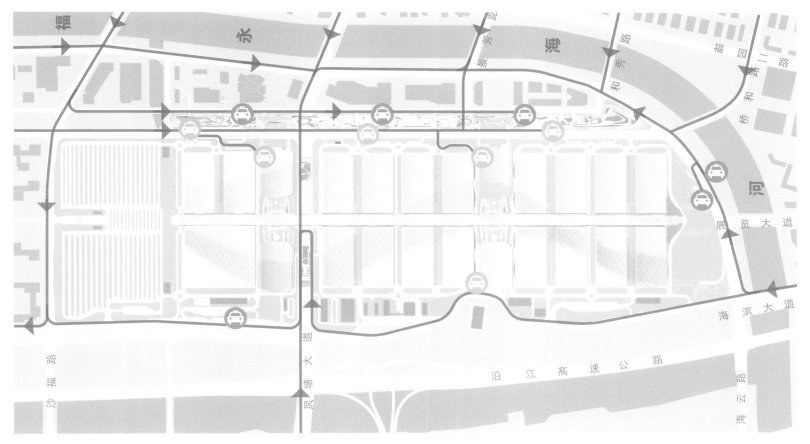

图 5-58 出租车驶入流线分析图
Figure 5-58 Taxi Access Circulation

图例
Legend

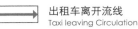 出租车到达流线
Taxi Arriving Circulation

 出租车停靠点
Taxi Stops

出租车离开流线
Taxi leaving Circulation

 出租车排队轮候区
Taxi Queing Area

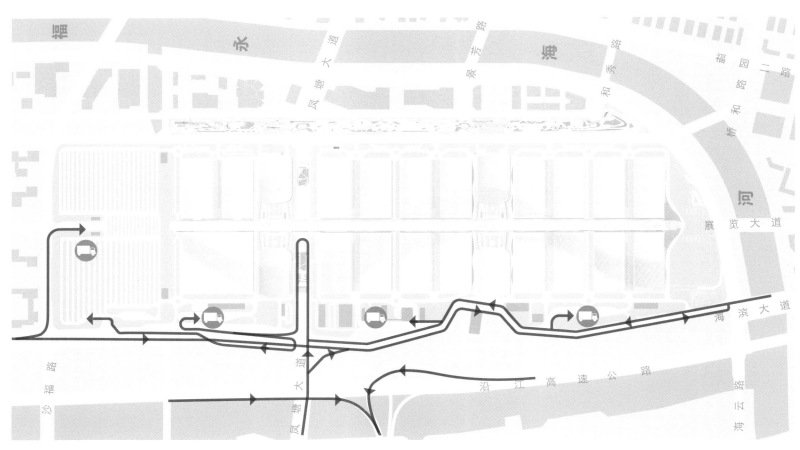

图 5-59　会展外部货运进入流线分析图
Figure 5-59　Circulation of Freight Transportation

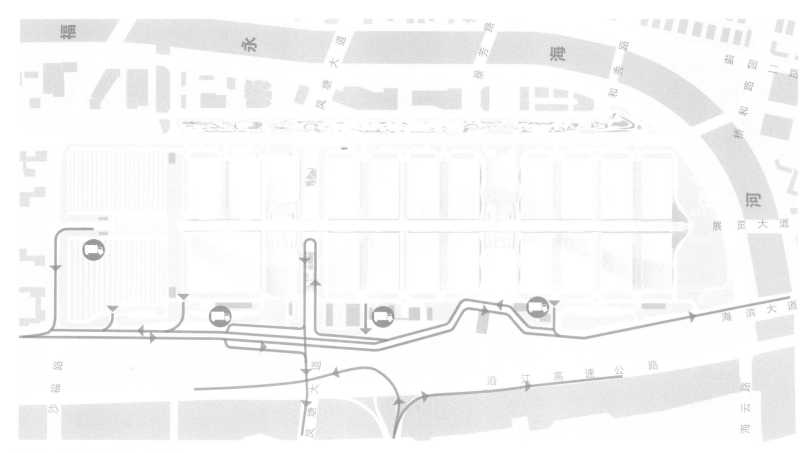

图 5-60　会展外部货运离开流线分析图
Figure 5-60　Analysis of the Circulation of the External Freight Transport

图例
Legend

货车离开流线
Truck leaving Circulation

货车到达流线
Truck arrival Circulation

货运停车区
Freight Parking Area

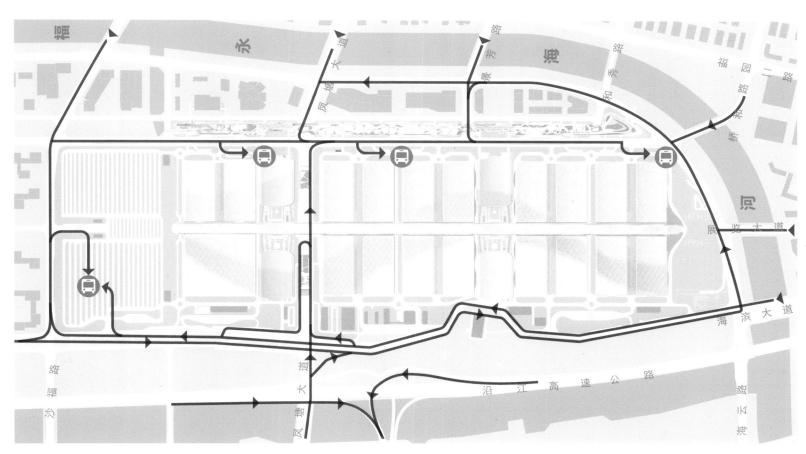

图 5-61 商务巴士进入流线分析图
Figure 5-61 Analysis diagram of Business Bus Arrival Circulation

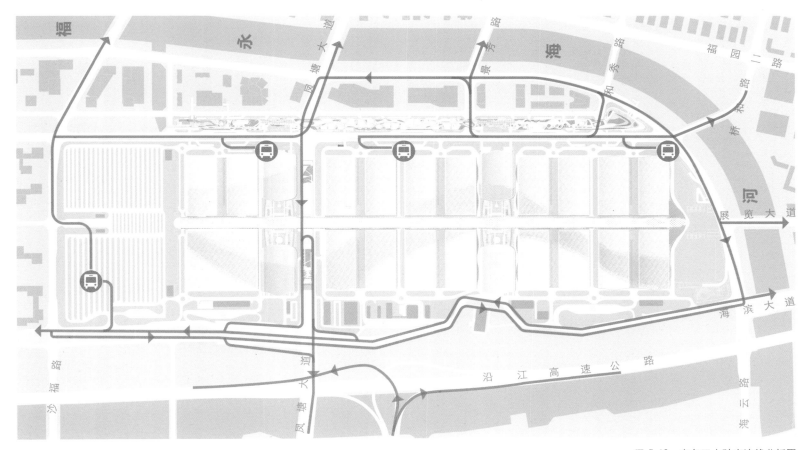

图 5-62 商务巴士驶出流线分析图
Figure 5-62 Analysis Diagram of Business Bus Exit Circulation

图例
Legend

 商务巴士到达流线
Business Bus Arrival Circulation

 商务巴士停车区
Business Bus Parking Area

 商务巴士离开流线
Business Bus Exit Circulation

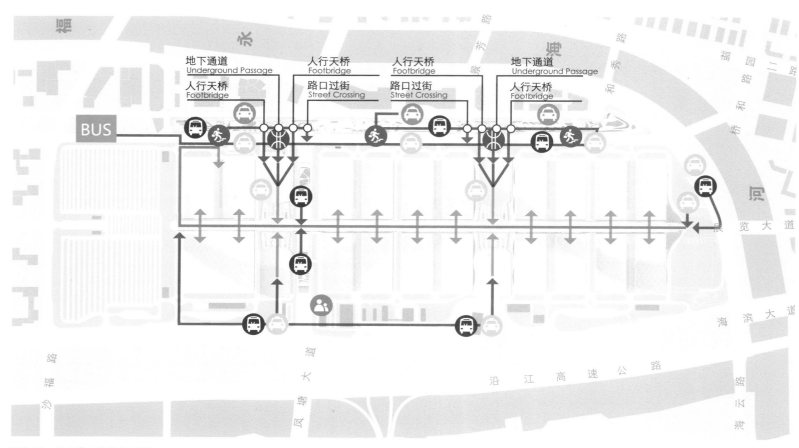

图 5-63　行人进入流线分析图
Figure 5-63　Analysis of Pedestrian Circulation

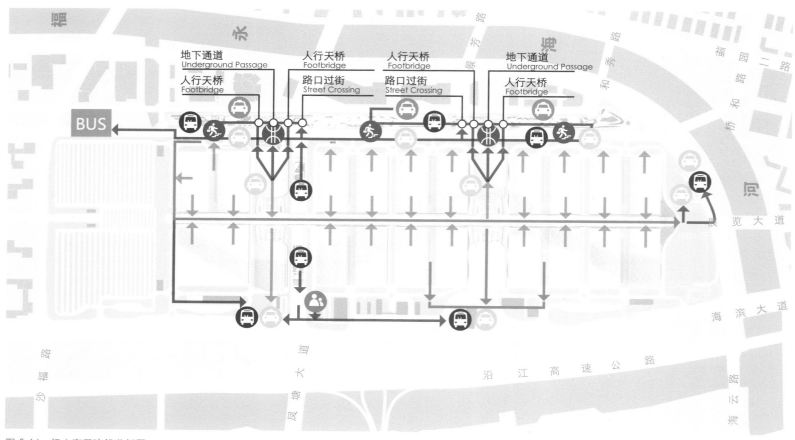

图 5-64　行人离开流线分析图
Figure 5-64　Analysis of Pedestrian Leaving Circulation

图例
Legend

 内部人行流线 Internal Pedestrian Circulation	公交车停靠站 Bus Station	出租车站台 Taxi Platform	地铁出入口 Subway Entrance	员工出入口 Staff Access
外部人行流线 External Pedestrian Circulation	公交车首末站 Bus Terminal	出租车轮候区 Taxi Waiting area	 人行过街天桥 Overpass	

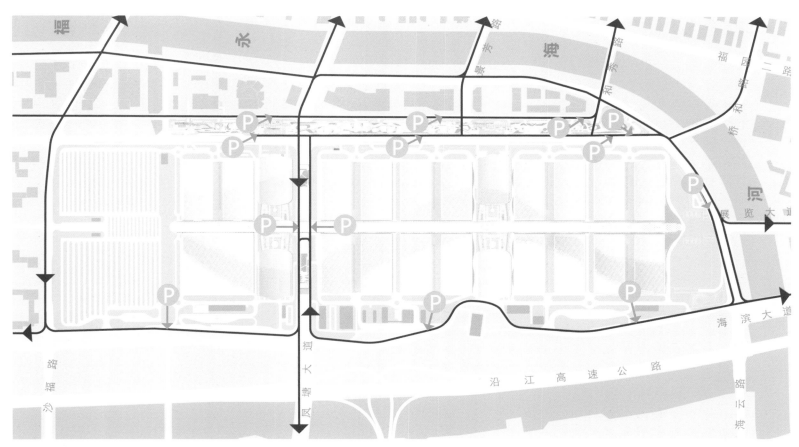

图 5-65　小汽车车行流线分析图
Figure 5-65　Analysis of Car Circulation

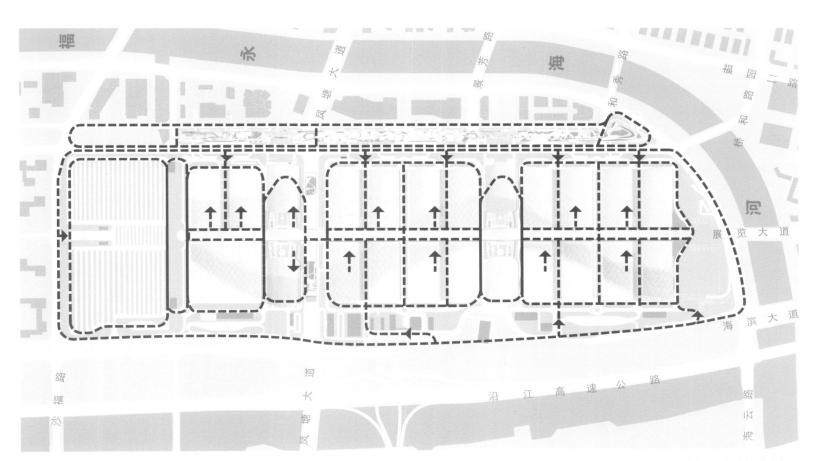

图 5-66　消防交通组织流线分析图
Figure 5-66　Analysis of Fire Traffic Organization Circulation

图例
Legend

 地下车库出口流线
Underground Parking Exit Circulation

 小汽车流线
Car Circulation

Ⓟ 小汽车地下停车场
Underground Car Parking

⬤ 消防交通组织
Fire Traffic Organization

项目特色
Project Features

① 在国内率先实现超大型会展中心全过程交通规划设计咨询服务，有效保障规划设计及建设运营的系统性、落地性。

会展中心及其配套设施以及周边商业区开发量总计超 300 万平方米，世界罕见，由于过硬的技术水准和客观公正的态度，项目深得业主信任，项目涵盖规划、设计、建设、运营等各阶段，周期长达 5 年，包括前期规划、内外交通设计、交通评估、运营服务等内容，在国内率先实现超大型公建全过程、全方位的交通规划设计咨询服务，主动引导相关工作的推进，有效保障规划设计及建设运营的系统性、延续性及可实施性。一般咨询服务难以介入运营阶段，由于会展中心配套交通设施尤其是地铁建设滞后，会展运营团队无类似规模展会运营经验，为避免出现同样规模的上海国展中心开业不久即引起交通瘫痪的情况，项目组强势介入，搭建了交通主管部门与运营单位的对话平台，同时配合交通主管部门及运营单位，协调相关单位落实责任范围内交通保障措施，制订运营初期具体展会的详细交通组织方案，制订并配合实施公交接驳方案，做好公共交通与会展内部交通的衔接（图 5-67 路径校核与需求分析图）。

② 交通专业作为独立一方，真正意义上主导交通规划设计，多专业、多部门开放探讨与协调，充分利用行政手段，确保交通规划设计"不走样"。

会展中心涉及众多管理部门、设计单位、建设单位及运营单位，各单位对交通均有不同诉求与想法，尤其是运营单位过多关注会展利益而忽视公共利益，而传统建筑项目规划设计中，建筑专业比较强势，交通更多是配合。在本项目中，鉴于会展交通的重要性、问题的复杂性，交通专业作为独立的一方总揽交通规划设计，统筹考虑会展与周边交通系统而不局限于会展中心，成功构建交通专业参与建筑设计的沟通平台，与建筑、景观、结构、机电等多专业合作，协调规划、交通、会展运营等多部门，在大尺度的空间下，充分发挥交通专业的宏观把控能力，确保交通规划设计"不走样"。

③ 提出深圳国际会展中心与欧美展馆交通疏解不具可比性，顶住压力，以公共交通为主导，抑制会展对非公共交通的诉求，缓解交通压力并有效减小对城市交通的冲击。

由于会展具有客流规模大、高峰集中等特征，欧美大型展馆选址偏远，交通设施充足，而深圳会展中心选址城市密集功能区，交通系统难以承担过多的非公共交通。项目组顶住会展行业主管部门、欧洲会展设计专家及美国运营商的压力，以公共交通为主导的理念配置交通设施，最大限度地抑制会展对非公共交通的诉求，大大缓解后续运营阶段的交通压力，同时有效避免会展高峰交通对城市交通的过度冲击（表 5-10）。以停车位配置为例，参照欧洲同规模会展中心需配置 3 万～5 万个停车位，项目组通过分析论证并多方协调，兼顾路网容量及会展运营需求，将车位控制在 9500 个，最终多方达成一致。

④ 整体考虑会展中心片区交通经济效益，采用日常配套与临时措施结合、配套商业区共享会展交通资源等手段，最大限度节省资源、集约空间、提高效率（图 5-67）。

由于会展中心淡旺季、大小展等特征，如果一味按照大型展会配置交通设施将占用大量空间及设施资源。因此项目采用日常配套与临时应急措施相结合的办法，其中日常配套应对日常高峰客流，临时应急措施应对极端高峰客流，以此大幅减小设施配套。同时，充分利用货运交通与客运交通在时间上存在分离的特性，分时共享场地空间，进一步集约用地。通过地下车库互联互通、立体慢行交通系统联通地铁站、公交场站的手段，实现会展配套商业区充分共享会展中心充足的交通资源，实现片区交通的融合发展。通过以上手段，最大限度地集约空间、节省资源，整体上提高会展中心片区交通经济效益（图表 5-1）。

表 5-10 欧洲非公共交通发达、国内依赖公交疏解统计表
Table 5-10   European Non-public Transportation is Advanced, But Domestic Transportation is Dependent on relieved public transport

	建筑面积（万平米） Construction area (ten thousand kilometers)	展览面积（万平米） Exhibition area (ten thousand kilometers)	轨道交通 Rail	小汽车分担率 （高峰时段） Car sharing rate (Peak hours)	停车位（个） Parking spaces	停车配建（个/百平） Parking configuration (a / 100 square meters)
国际会展中心（上海） International Convention and Exhibition Center (Shanghai)	147	40	3线4站 3 lines and 4 stations	10%~20%	11000	2.75
琶洲国际会展中心 International Convention and Exhibition Center (Shanghai)	110	33.8	3线4站 3 lines and 4 stations	10%~20%	7200	2.13
汉诺威会展中心 International Convention and Exhibition Center (Shanghai)	100	46.6	3线3站 3 lines and 3 stations	10%~20%	50000	10.73
米兰新国际会展中心 International Convention and Exhibition Center (Shanghai)	100	34.5	2线2站 2 lines and 2 stations	10%~20%	31000	8.99

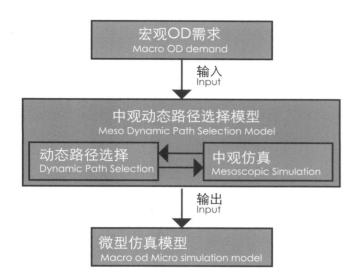

路径校核
Path Checking

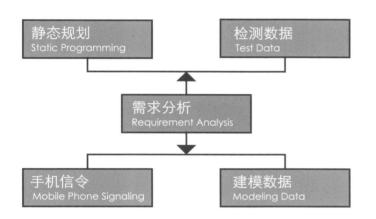

需求分析
Requirement Analysis

图 5-67 路径校核与需求分析图
Figure 5-67 Path Verification and Demand Analysis

图表 5-1 大数据及模型技术结合分析表
Chart 5-1 Combination of Big data and Model Technology

① The first case in China to provide full-process transportation planning and design consulting services for ultra-large exhibition centers, and effectively ensure the systematization and implementation of planning, design, construction and operation.

The Shenzhen World and its supporting facilities and surrounding business district construction has a total area of over 3 million square meters, which is rare in the world. Due to the high technical level, objective and fair attitude, the project is deeply trusted by the owners. This project has several stages of planning, design, construction, and operation, with a total period of 5 years, including preliminary scheme, internal and external traffic design, traffic evaluation, operation service, etc. We managed to provide a series of traffic planning and design consulting services in all aspects throughout the whole process for such a super large public project, which is the first case in the country. We also actively guided the advancement of related work, effectively ensure the planning design and construction operations systemic, continuity, and practicity.

In general, it is very difficult for consultant to engage operation. However, due to the delay of the construction, and especial the metro, and the lack of experience of operator for such large project. To avoid the traffic paralysis similar to what happened after the opening of the Shanghai International convention and exhibition center of same scale, the project team got involved and set up a the dialogue platform with the transportation departments and operation units, and cooperated with them to coordinate the relevant units to carry out traffic safeguard measures within the scope of responsibility, formulated detailed traffic organization plan for specific exhibition at the beginning of the operation, and made a shuttle bus plan and assisted the implementation, to guarantee good connection between public transportation and exhibition internal transportation(Figure 5-67).

② As an independent party, the transportation dominates the traffic planning and design, with open discussion and coordination of multi-profession and multi-department, make full use of administrative means to guarantee the traffic planning and unchanged design.

The Shenzhen World involves many management departments, design units, construction units and operating units, all of which have different demands and ideas on transportation. In particular, the operating units pay too much attention to the interests of the exhibitor and neglect the public needs. In the planning and design of traditional architectural projects, the architecture specialty is strong, with the transportation is more of a cooperative support. In this project, given the importance of the exhibition transportation, the complexity of the problem, the transportation specialty gave overall consideration of the traffic system of the exhibition center as well as its surrounding, successfully build traffic communication platform which the transportation get involved in architectural design, and cooperated with other special design units such as architecture, landscape, structure, mechanical and electrical team, and coordinated different departments of planning, traffic, operation, gave full play to macro control of the transportation specialty, and ensured that the traffic planning and unchanged design.

项目特色
Project Features

⑤ 践行"以人为本"的设计理念，在超大空间尺度上，提出"会展中心可容忍步行距离"的概念，采用多样化手段缩减步行距离。

践行"以人为本"的理念，最大限度实现人车分离以提高慢行安全，人行交通组织注重步行可容忍距离，公共交通设施与展馆出入口布置充分协调，布置充足自动步道、穿梭巴士等代步工具，地下车库构建与车流分离的专用人行通道以提高步行安全与品质，车库人行安检区实现 90 秒步行半径全覆盖。

⑥ 在超大范围内开展建筑与交通一体化微观仿真模拟，提出静态分析和动态仿真融合的微创新，为对策研究与方案评估提供有效支撑。

在会展中心周边道路即将启动建设前，以会展中心为核心，罕见地在 10 平方公里巨大范围内开展会展建筑及周边交通一体化微观仿真模拟。构建了宏观需求分析、中观路径选择、微观精细化模拟于一体的技术架构。通过车道级动态、定量的分析寻找细节交通问题，直接指导周边道路施工图的完善，增强交通疏解能力，减少局部交通拥堵风险。紧跟大数据时代潮流，模型技术微创新，将静态规划分析和动态车流仿真融合，通过交通大数据挖掘分析，支撑宏观交通需求分析、出行特征识别，为对策研究与方案评估提供支撑。有效推动了凤塘大道、国展立交等重要节点共计 18 项优化方案的落地实施（图 5-68）。

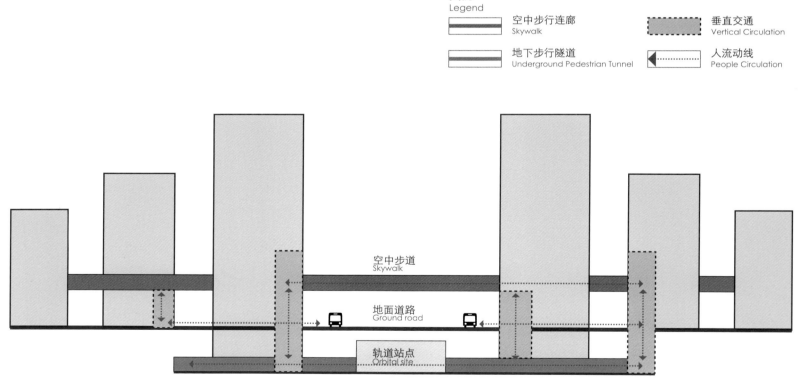

图 5-68　配套区通过立体慢行设施共享会展公共交通分析图
Figure 5-68　Supporting Area Shares Public Transportation Through Three-dimensional Slow-moving Facilities

③ The traffic relief of the Shenzhen World is incomparable with that of European and American. With this high pressure, we take public transportation as the leading factor to curb the appeal of exhibition to non-public transportation, so as to alleviate the traffic pressure and effectively reduce its impact on urban traffic.

In Europe and the United States, large exhibition center is located in remote place with sufficient transportation facilities. However, the Shenzhen World is located in intensive functional urban area, the transport system cannot afford too much non-public transportation. Under the pressure from the competent department of exhibition industry, the exhibition design experts from Europe, and the operators from the United States, we configured transport facilities in the concept of public transportation-oriented, greatly restrained the appeal of exhibition to non-public transportation, alleviated the traffic pressure of the subsequent operation, and effectively avoided the excessive impact of exhibition traffic peak on urban traffic (Table 5-11). Taking the parking space configuration as an example, referring to the European exhibition center of the same scale, 30,000 to 50,000 parking spaces need to be configured. Through analysis and coordination, and taking into account of the network capacity and exhibition operation needs, we managed to control the number of parking spaces within 9,500, and finally reached an agreement among all parties.

④ Given the overall consideration to the economic benefits of transportation in the area of the exhibition center, we took measures such as combination of daily supporting and temporary measures, and co-sharing of transportation resources with supporting business area, so as to furthest save resource, intensify space and improve efficiency(Figure 5-67).

There are low season and peak season, and large exhibitions and small exhibitions, it will waste a lot of space and facilities resources if the transportation is deployed according to a large-scale exhibition configuration. Therefore, we combined daily needs with temporary, emergency measures. Daily needs to deal with daily passenger flow peak, and temporary emergency measures to cope with the extreme passenger flow peak, in this way to sharply reduce the usage of facilities. At the same time, due to the separation of freight traffic and passenger traffic, the site can be time-share for intensive land use. By interconnecting the underground parking space and connecting subway stations and bus stations with the slow traffic system, the supporting commercial area can fully share the sufficient transportation resources of the exhibition center and achieve the integrated development of the area's transportation. Through the above means, the exhibition center can maximize the intensive use of space, resources saving, and improve the overall economic benefits of transportation (Chart 5-1).

⑤ Practice the design concept of "human touch", put forward the concept of "tolerable walking distance according to the super-large space scale, and adopt diversified means to reduce the walking distance.

Practice "human touch " concept, and separate people and the vehicle to improve the safety of slow zone. The pedestrian organization are emphasized by tolerate walking distance. Coordinate the public transport facilities and the entrance layout of exhibition hall to suarantee sufficient passenger belts, shuttle buses and other transportations. The pedestrian in the underground parking space is dedicated to improve safety and quality. The passenger security area in the underground parking space is within a radius of 90 seconds walk.

⑥ Carry out micro-simulation of building and transportation integration in a large scale, and propose micro-innovation of fusing static analysis and dynamic simulation, providing effective support for countermeasure research and program evaluation.

Before the construction of the surrounding roads, we took the Shenzhen World as the core and carried out an integrated micro-simulation of the exhibition building and surrounding transportation within a huge area of 10 square kilometers, which is rare. We also constructed a technical framework of macro demand analysis, middle path selection and micro refined simulation. Through lane-level dynamics and quantitative analysis, we found the details of traffic problems and directly guide the improvement of surrounding road construction drawings to enhance the ability of traffic relief and to reduce the risk of local traffic congestion. Following the trend of big data, and through the micro-innovation of model technology, we integrated the static planning analysis and dynamic traffic flow simulation; and through the digging and analysis of traffic big data, support the macro traffic demand analysis and the traveling feature identification, providing support for countermeasure research and program evaluation. Effectively promoted the implementation of 18 optimization schemes for Fengtang avenue, Guozhan interchange and other important nodes(Figure 5-68).

幕墙设计
Curtain Wall Design
内容摘自 ：华纳工程咨询（北京）有限公司
Warner Engineering Consulting (Beijing) Co., Ltd

幕墙的初步设计方案先明确了项目的设计理念，进而设计不同的方案并选择了不同的材料进行对比分析，经过多轮方案探讨及材料比选，达成共识后最终选用优化的设计方案及幕墙材料。

The preliminary design scheme of the curtain wall firstly clarified the design concept of the project, and then designed different schemes and choose different materials for comparative analysis. After several rounds of discussion and material comparison, the optimized design scheme and curtain wall materials were finally selected.

展厅幕墙设计
Curtain Wall Design of Exhibition Hall

系统介绍

展厅幕墙系统包括明框玻璃幕墙，竖明横隐玻璃幕墙、展厅通中央廊道及会议厅玻璃幕墙及轻质屋面，铝合金防风雨百叶、铝蜂窝板吊顶，玻璃中心吊门系统，电动滑移门，铝板（不锈钢）门，铝饰面防火门，超大门，电动遮阳卷帘，气动（电动）开窗器，穿孔铝板，特殊立面的立体穿孔铝板，展厅间电动格栅，室内穿孔铝板隔音幕墙，室内装饰百叶，甲级防火窗，室内外蜂窝板吊顶系统等。

System Introduction

The curtain wall system of exhibition hall includes: exposed framing glass curtain wall, semi-exposed framing glass curtain wall, glass curtain wall and life roof from exhibition hall to central concourse and conference hall, aluminum alloy weather-proof shutter, aluminum honeycomb suspended ceiling, glass center overhang-door system, ultra-large gate, electric sliding door, aluminum (stainless steel) door, aluminum coating fire door, over. Large door, electric roller blind, pneumatic (electric) window opener, perforated aluminum plate, tridimensional perforated aluminum plate with special facade, electric grid between exhibition halls, indoor perforated aluminum insulation curtain wall, indoor decorative shutter, class A fire window, indoor and outdoor honeycomb suspended ceiling systems, etc.

主要系统方案设计优化

项目设计阶段，我公司经过考察不同项目成功案例及总结类似项目经验教训及项目工期的紧迫性，与建筑师及业主方多轮汇报方案比选，并与不同材料厂家交流咨询后，得出以下最终优化效果原则：

项目尽量单元化完成，减少现场安装强度。因此玻璃幕墙、开缝式铝板系统均采用半单元的构造，以提高现场的安装速度及劳动强度。4.5m宽x1.0m 高，1.5m 宽x6m 高的铝板为保证平整度采用5mm 面板背衬铝合金加结构胶焊钉的综合连接措施。同时增加板材运输安装的可靠度。门、窗、百叶、排烟窗均采用成品厂家产品（表 5-11）。

Design Optimization of The Main System

In the design stage, we have investigated different successful cases and learnt the experience and lessons of similar projects. After multiple rounds of reporting to the architects and owners, and communicated and consulted with different material manufacturers, we finally got optimal results as follows:
Reduce the installation intensity on site by unitization. Therefore, the glass curtain wall and the slit aluminum plate system all adopt the half-unit structure to improve the installation speed and labor intensity. In order to ensure the smoothness, the high aluminum plate of 4.5m wide x 1.0m high, 1.5m wide x6m adopted a connection measure of 5mm panels backing aluminum alloy bonding with plastic welding nail, which could increase the reliability of the plate transportation and installation. The doors, windows, louvers, smoke exhaust windows are all finished products directly from the manufacturers.(Table 5-11)

表 5-11 标准展厅幕墙设计配置表
Table 5-11 The design and configuration of the standard exhibition hall curtain wall are as follows

序号 Serial number	项目名称 Project name	（宽 X 高）mm (W x H) mm	计算标高 Calculated elevation
1	HS8+1.52pvb+HS8low-e+12A+TP12mm夹胶中空超白玻璃 HS8 + 1.52pvb + HS8low-e + 12a + TP12mm laminated hollow ultra white glass	1500X4000	20.0 mm
2	HS8+1.52pvb+HS8low-e+12A+TP12mm夹胶中空超白玻璃 HS8 + 1.52pvb + HS8low-e + 12a + TP12mm laminated hollow ultra white glass	2250X5000	15.0 mm
3	5 mm厚穿孔铝单板 5mm thick perforated aluminum veneer	2250X5000	20.0 mm
4	5 mm厚铝单板 5mm thick aluminum veneer	1500X6000 2950X5000	28.0 mm
5	28 mm厚不锈钢蜂窝板 28 mm thick stainless steel honeycomb panel	1500X3000	6.0 mm
6	40 mm厚深灰色铝蜂窝板 HS8 + 1.52pvb + HS8low-e + 12a + TP12mm laminated hollow ultra white glass	2250X5000	6.0 mm
7	平开超大门 Level opening super gate	2250X5000	6.0 mm
8	电动推拉门 Electric sliding door	2250X5000	6.0 mm
9	电动推拉门 Electric sliding door	2250X5000	12.0 mm

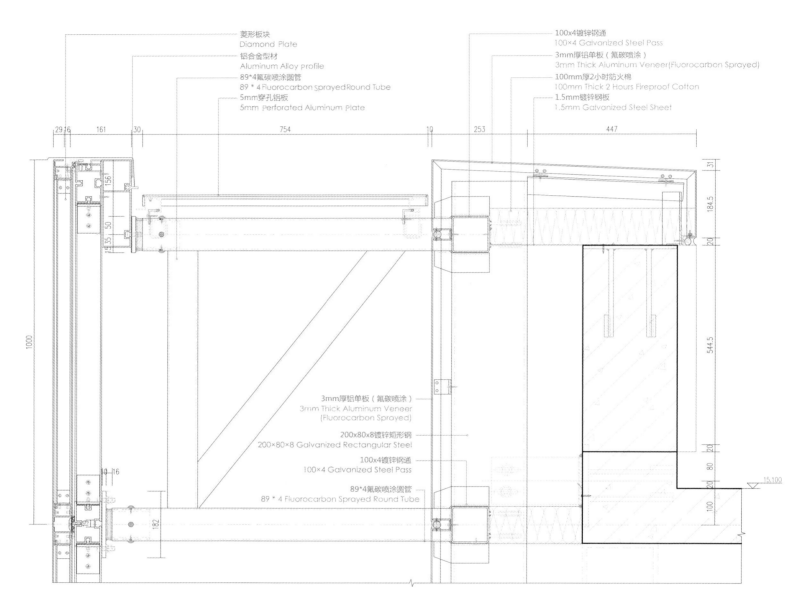

菱形板块
Diamond Plate

铝合金型材
Aluminum Alloy profile

89*4氟碳喷涂圆管
89 * 4 Fluorocarbon Sprayed Round Tube

5mm穿孔铝板
5mm Perforated Aluminum Plate

100x4镀锌钢通
100×4 Galvanized Steel Pass

3mm厚铝单板（氟碳喷涂）
3mm Thick Aluminum Veneer(Fluorocarbon Sprayed)

100mm厚2小时防火棉
100mm Thick 2 Hours Fireproof Cotton

1.5mm镀锌钢板
1.5mm Galvanized Steel Sheet

3mm厚铝单板（氟碳喷涂）
3mm Thick Aluminum Veneer
(Fluorocarbon Sprayed)

200x80x8镀锌矩形钢
200×80×8 Galvanized Rectangular Steel

100x4镀锌钢通
100×4 Galvanized Steel Pass

89*4氟碳喷涂圆管
89 * 4 Fluorocarbon Sprayed Round Tube

图 5-69 特殊立面构造图
Figure 5-69 Special Facade Structure

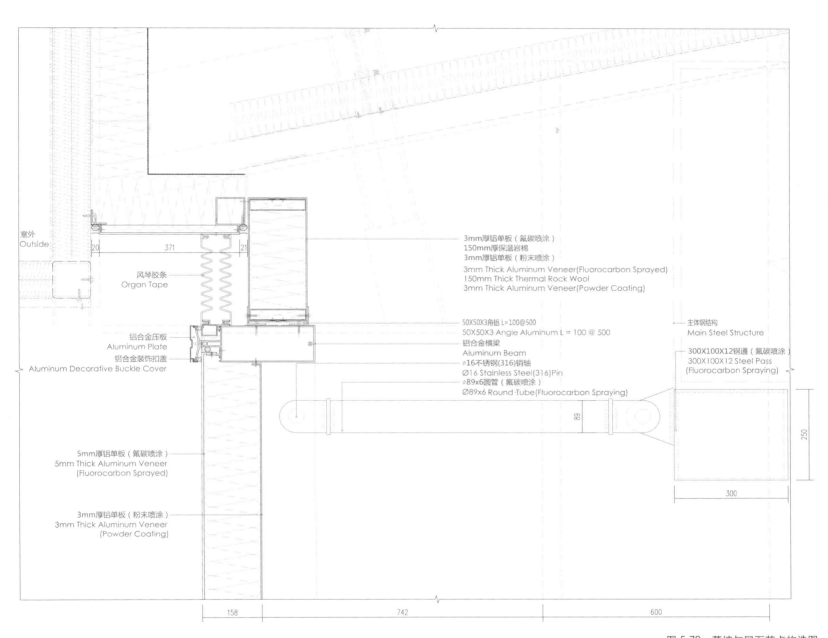

室外
Outside

风琴胶条
Organ Tape

铝合金压板
Aluminum Plate
铝合金装饰扣盖
Aluminum Decorative Buckle Cover

5mm厚铝单板（氟碳喷涂）
5mm Thick Aluminum Veneer
(Fluorocarbon Sprayed)

3mm厚铝单板（粉末喷涂）
3mm Thick Aluminum Veneer
(Powder Coating)

3mm厚铝单板（氟碳喷涂）
150mm厚保温岩棉
3mm厚铝单板（粉末喷涂）
3mm Thick Aluminum Veneer(Fluorocarbon Sprayed)
150mm Thick Thermal Rock Wool
3mm Thick Aluminum Veneer(Powder Coating)

50X50X3角铝 L=100@500
50X50X3 Angle Aluminum L = 100 @ 500
铝合金横梁
Aluminum Beam
∅16不锈钢(316)销轴
Ø16 Stainless Steel(316)Pin
∅89x6圆管（氟碳喷涂）
Ø89x6 Round Tube(Fluorocarbon Spraying)

主体钢结构
Main Steel Structure

300X100X12钢通（氟碳喷涂）
300X100X12 Steel Pass
(Fluorocarbon Spraying)

20   371   21

158   742   600

89   250   300

图 5-70　幕墙与屋面节点构造图
Figure 5-70　Node Structure of Curtain Wall and Roof

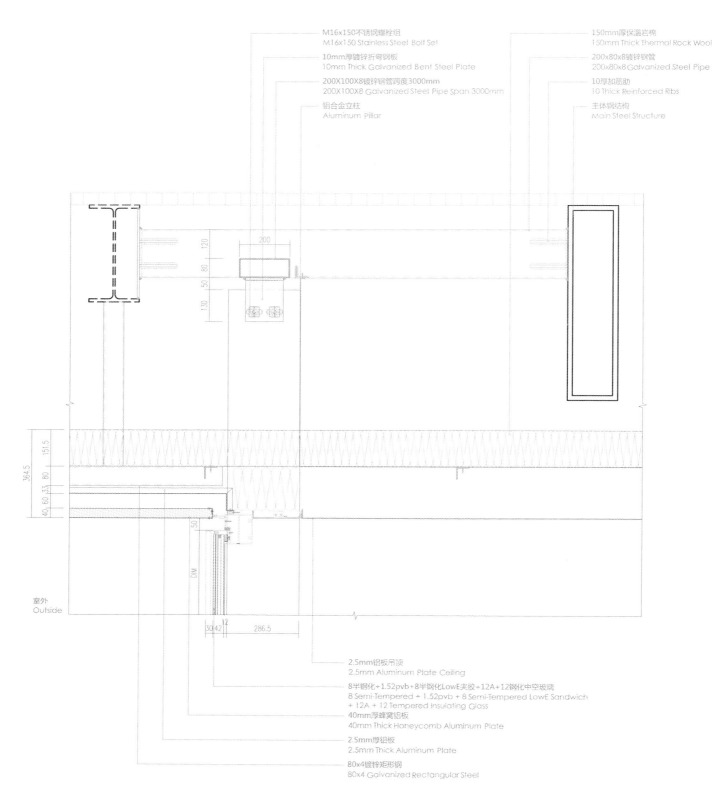

M16x150不锈钢螺栓组
M16x150 Stainless Steel Bolt Set

10mm厚镀锌折弯钢板
10mm Thick Galvanized Bent Steel Plate

200X100X8镀锌钢管跨度3000mm
200X100X8 Galvanized Steel Pipe Span 3000mm

铝合金立柱
Aluminum Pillar

150mm厚保温岩棉
150mm Thick Thermal Rock Wool

200x80x8镀锌钢管
200x80x8 Galvanized Steel Pipe

10厚加筋肋
10 Thick Reinforced Ribs

主体钢结构
Main Steel Structure

室外
Outside

2.5mm铝板吊顶
2.5mm Aluminum Plate Ceiling

8半钢化+1.52pvb+8半钢化LowE夹胶+12A+12钢化中空玻璃
8 Semi-Tempered + 1.52pvb + 8 Semi-Tempered LowE Sandwich + 12A + 12 Tempered Insulating Glass

40mm厚蜂窝铝板
40mm Thick Honeycomb Aluminum Plate

2.5mm厚铝板
2.5mm Thick Aluminum Plate

80x4镀锌矩形钢
80x4 Galvanized Rectangular Steel

图 5-71　铝蜂窝板吊顶节点构造图
Figure 5-71　Joint Detail of Aluminum Honeycomb Ceiling

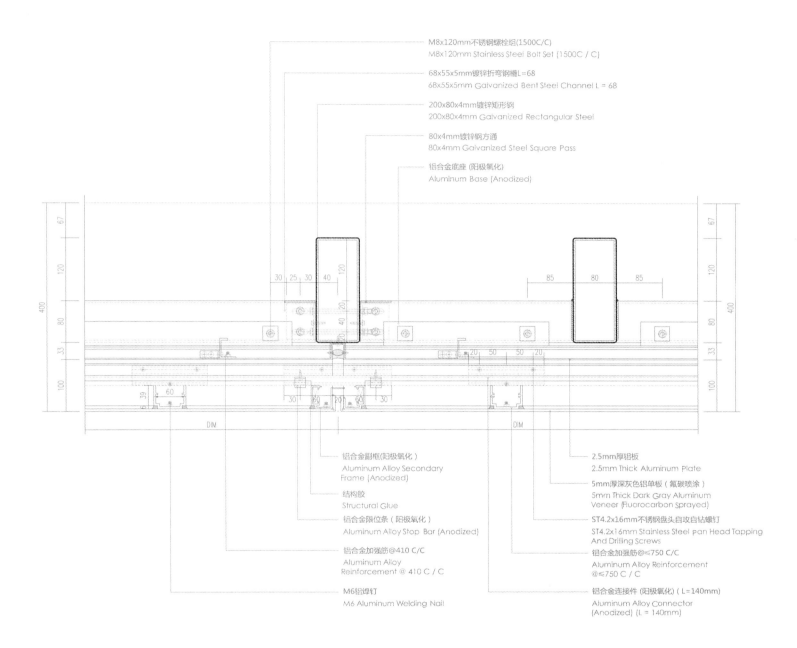

M8x120mm不锈钢螺栓组(1500C/C)
M8x120mm Stainless Steel Bolt Set (1500C / C)

68x55x5mm镀锌折弯钢槽L=68
68x55x5mm Galvanized Bent Steel Channel L = 68

200x80x4mm镀锌矩形钢
200x80x4mm Galvanized Rectangular Steel

80x4mm镀锌钢方通
80x4mm Galvanized Steel Square Pass

铝合金底座 (阳极氧化)
Aluminum Base (Anodized)

铝合金副框(阳极氧化)
Aluminum Alloy Secondary Frame (Anodized)

结构胶
Structural Glue

铝合金限位条（阳极氧化 ）
Aluminum Alloy Stop Bar (Anodized)

铝合金加强筋@410 C/C
Aluminum Alloy Reinforcement @ 410 C / C

M6铝焊钉
M6 Aluminum Welding Nail

2.5mm厚铝板
2.5mm Thick Aluminum Plate

5mm厚深灰色铝单板（氟碳喷涂 ）
5mm Thick Dark Gray Aluminum Veneer (Fluorocarbon Sprayed)

ST4.2x16mm不锈钢盘头自攻自钻螺钉
ST4.2x16mm Stainless Steel pan Head Tapping And Drilling Screws

铝合金加强筋@≤750 C/C
Aluminum Alloy Reinforcement @≤750 C / C

铝合金连接件 (阳极氧化) (L=140mm)
Aluminum Alloy Connector (Anodized) (L = 140mm)

图 5-72　板幕墙节点构造图
Figure 5-72　Node Structure Detail of Aluminum Panel Curtain Wall

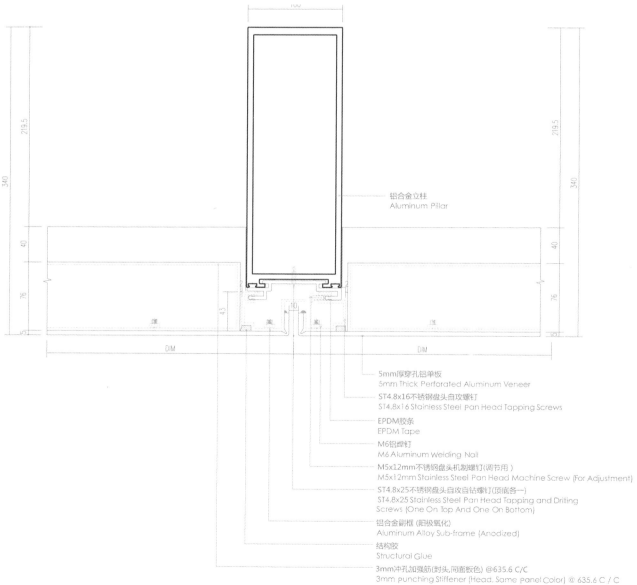

铝合金立柱(氟碳喷涂)
Aluminum Pillar(Fluorocarbon Spraying)

8半钢化+1.52pvb+8半钢化LowE夹胶+12A+12钢化中空玻璃
8 Semi-Tempered + 1.52pvb + 8 Semi-Tempered LowE
Sandwich + 12A + 12 Tempered Insulating Glass

EPDM胶条
EPDM Tape

Ø15泡沫棒&密封胶
Ø15 Foam Stick & Sealant

M6X25不锈钢螺栓组 @350配一弹垫一螺母
Grouting Hole

铝合金压条
Grouting Hole

图 5-73　玻璃幕墙节点构造图
Figure 5-73　Node Structure Detail of Glass Curtain Wall

铝合金立柱
Aluminum Pillar

5mm厚穿孔铝单板
5mm Thick Perforated Aluminum Veneer

ST4.8x16不锈钢盘头自攻螺钉
ST4.8x16 Stainless Steel Pan Head Tapping Screws

EPDM胶条
EPDM Tape

M6铝焊钉
M6 Aluminum Welding Nail

M5x12mm不锈钢盘头机制螺钉(调节用)
M5x12mm Stainless Steel Pan Head Machine Screw (For Adjustment)

ST4.8x25不锈钢盘头自攻自钻螺钉(顶底各一)
ST4.8x25 Stainless Steel Pan Head Tapping and Drilling
Screws (One On Top And One On Bottom)

铝合金副框 (阳极氧化)
Aluminum Alloy Sub-frame (Anodized)

结构胶
Structural Glue

3mm冲孔加强筋(封头,同面板色) @635.6 C/C
3mm punching Stiffener (Head, Same panel Color) @ 635.6 C / C

图 5-74　屋顶穿孔板构造图
Figure 5-74　Roof Perforated Panel Detail

登录厅、中央廊道屋面系统设计

① 系统介绍

登录厅、中央廊道屋面系统构造层次基本一致（图 5-75，图 5-76），只是建筑外形略有区别。

登录厅、中央廊道屋面系统分别由钢化夹胶玻璃面板、铝板＋降噪岩棉＋铝板（以下简称金属面板）、铝合金吊顶格栅、不锈钢天沟（不含虹吸系统）、太阳能光伏板（仅限登录厅屋面）、竖向导雨百叶（仅限中央廊道屋面）、檐口外包铝板、上人检修口、屋面防雷系统、防坠落系统等组成。

Roof System Design of the Arrival Hall and the Central Concourse

① System Introduction

The roof system structure of the arrival hall and the central concourse is basically the same (Figure 5-75, 5-76), which is only slightly different on the appearance of the building.

The roof system of the arrival hall and the central concourse is consisted of tempered laminated glass panel, aluminum plate+ denoised rock woll+ aluminum plate (hereinafter referred to as the metal plate), aluminum alloy suspended ceiling grilling, stainless steel gutter (excluding siphon system), solar photovoltaic panels (only for the roof of landing hall), vertical rainproof shutter (only for the roof of the central corridor), cornice with exterior aluminum , access hole for people, roof lightning protection system, fall prevention system, etc.

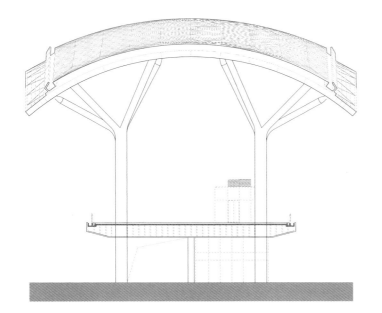

图 5-75　中央廊道横剖面图
Figure 5-75　Central Concourse Cross Section

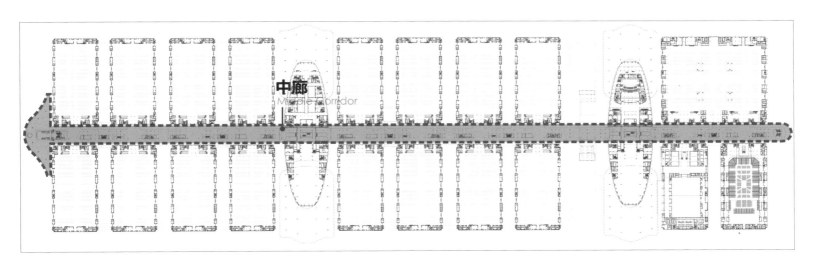

图 5-76　中央廊道图
Figure 5-76　Central Concourse

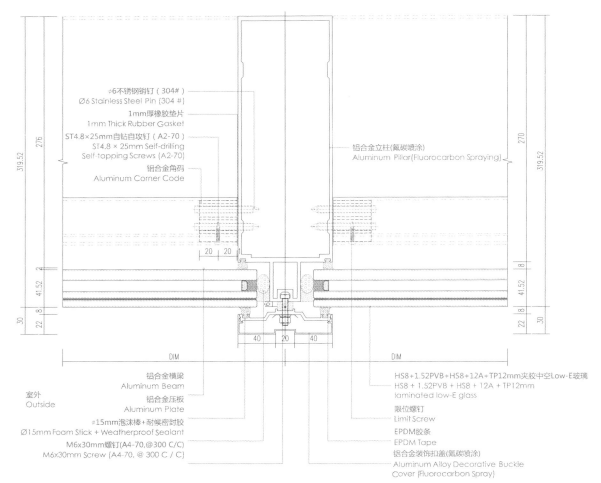

ø6不锈钢销钉（304#）
Ø6 Stainless Steel Pin (304 #)

1mm厚橡胶垫片
1mm Thick Rubber Gasket

ST4.8×25mm自钻自攻钉（A2-70）
ST4.8 × 25mm Self-drilling
Self-tapping Screws (A2-70)

铝合金角码
Aluminum Corner Code

铝合金立柱(氟碳喷涂)
Aluminum Pillar(Fluorocarbon Spraying)

铝合金横梁
Aluminum Beam

室外
Outside

铝合金压板
Aluminum Plate

ø15mm泡沫棒+耐候密封胶
Ø15mm Foam Stick + Weatherproof Sealant

M6x30mm螺钉(A4-70,@300 C/C)
M6x30mm Screw (A4-70, @ 300 C / C)

HS8+1.52PVB+HS8+12A+TP12mm夹胶中空Low-E玻璃
HS8 + 1.52PVB + HS8 + 12A + TP12mm
laminated low-E glass

限位螺钉
Limit Screw

EPDM胶条
EPDM Tape

铝合金装饰扣盖(氟碳喷涂)
Aluminum Alloy Decorative Buckle
Cover (Fluorocarbon Spray)

图 5-77　玻璃幕墙节点构造图
Figure 5-77　Node Structure Detail of Glass Curtain Wall

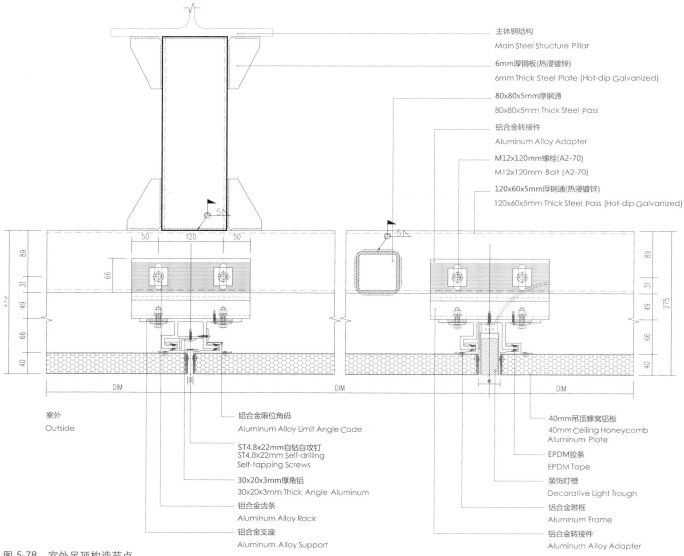

主体钢结构
Main Steel Structure Pillar

6mm厚钢板(热浸镀锌)
6mm Thick Steel Plate (Hot-dip Galvanized)

80x80x5mm厚钢通
80x80x5mm Thick Steel Pass

铝合金转接件
Aluminum Alloy Adapter

M12x120mm螺栓(A2-70)
M12x120mm Bolt (A2-70)

120x60x5mm厚钢通(热浸镀锌)
120x60x5mm Thick Steel Pass (Hot-dip Galvanized)

室外
Outside

铝合金限位角码
Aluminum Alloy Limit Angle Code

ST4.8x22mm自钻自攻钉
ST4.8x22mm Self-drilling
Self-tapping Screws

30x20x3mm厚角铝
30x20x3mm Thick Angle Aluminum

铝合金齿条
Aluminum Alloy Rack

铝合金支座
Aluminum Alloy Support

40mm吊顶蜂窝铝板
40mm Ceiling Honeycomb
Aluminum Plate

EPDM胶条
EPDM Tape

装饰灯槽
Decorative Light Trough

铝合金附框
Aluminum Frame

铝合金转接件
Aluminum Alloy Adapter

图 5-78　室外吊顶构造节点
Figure 5-78　Outdoor Ceiling Node Structure Detail

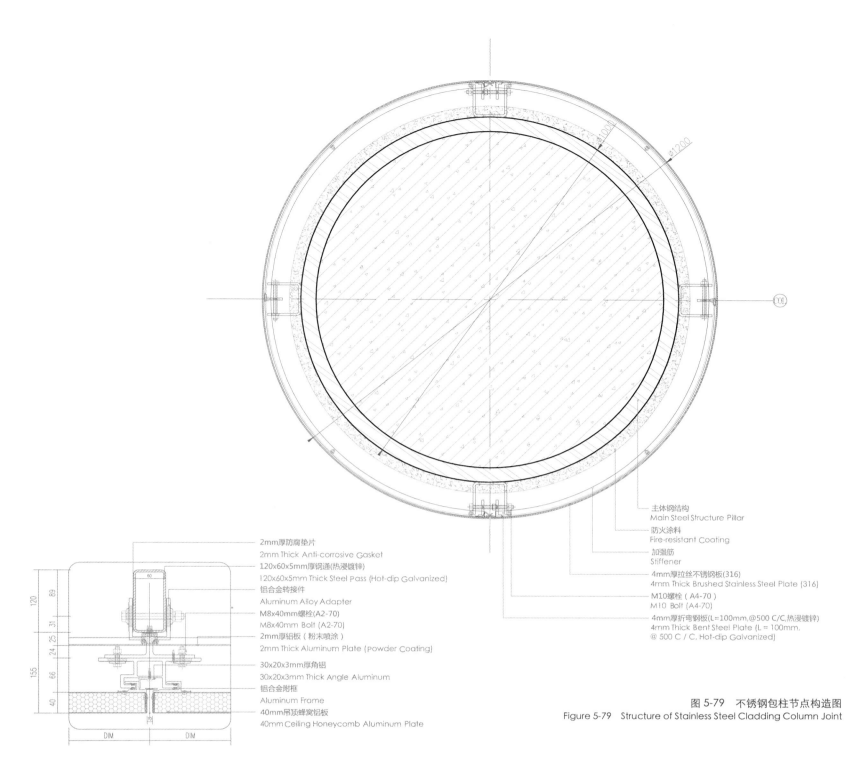

2mm厚防腐垫片
2mm Thick Anti-corrosive Gasket
120x60x5mm厚钢通(热浸镀锌)
120x60x5mm Thick Steel Pass (Hot-dip Galvanized)
铝合金转接件
Aluminum Alloy Adapter
M8x40mm螺栓(A2-70)
M8x40mm Bolt (A2-70)
2mm厚铝板（粉末喷涂）
2mm Thick Aluminum Plate (Powder Coating)
30x20x3mm厚角铝
30x20x3mm Thick Angle Aluminum
铝合金附框
Aluminum Frame
40mm吊顶蜂窝铝板
40mm Ceiling Honeycomb Aluminum Plate

主体钢结构
Main Steel Structure Pillar
防火涂料
Fire-resistant Coating
加强筋
Stiffener
4mm厚拉丝不锈钢板(316)
4mm Thick Brushed Stainless Steel Plate (316)
M10螺栓（A4-70）
M10 Bolt (A4-70)
4mm厚折弯钢板(L=100mm,@500 C/C,热浸镀锌)
4mm Thick Bent Steel Plate (L = 100mm,
@ 500 C / C, Hot-dip Galvanized)

图 5-79　不锈钢包柱节点构造图
Figure 5-79　Structure of Stainless Steel Cladding Column Joint

图 5-80　A – A 剖面图
Figure 5-80　Section A – A

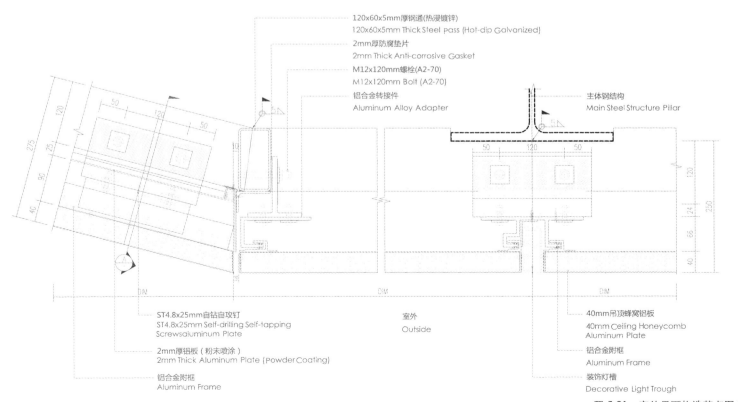

120x60x5mm厚钢通(热浸镀锌)
120x60x5mm Thick Steel pass (Hot-dip Galvanized)
2mm厚防腐垫片
2mm Thick Anti-corrosive Gasket
M12x120mm螺栓(A2-70)
M12x120mm Bolt (A2-70)
铝合金转接件
Aluminum Alloy Adapter

主体钢结构
Main Steel Structure Pillar

ST4.8x25mm自钻自攻钉
ST4.8x25mm Self-drilling Self-tapping
Screwsaluminum Plate
2mm厚铝板（粉末喷涂）
2mm Thick Aluminum Plate (Powder Coating)
铝合金附框
Aluminum Frame

室外
Outside

40mm吊顶蜂窝铝板
40mm Ceiling Honeycomb
Aluminum Plate
铝合金附框
Aluminum Frame
装饰灯槽
Decorative Light Trough

图 5-81　室外吊顶构造节点图 2
Figure 5-81　Outdoor Ceiling Structure Node

登录厅幕墙系统设计
Curtain Wall Design of Arrival Hall

系统介绍

幕墙系统主要包括：
明框玻璃幕墙＋横向装饰百叶系统（有主体结构）、明框玻璃幕墙＋横向装饰百叶系统（无主体结构）、拉索玻璃幕墙系统、明框玻璃幕墙系统、蜂窝一体保温板系统，玻璃栏板系统、玻璃中心吊门系统、铝合金门系统、铝饰面防火门系统、超大门系统、气动排烟窗系统等。

登录大厅幕墙立面（图 5-82 登录大厅幕墙设计图）使用材料：
① 10+2.28PVB+10+12A+10+2.28PVB+10mm 双银半钢化 LOW-E 中空双夹胶玻璃（四片超白）------ 用于立面
② 15+2.28pvb+15 +12A+15mm 双银钢化夹胶 LOW-E 中空超白玻璃 ------ 用于入口拉索幕墙
③ 25mm 蜂窝铝板 ------ 用于立面
④拉索幕墙玻璃最大分格为 1.78m 宽 X1.78m 高（图 5-82）

System Introduction

The curtain wall system mainly includes: exposed framing glass curtain wall + horizontal decorative shutter system (with main structure), exposed framing glass curtain wall + horizontal decorative shutter system (without main structure), cable glass curtain wall system, exposed framing glass curtain wall system, honeycomb integrated insulation board, balustrade with glass panel, glass center hanging door system, aluminum alloy door system, aluminum facing fire door system, ultra-l-arge gate system, pneumatic exhaust smoke window system, etc.

Materials used for the facade of the curtain wall of the arrival hall:(Figure 5-82)
① 10+2.28PVB+10+12A+10+2.28PVB+10mm double silver semi-tempered LOW-E hollow double-laminated glass (four pieces of ultra-white), for the facade
② 15+2.28 PVB +15 +12A+15mm double silver semi-tempered laminated LOW-E hollow ultra-white glass, for the cable curtain wall at the entrance
③ 25mm honeycomb aluminum plate, for the facade
④ The maximum partition of inhaul cable curtain wall glass is 1.78m wide x 1.78m high(Figure 5-82)

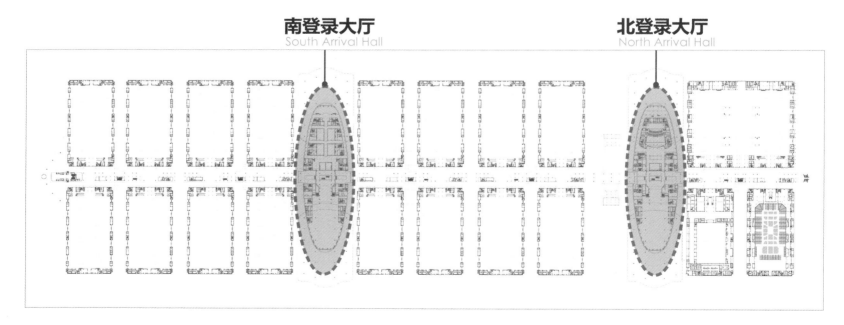

南登录大厅 South Arrival Hall　　　**北登录大厅** North Arrival Hall

图 5-82　登录大厅幕墙设计图
Figure 5-82　Arrival Hall Curtain Wall Design

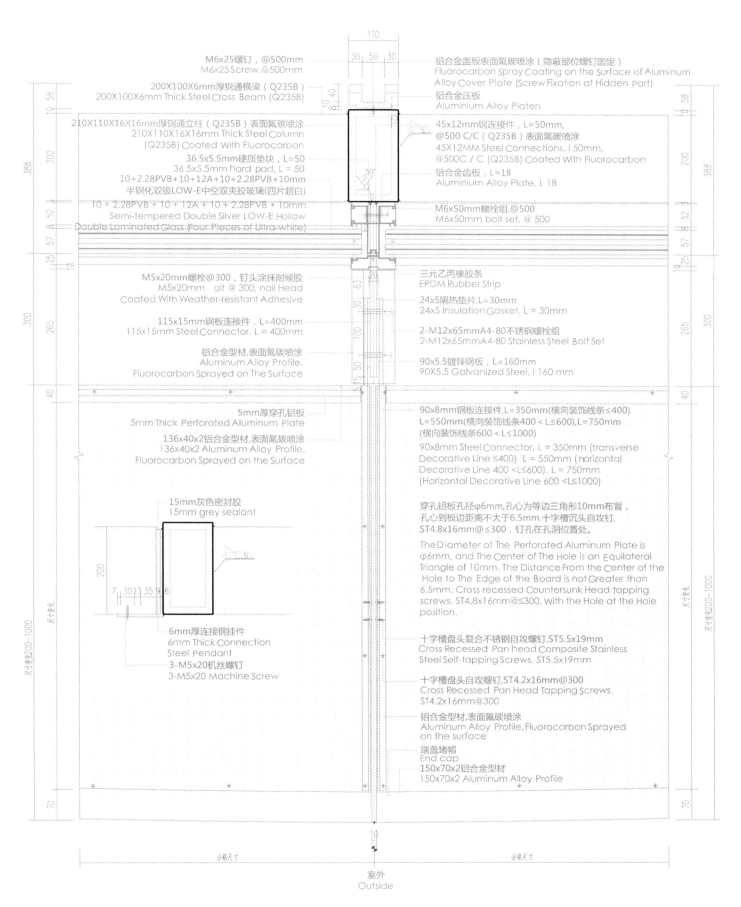

M6x25螺钉，@500mm
M6x25 Screw,@500mm

200X100X6mm厚钢通横梁（Q235B）
200X100X6mm Thick Steel Cross Beam (Q235B)

210X110X16X16mm厚钢通立柱（Q235B）表面氟碳喷涂
210X110X16X16mm Thick Steel Column
(Q235B) Coated With Fluorocarbon

36.5x5.5mm硬质垫块，L=50
36.5x5.5mm hard pad, L = 50

10+2.28PVB+10+12A+10+2.28PVB+10mm
半钢化双银LOW-E中空双夹胶玻璃(四片超白)

10+2.28PVB+10+12A+10+2.28PVB+10mm
Semi-tempered Double Silver LOW-E Hollow
Double Laminated Glass (Four Pieces of Ultra-white)

M5x20mm螺栓@300，钉头涂抹耐候胶
M5x20mm olt @ 300, nail Head
Coated With Weather-resistant Adhesive

115x15mm钢板连接件，L=400mm
115x15mm Steel Connector, L = 400mm

铝合金型材,表面氟碳喷涂
Aluminum Alloy Profile,
Fluorocarbon Sprayed on The Surface

5mm厚穿孔铝板
5mm Thick Perforated Aluminum Plate

136x40x2铝合金型材,表面氟碳喷涂
136x40x2 Aluminum Alloy Profile,
Fluorocarbon Sprayed on the Surface

15mm灰色密封胶
15mm grey sealant

6mm厚连接钢挂件
6mm Thick Connection
Steel Pendant

3-M5x20机丝螺钉
3-M5x20 Machine Screw

铝合金盖板表面氟碳喷涂（隐蔽部位螺钉固定）
Fluorocarbon Spray Coating on the Surface of Aluminum
Alloy Cover Plate (Screw Fixation at Hidden Part)

铝合金压板
Aluminium Alloy Platen

45x12mm钢连接件，L=50mm,
@500 C/C（Q235B）表面氟碳喷涂
45X12MM Steel Connections, I 50mm,
@500C / C (Q235B) Coated With Fluorocarbon

铝合金齿板，L=18
Aluminium Alloy Plate, L 18

M6x50mm螺栓组,@500
M6x50mm bolt set, @ 500

三元乙丙橡胶条
EPDM Rubber Strip

24x5隔热垫片,L=30mm
24x5 Insulation Gasket, L = 30mm

2-M12x65mmA4-80不锈钢螺栓组
2-M12x65mmA4-80 Stainless Steel Bolt Set

90x5.5镀锌钢板，L=160mm
90x5.5 Galvanized Steel, l 160 mm

90x8mm钢板连接件,L=350mm(横向装饰线条≤400)
L=550mm(横向装饰线条400<L≤600),L=750mm
(横向装饰线条600<L≤1000)
90x8mm Steel Connector, L = 350mm (transverse
Decorative Line ≤400) L = 550mm (horizontal
Decorative Line 400 <L≤600), L = 750mm
(Horizontal Decorative Line 600 <L≤1000)

穿孔铝板板孔径φ6mm，孔心为等边三角形10mm布置，
孔心到板边距离不大于6.5mm.十字槽沉头自攻钉，
ST4.8x16mm@≤300，钉孔在孔洞位置处。
The Diameter of The Perforated Aluminum Plate is
φ6mm, and The Center of The Hole is an Equilateral
Triangle of 10mm. The Distance From the Center of the
Hole to The Edge of the Board is not Greater than
6.5mm. Cross recessed Countersunk Head Tapping
screws, ST4.8x16mm@≤300, With the Hole at the Hole
position.

十字槽盘头复合不锈钢自攻螺钉,ST5.5x19mm
Cross Recessed Pan head Composite Stainless
Steel Self-tapping Screws, ST5.5x19mm

十字槽盘头自攻螺钉,ST4.2x16mm@300
Cross Recessed Pan Head Tapping Screws,
ST4.2x16mm@300

铝合金型材,表面氟碳喷涂
Aluminum Alloy Profile,Fluorocarbon Sprayed
on the surface

端盖堵帽
End cap

150x70x2铝合金型材
150x70x2 Aluminum Alloy Profile

分格尺寸

室外
Outside

分格尺寸

图 5-83　登录大厅玻璃幕墙节点构造图
Figure 5-83　Node Structure of Glass Curtain Wall in the Arrival Hall

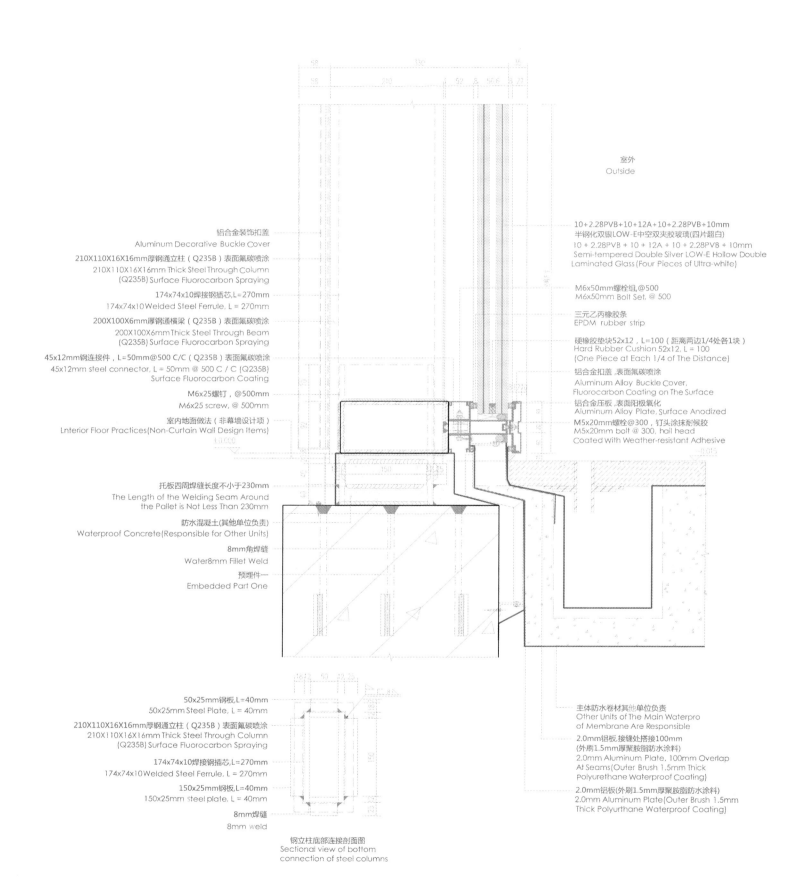

铝合金装饰扣盖
Aluminum Decorative Buckle Cover

210X110X16X16mm厚钢通立柱（Q235B）表面氟碳喷涂
210X110X16X16mm Thick Steel Through Column
(Q235B) Surface Fluorocarbon Spraying

174x74x10焊接钢插芯，L=270mm
174x74x10 Welded Steel Ferrule, L = 270mm

200X100X6mm厚钢通横梁（Q235B）表面氟碳喷涂
200X100X6mm Thick Steel Through Beam
(Q235B) Surface Fluorocarbon Spraying

45x12mm钢连接件，L=50mm@500 C/C（Q235B）表面氟碳喷涂
45x12mm steel connector, L = 50mm @ 500 C / C (Q235B)
Surface Fluorocarbon Coating

M6x25螺钉，@500mm
M6x25 screw, @ 500mm

室内地面做法（非幕墙设计项）
Lnterior Floor Practices(Non-Curtain Wall Design Items)

托板四周焊缝长度不小于230mm
The Length of the Welding Seam Around
the Pallet is Not Less Than 230mm

防水混凝土(其他单位负责)
Waterproof Concrete(Responsible for Other Units)

8mm角焊缝
Water8mm Fillet Weld

预埋件一
Embedded Part One

室外
Outside

10+2.28PVB+10+12A+10+2.28PVB+10mm
半钢化双银LOW-E中空双胶玻璃(四片超白)
10 + 2.28PVB + 10 + 12A + 10 + 2.28PVB + 10mm
Semi-tempered Double Silver LOW-E Hollow Double
Laminated Glass (Four Pieces of Ultra-white)

M6x50mm螺栓组,@500
M6x50mm Bolt Set, @ 500

三元乙丙橡胶条
EPDM rubber strip

硬橡胶垫块52x12，L=100（距离两边1/4处各1块）
Hard Rubber Cushion 52x12, L = 100
(One Piece at Each 1/4 of The Distance)

铝合金扣盖,表面氟碳喷涂
Aluminum Alloy Buckle Cover,
Fluorocarbon Coating on The Surface

铝合金压板,表面阳极氧化
Aluminum Alloy Plate, Surface Anodized

M5x20mm螺栓@300，钉头涂抹耐候胶
M5x20mm bolt @ 300, hail head
Coated With Weather-resistant Adhesive

主体防水卷材其他单位负责
Other Units of The Main Waterpro
of Membrane Are Responsible

2.0mm铝板,接缝处搭接100mm
(外刷1.5mm厚聚脂胶防水涂料)
2.0mm Aluminum Plate, 100mm Overlap
At Seams(Outer Brush 1.5mm Thick
Polyurethane Waterproof Coating)

2.0mm铝板(外刷1.5mm厚聚脂胶防水涂料)
2.0mm Aluminum Plate(Outer Brush 1.5mm
Thick Polyurthane Waterproof Coating)

50x25mm钢板,L=40mm
50x25mm Steel Plate, L = 40mm

210X110X16X16mm厚钢通立柱（Q235B）表面氟碳喷涂
210X110X16X16mm Thick Steel Through Column
(Q235B) Surface Fluorocarbon Spraying

174x74x10焊接钢插芯,L=270mm
174x74x10 Welded Steel Ferrule, L = 270mm

150x25mm钢板,L=40mm
150x25mm steel plate, L = 40mm

8mm焊缝
8mm weld

钢立柱底部连接剖面图
Sectional view of bottom
connection of steel columns

图 5-84　钢立柱底部连接剖面图
Figure 5-84　Steel Column Bottom Connection Section

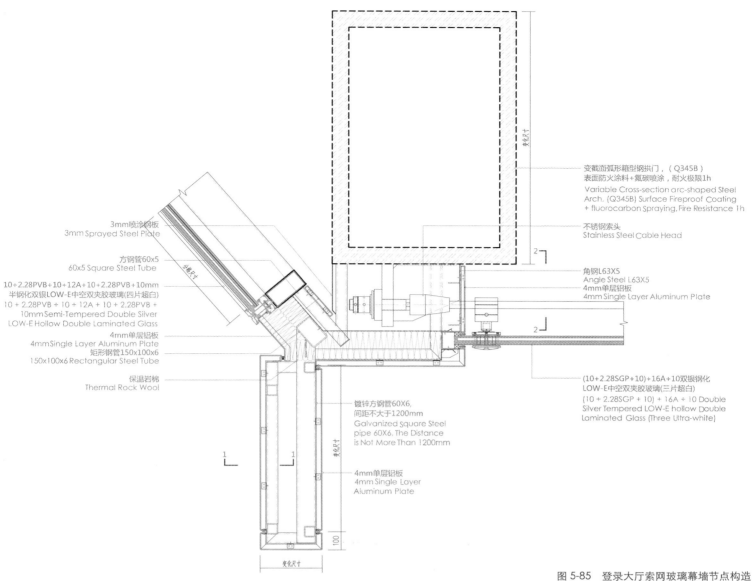

3mm喷涂钢板
3mm Sprayed Steel Plate

方钢管60x5
60x5 Square Steel Tube

10+2.28PVB+10+12A+10+2.28PVB+10mm
半钢化双银LOW-E中空双夹胶玻璃(四片超白)
10 + 2.28PVB + 10 + 12A + 10 + 2.28PVB +
10mm Semi-Tempered Double Silver
LOW-E Hollow Double Laminated Glass

4mm单层铝板
4mm Single Layer Aluminum Plate
矩形钢管150x100x6
150x100x6 Rectangular Steel Tube

保温岩棉
Thermal Rock Wool

镀锌方钢管60X6,
间距不大于1200mm
Galvanized Square Steel
pipe 60X6, The Distance
is Not More Than 1200mm

4mm单层铝板
4mm Single Layer
Aluminum Plate

变截面弧形箱型钢拱门,(Q345B)
表面防火涂料+氟碳喷涂,耐火极限1h
Variable Cross-section arc-shaped Steel
Arch. (Q345B) Surface Fireproof Coating
+ fluorocarbon Spraying,Fire Resistance 1h

不锈钢索头
Stainless Steel Cable Head

角钢L63X5
Angle Steel L63X5
4mm单层铝板
4mm Single Layer Aluminum Plate

(10+2.28SGP+10)+16A+10双银钢化
LOW-E中空双夹胶玻璃(三片超白)
(10 + 2.28SGP + 10) + 16A + 10 Double
Silver Tempered LOW-E hollow Double
Laminated Glass (Three Ultra-white)

图 5-85　登录大厅索网玻璃幕墙节点构造
Figure 5-85　Detail of Cable Net Glass Curtain Wall Joint in the Arrival Hall

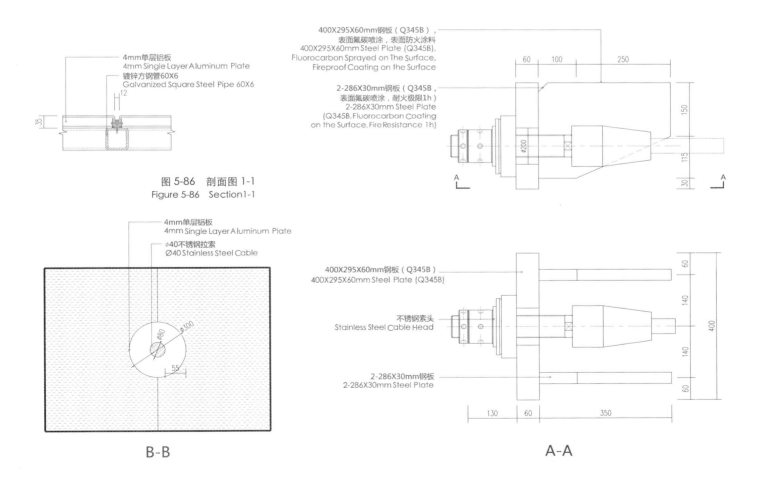

4mm单层铝板
4mm Single Layer Aluminum Plate
镀锌方钢管60X6
Galvanized Square Steel Pipe 60X6

图 5-86　剖面图 1-1
Figure 5-86　Section1-1

400X295X60mm钢板(Q345B),
表面氟碳喷涂,表面防火涂料
400X295X60mm Steel Plate (Q345B),
Fluorocarbon Sprayed on The Surface,
Fireproof Coating on the Surface

2-286X30mm钢板(Q345B,
表面氟碳喷涂,耐火极限1h)
2-286X30mm Steel Plate
(Q345B,Fluorocarbon Coating
on the Surface,Fire Resistance 1h)

4mm单层铝板
4mm Single Layer Aluminum Plate
∅40不锈钢拉索
Ø40 Stainless Steel Cable

400X295X60mm钢板(Q345B)
400X295X60mm Steel Plate (Q345B)

不锈钢索头
Stainless Steel Cable Head

2-286X30mm钢板
2-286X30mm Steel Plate

B-B

A-A

图 5-87　剖面图 B-B
Figure 5-87　SectionB-B

图 5-88　剖面图 A-A
Figure 5-88　SectionA-A

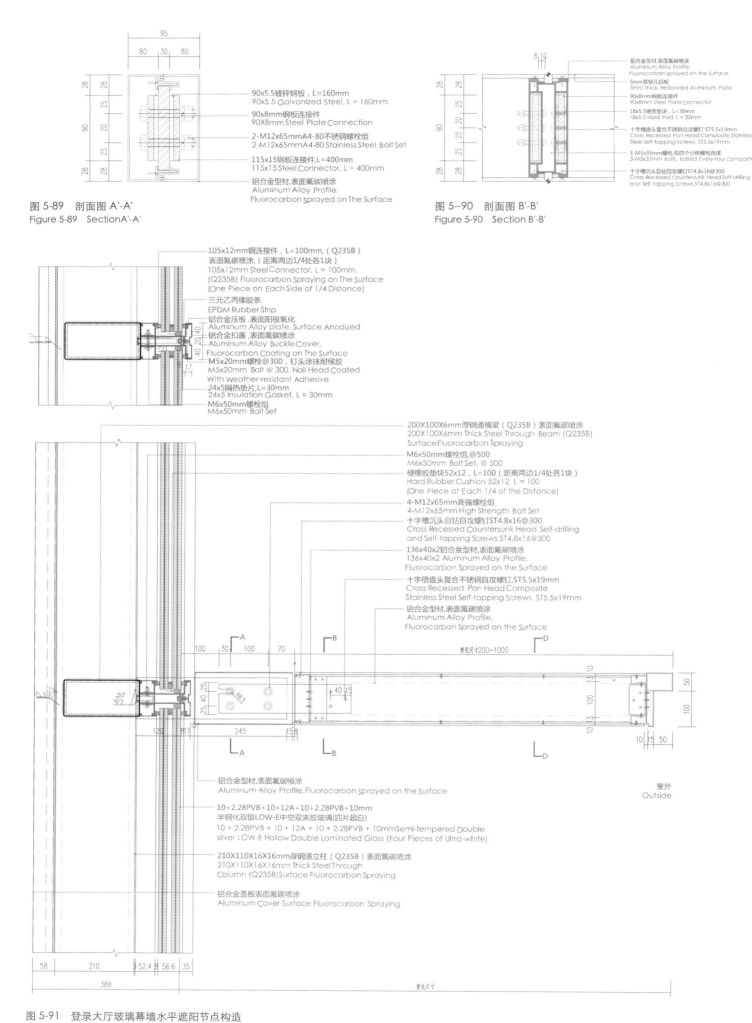

95
80 30 80

90x5.5镀锌钢板，L=160mm
90x5.5 Galvanized Steel, L = 160mm

90x8mm钢板连接件
90X8mm Steel Plate Connection

2-M12x65mmA4-80不锈钢螺栓组
2-M12x65mmA4-80 Stainless Steel Bolt Set

115x15钢板连接件,L=400mm
115x15 Steel Connector, L = 400mm

铝合金型材,表面氟碳喷涂
Aluminum Alloy Profile,
Fluorocarbon Sprayed on The Surface

图 5-89　剖面图 A'-A'
Figure 5-89　SectionA'-A'

铝合金型材,表面氟碳喷涂
Aluminum Alloy Profile,
Fluorocarbon Sprayed on the Surface

5mm厚穿孔铝板
5mm Thick Perforated Aluminum Plate

90x8mm钢板连接件
90X8mm Steel Plate Connector

18x5.5硬质垫块，L=30mm
18x5.5 Hard Pad, L = 30mm

十字槽盘头复合不锈钢自攻螺钉,ST5.5x19mm
Cross Recessed Pan Head Composite Stainless
Steel Self-tapping Screws, ST5.5x19mm

3-M5x35mm螺栓,每四个分格栓连接
3-M5x35mm Bolts, Bolted Every Four Compartments

十字槽沉头自钻自攻螺钉ST4.8x16@300
Cross Recessed Countersunk Head Self-drilling
and Self-tapping Screws ST4.8x16@300

图 5--90　剖面图 B'-B'
Figure 5-90　Section B'-B'

105x12mm钢连接件，L=100mm,（Q235B）
表面氟碳喷涂,（距离两边1/4处各1块）
105x12mm Steel Connector, L = 100mm,
(Q235B) Fluorocarbon Spraying on The Surface
(One Piece on Each Side of 1/4 Distance)

三元乙丙橡胶条
EPDM Rubber Strip

铝合金压板,表面阳极氧化
Aluminum Alloy plate, Surface Anodized

铝合金扣盖,表面氟碳喷涂
Aluminum Alloy Buckle Cover,
Fluorocarbon Coating on The Surface

M5x20mm螺栓@300，钉头涂抹耐候胶
M5x20mm Bolt @ 300, Nail Head Coated
With weather-resistant Adhesive

24x5隔热垫片,L=30mm
24x5 Insulation Gasket, L = 30mm

M6x50mm螺栓组
M6x50mm Bolt Set

200X100X6mm厚钢通横梁（Q235B）表面氟碳喷涂
200X100X6mm Thick Steel Through Beam (Q235B)
Surface Fluorocarbon Spraying

M6x50mm螺栓组,@500
M6x50mm Bolt Set, @ 500

硬橡胶垫块52x12，L=100（距离两边1/4处各1块）
Hard Rubber Cushion 52x12, L = 100
(One Piece at Each 1/4 of the Distance)

4-M12x65mm高强螺栓组
4-M12x65mm High Strength Bolt Set

十字槽沉头自钻自攻螺钉ST4.8x16@300
Cross Recessed Countersunk Head Self-drilling
and Self-tapping Screws ST4.8x16@300

136x40x2铝合金型材,表面氟碳喷涂
136x40x2 Aluminum Alloy Profile,
Fluorocarbon Sprayed on the Surface

十字槽盘头复合不锈钢自攻螺钉,ST5.5x19mm
Cross Recessed Pan Head Composite
Stainless Steel Self-tapping Screws, ST5.5x19mm

铝合金型材,表面氟碳喷涂
Aluminum Alloy Profile,
Fluorocarbon Sprayed on the Surface

变化尺寸200~1000

室外
Outside

铝合金型材,表面氟碳喷涂
Aluminum Alloy Profile,Fluorocarbon Sprayed on the Surface

10+2.28PVB+10+12A+10+2.28PVB+10mm
半钢化双银LOW-E中空双夹胶玻璃(四片超白)
10 + 2.28PVB + 10 + 12A + 10 + 2.28PVB + 10mmSemi-tempered Double
silver LOW-E Hollow Double Laminated Glass (Four Pieces of Ultra-white)

210X110X16X16mm厚钢通立柱（Q235B）表面氟碳喷涂
210X110X16X16mm Thick Steel Through
Column (Q235B)Surface Fluorocarbon Spraying

铝合金盖板表面氟碳喷涂
Aluminum Cover Surface Fluorocarbon Spraying

58　210　52.4　56.6　35
388
变化尺寸

图 5-91　登录大厅玻璃幕墙水平遮阳节点构造
Figure 5-91　Construction of Horizontal Shading Node for Glass Curtain Wall of Arrival Hall

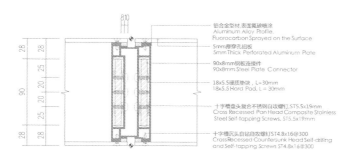

铝合金型材,表面氟碳喷涂
Aluminum Alloy Profile,
Fluorocarbon Sprayed on the Surface

5mm厚穿孔铝板
5mm Thick Perforated Aluminum Plate

90x8mm钢板连接件
90x8mm Steel Plate Connector

18x5.5硬质垫板,L=30mm
18x5.5 Hard Pad, L=30mm

十字槽盘头复合不锈钢自攻螺钉,STS.5x19mm
Cross Recessed Pan Head Composite Stainless
Steel Self-tapping Screws, STS.5x19mm

十字槽沉头自钻自攻螺钉ST4.8x16@300
Cross Recessed Countersunk Head Self-drilling
and Self-tapping Screws ST4.8x16@300

图 5-92 剖面图 B-B
Figure 5-92 Section B-B

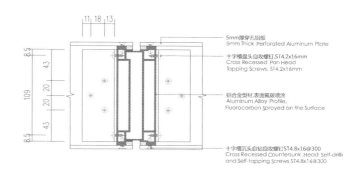

5mm厚穿孔铝板
5mm Thick Perforated Aluminum Plate

十字槽盘头自攻螺钉ST4.2x16mm
Cross Recessed Pan Head
Tapping Screws, ST4.2x16mm

铝合金型材,表面氟碳喷涂
Aluminum Alloy Profile,
Fluorocarbon Sprayed on the Surface

十字槽沉头自钻自攻螺钉ST4.8x16@300
Cross Recessed Countersunk Head Self-drilling
and Self-tapping Screws ST4.8x16@300

图 5-93 剖面图 D-D
Figure 5-93 Section D-D

2-295X20钢板,(Q345B)表面
防火涂料+氟碳喷涂,耐火极限1h
2-295X20 Steel Plate, (Q345B) Surface
Fireproof Coating + Fluorocarbon Spraying,
Fire Resistance Limit 1h

Φ18不锈钢拉索
Ø18 Stainless Steel Cable

315X300X30钢板,(Q345B)表面
防火涂料+氟碳喷涂,耐火极限1h
315X300X30 Steel Plate, (Q345B) Surface
Fireproof Coating + Fluorocarbon Spraying,
Fire Resistance Limit 1h

图 5-94 剖面图 A-A
Figure 5-94 Section A-A

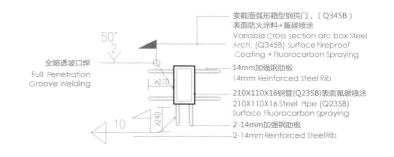

变截面弧形箱型钢拱门,(Q345B)
表面防火涂料+氟碳喷涂
Variable Cross Section arc box Steel
Arch, (Q345B) Surface Fireproof
Coating + Fluorocarbon Spraying

14mm加强钢肋板
14mm Reinforced Steel Rib

210X110X16钢管(Q235B)表面氟碳喷涂
210X110X16 Steel Pipe (Q235B)
Surface Fluorocarbon spraying

2-14mm加强钢肋板
2-14mm Reinforced Steel Rib

全熔透坡口焊
Full Penetration
Groove Welding

图 5-95 剖面图 C-C
Figure 5-95 Section C-C

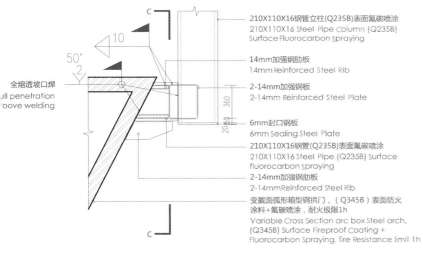

210X110X16钢管立柱(Q235B)表面氟碳喷涂
210X110X16 Steel Pipe Column (Q235B)
Surface Fluorocarbon spraying

14mm加强钢肋板
14mm Reinforced Steel Rib

2-14mm加强钢板
2-14mm Reinforced Steel Plate

6mm封口钢板
6mm Sealing Steel Plate

210X110X16钢管(Q235B)表面氟碳喷涂
210X110X16 Steel Pipe (Q235B) Surface
fluorocarbon spraying

2-14mm加强钢肋板
2-14mm Reinforced Steel Rib

变截面弧形箱型钢拱门,(Q345B)表面防火
涂料+氟碳喷涂,耐火极限1h
Variable Cross Section arc box Steel arch,
(Q345B) Surface Fireproof Coating +
Fluorocarbon Spraying, fire Resistance limit 1h

全熔透坡口焊
full penetration
groove welding

图 5-96 大面幕墙底部连接节点图
Figure 5-96 Large-scale Curtain Wall Bottom Connection Node

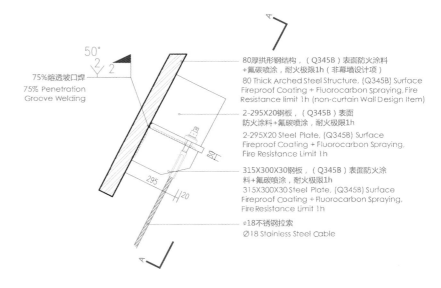

80厚拱形钢结构,(Q345B)表面防火涂料
+氟碳喷涂,耐火极限1h(非幕墙设计项)
80 Thick Arched Steel Structure, (Q345B) Surface
Fireproof Coating + Fluorocarbon Spraying, Fire
Resistance limit 1h (non-curtain Wall Design Item)

2-295X20钢板,(Q345B)表面
防火涂料+氟碳喷涂,耐火极限1h
2-295X20 Steel Plate, (Q345B) Surface
Fireproof Coating + Fluorocarbon Spraying,
Fire Resistance Limit 1h

315X300X30钢板,(Q345B)表面防火涂
料+氟碳喷涂,耐火极限1h
315X300X30 Steel Plate, (Q345B) Surface
Fireproof Coating + Fluorocarbon Spraying,
Fire Resistance Limit 1h

Φ18不锈钢拉索
Ø18 Stainless Steel Cable

75%熔透坡口焊
75% Penetration
Groove Welding

图 5-97 Φ18 拉索支座示意图
Figure 5-97 Schematic Diagram of Φ18 Cable Support

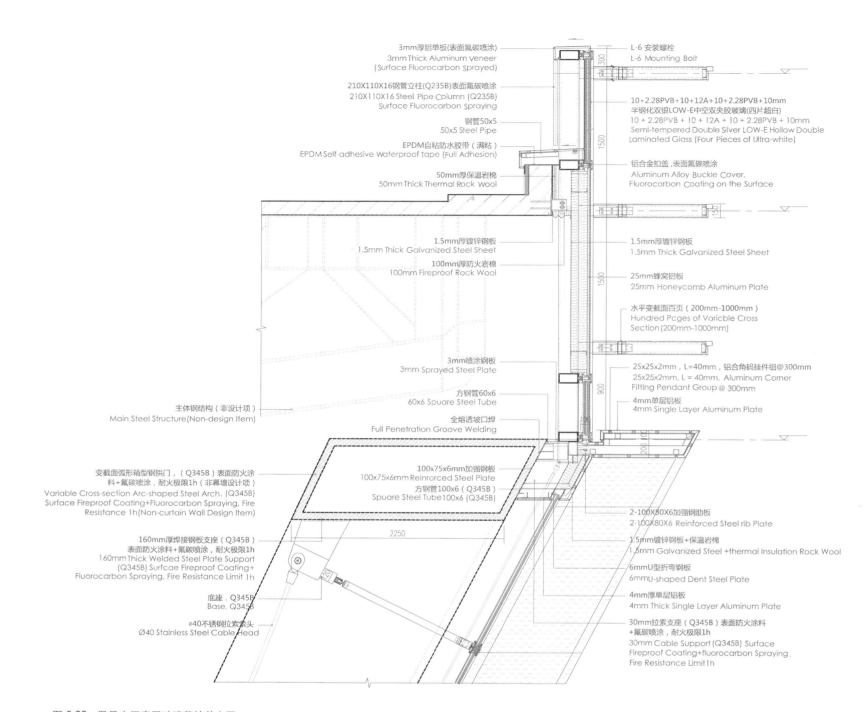

3mm厚铝单板(表面氟碳喷涂)
3mm Thick Aluminum Veneer (Surface Fluorocarbon Sprayed)

210X110X16钢管立柱(Q235B)表面氟碳喷涂
210X110X16 Steel Pipe Column (Q235B) Surface Fluorocarbon Spraying

钢管50x5
50x5 Steel Pipe

EPDM自粘防水胶带（满粘）
EPDM Self-adhesive Waterproof Tape (Full Adhesion)

50mm厚保温岩棉
50mm Thick Thermal Rock Wool

1.5mm厚镀锌钢板
1.5mm Thick Galvanized Steel Sheet

100mm厚防火岩棉
100mm Fireproof Rock Wool

3mm喷涂钢板
3mm Sprayed Steel Plate

方钢管60x6
60x6 Spuare Steel Tube

主体钢结构（非设计项）
Main Steel Structure(Non-design Item)

全熔透坡口焊
Full Penetration Groove Welding

变截面弧形箱型钢拱门，（Q345B）表面防火涂料+氟碳喷涂，耐火极限1h（非幕墙设计项）
Variable Cross-section Arc-shaped Steel Arch. (Q345B) Surface Fireproof Coating+Fluorocarbon Spraying, Fire Resistance 1h(Non-curtain Wall Design Item)

160mm厚焊接钢板支座（Q345B）表面防火涂料+氟碳喷涂，耐火极限1h
160mm Thick Welded Steel Plate Support (Q345B) Surfcae Fireproof Coating+ Fluorocarbon Spraying, Fire Resistance Limit 1h

底座，Q345B
Base. Q345B

Ø40不锈钢拉索索头
Ø40 Stainless Steel Cable Head

L-6 安装螺栓
L-6 Mounting Bolt

10+2.28PVB+10+12A+10+2.28PVB+10mm
半钢化双银LOW-E中空双夹胶玻璃(四片超白)
10 + 2.28PVB + 10 + 12A + 10 + 2.28PVB + 10mm
Semi-tempered Double Silver LOW-E Hollow Double Laminated Glass (Four Pieces of Ultra-white)

铝合金扣盖，表面氟碳喷涂
Aluminum Alloy Buckle Cover, Fluorocarbon Coating on the Surface

1.5mm厚镀锌钢板
1.5mm Thick Galvanized Steel Sheet

25mm蜂窝铝板
25mm Honeycomb Aluminum Plate

水平变截面百页（200mm-1000mm）
Hundred Pcges of Varicble Cross Section (200mm-1000mm)

25x25x2mm，L=40mm，铝合角码挂件组@300mm
25x25x2mm, L = 40mm, Aluminum Corner Fitting Pendant Group @ 300mm

4mm单层铝板
4mm Single Layer Aluminum Plate

2-100X80X6加强钢肋板
2-100X80X6 Reinforced Steel rib Plate

1.5mm镀锌钢板+保温岩棉
1.5mm Galvanized Steel +thermal Insulation Rock Wool

6mmU型折弯钢板
6mmU-shaped Dent Steel Plate

4mm厚单层铝板
4mm Thick Single Layer Aluminum Plate

30mm拉索支座（Q345B）表面防火涂料+氟碳喷涂，耐火极限1h
30mm Cable Support (Q345B) Surface Fireproof Coating+fluorocarbon Spraying, Fire Resistance Limit 1h

100x75x6mm加强钢板
100x75x6mm Reinforced Steel Plate

方钢管100x6（Q345B）
Spuare Steel Tube 100x6 (Q345B)

图 5-98　登录大厅索网玻璃幕墙节点图
Figure 5-98　The Cable Net Glass Curtain Wall Node Diagram of the Arrival Hall

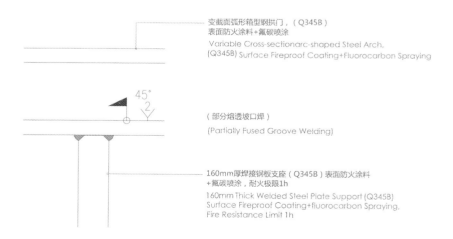

变截面弧形箱型钢拱门，（Q345B）表面防火涂料+氟碳喷涂
Variable Cross-sectionarc-shaped Steel Arch, (Q345B) Surface Fireproof Coating+Fluorocarbon Spraying

（部分熔透坡口焊）
(Partially Fused Groove Welding)

160mm厚焊接钢板支座（Q345B）表面防火涂料+氟碳喷涂，耐火极限1h
160mm Thick Welded Steel Plate Support (Q345B) Surface Fireproof Coating+fluorocarbon Spraying, Fire Resistance Limit 1h

图 5-99　160mm 厚拉锁耳板焊接示意图
Figure 5-99　Welding Schematic Diagram of 160mm Thick Zipper Lug

展厅屋面系统设计
Exhibition Hall Ro of System Design

系统介绍

展厅屋面系统采用一级防水屋面构造：镀铝锌钢直立锁边防水屋面系统加装饰面板的构造系统。构造层次由内到外分别为：0.8mm 镀铝锌穿孔压型钢板；吸音玻璃纤维布；100mm 吸音棉屋面次檩条容重 32K；1.0mm 镀铝锌压型钢板；0.3mm 隔气膜；100mm 保温岩棉容重 180K；1.5mm 镀锌钢板；1.5mmTPO（二次防水层）；0.8mm 镀铝锌氟碳预辊涂直立锁边钢屋面板（防水层）；铝合金支撑檩条及夹具；3.0mm 外装饰铝板（装饰层）；

项目位处海边，加上深圳基本风压为 0.75kpa。项目所在地的地面粗糙度为 A 类。由于项目的重要性，业主和我公司均建议项目的风荷载重现期按 100 年一遇考虑。但项目建筑众多（16 个标准展厅及一个非标展厅）组相互影响较大，通过风动试验确认的风荷载值在边角风荷载最大吸力约有 6.5KPa（大大高于规范的标准取值 5.0KPa）；为了抵抗抗风掀能力，在边角部我们也采用了加强的设计措施保证屋面系统的整体，局部均满足项目的抗风性能需求（图 5-100）。

项目立项之初，屋面系统考虑在铝合金屋面系统和钢屋面系统中选取。后来综合若干次专家会议的意见，及相关项目的考察，最终项目决定采用抗风掀能力更好的钢屋面系统。其在施工期间也经历了台风"山竹"的考验，证明了系统抗风掀能力的可靠性。

System Introduction

The roof system of exhibition hall adopts the class A waterproof roof structure: aluminum zinc steel standing seam waterproof roof system with decorative panel construction. The structural layers from inside to outside are: 0.8mm aluminum-zinc perforated pressed steel plate; sound-absorbing glass fiber cloth; 100mm sound-absorbing cotton roof secondary purlin with unit weight of 32K; 1.0mm aluminum-zinc pressed steel plate; 0.3mm gas isolation film; 100mm heat preservation rock wool with unit weight of 180K; 1.5mm aluminum-zinc steel plate; 1.5mmTPO (secondary waterproof layer); 0.8mm aluminum zinc standing seam steel roofing board (waterproof layer); aluminum alloy supporting purlin and clamp; 3.0mm outer decorative aluminum plate (decorative layer).

The project is located at the seaside and the basic wind pressure in Shenzhen is 0.75kpa. The surface roughness of the project site is class A. Due to the importance of the project, both the owner and our company suggest that the current wind-load period of the project should be considered on a 100-year basis. However, due to the number of the buildings (16 standard exhibition halls and one non-standard exhibition hall) and their great mutal influence on each other, the maximum suction of wind load confirmed by the wind dynamic test is about 6.5kpa (significantly higher than the standard value of 5.0kpa). In order to resist the wind, we also adopted strengthened design measures at the corners to guarantee the ablity of wind resistance of the whole roof system. The following figure shows the wind load distribution of the project roof(Figure 5-100).

In the early stage of this project, we considered to select the roof system between aluminum alloy roof system and steel roof system. Later, based on the opinions of several expert meetings and the investigation of related projects, the steel roof system with better wind resistance was adopted. During the construction, it also experienced the test of typhoon "Mangosteen", which proved the reliability of the system against the wind.

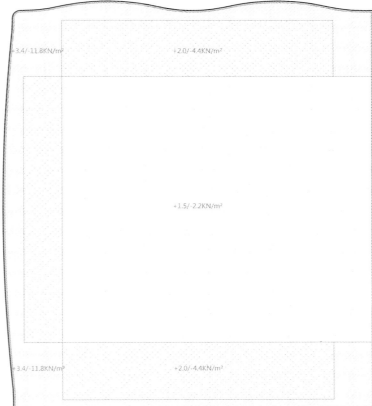

图 5-100　屋顶风洞试验风压平面图（100 年一遇）
Figure 5-100　Roof Wind Tunnel Test: Wind Pressure Plan (Once Every 100 Years)

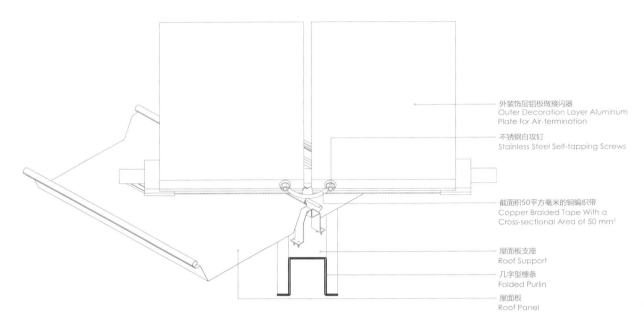

外装饰层铝板做接闪器
Outer Decoration Layer Aluminum
Plate for Air-termination

不锈钢自攻钉
Stainless Steel Self-tapping Screws

截面积50平方毫米的铜编织带
Copper Braided Tape With a
Cross-sectional Area of 50 mm²

屋面板支座
Roof Support

几字型檩条
Folded Purlin

屋面板
Roof Panel

图 5-101 展厅屋面 A 防雷节点图
Figure 5-101 A Lightning Protection Joint on the Roof of the Exhibition Hall

图 5-103 剖面图 B-B
Figure 5-103 Section B-B

3.0mm 厚外装饰层铝板（外装饰层）
3.0mm Thick Outer Ddecorative Layer
Aluminum Plate (Outer Decorctive Layer)

铝合金支承檩条及连接夹具（外装饰支承层）
Aluminum Alloy Supporting Purlin And Connecting
Fixture (External Decorction Supporting Layer)

0.8 毫米厚镀铝锌 PVDF 扣合式屋面防水板（防水层）
0.8mm Thick Aluminum Zinc Coated PVDF Snap
on Roof Waterproof Board (Waterproof Layer)

1.5mm TPO 防水卷材（二次防水层）
1.5mm tpo Waterproof roll (Secondary Waterproof Layer)

1.5mm 镀锌平钢板（隔声降噪层）
1.5mm Galvanized Flat Steel Plate
(Sound Insulation and Noise Reduction Layer)

50 保温岩棉（容重 180kg/m³）+ 50 保温岩棉（容重 120kg/m³）
50 Insulating Rock Wool (Volume Weight 180kg / m³)
+ 50 Insulating Rock Wool (Volume Weight 120kg / m³)

0.3mm 隔汽膜（防潮层）
0.3mm Vapor Barrier (Moisture Barrier)

1.0mm 厚镀铝锌压型钢板（支撑层）
1.0mm Thick Aluminum Zinc Coated Profiled Steel Plate (Support Layer)

屋面次檩条镀锌矩形钢管 120x60x4mm
Roof Secondary Purlin Galvanized Rectangular Steel Pipe 120x60x4mm

100mm 吸音棉（吸音层，容重 32kg/m³）
100mm Sound-absorbing Cotton (Sound-absorbing Layer, Unit Weight 32kg /

吸音玻璃纤维布（兼做防尘层）
Sound Absorbing Fiberglass Cloth (Also Used as Dust-proof Layer)

0.8mm 厚镀铝锌压型钢板（吸声吊顶支承层）
0.8mm Thick Aluminum Zinc Plated Profiled Steel Plate
(acoustic ceiling supporting layer)

3.0mm 厚外装饰层铝板（镂空区域）
3.0mm Thick Outer Decorative Aluminum Plate (Hollow Area)

25mm 檐口蜂窝铝板
25mm Cornice Honeycomb Aluminum Plate

1.5mmTPO 防水卷材
1.5mmtpo Waterproof Roll

3mm 厚不锈钢天沟
3mm Thick Stainless Steel Gutter

4*50mm 厚保温岩棉
（上层 50mm 容重 180kg/m³，下层 3*50mm 容重 120kg/m³）
4 * 50mm Thick Insulation Rock Wool
(Volume Weight of Upper Layer 50mm 180kg / m³,
Volume Weight of Lower Layer 3 * 50mm 120kg / m³)

0.3mm 隔汽膜（防潮层）
0.3mm Vapor Barrier (Moisture Barrier)

1.0mm840 镀铝锌压型钢底板
1.0mm840 Aluminum Plated Zinc Profiled Steel Base Plate

1684 1318

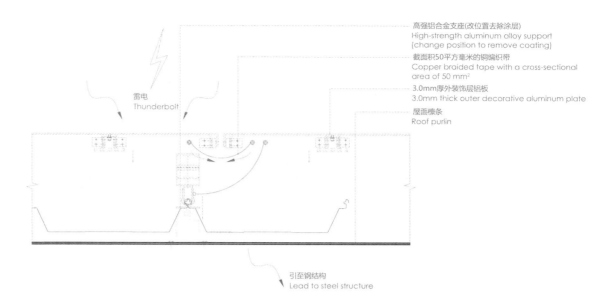

高强铝合金支座(改位置去除涂层)
High-strength aluminum alloy support
(change position to remove coating)

截面积50平方毫米的铜编织带
Copper braided tape with a cross-sectional
area of 50 mm²

3.0mm厚外装饰层铝板
3.0mm thick outer decorative aluminum plate

屋面檩条
Roof purlin

雷电
Thunderbolt

引至钢结构
Lead to steel structure

图 5-102 展厅屋面 A 防雷节点图 2
Figure 5-102 A lightning Protection Joint on the Roof of the Exhibition Hall

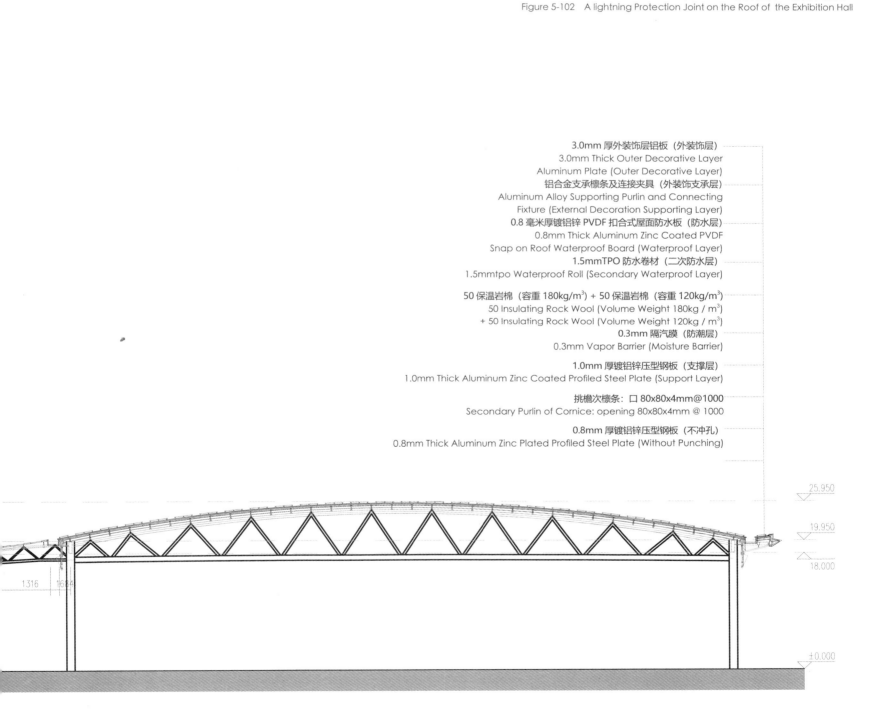

3.0mm 厚外装饰层铝板（外装饰层）
3.0mm Thick Outer Decorative Layer
Aluminum Plate (Outer Decorative Layer)

铝合金支承檩条及连接夹具（外装饰支承层）
Aluminum Alloy Supporting Purlin and Connecting
Fixture (External Decoration Supporting Layer)

0.8 毫米厚镀铝锌 PVDF 扣合式屋面防水板（防水层）
0.8mm Thick Aluminum Zinc Coated PVDF
Snap on Roof Waterproof Board (Waterproof Layer)

1.5mmTPO 防水卷材（二次防水层）
1.5mmtpo Waterproof Roll (Secondary Waterproof Layer)

50 保温岩棉（容重 180kg/m³）+ 50 保温岩棉（容重 120kg/m³）
50 Insulating Rock Wool (Volume Weight 180kg / m³)
+ 50 Insulating Rock Wool (Volume Weight 120kg / m³)

0.3mm 隔汽膜（防潮层）
0.3mm Vapor Barrier (Moisture Barrier)

1.0mm 厚镀铝锌压型钢板（支撑层）
1.0mm Thick Aluminum Zinc Coated Profiled Steel Plate (Support Layer)

挑檐次檩条：口 80x80x4mm@1000
Secondary Purlin of Cornice: opening 80x80x4mm @ 1000

0.8mm 厚镀铝锌压型钢板（不冲孔）
0.8mm Thick Aluminum Zinc Plated Profiled Steel Plate (Without Punching)

25.950

19.950

18.000

1316 16

±0.000

图 5-104　展厅屋面 A 节点详图
Figure 5-104　Detail of Node A of Exhibition Hall Roof

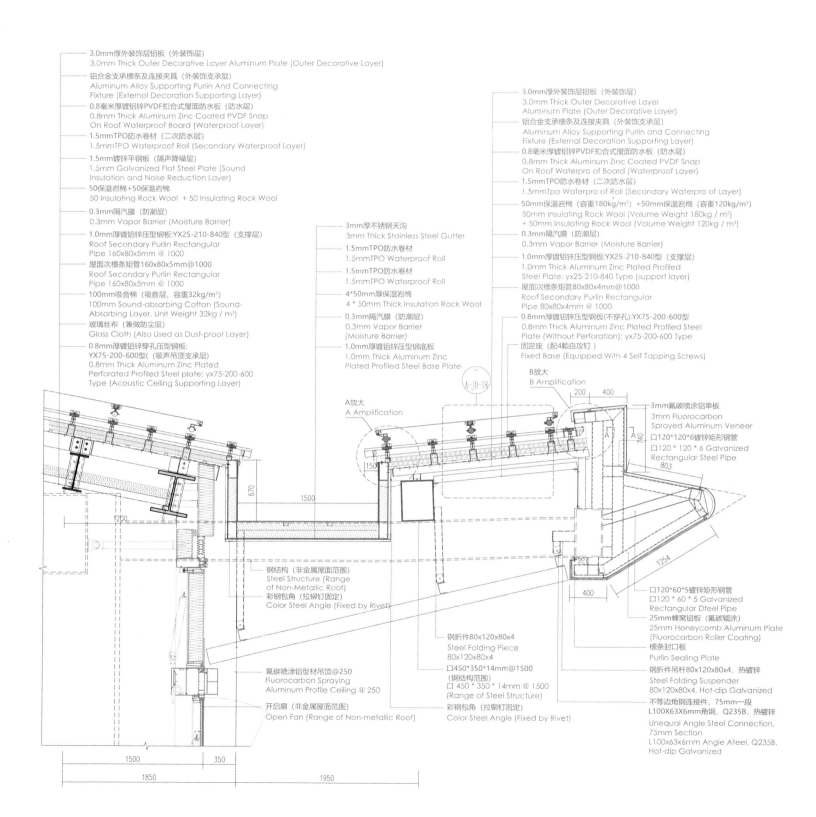

3.0mm厚外装饰层铝板（外装饰层）
3.0mm Thick Outer Decorative Layer Aluminum Plate (Outer Decorative Layer)

铝合金支承檩条及连接夹具（外装饰支承层）
Aluminum Alloy Supporting Purlin And Connecting Fixture (External Decoration Supporting Layer)

0.8毫米厚镀锌PVDF扣合式屋面防水板（防水层）
0.8mm Thick Aluminum Zinc Coated PVDF Snap On Roof Waterproof Board (Waterproof Layer)

1.5mmTPO防水卷材（二次防水层）
1.5mmTPO Waterproof Roll (Secondary Waterproof Layer)

1.5mm镀锌平钢板（隔声降噪层）
1.5mm Galvanized Flat Steel Plate (Sound Insulation and Noise Reduction Layer)

50保温岩棉+50保温岩棉
50 Insulating Rock Wool + 50 Insulating Rock Wool

0.3mm隔汽膜（防潮层）
0.3mm Vapor Barrier (Moisture Barrier)

1.0mm厚镀锌铝型压型钢板:YX25-210-840型（支撑层）
Roof Secondary Purlin Rectangular Pipe 160x80x5mm @ 1000

屋面次檩条矩管160x80x5mm@1000
Roof Secondary Purlin Rectangular Pipe 160x80x5mm @ 1000

100mm吸音棉（吸音层，容重32kg/m³)
100mm Sound-absorbing Cotton (Sound-Absorbing Layer, Unit Weight 32kg / m³)

玻璃丝布（兼做防尘层）
Glass Cloth (Also Used as Dust-proof Layer)

0.8mm厚镀锌穿孔压型钢板:
YX75-200-600型（吸声吊顶支承层）
0.8mm Thick Aluminum Zinc Plated Perforated Profiled Steel plate: yx75-200-600 Type (Acoustic Ceiling Supporting Layer)

3mm厚不锈钢天沟
3mm Thick Stainless Steel Gutter

1.5mmTPO防水卷材
1.5mmTPO Waterproof Roll

1.5mmTPO防水卷材
1.5mmTPO Waterproof Roll

4*50mm厚保温岩棉
4 * 50mm Thick Insulation Rock Wool

0.3mm隔汽膜（防潮层）
0.3mm Vapor Barrier (Moisture Barrier)

1.0mm厚镀锌铝型压型钢底板
1.0mm Thick Aluminum Zinc Plated Profiled Steel Base Plate

3.0mm厚外装饰层铝板（外装饰层）
3.0mm Thick Outer Decorative Layer Aluminum Plate (Outer Decorative Layer)

铝合金支承檩条及连接夹具（外装饰支承层）
Aluminum Alloy Supporting Purlin and Connecting Fixture (External Decoration Supporting Layer)

0.8毫米厚镀锌PVDF扣合式屋面防水板（防水层）
0.8mm Thick Aluminum Zinc Coated PVDF Snap On Roof Waterpro of Board (Waterproof Layer)

1.5mmTPO防水卷材（二次防水层）
1.5mmTpo Waterpro of Roll (Secondary Waterpro of Layer)

50mm保温岩棉（容重180kg/m³) +50mm保温岩棉（容重120kg/m³)
50mm Insulating Rock Wool (Volume Weight 180kg / m³) + 50mm Insulating Rock Wool (Volume Weight 120kg / m³)

0.3mm隔汽膜（防潮层）
0.3mm Vapor Barrier (Moisture Barrier)

1.0mm厚镀锌铝型压型钢板:YX25-210-840型（支撑层）
1.0mm Thick Aluminum Zinc Plated Profiled Steel Plate: yx25-210-840 Type (support layer)

屋面次檩条矩管80x80x4mm@1000
Roof Secondary Purlin Rectangular Pipe 80x80x4mm @ 1000

0.8mm厚镀锌铝锌压型钢板(不穿孔):YX75-200-600型
0.8mm Thick Aluminum Zinc Plated Profiled Steel Plate (Without Perforation): yx75-200-600 Type

固定座（配4颗自攻钉）
Fixed Base (Equipped With 4 Self Tapping Screws)

A放大
A Amplification

B放大
B Amplification

3mm氟碳喷涂铝单板
3mm Fluorocarbon Sprayed Aluminum Veneer

口120*120*6镀锌矩形钢管
口120 * 120 * 6 Galvanized Rectangular Steel Pipe

钢结构（非金属屋面范围）
Steel Structure (Range of Non-Metallic Roof)

彩钢包角（拉铆钉固定）
Color Steel Angle (Fixed by Rivet)

钢折件80x120x80x4
Steel Folding Piece 80x120x80x4

口450*350*14mm@1500（钢结构范围）
口 450 * 350 * 14mm @ 1500 (Range of Steel Structure)

彩钢包角（拉铆钉固定）
Color Steel Angle (Fixed by Rivet)

氟碳喷涂铝型材吊顶@250
Fluorocarbon Spraying Aluminum Profile Ceiling @ 250

开启扇（非金属屋面范围）
Open Fan (Range of Non-metallic Roof)

口120*60*5镀锌矩形钢管
口120 * 60 * 5 Galvanized Rectangular Dteel Pipe

25mm蜂窝铝板（氟碳辊涂）
25mm Honeycomb Aluminum Plate (Fluorocarbon Roller Coating)

檩条封口板
Purlin Sealing Plate

钢折件吊杆80x120x80x4，热镀锌
Steel Folding Suspender 80x120x80x4, Hot-dip Galvanized

不等边角钢连接件，75mm一段
L100X63X6mm角钢，Q235B，热镀锌
Unequal Angle Steel Connection, 75mm Section
L100x63x6mm Angle Ateel, Q235B, Hot-dip Galvanized

图 5-105 A 节点放大图
Figure 5-105 Node A Detail

不锈钢铆钉@300
Stainless Steel Rivets @ 300

ST5.5X25不锈钢复合自攻自钻钉 @500
ST5.5X25 Stainless Steel Composite
Self-drilling Self-drilling Nails @ 500

檐口滴水片（通长）
Cornice Drip Sheet (Through Length)

0.8mm镀铝锌泛水板（材质同屋面板）
0.8mm Aluminized Zinc Flooding Board
(Same Material as Roof Panel)

2.0mm厚镀锌钢折件（通长）
2.0mm Thick Galvanized Steel Folding
Piece (through length)

铝合金型材(氟碳喷涂)
Aluminum Alloy Profile (Fluorocarbon Spray)

25mm蜂窝铝板
25mm Honeycomb Aluminum Plate

2mm厚橡胶垫片
2mm Thick Rubber Gasket

ST5.5*22不锈钢自攻自钻钉
ST5.5 * 22 Stainless Steel Self-tapping
and Self-drilling nails

口120*60*5镀锌矩形钢管
120 * 60 * 5 Galvanized Rectangular
Steel Pipe

图 5-106 剖面图 B-B
Figure 5-106 Section B-B

3mm厚铝单板（氟碳喷涂）
3mm Thick Aluminum Veneer
(Fluorocarbon Sprayed)

铝合金背筋@500mm
Aluminum Back Bar @ 500mm

图 5-107 加强筋布置图
Figure 5-107 Layout of Reinforcement

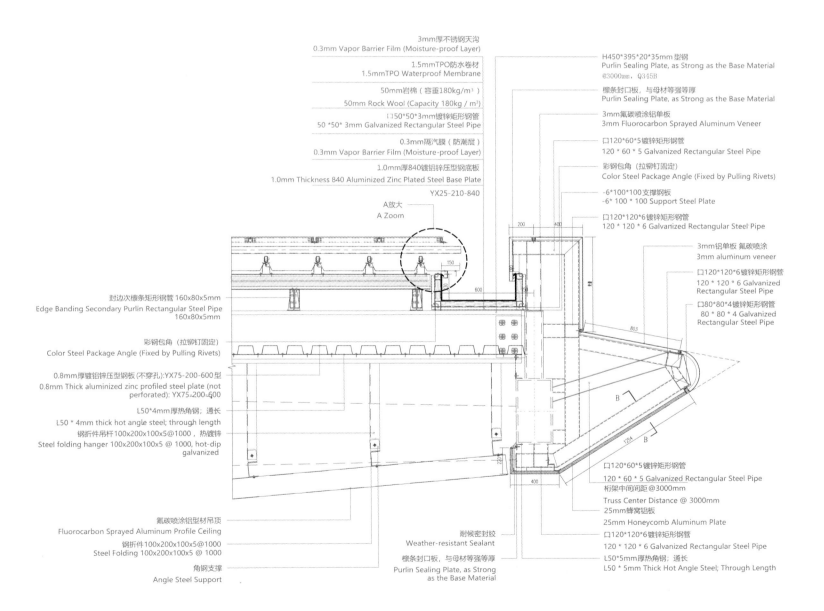

3mm厚不锈钢天沟
0.3mm Vapor Barrier Film (Moisture-proof Layer)

1.5mmTPO防水卷材
1.5mmTPO Waterproof Membrane

50mm岩棉（容重180kg/m³）
50mm Rock Wool (Capacity 180kg / m³)

口150*50*3镀锌矩形钢管
50 *50* 3 Galvanized Rectangular Steel Pipe

0.3mm隔汽膜（防潮层）
0.3mm Vapor Barrier Film (Moisture-proof Layer)

1.0mm厚840镀铝锌压型钢底板
1.0mm Thickness 840 Aluminized Zinc Plated Steel Base Plate
YX25-210-840

A放大
A Zoom

封边次檩条矩形钢管 160x80x5mm
Edge Banding Secondary Purlin Rectangular Steel Pipe
160x80x5mm

彩钢包角（拉铆钉固定）
Color Steel Package Angle (Fixed by Pulling Rivets)

0.8mm厚镀锌铝压型钢板（不穿孔）:YX75-200-600型
0.8mm Thick aluminized zinc profiled steel plate (not
perforated): YX75-200-600

L50*4mm厚热角钢；通长
L50 * 4mm thick hot angle steel; through length
钢折件吊杆100x200x100x5@1000，热镀锌
Steel folding hanger 100x200x100x5 @ 1000, hot-dip
galvanized

氟碳喷涂铝型材吊顶
Fluorocarbon Sprayed Aluminum Profile Ceiling

钢折件100x200x100x5@1000
Steel Folding 100x200x100x5 @ 1000

角钢支撑
Angle Steel Support

耐候密封胶
Weather-resistant Sealant

檩条封口板，与母材等强等厚
Purlin Sealing Plate, as Strong
as the Base Material

H450*395*20*35mm 型钢
Purlin Sealing Plate, as Strong as the Base Material
@3000mm，Q345B

檩条封口板，与母材等强等厚
Purlin Sealing Plate, as Strong as the Base Material

3mm氟碳喷涂铝单板
3mm Fluorocarbon Sprayed Aluminum Veneer

口120*60*5镀锌矩形钢管
120 * 60 * 5 Galvanized Rectangular Steel Pipe

彩钢包角（拉铆钉固定）
Color Steel Package Angle (Fixed by Pulling Rivets)

-6*100*100支撑钢板
-6* 100 * 100 Support Steel Plate

口120*120*6镀锌矩形钢管
120 * 120 * 6 Galvanized Rectangular Steel Pipe

3mm铝单板 氟碳喷涂
3mm aluminum veneer

口120*120*6镀锌矩形钢管
120 * 120 * 6 Galvanized
Rectangular Steel Pipe

口80*80*4镀锌矩形钢管
80 * 80 * 4 Galvanized
Rectangular Steel Pipe

口120*60*5镀锌矩形钢管
120 * 60 * 5 Galvanized Rectangular Steel Pipe
桁架中间间距@3000mm
Truss Center Distance @ 3000mm

25mm蜂窝铝板
25mm Honeycomb Aluminum Plate

口120*120*6镀锌矩形钢管
120 * 120 * 6 Galvanized Rectangular Steel Pipe

L50*5mm厚热角钢；通长
L50 * 5mm Thick Hot Angle Steel; Through Length

图 5-108 展厅檐口排水天沟节点构造2
Figure 5-108 Detail of Gutter Joint in the Eaves of Exhibition Hall 2

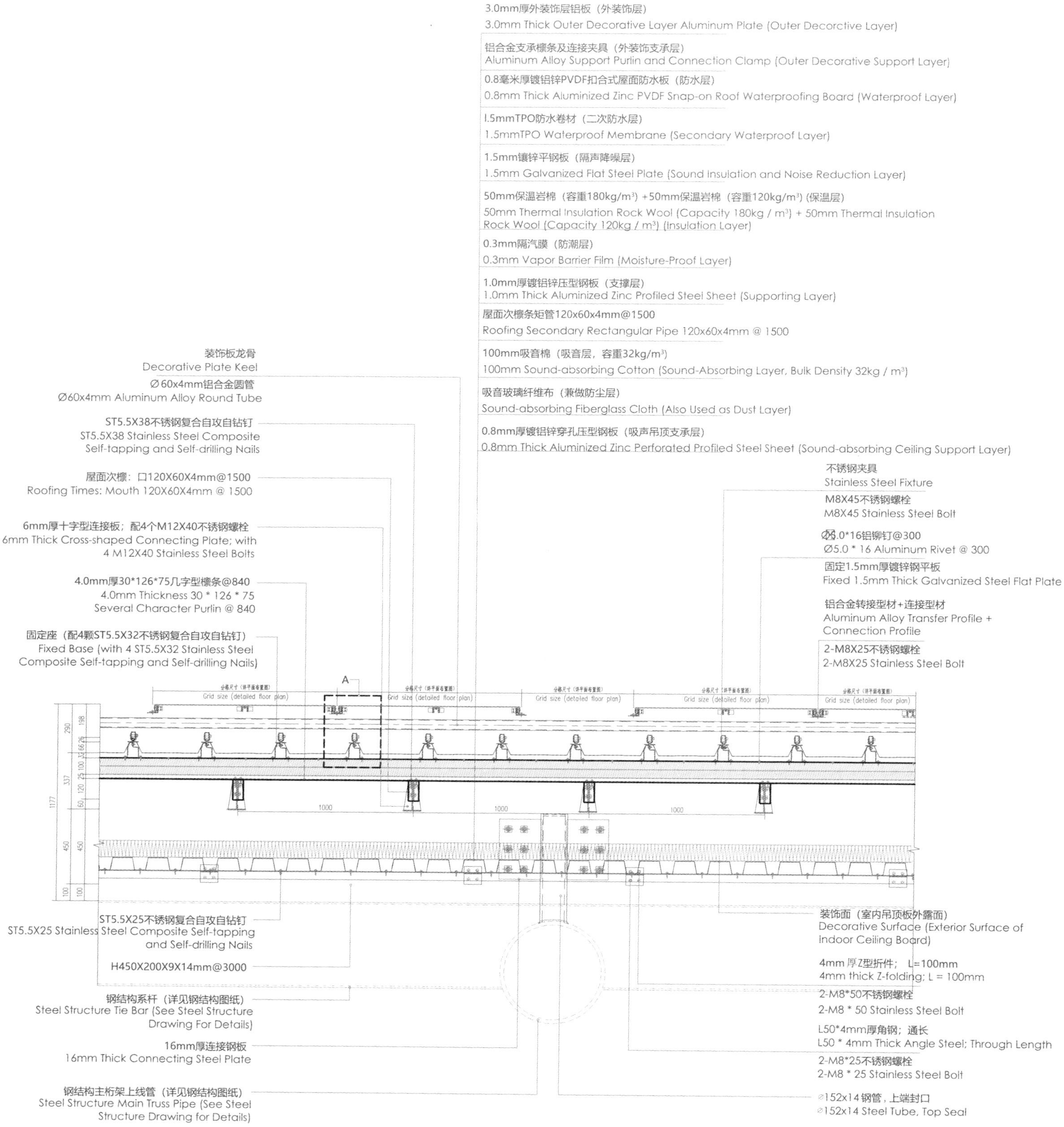

3.0mm厚外装饰层铝板（外装饰层）
3.0mm Thick Outer Decorative Layer Aluminum Plate (Outer Decorctive Layer)

铝合金支承檩条及连接夹具（外装饰支承层）
Aluminum Alloy Support Purlin and Connection Clamp (Outer Decorative Support Layer)

0.8毫米厚镀铝锌PVDF扣合式屋面防水板（防水层）
0.8mm Thick Aluminized Zinc PVDF Snap-on Roof Waterproofing Board (Waterproof Layer)

1.5mmTPO防水卷材（二次防水层）
1.5mmTPO Waterproof Membrane (Secondary Waterproof Layer)

1.5mm镀锌平钢板（隔声降噪层）
1.5mm Galvanized Flat Steel Plate (Sound Insulation and Noise Reduction Layer)

50mm保温岩棉（容重180kg/m³）+50mm保温岩棉（容重120kg/m³）（保温层）
50mm Thermal Insulation Rock Wool (Capacity 180kg / m³) + 50mm Thermal Insulation Rock Wool (Capacity 120kg / m³) (Insulation Layer)

0.3mm隔汽膜（防潮层）
0.3mm Vapor Barrier Film (Moisture-Proof Layer)

1.0mm厚镀铝锌压型钢板（支撑层）
1.0mm Thick Aluminized Zinc Profiled Steel Sheet (Supporting Layer)

屋面次檩条矩管120x60x4mm@1500
Roofing Secondary Rectangular Pipe 120x60x4mm @ 1500

100mm吸音棉，容重32kg/m³）
100mm Sound-absorbing Cotton (Sound-Absorbing Layer, Bulk Density 32kg / m³)

吸音玻璃纤维布（兼做防尘层）
Sound-absorbing Fiberglass Cloth (Also Used as Dust Layer)

0.8mm厚镀铝锌穿孔压型钢板（吸音吊顶支承层）
0.8mm Thick Aluminized Zinc Perforated Profiled Steel Sheet (Sound-absorbing Ceiling Support Layer)

装饰板龙骨
Decorative Plate Keel

Ø60x4mm铝合金圆管
Ø60x4mm Aluminum Alloy Round Tube

ST5.5X38不锈钢复合自攻自钻钉
ST5.5X38 Stainless Steel Composite Self-tapping and Self-drilling Nails

屋面次檩：口120X60X4mm@1500
Roofing Times: Mouth 120X60X4mm @ 1500

6mm厚十字型连接板；配4个M12X40不锈钢螺栓
6mm Thick Cross-shaped Connecting Plate; with 4 M12X40 Stainless Steel Bolts

4.0mm厚30*126*75几字型檩条@840
4.0mm Thickness 30 * 126 * 75 Several Character Purlin @ 840

固定座（配4颗ST5.5X32不锈钢复合自攻自钻钉）
Fixed Base (with 4 ST5.5X32 Stainless Steel Composite Self-tapping and Self-drilling Nails)

不锈钢夹具
Stainless Steel Fixture

M8X45不锈钢螺栓
M8X45 Stainless Steel Bolt

Ø5.0*16铝铆钉@300
Ø5.0 * 16 Aluminum Rivet @ 300

固定1.5mm厚镀锌钢平板
Fixed 1.5mm Thick Galvanized Steel Flat Plate

铝合金转接型材+连接型材
Aluminum Alloy Transfer Profile + Connection Profile

2-M8X25不锈钢螺栓
2-M8X25 Stainless Steel Bolt

A

分格尺寸（详平面布置图）
Grid size (detailed floor plan)

ST5.5X25不锈钢复合自攻自钻钉
ST5.5X25 Stainless Steel Composite Self-tapping and Self-drilling Nails

H450X200X9X14mm@3000

钢结构系杆（详见钢结构图纸）
Steel Structure Tie Bar (See Steel Structure Drawing For Details)

16mm厚连接钢板
16mm Thick Connecting Steel Plate

钢结构主桁架上线管（详见钢结构图纸）
Steel Structure Main Truss Pipe (See Steel Structure Drawing for Details)

装饰面（室内吊顶板外露面）
Decorative Surface (Exterior Surface of Indoor Ceiling Board)

4mm厚Z型折件；L=100mm
4mm thick Z-folding; L = 100mm

2-M8*50不锈钢螺栓
2-M8 * 50 Stainless Steel Bolt

L50*4mm厚角钢；通长
L50 * 4mm Thick Angle Steel; Through Length

2-M8*25不锈钢螺栓
2-M8 * 25 Stainless Steel Bolt

⌀152x14钢管，上端封口
⌀152x14 Steel Tube, Top Seal

图 5-109 展厅屋面基本节点构造图
Figure 5-109 Basic Node Structure of the Exhibition Hall Roof

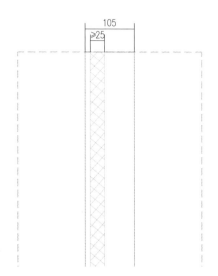

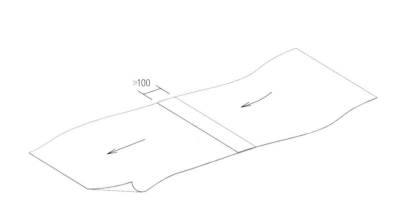

图 5-110　TPO 防水卷材搭接节点图
Figure 5-110　TPO Waterproof Coil Overlap Node

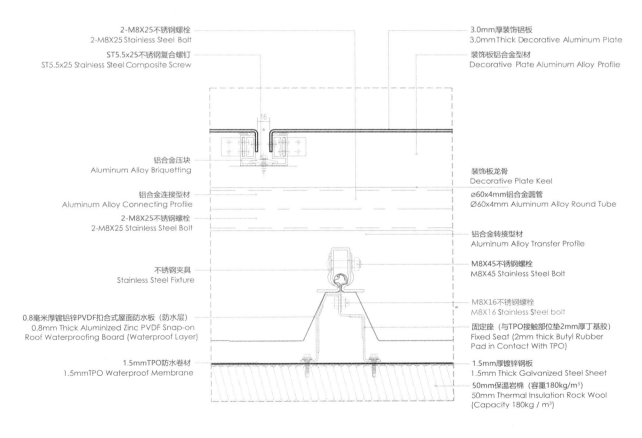

A 放大

A zoom

图 5-111　展厅屋面基本节点构造 A
Figure 5-111　Basic Structure Node of Exhibition Hall Roof A

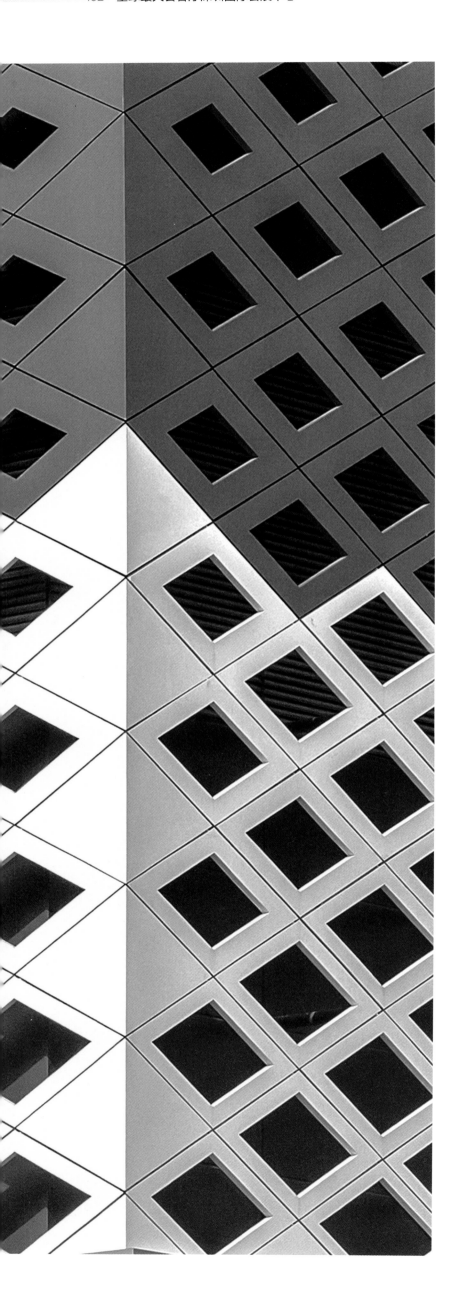

标识设计
Signage System Design
内容摘自：深圳市西利标识设计制作有限公司
Shenzhen Xili Logo Design and Production Co., Ltd

深圳国际会展中心作为具有展览功能及会议功能的场所，它标识系统的规划设计需要清晰地展示场馆、会议室以及配套设施位置的视觉形象，以便参展人员、布展人员、观展人员尽快找到自己的目的地。

深圳国际会展中心的标识系统规划设计，重点在于解决会展所有业态：展馆、登录厅、多功能厅、停车场以及众多会展配套设施之间的关系，并且要解决会展业态之间的名称、位置、通道以及受众群体在会展区域流动对标识系统的需求。

The Shenzhen World is an exhibition center with conference functions, thus the signage system should clearly display the visual image system of the exhibition halls, conference rooms and supporting facilities, so that exhibitors, organizers and visitors can find their destinations as soon as possible.

The signage system design of the Shenzhen World focuses on all kinds of functional space and facilities within: the exhibition hall, arrival hall, multi-function hall, parking space, and many other facilities and their relationship with each other. In addition, it also need to display the name, location, passage clearly.

系统全面，指示一目了然
Comprehensive System with Clear Instructions

标识系统规划在设计初期，必须要明确会展的建筑总体规划以及会展建设和运营的预期效果，并依据建筑设计、室内设计、景观设计、机电设计以及会展运营管理团队的要求，首先要做好会展标识系统的总体规划，解决外围交通系统，内部交通循环系统，合理规划高速公路、城市道路、轨道交通站点、公交站点、的士停靠点、网约车停靠点等等为会展服务的设施，重点考虑（图 5-112）。

At the beginning of the design, we must clearly understand the overall planning of the building and clear the expectation of the construction and operation. According to the requirements of the architectural design, interior design, landscape design, electrical design and exhibition operation management team, the primary step is to make an overall plan for the signage system, provide signage system for the periphery traffic system, the internal traffic circulation system, highway, city road, rail transit stations, bus stops, taxi stands, shared car stands,etc.(Figure 5-112)

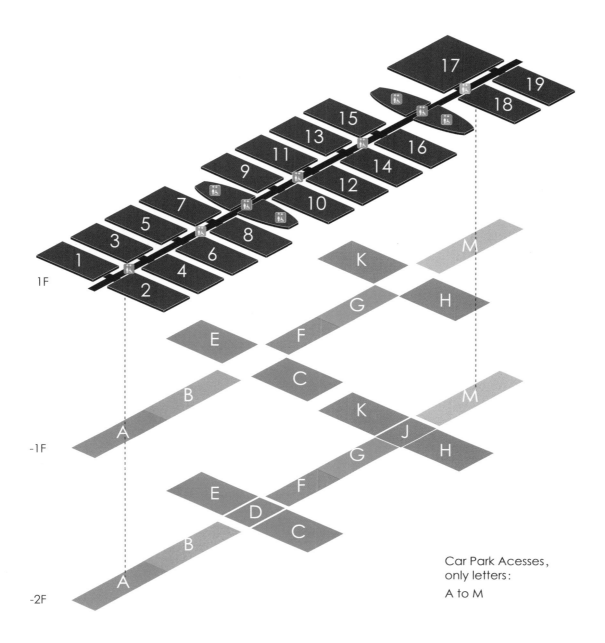

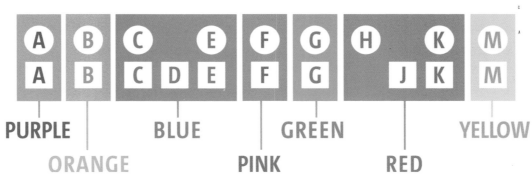

图 5-112　色彩图形分区图
Figure 5-112　Colored Partition Diagram

观展系统

观展人群通过各种交通设施进出展馆区域的清晰引导，主要指具有一定展览功能的场馆以及具有一定会议功能的场所，在人们使用时能清晰地展示场馆、会议室以及配套设施位置的视觉形象系统。方便参展人员、布展人员、观展人员尽快找到自己的目的地（图5-113）。主要包括停车场、登录厅、展馆、展馆服务、会议室及配套服务设施（餐厅、洗手间等）区域。

Visitors System

Visitors need clear guidance to enter and leave the exhibition area through various traffic facilities, and also need to find the locations of all the exhibition halls, conference rooms and other supporting facilities.(Figure 5-113) This system mainly includes parking space, arrival hall, exhibition hall, exhibition hall service center, conference room and supporting service facilities (restaurant, toilet, etc.).

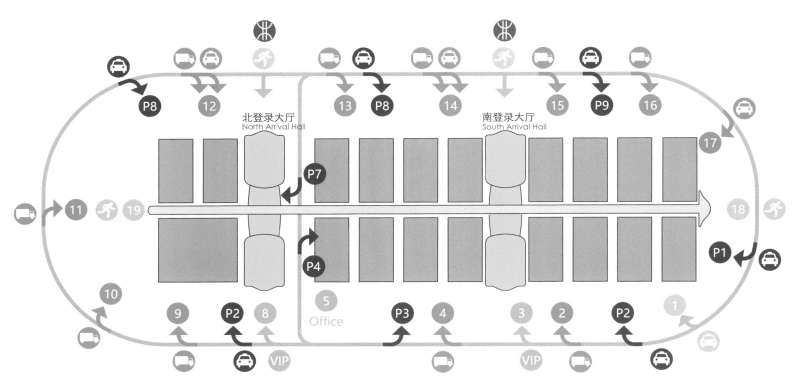

图 5-113　项目交通示意图
Figure 5-113　Project Traffic Diagram

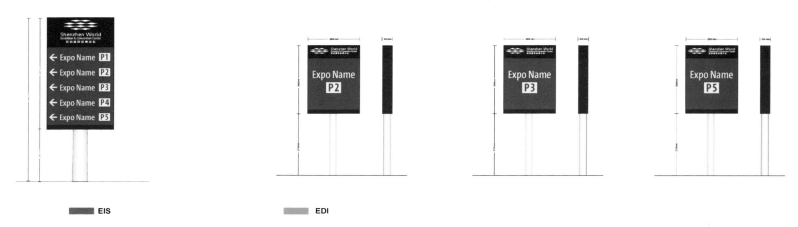

图 5-114　地下停车场指引标识图
Figure 5-114　Underground Parking Signage

图 5-115　外围出入口标识图
Figure 5-115　Peripheral Entrance and Exit Signage

布展撤展系统

该系统包括会展外围的进入通道，会展区域的展馆位置指示系统，货车的停靠位置，以及参展人员的进出方位需求等（图5-116）。

This system includes the access passage outside the exhibition center, provide guidance for the exhibition hall within the exhibition area, the stop position for trucks, as well as the guidance for exhibitors(Figure 5-116).

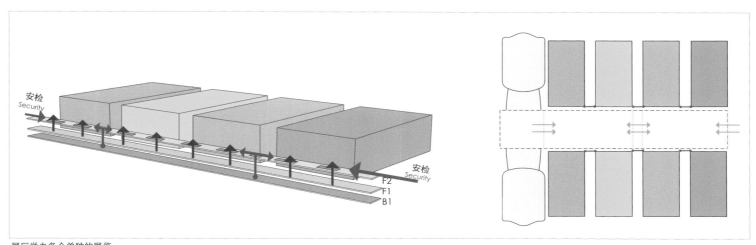

展厅举办多个单独的展览
The exhibition hall holds several separate exhibitions

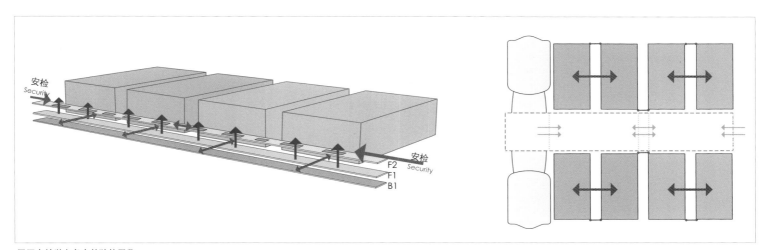

展厅合并举办多个单独的展览
Combined exhibition halls to host multiple separate exhibitions

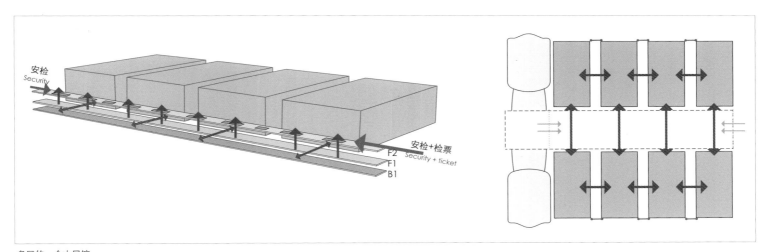

多区的一个大展馆
A large Exhibition hall in multiple districts

图 5-116 展览策略分析图
Figure 5-116 Exhibition Strategy Analysis

图例
Legend

 竖向交通 Vertical Traffic　　 中央廊道控制出入口 Central Concourse Controlled Access　　 临时围栏 Temporary Fence　　 入口 Access　　 展厅间连接，服务于同一展馆 Exhibition Connection(Serve the Same Exhibition Hall)

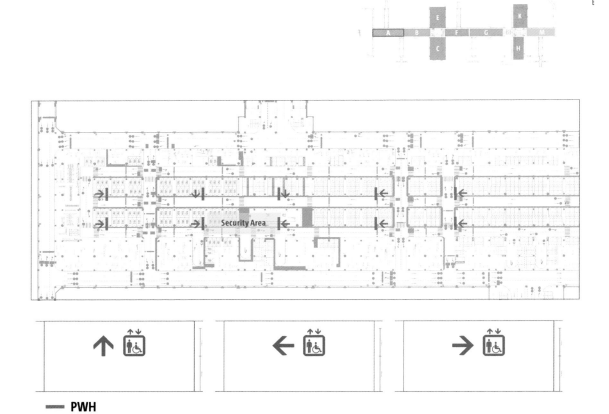

图 5-117　电梯厅指引图
Figure 5-117　Elevator Hall Guide Map

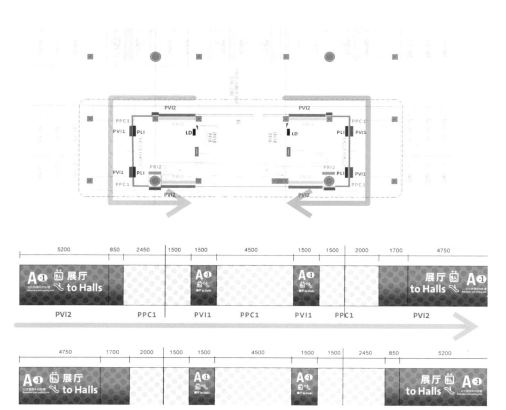

图 5-118　电梯厅核心筒外墙标识图
Figure 5-118　Elevator Hall Core Tube Exterior Wall Signage

简洁明快，彰显地域特色
Simple and lively with Regional Characteristics

深圳国际会展中心标识系统的造型设计要新颖、明快、醒目、富有个性，要强调会展文化、深圳特色以及世界级会展的要求，交通标识需严格按照国家标准以及世界通行的基本要求，其他标识在造型上一定要全方位考虑会展建筑特征及公共空间，充分满足参会人员的观展需求，尽量满足人们的快捷需求，可以彰显世界级现代化会展的标识服务体系（图5-119）。

在确保标识牌结构合理安全稳定、指示到位、功能明确、线条明快、美观大方的同时，设计过程中也要把握标识系统的创意设计、视觉设计、图形设计、文字设计、色彩设计、结构设计、材质设计、工艺设计以及安装设计，努力将会展中心完美、合适的标识系统展现出来。

The style design of the signage system should be unique, bright, bold, special, which should manifest the exhibition culture and urban features of Shenzhen. Traffic signs must be strictly comply with the national standard and basic requirements around the world, other signs must be comprehensive combined with the characteristics of exhibition building and the public space, fully meet the demand of the exhibition participants.(Figure 5-119)

On the premise of safe and stability, the signage system should present clear instructions with simple, lively and beautiful design. In the process of the design, the creative design, visual design, graphic design, character design, color design, structural design, material design, process design, installation design should also be considered carefully, and altogether, present a perfect signage system for the Shenzhen World.

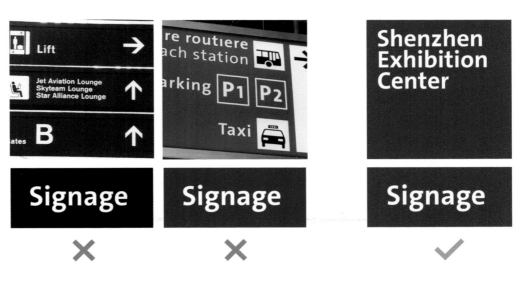

图 5-119　标识设计来源1
Figure 5-119　Signage Design Inspiration 1

会展紫，跳出传统

公共建筑项目的标识追求的是简洁明了，造型方面总是设计的非常简约，而我们希望在深圳国际会展中心项目中的标识能给人带来更深的印象，首先考虑从色彩入手，而考察现有会展或公建项目中的标识大多使用黑色与蓝色，这些颜色都过于传统，不适用于世界第一大的会展中心这样的特殊项目，于是经过研究中国文化与深圳地区特色，最终采用了中国传统红色与深圳市花"勒杜鹃"相结合而形成的特殊色彩——"会展紫"（图 5-120）。

Purple: Unconventional Color Selection

The signage design for public building is usually concise and clear with a simple style. We hope that the signage design of the Shenzhen World can impress people, so we decided to chose a main color first. Most existing exhibition centers and public building use black or blue for signage design which are too conventional for world-class exhibition center like the Shenzhen World. Through a study on Chinese culture and the characteristics of Shenzhen, we finally use a special purple-color, which is a combination from China red and the color of the bougainvillea (Shenzhen's city flower) fuchsia.(Figure 5-120)

海浪元素，活力会展

深圳是一个滨海城市，深圳国际会展中心也建立在海边，建筑与标志都运用了海浪元素，在标识的图形设计上，也沿用了这些元素，契合建筑空间理念，为指向标识增加了一丝独特的韵味。
同样，不止于勒杜鹃的色彩，它的造型也被运用，主要表现在图形符号的设计上，在国际通用符号易识别的基础上，与勒杜鹃提炼的元素结合，使标识更加新颖独特（图 5-121）。
在贯通整个项目的中央廊道，展厅的数字编号是一个非常重要的信息，我们使用了一个特殊的造型来承载这些展厅编号的信息，概念依旧来自于海浪，水波的微妙运动给整个项目带来了活力。

Wave: The Element of Sea

Shenzhen is a coastal city, the Shenzhen World is also located on the seaside, both the architectural design and the LOGO design applied the elements of waves. Therefore, it is an extension of the concept to use this element in the graphic design of the signage system.

In addition, not only the color of bougainvillea was introduced in the signage design but also its shape, which made the signage design innovative and unique.(Figure 5-121)

Throughout the central concourse, the serial number of the exhibition hall is a very important information. We use a special shape to display, which was inspired by the concept of ocean wave. The subtle movement of water wave brings vitality to the design.

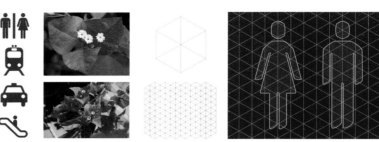

但是，符号标志也应该有新颖独特的细节反映园区鲜明的身份特征。
However, symbol signs should also have novel and unique details to reflect the distinctive identity characteristics of the park.
该模式为符号创造了新的形状，同时也给传统的人行符号带来新的一面。
This mode creates new shapes for symbols and brings new aspects to traditional human symbols.

设计灵感来自于水的微妙运动
Inspired by the subtle movement of water

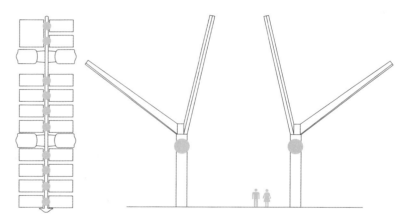

在走廊里，大厅的数字是这个空间的一个非常重要的元素。
In the corridor, the number of the hall is a very important element of the space.

图 5-120　标识设计来源 2
Figure 5-120　Signage Design Inspiration 2

图 5-121　中央廊道指示牌
Figure 5-121　Central Concourse Signage

图 5-122 中央廊道指示牌效果图
Figure 5-122 Central Concourse Signage Rendering

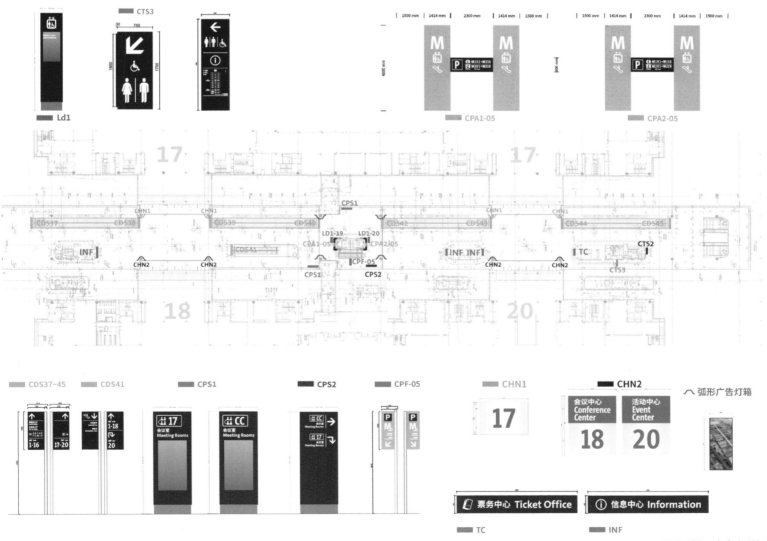

图 5-123 中央廊道标识图
Figure 5-123 Central Concourse Signage

A11展厅 ?~? 轴立面图

图 5-124 展厅外立面标示图
Figure 5-124 Exhibition Hall Facade Signage

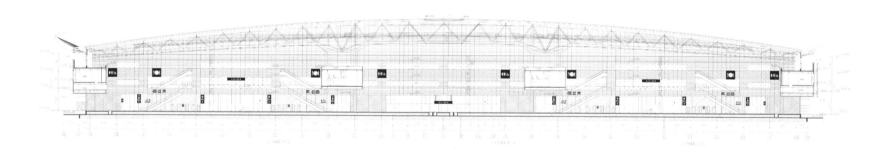

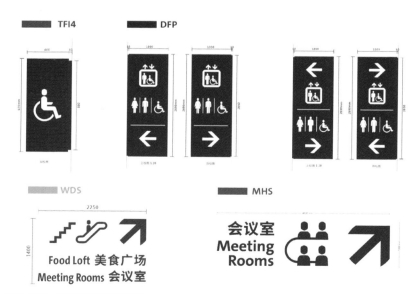

图 5-125 展厅内立面标示图
Figure 5-125 Exhibition Hall Interior Facade Signage

室内设计
Interior Design
内容摘自：深圳市广田建筑装饰设计研究院
Shenzhen Guangtian Architectural Decoration
Design and Research Institute

作为一家旨在打造世界体量最大，功能最完善，体验最优质的国际性展览中心，深圳国际会展中心项目的建筑目标对于室内设计而言有着不少挑战，其中包括面积体量巨大、时间紧迫、专业交叉性高，新技术工艺、新材料涉及运营多等问题，需反复推敲论证等。

As the world's largest exhibition center with the goal of impeccable functions and best experience, the interior design of the Shenzhen World has also presents lots of challenges, such as: huge volume, tight schedule, multi professional cooperation, and new technology and new material that need to be tested.

室内精装设计特色
Interior Design Features

国际报告厅

位于北登陆大厅的国际报告厅有上下两层，可容纳 1900 人同时参会。

报告厅外围由 3000m 双曲面异形铝材围绕而成，每片铝板均通过独立加工制作而成，在曲面位置对接难度极大，表面覆膜难度极大。对于安装的精准度要求极高。此类做法（横向异形铝材联通）的大型项目，在国内尚属首例。

国际报告厅内部同样是由数百块完全独立的铝板精密加工而成。通过特殊工艺加工，使曲面缝隙不易被察觉，形成连贯的曲线造型，视觉效果震撼。

而多种功能设备暗藏在层叠造型内部，配置丰富、全面。并且从观众视觉无法察觉，在满足超大会议功能的基础之上，保证了国际报告厅的会议视觉效果（图 5-126）。

报告厅内部空调系统均由地面"辐射型"出风，在节能的基础之上相比传统天花出风与墙面出风更环保与舒适。

图 5-126　国际报告厅效果图
Figure 5-126　International Lecture Hall Rendering

International Conference hall

The international conference hall located in the north arrival hall, has two floors in total, which can accommodate 1,900 people.
The outer part of the conference hall is made of 3000m of hyperboloid special-shaped aluminum plate, and each piece of the aluminum is processed independently. It is extremely difficult to weld on the curved surface and cover the surface with film, and it requires high accuracy in installation. This is the first case in China.
The interior of the international conference hall is also made of hundreds of completely independent aluminum plates. Through special processing, the surface gap is not easy to be noticed, which guarantees a beautiful smooth curve shape with continuous stroke and a great visual effect.
A variety of functional equipment hidden inside the cascade shape, with a rich and comprehensive configuration, and the audience would not notice, which guarantees the basic needs of this conference hall without affecting any visual effect.(Figure 5-126)
The air conditioning system supply air from the ground, which is more environmentally friendly and comfortable than traditional ceiling or wall air supply.

图 5-127　宴会厅效果图
Figure 5-127　Banquet Hall Renderings

图 5-128　多功能厅效果图
Figure 5-128　Multifunctional Hall Rendering

中大型会议室

在本项目中，对外开放的中大型会议室几乎贯穿所有场馆（含标准展厅）

其中中型会议室面积约为 80 平方米、大型会议室从 400 平方米 - 800 平方米不等。

大小会议室均按照行业标准配备多媒体影音设备与声学处理。

大型会议室与多功能厅类似，可切分为 3 个独立会议室使用（多媒体设备独立控制），满足多功能需要（图 5-130）。

在会议室内均配备 P1.9 高清 LED 显示屏，无需临时租借搭建。
在大型会议室天花内暗藏搞荷载电动吊挂设备，即使需要搭建舞台，也无需提前搭建钢架，极大的缩短了会议准备时间。

图 5-130　大会议室效果图
Figure 5-130　Large Conference Room Rendering

超大型多功能厅

南登录多功能厅总长 80m，宽度 40m，总面积 3200 平方米。

影音配备、声学处理均可达到标准大型会议、宴会的标准。（多媒体音响灯光、设备控制室、追光室、同声传译室等）
宴会模式下可容纳近 130 桌 /1500 人，会议模式下可容纳近 2000 人。
同时，多功能厅可以通过半自动活动隔断被切分为三个空间同时使用，
其采用国际领先的"超轻型、超隔音"隔断，在保证结构安全的同时阻隔两空间的声音干扰，达到真正意义上的三空间独立使用（图 5-128）。

图 5-131　多功能实景效果图
Figure 5-131　Small Conference Room Rendering

图 5-129　小型会议室效果图
Figure 5-129　Small Conference Room Rendering

Medium and Large Meeting Room

In this project, large and medium-sized conference rooms open to the public are allocated in almost all the exhibition halls (including the standard exhibition hall).

Among them, medium-sized meeting room is about 80 square meters, and the large ones varies from 400 square meters to 800 square meters.

All the meeting rooms are equipped with multimedia audio equipment and acoustic treatment in accordance with the industry standards.

Similar to the multi-function hall, the large meeting room can be divided into three independent meeting rooms (each with independent multimedia equipmens) (Figure 5-130).

All meeting rooms are equipped with P1.9 hd LED display, no need for temporary rental.

In the ceiling of the large meeting room, there is a hidden load of electric hanging equipment, therefore it can build a stage without building a steel frame in advance, which greatly shortens the preparation time for the meeting.

Ultra-large Multi-function Hall

The south arrival hall multi-functional room has a total length of 80 meters, a width of 40 meters and a total area of 3200 square meters.

The standard of the audio and video equipment and the acoustic treatment is the same as those of large conferences and banquets. (multimedia sound and lighting, equipment control room, the spotlight chambor, simultaneous interpretation room, etc.)

For banquet, it can accommodate nearly 130 tables /1500 people, for meeting, and accommodate nearly 2000 people for conference.

Meanwhile, the multi-function hall can be divided into three spaces at the same time by semi-automatic partition, which is worldclass "ultra-light, ultra-sound insulation" to guarantee the safety of the structure and the sound proof capacity, which will transferring the ultra-Large hall into three independent space for different needs. (Figure 5-128)

全国最大的单体无柱多功能厅 / 宴会厅

C11 多功能厅总长 100m，宽度 70m，总面积 7000 平方米（图 5-131 多功能厅效果图）。

影音配备、声学处理均可达到标准大型会议、宴会的标准。（多媒体音响灯光、设备控制室、追光室、同声传译室等）除作为展厅之外，还可以举办超大型会议与宴会（图 5-127）。宴会模式下可容纳近 280 桌 /3000 人，会议模式下可容纳近 4000 人。

同时，宴会厅可以通过半自动活动隔断切分为两个空间同时使用，其采用国际领先的"超轻型，超隔音"隔断，保证结构安全的同时阻隔两空间的声音干扰，达到真正意义上的双空间独立使用。

图 5-132　登录大厅效果图
Figure 5-132　Arrival Hall Rendering

登录大厅

超大型多曲面铝制天花。

① 整个近万平方的登录大厅天花均由非复制性的定制铝板焊接而成。 在两个登录大厅的主入口，巨型海浪造型充分展现了深圳海洋文化的主体性与深圳敢闯敢拼破浪前行的意志（图 5-132）。

② 超长巨幅 LED 屏幕，整体 LED 屏幕长 72m，宽 5m，以环抱形态面向入口，整体视觉效果震撼，国内罕见（图 5-133）。

③ 在登录大厅曲面天花之内暗藏电动智控广告悬挂轴，通过遥控便可落下安装广告条幅。 操作便捷，极大缩短会展布置时间（图 5-134）。

Arrival Hall

The Arrival hall features a super large polycurved aluminum ceiling.

① The entire ceiling of the Arrival hall covers with nearly 10,000 square meters of non-replicative customized aluminum plates which are welded together. At the main entrance of the two arrival halls, the giant wave shape fully demonstrates the marine culture of Shenzhen and its spirit to break through the waves.(Figure 5-132)

② the super long and large LED screen with a length of 72 meters long and width of 5 meters, is facing the entrance in the form of an embrace, demonstrating an eye-catching visual effect which is rare in China.(Figure 5-133)

③ Hidden in the curved ceiling of the arrival hall, there is an electrically controlled advertising suspension shaft, which can be lowered by remote control and then being installed, which is convenient for operation and greatly reduces the time for exhibition preparation.(Figure 5-134)

Banquet Hall

The C11 multifunctional hall is 100 meters long, 70 meters wide, and 7,000 square meters in total.(Figure 5-131)

The standard of the audio and video equipment, acoustic treatment is the same as those of large conferences and banquets. (multimedia sound and lighting, equipment control room, dome, simultaneous interpretation room, etc.)
It can be used as an exhibition hall, or for super large conferences and banquets. (Figure 5-127)
For banquet, it can accommodate nearly 280 tables /3000 people, and for meeting, it can accommodate nearly 4000 people.
Meantime, the banquet hall can be divided into two spaces at the same time by semi-automatic partition, which is world-class "ultra-light, ultra-sound insulation" to guarantee the safety of the structure and sound proof capacity, and this would transfer the banquet hall into two independent space for different needs.

图 5-133　登录大厅实景图
Figure 5-133　Arrival Hall Real Scene

图 5-134　登录大厅实景图
Figure 5-134　Arrival Hall Real Scene

垂直交通

Vertical Transportation
SZEXPO – 垂直交通顾问
Vertical Transport Consultant-SZEXPO

垂直交通在研究入口、楼梯、廊道位置，电梯、扶梯、步道位置及配置，以及人员分布情况和流转空间基础上，根据缩短客流、货流动线或移动时间、避免客流与货流动线交叉、瓶颈等原则进行规划设计。由于深圳国际会展中心建筑体量庞大，跨度巨大，人流动线与流量分析，成为垂直交通设计的最大难点。

The scope of considerations for vertical transportation includes: entrance, staircase, corridor location, elevator, escalator, visitors location and configuration, passenger flow and distribution,circulation spacing. And the design principles are: shorten the passenger flow, logistic circulation and travel time, prevent crossing of passenger and logistic circulation; prevent traffic bottle neck. Due to the large volume and span of the building, the circulation and circulation analysis are the most difficult in the vertical transportation design.

人流动线
People Circulation

经过与交通顾问及建筑师多次沟通，我们定义了各个入口到展馆的主要人流动线，主要人流动线由多个交通核组成，点与点之间为大量的人流辅线，我们称之为动线分段，把所有动线分段规划后，便基本完成了人流动线的定义（表 5-12）。

After communicating with the traffic consultant and the architect for many times, we defined the main people circulation from each entrance to the exhibition hall. The main people circulation is composed of multiple traffic cores, and there are a large number of auxiliary lines between these traffic cores, which is called moving line sections. After the planning of moving line sections, the basic people circulation is done.(Table 5-12)

登录大厅至展览厅的动线分段

① 展览厅 1F 至中央廊道 2F
② 中央廊道 2F 至登录大厅 2F
③ 登录大厅 2F 至登录大厅 1F
④ 垂直交通系统建议配置汇总

Moving line Section from Arrival Hall to Exhibition Hall

① Exhibition hall 1F to central concourse 2F
② Central concourse 2F to arrival hall 2F
③ Arrival hall 2F to arrival hall 1F
④ Suggested configuration summary of vertical circulation system

车库至中央廊道的动线分段

① 车库 B3 至中央廊道 L2
② 垂直交通系统建议配置汇总

Moving line Sections from Parking Space to Central Concourse

① parking B3 to central concourse L2
② suggested configuration summary of vertical traffic system

南、北登录大厅首层至会议厅及餐饮层动线的动线分段

① 南北登录大厅会议厅 L2 至餐饮层 L3
② 南北登录大厅首层会议厅 L1 至南北登录大厅首层会议厅 L2
③ 垂直交通系统建议配置汇总

此外，展厅内部、体育中心内部亦细分了动线分段。

Moving line Sections from Arrival Hall L1 to Conference Room and Catering Area

① Conference Hall L2 to Catering Floor L3
② Conference hall L1 to Conference Hall L2
③ Suggested Configuration Summary of Vertical Transportation System

In addition, there are also moving line sections inside the exhibition hall and the sports center.

流量分析

各个动线分段主要通过扶梯或自动人行步道处理人流，根据各分段所能到达的功能区面积与特性，结合运营方式、使用时间属性等，我们建立了人流量模型，并对各分段运能进行分析，经过行业软件模拟分析，对各分段的垂直交通设备提供配置建议。

Flow Analysis

Each line section processes people circulation mainly by escalator or automatic pedestrian walk way. According to the function area and the characteristics of each section, and combining with way of operation and temporalities, we have built the people circulation model to analyze each section's capacity, and provided configuration advices for the vertical transportation equipment of each section.

设计标准

为清晰评价本项目垂直交通运输系统的运行概况，需要设置合理、客观的设计标准对其运行概况进行量化表达，并为配合垂直交通运输系统的规划与设计提供基准参考。通常情况下垂直交通运输系统运行指标描述由以下参数组成：
① 人均面积：展览厅每人 0.9 平方米，餐厅每人 1.5 平方米
② 五分钟处理能力 与 五分钟客流量完全匹配
③ 间隔时间 ≤ 40s
④ 轿厢装载率 ≤ 60%

Design Criteria

In order to clearly evaluate the operation profile of the vertical transportation system of this project, it is necessary to set reasonable and objective design standards to quantitatively express the operation profile, and provide reference for the planning and design of the vertical transportation system. In general, the performance index description of vertical transportation system consists of the following parameters:
① per capita area: 0.9 square meter per person in the exhibition hall and 1.5 square meters per person in the catering area
② 5-minute processing capacity perfectly matches the 5-minute passenger flow
③ interval time ≤40s
④ cage loading rate ≤60%

楼梯与交通系统客流分配比例设计依据 Proportion of usage of stairs and lift	Table 2.13 Comparison of usage of stairs and life

Table 2.13 Comparison of usage of stairs and life

Floors travelled	Usage (stair : lift)	
	Up (%)	Down (%)
1	10 : 90	15 : 85
2	5 : 95	10 : 90

电梯与扶梯客流分配比例设计依据 Proportion of usage of lift and escalator	Table 2.14 Likely division of traffic between lifts and escalators

Table 2.14 Likely division of traffic between lifts and escalators

Floors travelled	Escalator (%)	Lift (%)
1	90	10
2	75	25
3	50	50
4	25	75
5	10	90

走道客流计算依据
Corridor passenger flow rate standard

Table 2.10 Actual mall pedestrian flow rates per metre width of mall

	Uncrowed (0.2 person/m²)		Crowded (0.2 person/m²)	
	Speed (m/s)	Flow rate (person/h)	Speed (m/s)	Flow rate (person/h)
All shoppers	13	936	1.0	1620

扶梯运载能力计算依据
Corridor passenger flow rate standard

Table 2.10 Theoretical handing capacities as stated in BS EN 115-1: 1998 and BS EN 115-1: 2008

Step wide (mm)	Handing capacity (person/h) for stated contract speed (m/s)		
	0.5	0.65	0.75
(a) BS EN 115-1:1998: clause 3.8			
600	4500	5850	6750
800	6750	8775	10125
1000	9000	11700	13500
(b) BS EN 115-1:2008: Annex H			
600	3600	4400	4900
800	6750	8775	10125
1000	9000	11700	13500

Table 2.9 Practical escalator handing capacity (BS 5656-2: 2004)

Speed	Step width					
	600 mm		800 mm		1000 mm	
	Person/ min.	Person/ hour	Person/ min.	Person/ hour	Person/ min.	Person/ hour
0.50	37	2250	57	3375	75	4500
0.65	49	2925	73	4388	97	5850
0.75	57	3375	85	5063	113	6750

自动步道运载能力计算依据
Travellator capacity standard

Table 2.10 Practical handing capacities of moving walks and ramps (source:BS 5656-2(BSI,2004))

Inclination (degree)	Speed (m/s)	Nominal width					
		600 mm		800 mm		1000 mm	
		Person/ min.	Person/ hour	Person/ min.	Person/ hour	Person/ min.	Person/ hour
Horizontal moving walks :							
0	0.50	48	2880	60	3600	84	5040
0	0.65	62	3648	78	4560	106	6350
0	0.75	72	4320	90	5400	126	7560
Inclined moving walks :							
6	0.50	48	2880	60	3600	84	5040
10	0.50	48	2880	60	3600	--	--
12	0.50	48	2880	60	3600	--	--

表 5-12 分析及设计依据表
Table 5-12 Analysis and Design Basis Table

运能分析

① 中央廊道 2F 至登录大厅 2F

本动线分段为水平交通，中央廊道同时连接 8 个展览厅，最不利情况考虑同时使用概率为 0.7。

考虑 2 小时内展馆人数达到目标人次，客流分别从南登录大厅及登录大厅（分配比例为 7：3），从中央廊道采用走廊及自动步道到达各展览厅二层入口（图 5-135）。

Transportation Capacity Analysis

① 2F Central Concourse to 2F Exhibition Hall

This segment is mainly horizontal traffic. Focusing on central concourse between SE portal to SS portal, central concourse is connecting eight exhibition halls, consider the worst scenario 70% of peak occupancy, exhibition hall reaches targeted number of visitors in two hours. Entrance bias of SE portal and SS portal as 7 ： 3. Passenger would be served by both concourse walkway and travellator(Figure 5-135).

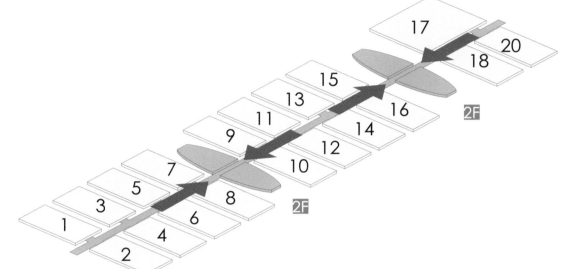

图 5-135　客流从中廊到登录大厅示意图
Figure 5-135　Diagram of Passenger Circulation from Central Concourse to Arrival Hall

动线分段 Flow segment	客流流转量 Visitor Traffic Capacity					客流分配 Distribution		
	每展厅人次 Number of visitors per exhibition hall	展厅数量 Number of showrooms	同时使用系数 Diversity factor	时间 Time	五分钟客流量 5 mins traffic flow	登录口 Entrance	分配比 Ratio	人数 Person
中央廊道2F至登录大厅2F Central Concourse2F to Arrival Hall 2F	19500	8	0.7	2	4550	南登录大厅 South Arrival Hall	70%	3185
						北登录大厅 North Arrival Hall	30%	1365

表 5-13　客流流转量和客流分配分析表
Table 5-13　Analysis of Passenger Flow and Distribution

设备 Equipment	五分钟运能需求 5 mins capacity demand 人 /person	每小时处理能力 Hourly handling capacity 人 /person	五分钟运载能力 5 mins handling capacity 人 /person	配置数量 Required number
A 区域步道 Zone A travellator	1792	7560	630	3
B 区域步道 Zone B travellator	1792	7560	630	3
C 区域步道 Zone C travellator	1393	7560	630	3
D 区域步道 Zone D travellator	796	7560	630	2
E 区域步道 Zone E travellator	597	7560	630	1
注1：小时处理能力按1400mm/0.75ms 考虑，见3.7.5 3.7.5 consideration, see 1400mm / 0.75ms: 1 hour processing capacity				

表 5-14　步道运能分析表 1
Table 5-14　Footpath Capacity Analysis 1

客流分类 Flow type	五分钟客流量 5 mins traffic flow	通行方式 Mode of transportation	分配比例 Ratio ％	客流分配 Person	A 步道 使用人数 Travellator A	B 步道 使用人数 Travellator B	C 步道 使用人数 Travellator C	D 步道 使用人数 Travellator D	E 步道 使用人数 Travellator E
南登录大厅至1#厅 South Arrival Hall to Hall 1	796.25	走廊 Walkway	0	0					
		步道 Travellator	100	796	796	796	796	796	
南登录大厅至 2#厅 South Arrival Hall to Hall 2	796.25	走廊 Walkway	25	199					
		步道 Travellator	75	597	597	597	597		
南登录大厅至 3#厅 South Arrival Hall to Hall 3	796.25	走廊 Walkway	50	398					
		步道 Travellator	50	398	398	398			
南登录大厅至 4#厅 South Arrival Hall to Hall 4	796.25	走廊 Walkway	100	796					
		步道 Travellator	0	0	0				
登录大厅至 1#厅 South Arrival Hall to Hall 1	341.25	走廊 Walkway	100	341					
		步道 Travellator	0	0					
登录大厅至 2#厅 Arrival Hall to Hall 2	341.25	走廊 Walkway	75	256					
		步道 Travellator	25	85					85
登录大厅至 3#厅 Arrival Hall to Hall 3	341.25	走廊 Walkway	50	171					
		步道 Travellator	50	171					171
登录大厅至 4#厅 Arrival Hall to Hall 4	341.25	走廊 Walkway	0	0					
		步道 Travellator	100	341					341
运能需求总计 Capacity demand					1792	1792	1393	796	597

表 5-15　步道运能分析表 2
Table 5-15　Footpath Capacity Analysis 2

动线分段 Flow segment	客流流转量 Traffic					客流分配 Distribution				
	每展厅人次 Person/hall	展厅数量 No. of hall	同时使用系数 Diversity factor	时间 Time/hr	五分钟客流量 5 mins traffic flow 人 /person	登录口 Portal	分配比 Ratio ％	设备 Equipment	分配比 Ratio％	人数 Persons
登录大厅2F至登录大厅1F Arrival Hall 2F to Arrival Hall 1F	19500	20	0.5	2	7412.5	南登录大厅 South Arrival Hall	40	扶梯 Escalator	90	2669
								电梯 Elevator	5	148
								楼梯 Stair	5	148
						北登录大厅 North Arrival Hall	30	扶梯 Escalator	90	2001
								电梯 Elevator	5	111
								楼梯 Stair	5	111
						登录大厅 Arrival Hall	27	扶梯 Escalator	90	1801
								电梯 Elevator	5	100
								楼梯 Stair	5	100
						北侧扶梯平台 North Escalator Platform	3	扶梯 Escalator	95	211
								楼梯 Stair	5	11

表 5-16　客流流转量和客流分配分析表 2
Table 5-16　Analysis of Passenger Flow and Distribution 2

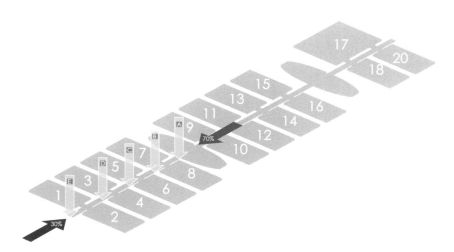

图 5-136　客流从南登录大厅到达展厅示意图
Figure 5-136　Diagram of Passenger Circulation from South Arrival Hall to Exhibition Hall

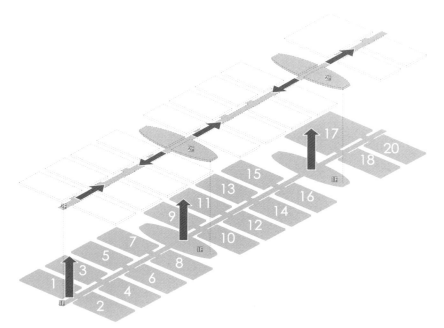

图 5-137　客流从登录大厅通往 2F 示意图
Figure 5-137　Diagram of Passenger Circulation from Arrival Hall to 2F

运能分析
Transportation Capacity Analysis

① 登录大厅 2F 至登录大厅 1F
建筑物主要有 6 个出入口，东、南、北侧共 4 个，西侧共 2 个，其中由于所有大型公共交通枢纽皆设于东侧，故西侧人流量较少。4 个登录口连接 20 个展览厅，最不利情况考虑为同时使用概率为 0.5（表 5-16）。
考虑 2 小时内展馆人数达到目标人次，客流分别从南侧登陆大厅 1F，东侧南登陆大厅 1F 及北登录大厅 1F，北侧扶梯平台 1F（分配比例为 3：8：8：1），从各口部采用扶梯、电梯、楼梯（分配比例为 18：1：1）到达各登录大厅及扶梯平台 2F（表 5-17）。另约 17000 人经车库到达后通过车库扶梯及电梯到达中央廊道，该部分人数将于以下计算中予以扣除（表 5-18）。
针对西侧 2 个口部，由于 5.1.3 部分仅考虑参展访客，不考虑会议厅访客，同时，会议厅访客流量远大于参展访客，故 NW、SW 西侧垂直交通配置以 5.3.2 分析结果为准（表 5-19）。

① Arrival Hall F2 to Arrival Hall 1F
Visitors entering the building from six portals, for the existence of mass transit, and the transportation hub is located on the east, the passenger flow from portal on the west could be neglected. There are four portals, respectively on east, south and north, which are connecting twenty exhibition halls, consider worst situation as 50% of peak occupancy, exhibition hall reaches targeted visitors capacity in two hours. Entrance bias of north portal, south portal, north escalator platform and portal is 3 : 8 : 8 : 1. Passenger would be served by escalator, elevator and staircase. Approx. 17000 visitors would enter from carpark through carpark escalator and elevator, which will be considered in the traffic estimation below(Table 5-18).

设备 Equipment	五分钟运能需求 5 mins capacity demand (person)	每小时处理能力 Hourly handling capacity (person)	五分钟运载能力 5 mins handling capacity (person)	配置数量 Required number
南登录大厅扶梯 South Arrival Hall Escalator	2669	7300	608	5
北登录大厅扶梯 North Arrival Hall Escalator	2002	7300	608	4
登录大厅扶梯 Arrival Hall Escalator	1802	7300	608	3
北侧扶梯平台扶梯 North Escalator Platform Escalator	212	7300	608	1

注 1：小时处理能力按 1000mm/0.65ms 考虑，见 3.7.4
Note 1: the hourly processing capacity is considered as 1000mm / 0.65ms, see 3.7.4

表 5-17　扶梯运能分析表
Table 5-17　Escalator Traffic Analysis

设备 Equipment	五分钟运能需求 5 min Capacity Demand （/person）	梯速 Speed （m/s）	载重 Capacity （kg）	数量 Numbers	间隔时间 Interval （s）	装载率 Capacity factor （%）
南、北登录大厅电梯 South/North Arrival Hall Elevator	上行：148 UP :148	1	1000	3	23.4	97.6
			1350		23.4	70.5
			1600		23.4	60.4
			1800		23.4	52.9
			1000	4	14.5	60.6
			1350		14.5	43.8
			1600		14.5	37.5
			1800		14.5	32.8
			1000	5	10.5	43.8
			1350		10.5	31.6
			1600		10.5	27.1
			1800		10.5	23.7
			1000	6 （现方案） (current plan)	8.1	34
			1350		8.1	24.5
			1600		8.1	21
			1800		8.1	18.4

注 1：10 月 31 日更新建筑方案中，南北登录大厅电梯同时服务L1，L2-3 餐饮动线，关于中午餐饮高峰运能分析请见 5.3.1 。
Note 1: in the updated building scheme on October 31, the elevators in the North-South landing hall serve L1 and L2-3 catering lines at the same time. Please refer to 5.3.1 for the analysis of peak capacity of catering at noon.

表 5-18　电梯运能分析表
Table 5-18　Elevator Traffic Analysis

	区域 region	设备 Equipment	建议数量 Suggested number	建议规格 Suggested spec.	备注 Remark
A	展览厅 A1-10 Exhibition Hall A1-10 展览厅 C1-10 Exhibition Hall C1-10	扶梯 Escalator	2 上 2 下，共4台 2 up and 2 down, 4 sets in total	30°/1000mm/0.5ms	Refer to page 7
		电梯 Elevator	2+2	1000kg/1.0ms	Refer to page 8
B	中央廊道 Central Concourse	步道 Travellator	3	1400mm/0.75ms	Refer to page 11
C			3	1400mm/0.75ms	
D			3	1400mm/0.75ms	
E			2	1400mm/0.75ms	
F			1	1400mm/0.75ms	
G	南北登录大厅 South/North Arrival Hall	扶梯 Escalator	5 上 3 下，共8台 5 up and 3 down, 8 sets in total	30°/1000mm/0.5ms	Refer to page 14
		电梯 Elevator	3+3	1000kg/1.5ms	Refer to page 26
H	登录大厅 Arrival Hall	扶梯 Escalator	3 上 1 下，共4台 3 up and 1 down, 4 sets in total	30°/1000mm/0.5ms	Refer to page 14
		电梯 Elevator	2+2	1000kg/1.0ms	Refer to page 16
I	北侧扶梯平台 North Escalator Platform	扶梯 Escalator	1 上 1 下，共2台 1 up and 1 down, 2 sets in total	30°/1000mm/0.5ms	Refer to page 14

表 5-19　垂直交通系统建议配置表
Table 5-19　Vertical Traffic Suggested Configuration

声学设计
Acoustic Design

内容摘自：黄展春剧场建筑设计顾问（北京）有限公司
Huang Zhanchun Theater Architectural Design
Consultant (Beijing) Co., Ltd

深圳国际会展中心选址深圳空港新城，距 T3 航站楼 7 公里，距 T4 枢纽 3 公里，周边汇聚多条铁路、城际线、高速公路和地铁路线，交通便利的同时，也会因航空噪音和交通噪音造成较大的影响，须采取相应措施减小其影响。这个项目的建筑声学设计参考多个设计依据，本篇章重点阐述几个功能展厅的声学处理。

The Shenzhen World is located in the Shenzhen Airport New Town, which is 7km away from T3 terminal and 3 km away from T4. In addition, as an transportation hub, there are railway, intercity lines, highways and subway lines converged around. Therefore, both the aircraft noise and traffic noise will cause great influence on the Shenzhen World, and corresponding measures should be taken to reduce relative impact. The acoustical design of this project refers to several design codes. This chapter focuses on the acoustical treatment of several functional exhibition halls.

建声指标设定
Setting of building sound index

空间名称 Space name	混响时间 (中频带) Space name (Mid-band)	允许背景噪声 Allow background noise	固定墙体隔声量 Fixed wall sound insulation	活动隔声墙体隔声量 Space name	屋顶隔声量或楼版隔声量 Roof sound insulation or floor sound insulation	门隔声量（无声闸） Door sound insulation (Silent brake)	门隔声量（有声闸） Door sound insulation (with sound brake)	备注 Remark
展厅 Exhibition Hall	≤3.0s	NR-50	Rw+Ctr≥35		Rw+Ctr≥35	Rw+Ctr≥35	Rw+Ctr≥35	预估在大多数航班经过时，满足NR-50要求 Nr-50 is expected to be metwhen most flights pass
国际报告厅 International Lecture Hall	1.2s(±0.1)	NR-30	Rw+Ctr≥58		Rw+Ctr≥58	Rw+Ctr≥45	Rw+Ctr≥40	
多功能厅（南登录大厅） Multi-function Hall (South Arrival Hall)	1.2s(±0.1)	NR-30	Rw+Ctr≥58	Rw+Ctr≥52	Rw+Ctr≥58	Rw+Ctr≥45	Rw+Ctr≥40	
宴会厅(18) Banquet Hall (18)	≤1.5s	NR-35	Rw+Ctr≥50	Rw+Ctr≥52	Rw+Ctr≥50	Rw+Ctr≥45	Rw+Ctr≥40	
多功能场馆(20) Multi-purpose Stadium(20)	≤1.7s	NR-40	Rw+Ctr≥40		Rw+Ctr≥40	Rw+Ctr≥35	Rw+Ctr≥30	满席时,混响时间达要求值 When the seat is full, the reverberation time meets the requirements
登录大厅前厅 Arrival Hall Lobby		NR-40	Rw+Ctr≥40		Rw+Ctr≥40	Rw+Ctr≥35	Rw+Ctr≥30	预估在大多数航班经过时,满足NR-40要求 Nr-40 is expected to be metwhen most flights pass
会议室 Conference Room	≤0.6s	NR-30	Rw+Ctr≥50	Rw+Ctr≥52		Rw+Ctr≥35	Rw+Ctr≥30	
同声传译室 Simultaneous Interpretation Room	0.4s(±0.1s)	NR-30	Rw+Ctr≥50		Rw+Ctr≥50	Rw+Ctr≥45	Rw+Ctr≥40	
音控室 Sound Control Room	0.4s(±0.1s)	NR-30	Rw+Ctr≥50		Rw+Ctr≥50	Rw+Ctr≥45	Rw+Ctr≥40	
放映室 Projection Room			Rw+Ctr≥45		Rw+Ctr≥45	Rw≥45	Rw≥40	内部作吸声降噪处理 Internal sound absorption and noise reduction
行政或私人办公室 Administrative or Private Offices	≤0.6s	NR-35	Rw+Ctr≥50		Rw+Ctr≥50	Rw+Ctr≥35	Rw+Ctr≥30	与外部空间相邻墙面，隔声量Rw+Ctr≥40 Wall adjacent to external space, sound insulation RW + CTR ≥ 40
没有声掩蔽的开放办公室 An open plan office with No sound proof	≤0.8s	NR-35	Rw+Ctr≥50		Rw+Ctr≥50	Rw+Ctr≥35	Rw+Ctr≥30	与外部空间相邻墙面，隔声量Rw+Ctr≥40 Wall adjacent to external space, sound insulation RW + CTR ≥ 40
有声掩蔽的开放办公室 An open-plan office with sound proof	0.8s	NR-35	Rw+Ctr≥50		Rw+Ctr≥50	Rw+Ctr≥35	Rw+Ctr≥30	与外部空间相邻墙面，隔声量Rw+Ctr≥40 Wall adjacent to external space, sound insulation RW + CTR ≥ 40
接待室 Reception Room		NR-40	Rw+Ctr≥40			Rw+Ctr≥35	Rw+Ctr≥30	
酒吧 The Bar		NR-35	Rw+Ctr≥50		Rw+Ctr≥50	Rw+Ctr≥35	Rw+Ctr≥30	与外部空间相邻墙面，隔声量Rw+Ctr≥40 Wall adjacent to external space, sound insulation RW + CTR ≥ 40
员工室、休息室 Staff Room, Lounge		NR-35	Rw+Ctr≥50		Rw+Ctr≥50	Rw+Ctr≥35	Rw+Ctr≥30	与外部空间相邻墙面，隔声量Rw+Ctr≥40 Wall adjacent to external space, sound insulation RW + CTR ≥ 40
大厅、前厅 Lobby, Vestibule		NR-40	Rw+Ctr≥40			Rw+Ctr≥35	Rw+Ctr≥30	
空调机房 Air Conditioning Plant Room			Rw+Ctr≥55		Rw+Ctr≥55	Rw+Ctr≥45	Rw+Ctr≥40	
一般机房 General Engine Room			Rw+Ctr≥50		Rw+Ctr≥50	Rw+Ctr≥45	Rw+Ctr≥40	
餐厅 Restaurant		NR-40	Rw+Ctr≥45			Rw+Ctr≥35	Rw+Ctr≥30	与外部空间相邻墙面，隔声量Rw+Ctr≥40 Wall adjacent to external space, sound insulation RW + CTR ≥ 40
厨房和餐饮零售点 Kitchen and Food and Beverage Outlets		NR-40						
卸货台和服务车道 Loading Dock and Service Lane		NR-40						
更衣室、卫生间 Changing Room, Toilet		NR-40						

表 5-20　建声指标设定表
Table 5-20　Indicators of Architecture Acoustic Design

NR值 NR value	倍频带中心频率(Hz) Center Frequency of Octave Band (Hz)								
	31.5	63	125	250	500	1000	2000	4000	8000
NR-25	72	55	43	35	29	25	21	19	18
NR-30	76	59	48	39	34	30	26	25	23
NR-35	79	63	52	44	38	35	32	30	28
NR-40	82	67	56	49	43	40	37	35	33
NR-45	86	71	61	53	48	45	42	40	38
NR-50	89	75	66	59	54	50	47	45	44

表 5-21　参考表：噪声评价曲线 NR 对应的各倍频带声压级（DB）
Table 5-21　Reference table: Sound Pressure Level (DB) of Each Octave Band Corresponding to Noise Evaluation Curve NR

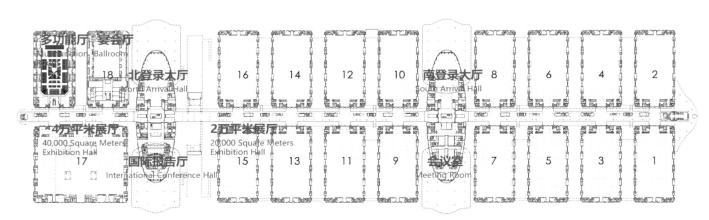

图 5-138　一层平面图
Figure 5-138　First Floor Plan

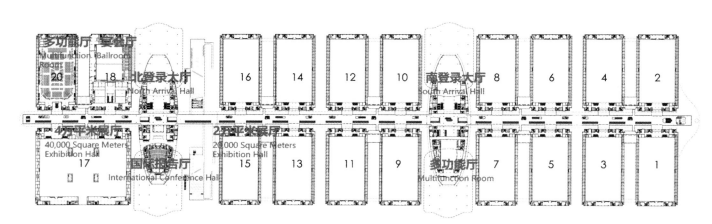

图 5-139　二层平面图
Figure 5-139　Second Floor Plan

噪声控制
Noise Control

环境噪声来源

① 航空噪声

飞机的噪声是全频带噪声，且低频（<250Hz）噪声大。飞机的噪声影响程度取决于飞机的起降次数、时刻、强度、飞机噪声的频谱分布、持续时间、距离和传播途径等。

本项目距离机场 T3 航站楼约 7 公里，距离规划中的 T4 航站楼约 3 公里。本项目之航空噪声应依据业主方提供的噪声监测相关资料，以及目前我方于现场量测之噪声资料进行分析，多数航班（中小型）之最大噪声值 Lmax 约为 75-80dB（A），大型飞机则可达到 80dB（A）以上，本项目以 80dB（A）作为基地航空噪声值进行噪声控制的依据，设定建筑围护结构的隔声等级，以满足主要空间（如展厅）的容许背声噪声要求。

② 雨噪声

深圳属于多雨地区，雨量充沛，而这对轻质屋盖的雨噪声影响也带来一定的考验。雨噪声主要与雨强、雨量、雨滴速度等有关系，而屋面做法将直接影响到室内雨噪声声压级的大小。本项目屋顶采用桁架结构金属屋面，屋面构造内设置减振隔声的构造以降低雨噪声的影响。

③ 交通噪声

深茂铁路、穗莞深城际线、深中通道均在本项目周边经过，规划与地铁 20 号线、12 号线，以及沿江高速、广深高速两条高速接驳。对于项目周边的交通噪声，须进行监测获得噪声数据，进行其数据分析，再采取相应措施以减少交通噪声的影响。其主要可通过提高建筑围护结构和隔声门窗的隔声水平、加强出入口隔声（如设置声闸）和室内吸声降噪等方式来降低交通噪声的影响。

④ 室内空调噪声

空调系统对于各空间的背景噪声有很大的影响，在声学要求高的空间中，其送回风系统均应采取相应的消声、减振措施，并且合理的控制系统内的气流速度，有效地降低气流的再生噪声。为了有效降低空调系统的风机噪声及气流噪声，可设置消声箱、消声弯及消声风管及减振垫等方式来降低空调噪声的影响，另外控制风道内的气流速度也是一个必要可行的方法。下表为不同噪声指标要求的空调系统气流速度允许值（表 5-22，表 5-23）。

Source of Ambient Noise

① Aircraft Noise

Aircraft noise is full-band noise, and the low frequency (<250Hz) noise is the major source. The influence of aircraft noise depends on the frequency, time, intensity, noise spectrum distribution, duration, distance and transmission of aircraft.

The Shenzhen World is about 7 kilometers from the airport T3 terminal and 3 kilometers from the planned T4 terminal. According to the noise monitoring, and based on the analysis of the noise we measured on scene, most of the aircrafts(small and medium size) have a maximum noise Lmax of approximately 75-80 db (A), large aircraft can reach above 80dB(A). For the Shenzhen World, we adopt the 80 db (A) value to set the sound insulation grades for the building envelope, so as to meet the requirements for the main space to handle noise.

② Rain Noise

Shenzhen is a rainy area with abundant rainfall, which also brings noise to the light roof. Rain noise is mainly related to the rain intensity, rainfall, rain drop velocity, etc., and the roof directly affects the sound pressure level (SPL) of the rain noise. The roof of the Shenzhen World is metal roof of truss structure, which has equipped with vibration control and sound insulation structure to reduce the influence of rain noise.

③ Traffic Noise

Shenmao railway, Guangzhou-Dongguang-Shenzhen intercity line and Shenzhong channel, all pass through the periphery of the Shenzhen World, which will be connected with metro line 20 and line 12, as well as riverside expressway and GuangShen expressway. For the traffic noise around the Shenzhen World, it is necessary to monitor, obtain and analyze the noise data, then take corresponding measures to reduce the impact of traffic noise. It's mainly by improving the sound insulation of building envelope and doors and windows, strengthening the sound insulation of entrances and exits (for example, setting sound gates) and indoor sound absorption and noise reduction.

④ Indoor Air Conditioning Noise

The air conditioning system has a great impact on the background noise of each space. In the space with high acoustic requirements, the ventilation systems should be able to eliminate noise and reduce vibration, and reasonably control the air flow speed in the system to effectively reduce the regenerative noise of air flow. In order to effectively reduce the fan noise and airflow noise of the air conditioning system, we can set muffler box, muffler bend, muffler windpipe and damping pad to reduce the influence of the air conditioning noise. In addition, it is also a necessary method to control the airflow speed in the air duct. The following table shows the allowable value of air flow velocity required by different noise indexes(Table 5-22, Table5-23).

	对应LA（dB） Corresponding LA (dB)	主管道风速（m/s） Wind speed of main pipeline (m/s)	支管道风速（m/s） Branch duct wind speed (m/s)	终端管道风速（m/s） Wind speed of terminal pipe (m/s)
NR20	25	≦4.5	≦3.5	1.5-2.0
NR25	30	≦5.0	≦4.5	2.0-2.5
NR30	35	≦6.5	≦5.5	≦3.3
NR35	40	≦7.5	≦6.0	≦4.0
NR40	45	≦8.5	≦6.5	≦5.0
NR45	50	≦9.5	≦7.0	≦5.5

表 5-22　参考表：不同噪声标准对应推荐流速控制值
Table 5-22　Recommended Flow Rate Control Values Corresponding to Different Noise Standards

序号 Serial number		说明 Explain	声学软件模拟计算值 Simulated value of acoustic software
1	建筑轻质屋面 Building Light Roof	1. 0.8mm厚PVDF镀铝锌压型钢板 　0.8mm thick PVDF aluminum zinc plated profiled steel plate 2. 1.5mmTPO防水卷材 　1.5mmtpo waterproof roll 3. 1.5mm镀锌钢板 　1.5mm galvanized steel plate 4. 50mm保温岩棉（容重180kg/m³）+ 　50mm保温岩棉（容重120kg/m³） 　50mm thermal insulation rock wool (volume weight 180kg/ m³)+ 　50mm thermal insulation rock wool (volume weight 120kg/ m³) 5. 1.0mm厚镀铝锌压型钢板 　以上屋面构造满足Rw+Ctr≥35的要求。 　1.0mm thick aluminum zinc plated profiled steel plate 　The above roof structure meets the requirements of Rw + Ctr ≥ 35.	Rw　52 dB C　　-4 dB Ctr　-11dB DnTw　54 dB　[v:50m³] 　　　　　　　　[A:11m²]
2	建筑玻璃幕墙 Building Glass Curtain Wall	展厅与外部空间相邻之玻璃门扇及玻璃幕墙采用 6+1.52+6+12A+6mm中空夹胶玻璃， 可满足Rw+Ctr≥35的要求。 The glass door leaf and glass curtain wall adjacent to the exhibition hall and the external space 6 + 1.52 + 6 + 12a + 6mm hollow laminated glass, It can meet the requirements of Rw + Ctr ≥ 35.	Rw　41 dB C　　-2 dB Ctr　-6 dB DnTw　43 dB　[v:50m³] 　　　　　　　　[A:11m²]
3	建筑金属板墙体 Building Metal Plate Wall	展厅上部金属板墙体（标高约18米至20米范围内，建议由1mm不锈钢+150mm岩棉+1mm铝板改采用1mm不锈钢+10mm硅酸钙板+（厚度≥130mm岩棉容重≥120kg/m³）+10mm硅酸钙板+1mm铝板），可满足Rw+Ctr≥35的要求。 The metal plate wall at the upper part of the exhibition hall (the elevation is about 18-20 meters, it is suggested to change from 1m m stainless steel + 150 mm rock wool + 1 mm aluminum plate to 1mm stainless steel + 10 mm calcium silicate plate + (thickness ≥ 130 mm rock wool volume weight ≥ 120 kg/ m³) + 10 mM calcium silicate plate + 1 mm aluminum plate), which can meet the requirements of Rw + Ctr ≥ 35.	Rw　57 dB C　　-2 dB Ctr　-7 dB DnTw　59 dB　[v:50m³] 　　　　　　　　[A:11m²]
4	空调机房墙体 Wall of Air-conditioning Machine Room	空调机房邻展厅内部空间之隔墙，其隔墙构造须为200mm厚混凝土空心砖（密度＞1250KG/m³）+50mm厚空腔内填玻璃棉+双层硅酸钙板（12+10），可满足Rw+Ctr≥55的要求。 The partition wall of the air conditioning room adjacent to the interior space of the exhibition hall shall be made of 200mm thick concrete hollow brick (density > 1250kg/ m³) + 50mm thick hollow brick filled with glass wool + double-layer calcium silicate board (12 + 10), which can meet the requirements of Rw + Ctr ≥ 55.	Rw　58 dB C　　-1 dB Ctr　-3 dB DnTw　60 dB　[v:50m³] 　　　　　　　　[A:11m²]

表 5-23　主要隔声构造之建议与计算表
Table 5-23　Suggestions and Calculation Table of Main Sound Insulation Structures

室内混响时间控制
Control of Indoor Reverberation Time

展厅

展厅为会展中心内的主要空间，大型展厅举办展会时，语言交流特点有别于观演和体育馆等类型建筑，语言清晰度是评价展厅声环境好坏的一项重要指标。对展厅的语言清晰度的考察应包括两部分：一是人与人直接对话交流时的语言清晰度；二是使用扩声系统广播时的语言清晰度。本展厅为一大型展厅，为确保广播及交流时的语言清晰度，除降低室内的背景噪声外，须控制展厅的混响时间，降低混响声能，提高语言清晰度（图 5-140 展厅空间平面位置标示图）。

混响时间（中频带）：≤ 3 秒

关于展厅的混响时间建议值，国内规范并无相关规定。国内《展览建筑设计规范》JGJ218-2010 中建议展厅室内装修宜采取吸声措施。本展厅体积约为 47 万立方米，后经确定展厅之混响时间定为 ≤ 3 秒，混响时间的频率特性曲线允许低频有 15-30% 的提升，高频有 10-20% 的降低。

混响控制策略：经计算，整体吸声吊顶为铝冲孔吸音天花，为屋顶构造的一部分，其吸声降噪系数 NRC ≥ 0.8，墙面部分 ≥ 50% 面积设计为铝冲孔吸声墙，其降噪系数 NRC ≥ 0.7，声反射部位包括地面（合金骨料耐磨地坪）、玻璃幕墙及清水混凝土墙面等。

Exhibition Halls

The exhibition hall is the main space in the exhibition center. When exhibition is held in a large exhibition hall, the characteristic of language communication is different from the performance hall and the gymnasium. Clarity is an important indicator to evaluate the sound environment of the exhibition hall, which includes two parts: First, the language clarity of direct communication between people; Second, the language clarity of the public address system. The exhibition hall is a large one, in order to guarantee the language clarity during broadcasting and communication, and to reduce indoor background noise, reverberation time of the exhibition hall must be controlled so as to reduce the reverberation energy and improve the language clarity(Figure 5-140).

Reverberation time (medium frequency band): ≤3 seconds

There is no relevant norm of suggested reverberation for exhibition hall in China. According to the Design code for exhibition building JGJ218-2010, it's suggested that the indoor decoration of the exhibition hall should adopt sound absorption measures. The volume of this exhibition hall is about 470,000 cubic meters. It's confirmed that the reverberation time is ≤3 seconds, and the frequency curve of reverberation time allows 15-30% increase in low frequency and 10-20% decrease in high frequency.

Reverberation control strategy: through calculation, the overall piercing sound-absorbing ceiling is aluminum punching sound-absorbing ceiling, as part of the roof structure and its noise absorption coefficient is NRC ≥ 0.8, the area of the wall is 50%, and the aluminum perforated sound-absorbing wall with its noise absorption coefficient is NRC ≥0.7, the sound reflection area includes the ground (alloy aggregate wearproof floor), the glass curtain wall and the concrete walls, etc.

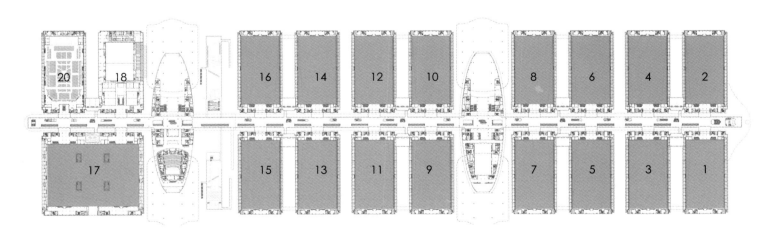

图 5-140　展厅空间平面位置标示图
Figure 5-140　Plan and Location of the Exhibition Hall

国际报告厅

功能定位：会议、讲座及文艺演出

国际报告厅主要作为会议使用的空间，良好的语言清晰度是声学的主要目标，为控制观众厅内有良好的音质环境，需要达到低的背景噪声，以及合适的混响时间（短混响时间），才能获得良好的语音清晰度。本厅体积约为 1.9 万立方米，容积约为 10m³/ 座。混响时间设定为 1.2 秒（±10%），混响时间的频率特性曲线允许低频有 15-30% 的提升，高频有 10-20% 的降低。

混响控制策略：经计算机模拟及计算，需设置较多吸声面积以达成设定混响时间。设计厅内后墙墙面均做强吸声处理，其降噪系数 NRC 应 ≥ 0.8，墙面（两侧墙）应有至少 2/3 的面积作吸声处理，其降噪系数 NRC 应 ≥ 0.7，位置建议在靠近观众席的位置均匀布置，后墙及侧墙材料可采用冲孔吸音板。地面主要为座椅 + 地毯，前区地面同样设置地毯，整体天花吊顶目前采用 3mm 厚的铝板，其背后空腔较深，预估可产生共振吸声 80Hz 以下之频带，墙面冲孔蜂窝铝板针对中低频起着较佳的吸声效果，而观众席及地毯则为中高频的吸声特点，因此能平衡各频带的吸声效果，获得较为平直的混响频率响应（图 5-141 国际报告厅平面标示图）。

舞台空间内，顶部全部吸声处理，其降噪系数 NRC 应 ≥ 0.8，侧墙及后墙于二层马道下方墙面吸声处理，其降噪系数 NRC 应 ≥ 0.7.

International Conference Hall

Function: conference, lecture and performance

The international conference hall is mainly used as a meeting space, and good speech articulation is the main goal for acoustics. In order to control the sound quality in the audience hall, low background noise and appropriate reverberation time (short reverberation time) are necessary to guarantee good speech articulation. The volume of this hall is about 19,000 cubic meters, and the volume of each seat is about 10 m³/seat. The reverberation time is set at 1.2 seconds (±10%), and the frequency characteristic curve of the reverberation time with a 15-30% allowed increase in low frequencies and a 10-20% decrease in high frequencies.

Reverberation Control Strategy: after computer simulation and calculation, the reverberation time needs more sound absorption area. Strong sound absorption treatment has been used for the back wall of the design hall , with the noise reduction coefficient NRC ≥0.8. At least 2/3 area of the wall surface (two sides) should be covered with sound absorption treatment, with the noise reduction coefficient NRC≥0.7, which should be installed close to the audience seat. The back wall and side wall materials can used perforated sound absorption board. On the ground, there is mainly seat and carpet, there is also carpet in the front area, the ceiling which uses 3 mm thick aluminum plate has a deep cavity behind, which would cause resonant sound absorbing with a frequency band under 80 hz. The perforated honeycomb aluminum plate on the wall is effective for sound absorption of low frequency, while the audience seat and the carpet for the sound absorption of high frequency, thus to balance the absorption of each frequency band, achieve the reverberation of flat frequency response(Figure 5-141).

In the stage space, all sound absorption treatment are done on the top, with the noise reduction coefficient NRC ≥0.8, the side wall also adopted absorption treatment, with the noise reduction coefficient NRC ≥0.7.

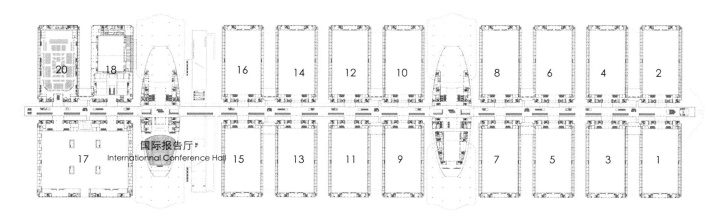

图 5-141　国际报告厅平面标示图
Figure 5-141　International Lecture Hall Plan

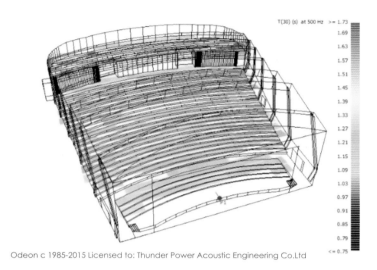

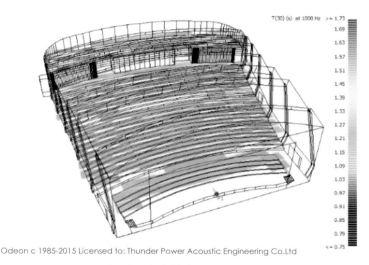

Odeon c 1985-2015 Licensed to: Thunder Power Acoustic Engineering Co.Ltd

Odeon c 1985-2015 Licensed to: Thunder Power Acoustic Engineering Co.Ltd

图 5-142　建筑声学计算机模拟结果 - 混响时间 T30 （500Hz）
Figure 5-142　Computer Simulation Results of Building Acoustics Reverberation Time T30 (500Hz)

说明:依模拟结果可知,中频带500-1000Hz的混响时间约为1.1-1.3秒，符合设计目标无明显的混响时间过长或过短的现象。

Note: According to the simulation results, the reverberation time in the mid-band 500-1000Hz is about 1.1-1.3 seconds, and there is no obvious reverberation time that is too long or too short in accordance with the design goals.

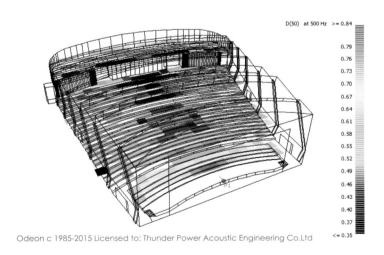

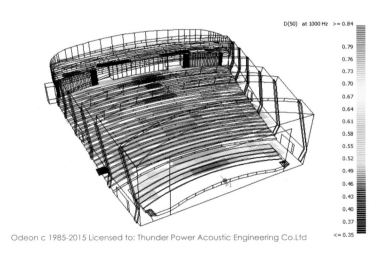

Odeon c 1985-2015 Licensed to: Thunder Power Acoustic Engineering Co.Ltd

Odeon c 1985-2015 Licensed to: Thunder Power Acoustic Engineering Co.Ltd

图 5-143　建筑声学计算机模拟结果 - 语言清晰度 /D50 （500Hz）
Figure 5-143　Computer Simulation Results of Building Acoustics - Speech Intelligibility / D50 (500Hz)

说明:由D50中频带的数值分析，整个厅里的语音清晰度主要分布在50%以上，对于不论是自然声的演出或是电声系统的使用，均能满足其需求。

Note: Based on the numerical analysis of the D50 mid-band, the speech intelligibility in the entire hall is mainly distributed above 50%. It can meet its needs for both natural sound performances and the use of electro-acoustic systems.

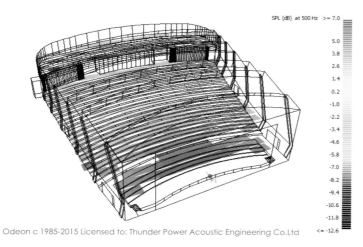 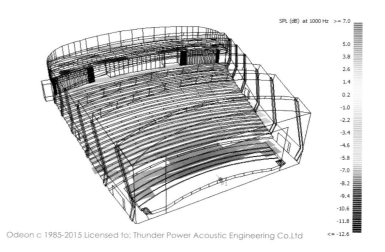

图 5-144 建筑声学计算机模拟结果 - 声压级分布 /SPL（500Hz）
Figure 5-144 Computer Simulation Results of Building Acoustics - Sound Pressure Level Distribution / SPL (500Hz)

说明:依模拟结果可知,各位置的声压级分布均匀，前区后区的声压级差距小，显示厅内的顶部及墙面早期反射声有效的反射至后区的座位区，同时音压级的均匀分布显示观众厅无明显声缺陷的现象，预计室内不均匀度可在±4dB内。

Note: According to the simulation results, the sound pressure levels are distributed evenly at each location, and the difference in sound pressure levels between the front and rear areas is small. The uniform distribution of pressure levels shows that there is no obvious acoustic defect in the auditorium, and the indoor unevenness is expected to be within ± 4dB.

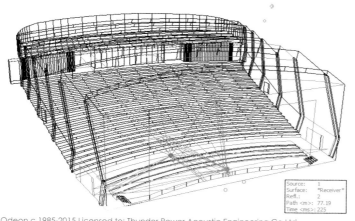 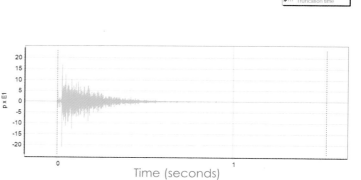

图 5-145 建筑声学计算机模拟结果 - 单点脉冲响应
Figure 5-145 Computer Simulation Results of Building Acoustics - Single Point Impulse Response

说明:各点位模拟之回声图（脉冲响应图），无产生回声之现象,本位置位于池座前区。

Note: The echo map (impulse response map) of each point simulation has no echo phenomenon. This position is located in the front area of the pool.

多功能会议厅及各会议室

功能定位：会议、宴会及展览等

声学指标参考依据

-BREEAM 声学舒适性评价要求室内空闲环境噪音水平满足 35-40dBL，本会议室背景噪声要求达到 NR30，符合 BREEAM 要求。

-LEED 的声学要求会议空间要有较高的语音清晰度，而高语音清晰度则需透过短混响，同时需要较高的讯噪比（S/N），故建声方面需控制较低的背景噪声以提高讯噪比（S/N），同时本会议空间须控制其短混响时间，混响时间的控制应为一平衡的概念，合适的混响时间使本空间得到良好的语言清晰度，同时又不使声音过于干涩，中频带混响时间为 ≤ 0.6 秒（图 5-146 多功能会议厅及各会议室平面位置标示图）。

Multi-functional Conference Hall and other Meeting Rooms

Function: conference, banquet, exhibition, etc.

Acoustic index reference

The BREEAM acoustic comfort evaluation requires that the indoor environment noise level should be within 35-40dbl, and the background noise of this space should be NR30, which fits the BREEAM requirements.

The LEED requires higher speech articulation, which should requires short reverberation and higher signal noise ratio (S/N), so the acoustic design should control low background noise to increase the signal noise ratio (S/N). At the same time this space needs to control its reverberation time. Applicable reverberation time guarantees good speech articulation without making the sound too dry, and the band reverberation time is ≤ 0.6 seconds.(Figure 5-146)

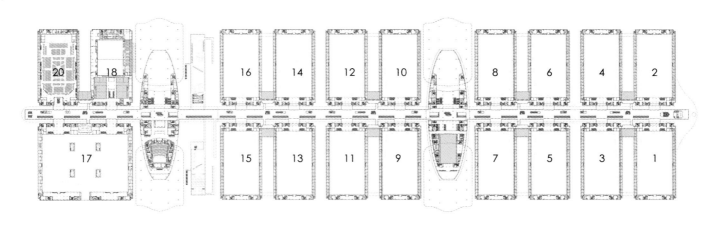

图 5-146　多功能会议厅及各会议室平面位置标示图
Figure 5-146　Multi-function Conference Hall and Meeting Rooms Location Plan

南、北登录大厅

设计依据：登录大厅除作为登录使用外也具有前厅空间的特质，前厅空间内宜做吸声降噪处理，以降低空间喧闹感，并得到合适的混响感，另外前厅空间也是作为隔绝外界噪声的空间，背景噪声需满足 NR-40[45dB(A)]。

建筑声学处理措施：天花部分调整为吸声构造，在入口大厅顶部之铝单板调整为铝冲孔吸音板，其降噪系数 NRC ≥ 0.7，而位于手扶梯上方的铝板天花，也应改为铝冲孔吸音天花，其降噪系数 NRC ≥ 0.7。

North and South Arrival Hall

Design basis: The arrival hall is not only used as the registration Lobby, but also has the characteristics of the vestibule. In the front hall space, sound absorption and noise reduction are needed to reduce the noise of the space and obtain the appropriate reverberation. In addition, the front hall space is also a space insulated from external noise, and the background noise should be NR-40 [45dB(A)].

Architectural acoustic treatment measures: the ceiling is adjusted to sound absorption structure. The aluminum veneer at the top of the entrance hall is adjusted to be aluminum perforated sound-absorbing plate, with the noise reduction coefficient NRC≥0.7, while the ceiling above the escalator should also be changed to aluminum perforated sound-absorbing ceiling with the noise reduction coefficient NRC≥0.7.

序号 Serial number	空间名称 Space Name	体积 Volume	混响时间(RT) Reverberation time(RT)	声学装修处理建议 Suggestions on acoustic decoration treatment
1	多功能厅 Multifunctional Hall	约3万立方米 About 30,000 cubic meters	<0.6秒 < 0.6 seconds	顶棚：蜂窝铝板、冲孔蜂窝铝板 Smallpox: honeycomb aluminum plate, punched honeycomb aluminum plate 地面：地毯 Floor: Carpet 墙面：冲孔蜂窝铝板、石材、亚克力板、玻璃 Wall surface: punched honeycomb aluminum plate, stone, acrylic plate, glass 冲孔蜂窝铝板,其降噪系数NRC≥0.8 Punched honeycomb aluminum plate, its noise reduction coefficient NRC≥0.8 冲孔铝板,其降噪系数NRC≥0.7 Perforated aluminum plate, its noise reduction coefficient NRC≥0.7 地毯,其降噪系数NRC≥0.35 Carpet with noise reduction coefficient NRC≥0.35
2	展厅之间会议室 Meeting Room Between Exhibition Halls	约3800立方米 About 3,800 cubic meters	<0.6秒 < 0.6 seconds	顶棚：铝板、冲孔蜂窝铝板、乳胶漆 Smallpox: aluminum plate, punched honeycomb aluminum plate, latex paint 地面：地毯 Floor: Carpet 墙面：铝板、冲孔铝板、冲孔蜂窝铝板 Wall surface: punched aluminum plate, punched honeycomb aluminum plate 冲孔蜂窝铝板,其降噪系数NRC≥0.8 Punched honeycomb aluminum plate, its noise reduction coefficient NRC≥0.8 冲孔铝板,其降噪系数NRC≥0.7 Perforated aluminum plate, its noise reduction coefficient NRC≥0.7 地毯,其降噪系数NRC≥0.35 Carpet with noise reduction coefficient NRC≥0.35
3	大会议室 （18号展厅二层） Large Conference Room (2nd floor of Hall 18)	约5300立方米 About 5,300 cubic meters	<0.6秒 < 0.6 seconds	顶棚：铝板、冲孔蜂窝铝板、乳胶漆 Smallpox: aluminum plate, punched honeycomb aluminum plate, latex paint 地面：地毯 Floor: Carpet 墙面：铝板、冲孔铝板、冲孔蜂窝铝板 Wall surface: punched aluminum plate, punched honeycomb aluminum plate 冲孔蜂窝铝板,其降噪系数NRC≥0.8 Punched honeycomb aluminum plate, its noise reduction coefficient NRC≥0.8 冲孔铝板,其降噪系数NRC≥0.7 Perforated aluminum plate, its noise reduction coefficient NRC≥0.7 地毯,其降噪系数NRC≥0.35 Carpet with noise reduction coefficient NRC≥0.35
4	小会议室（一层） Small Meeting Room (first floor)	约250立方米 About 250 cubic meters	<0.6秒 < 0.6 seconds	顶棚：铝板、冲孔蜂窝铝板、乳胶漆 Smallpox: aluminum plate, punched honeycomb aluminum plate, latex paint 地面：地毯 Floor: Carpet 墙面：铝板、冲孔铝板、冲孔蜂窝铝板 Wall surface: punched aluminum plate, punched honeycomb aluminum plate 冲孔蜂窝铝板,其降噪系数NRC≥0.8 Punched honeycomb aluminum plate, its noise reduction coefficient NRC≥0.8 冲孔铝板,其降噪系数NRC≥0.7 Perforated aluminum plate, its noise reduction coefficient NRC≥0.7 地毯,其降噪系数NRC≥0.35 Carpet with noise reduction coefficient NRC≥0.35

表 5-24　声学指标参考依据表
Table 5-24　Acoustic Index Reference

宴会厅

宴会厅的声学设计要求保证厅堂内的语言清晰度、丰满度，混响时间既要有利于语言类活动的使用，也须为音乐活动留出足够的可调范围（图 5-147 宴会厅平面位置标示图）。需无回音等声学缺陷，以保证电声系统得到良好的发挥。本宴会厅之背景噪声值设定为 NR-35，混响时间（500-1000Hz，满场）的设定为 ≤ 1.5 秒，另外辅助用房之音响控制室依 JGJ57-2016 剧场建筑设计规范之建议值，背景噪声满足 NR-35，中频带混响时间为 0.3-0.5s（平直），而辅助用房之同声传译室则依 ISO-2063 的规定，背景噪声满足 35dB（A），中频带混响时间为 0.3-0.5s。

混响时间（中频带）：≤ 1.5 秒

关于宴会厅的混响时间建议值，国内规范并无相关规定。本宴会厅体积约为 12 万立方米，后经确定宴会厅之混响时间定为 ≤ 1.5 秒，混响时间的频率特性曲线允许低频有 15-30% 的提升，高频有 10-20% 的降低。

混响控制策略：经计算，吊顶为钢格栅透声构造，于其背后的隔声吊顶面（全部）做吸声喷涂，其吸声降噪系数 NRC ≥ 0.7，墙面部分≥ 50% 面积设计为铝冲孔吸声墙，其降噪系数 NRC ≥ 0.7，侧墙反射扩散材料，面密度 ≥ 30kg/m²，地面铺设地毯，其降噪系数 NRC ≥ 0.35。

Banquet Hall

The acoustical design of the banquet hall requires the clarity and fullness of the speech, the reverberation time should not only be conducive to the use of language activities, but also need to leave enough adjustable range for music activities (Figure 5-147). No acoustic defects, such as echo, to ensure the sound system function well. The banquet hall of background noise values set to NR - 35, reverberation time (500-1000 hz, full) ≤1.5 seconds. In addition, according to the Design code for theater (JGJ57-2016), the background noise of the audio control room in the auxiliary room is NR - 35, and the band reverberation time is 0.3-0.5 s (straight), while according to the provisions of ISO - 2063, the background noise of simultaneous interpretation room is 35 db (A), and the band reverberation time is 0.3-0.5 s.

Reverberation time (medium frequency band) : ≤1.5 seconds

There is no relevant norm of suggested reverberation for banquet hall in China. The volume of this hall is about 120,000 cubic meters, the reverberation time is ≤1.5, and the frequency characteristic curve of the reverberation time allowed a 15-30% increase in low frequencies and a 10-20% decrease in high frequencies.

Reverberation control strategy: through calculation, the ceiling is steel grating structure, the surface behind (all) is sprayed with sound absorption material, its noise absorption coefficient is NRC≥0.7, the covered wall area is 50%, and the aluminum perforated sound-absorbing wall, with its noise absorption coefficient is NRC≥0.7. The side wall uses reflection diffusion material, with a surface density ≥30kg/m², the ground is covered by carpet, with the noise reduction coefficient NRC≥0.7.

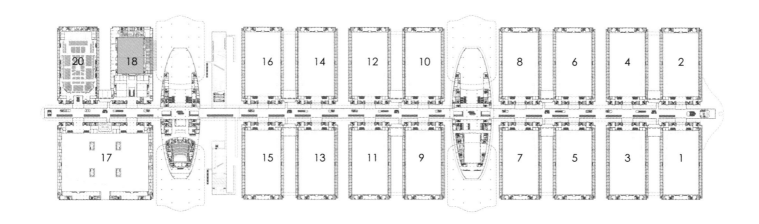

图 5-147　宴会厅平面位置标示图
Figure 5-147　Plan and Location of the Banquet Hall

C-12 多功能场馆

功能定位：展览、体育赛事、演唱会（图 5-148 C-12 多功能场馆平面位置标示图）。

依照《体育馆声学设计及测量规程 JGJ/T131-2012》要求，本多功能场馆体积约为 35 万立方米，依规范建议，本多功能场馆混响时间（500-1000Hz, 满场）的合适范围则介于约 1.9-2.1 秒之间，但为更有利于其他活动的进行，评估其功能以语言功能为主，故混响时间设为 ≤ 1.7 秒，混响时间的频率特性曲线允许低频有 15-30% 的提升，高频有 10-20% 的降低。另外辅助用房音响控制室依照 JGJ57-2016 剧场建筑设计规范建议值设计，中频带混响时间为 0.3-0.5s(平直)

混响控制策略：经计算，整体吸声吊顶为铝冲孔吸音天花，为屋顶构造的一部分，其吸声降噪系数 NRC ≥ 0.8，墙面部分 ≥ 55%，设计为铝冲孔吸声墙，其降噪系数 NRC ≥ 0.7，声反射部位包括地面（合金骨料耐磨地坪）、玻璃幕墙及少许混凝土墙面等（表 5-25　不同容积比赛大厅 500-1000HZ 满场混响时间）。

C-12 Multifunction Hall

Function: exhibition, sports events, concerts(Figure 5-148).

In accordance with the "Regulations for Acoustical Design and Measurement of Gymnasiums JGJ/T131-2012", the volume of this space is about 350,000 cubic meters, the suitable scope of the reverberation time (500-1000 hz,full) is between 1.9-2.1 seconds. However, for other activities that requires speech articulation, the reverberation time is set as ≤ 1.7 seconds, and the frequency characteristic curve of the reverberation time allowed a 15-30% increase in low frequencies and a 10-20% decrease in high frequencies. Additionally, according to the Design code for theater JGJ57-2016, the background noise of the audio control room in the auxiliary room is designed with the mid-frequency band reverberation time is 0.3-0.5 s (straight).

Reverberation control strategy: through calculation, the sound-absorbing ceiling is aluminum perforated sound-absorbing ceiling, as part of the roof structure, its noise absorption coefficient is NRC≥0.8, the wall area is 55% with aluminum perforated sound-absorbing wall, its noise absorption coefficient is NRC≥0.7, the sound reflection area includes the ground (alloy aggregate wearproof floor), glass curtain wall and concrete walls, etc.

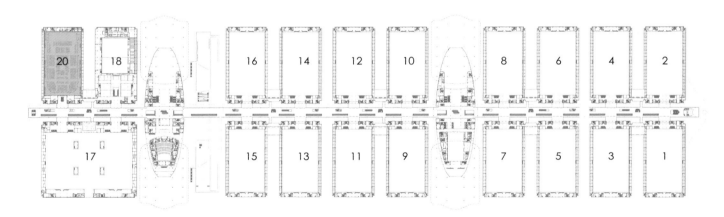

图 5-148　C-12 多功能场馆平面位置标示图
Figure 5-148　C-12 Multi Functional Venue Location Plan

不同容积比赛大厅500-1000Hz满场混响时间 Different Volume Competition Hall 500-1000hz full reverberation time				
容积（m³） Volume (m³)	< 40000	40000-8000	80000-160000	>160000
混响时间（S） Reverberation time (s)	1.3-1.4	1.4-1.6	1.6-1.8	1.9-2.1

表 5-25　不同容积比赛大厅 500-1000Hz 满场混响时间表
Table 5-25　Full Reverberation Time of 500-1000Hz in Competition Hall with Different Volume

绿建设计
Green Architecture Design
内容摘自：深圳市建筑科学研究院
Shenzhen Academy of Building Science

深圳国际会展中心是粤港澳大湾区的揭幕之作，也是深圳会展面向全球的展示窗口，作为全球绿建定位目标最高的会展类建筑，旨在打造第四代绿色会展标杆，并向世界展示深圳生态宜居的经营理念。

The Shenzhen World is the starting project of the GBA, and the showcase of Shenzhen exhibition industry to the world. With the highest positioning goal of global green architecture, it aims to build the fourth-generation green exhibition model and show the world the management concept of ecological living environment of Shenzhen.

技术框架
Technical Framework

图 5-149　绿色会展框架示意图
Figure 5-149　Green Exhibition Framework

最节能——全球能耗指标最低的会展中心

通过围护系统＋空调系统＋照明及设备系统的优化，实现单位面积能耗约 70.98kWh/m^2·a，较深圳大型公建降低 40%，远低于深圳市大型公建平均年能耗（图 5-150 建筑单位面积能耗对比分析图）。

Energy-saving - the Lowest Energy Consumption Index in the World

Through the optimization of envelope system + air conditioning system + lighting and equipment system, the energy consumption per unit area is about 70.98kWh/m^2·a, which is only 46% of the buildings with same scale in the United States, and much lower than the average annual energy consumption of large-scale public buildings in Shenzhen(Figure 5-150).

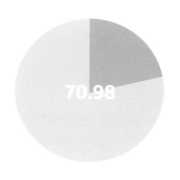

美国大型展馆
Large American Exhibition Pavilions

深圳大型公建
Large Public Constructions in Shenzhen

深圳国际会展中心
The Shenzhen World

备注：统计单位kWh/㎡·a美国大型展馆数据由底特律、丹佛、芝加哥、麦考密克展览中心的运营能耗值组成。
Remarks: The statistical unit kWh / ㎡ · a The data of large-scale exhibition halls in the United States are composed of the operating energy consumption values of the Detroit, Denver, Chicago, and McCormick Exhibition Centers.

图例
Legend

 美国大型展馆
Large American Exhibition Pavilion

 深圳大型公建
Large Public constructions in Shenzhen

 深圳国际会展中心
The Shenzhen World

图 5-150　建筑单位面积能耗对比分析图
Figure 5-150　Comparative Analysis Diagram of Energy Consumption per Unit Area of Building

最节水—— 大型单体建筑雨水收集系统

通过雨水收集回用＋市政中水＋节水器具的优化，每年降低市政用水 42.21 万吨，雨水替代率 3.1%，超过《深圳市海绵城市专项规划》近期目标（1.5%）（图 5-151 雨水收集系统图）。

Water-saving - Rainwater Collection System for Single Large-scale Building

Through the optimization of rainwater collection and reuse + municipal reclaimed water + water-saving appliances, 422,100 tons of municipal water would be saved every year, and the replacement rate of the rainwater would be 3.1%, exceeding the short-term goal (1.5%) of the Special Plan for Sponge City of Shenzhen(Figure 5-151).

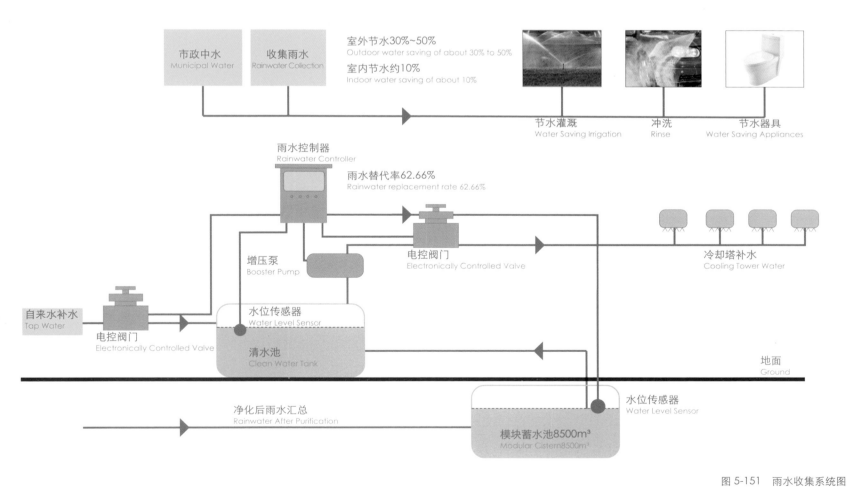

图 5-151 雨水收集系统图
Figure 5-151 Rainwater Collection System Diagram

最节材——全国装配率最高的大型公建

通过全钢结构 + 地上综合管廊 + 可回收材料的优化，降低施工费用（1470 万元）和建筑主体结构的运营维护费用（图 5-152 技术示意图）。

Material-Saving - the Largest Public Construction with the Highest Prefabriction Rate in the Country

Through the optimization of all-steel structure + overground utility tunnel + recyclable materials, the construction cost (14.7 million yuan) and the operation and maintenance cost of the main structure of the building were reduced(Figure 5-152).

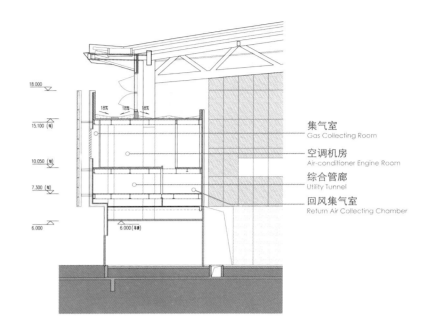

集气室
Gas Collecting Room

空调机房
Air-conditioner Engine Room

综合管廊
Utility Tunnel

回风集气室
Return Air Collecting Chamber

图 5-152　技术示意图
Figure 5-152　Technical Diagram

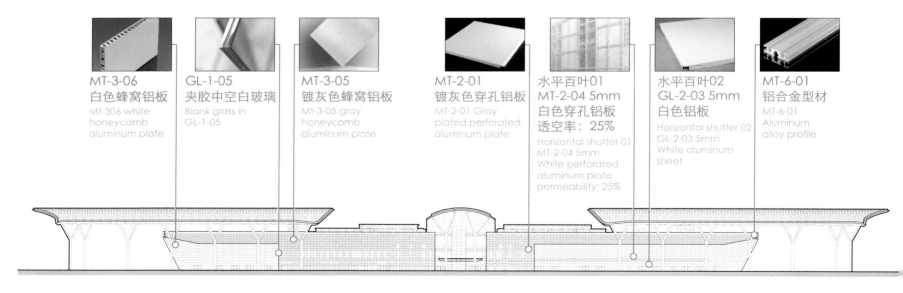

MT-3-06
白色蜂窝铝板
MT-306 white honeycomb aluminum plate

GL-1-05
夹胶中空白玻璃
Blank glass in GL-1-05

MT-3-05
镀灰色蜂窝铝板
MT-3-05 gray honeycomb aluminum plate

MT-2-01
镀灰色穿孔铝板
MT-2-01 Gray plated perforated aluminum plate

水平百叶01
MT-2-04 5mm
白色穿孔铝板
透空率：25%
Horizontal shutter 01 MT-2-04 5mm White perforated aluminum plate permeability: 25%

水平百叶02
GL-2-03 5mm
白色铝板
Horizontal shutter 02 GL-2-03 5mm White aluminum sheet

MT-6-01
铝合金型材
MT-6-01 Aluminum alloy profile

图 5-153　立面铝板示意图
Figure 5-153　Aluminum Plate Elevation

最低碳——国际先进的建筑全生命周期评估

通过 LEED+BREEAM 认证的全生命周期评估及优化，实现碳排放指数 12kgCO_2e/m^2·a，生态影响因子大大低于国际标准（图表 5-2 生态影响因子分析）。

Low-Carbon-International Advanced Full Building Life Cycle Assessment

Certified by LEED+BREEAM for the full life cycle assessment and optimization, achieving a carbon emission index of 12kgCO_2 e/m²·a, with ecological impact factor significantly lower than the international standard(Chart 5-2).

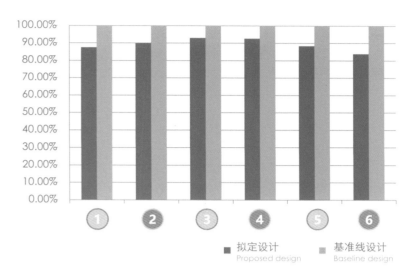

图表 5-2　生态影响因子分析
Chart 5-2　Analysis of Ecological Impact Factors

① 全球变暖潜力（温室气体）12.50%↓
Global warming potential (greenhouse gas) 12.50% ↓

② 平流层臭氧层损耗系数 10.02%↓
Stratospheric ozone layer loss coefficient 10.02% ↓

③ 土壤和水源酸化指数 7.04%↓
Soil and water acidification index 7.04% ↓

④ 富营养化影响 7.32%↓
Effect of eutrophication 7.32% ↓

⑤ 对流层臭氧形成指数 11.57%↓
Tropospheric ozone formation index 11.57% ↓

⑥ 不可再生能源资源的消耗量 16.08%↓
Consumption of non renewable energy resources 16.08% ↓

实现路径
Implementation Methods

图 5-154　四大绿色技术板块图
Figure 5-154　Four Main Green Technology Configuration

四大绿色技术板块，保障可持续发展
Four Green Technology Sectors to Guarantee Sustainable Development

① 综合规划

项目分析——因地制宜

• 屋面面积占比高

• 可绿化面积小

• 地面荷载要求高

• 雨水下渗空间有限

• 项目冷却水用水需求大

设计原则——应做尽做

• 最大限度利用绿地净化径流

• 满足雨水回用需求

• 局部适宜地面采用透水铺装（图 5-156 地面经流组织方案）

• 局部适宜屋顶设置绿色屋顶（图 5-155 屋面经流组织方案）

系统方案——有机结合（图 5-157 雨水组织方案）

• 5.72ha　下沉式绿地

• 1.15ha　雨水花园

• 2.2ha　透水混凝土

• 2.6ha　绿色屋顶

• 8500m^3　雨水回用池

Comprehensive planning

Project analysis - adapt to local conditions

• High proportion of roof area

• Small green area

• High ground load requirements

• Limited rainwater infiltration space

• Large demand for cooling water

Design principle – Try Best

• Maximum use of green space to purify runoff

• Meet the demand for rainwater reuse

• Suitable ground Partly applied permeable pavement (Figure 5-156)

• Suitable roof Partly applied green roof (Figure 5-155)

System solution - Organic combination (Figure 5-157)

• 5.72ha　sunken green space

• 1.15ha　rain garden

• 2.2ha　permeable concrete

• 2.6ha　green roof

• 8500m^3　rainwater reuse pool

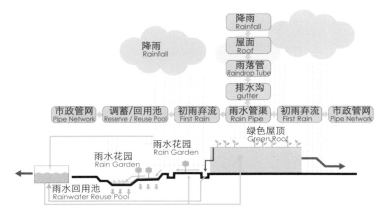

图 5-155　屋面经流组织方案
Figure 5-155　Roof Run-off Organization Scheme

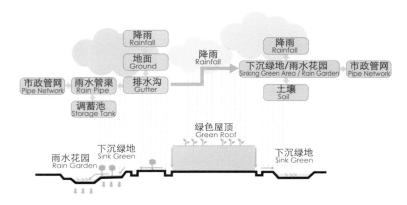

图 5-156　地面经流组织方案
Figure 5-156　Ground Surface Run off Organization Scheme

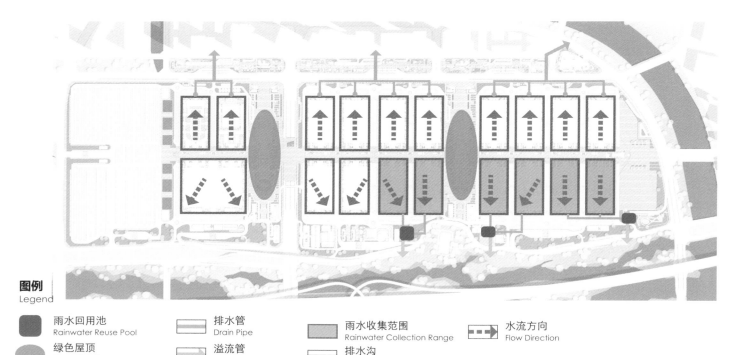

图 5-157　雨水组织方案
Figure 5-157　Rainwater Organization Plan

② 建筑本体

高效送冷
- 分布式冷站，大温差送冷，冷却循环水量减少 44%；
- 全部采用一级能效变频式冷水机组，调节灵活，按需取冷；
- 大功率机组采用 10kV 高压供电，降低启动和运行电流，节能的同时又节省了配电投资。

调峰蓄谷
- 深圳市峰谷电价鼓励调蓄，采用水蓄冷措施，白天用电量年节约 420 万 kWh，节费 244 万元 / 年；
- 供冷高峰时期有水蓄冷助力，避免冷水机组频繁变频，降低设备损耗。

智能控制
- 高大空间分层空调，舒适节能；
- 结合功能设置独立控制系统；
- 设置过渡季全新风运行，生态环保；
- 传感器反馈二氧化碳浓度，实现新风精准调节。

The Architecture Noumenon

Efficient Cooling System
- Distributed cold station, large temperature difference for cooling, reducing circulation cooling water by 44%；
- Use water chilling unit of variable-frequency with first-level energy efficiency, which can be adjusted flexibly and can be cooled as needed；
- The high-power unit adopts 10kV high-voltage power supply to reduce the starting and running current, saving both energy and power distribution investment at the same time.

Shaving and Storage
- Shenzhen peak-valley electricity price encourages regulation and storage, and adopts water cooling measures to save 4.2 million kWh and 2.44 million yuan per year for daytime electricity consumption；
- During the peak cooling period, water can be used to store cold power to avoid frequent frequency conversion of the water chilling unit and reduce equipment depletion.

Intelligent Control
- Tall space adopts layered air conditioning, which is comfortable and energy-saving；
- Independent control system sets according to functions；
- Set full fresh air operation during the transition of seasons to be ecological and environmental friendly；
- Sensor feedback of CO_2 concentration, guarantees accurate regulation of fresh air.

	环境温度 Ambient Temperature	人体衣物热阻 Body clothing Thermal Resistance	APMC分布云图1级 APMC Cloud Map Level 1	室内人行活动高度风速云图 Wind Speed Cloud Diagram of Indoor Pedestrian Activity Height	AGE空气龄云图 AGE Air Age Cloud Map	面积比例 Area Ratio
标准 展厅 Standard Exhibition Hall	17℃	CLO 1.5				92%
	18℃	CLO 1.5				100%
	19℃	CLO 1.5				100%
	20℃	CLO 1.5				92%
	21℃	CLO 1.5				64%
	25℃	CLO 1.5				84%

表 5-26　标准展厅热环境分析
Table 5-26　Standard Exhibition Hall Thermal Environment Analysis

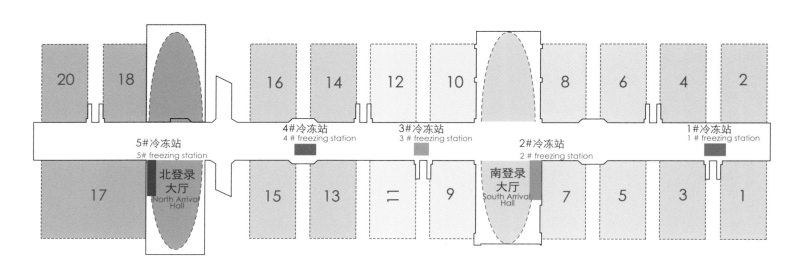

图 5-158　会展建筑冷冻站分布图
Figure 5-158　Standard Exhibition Hall Thermal Environment Analysis

③ 室内环境

光照效率
• 全面采用高效 LED 灯具，视觉舒适度 140lm/W，高于传统 LED 灯具 56%；
• 照明功率密度设计值比标准值低 60% 以上。

Dali 调光
• 采用控制讯号线与控制设备并接，智能寻址，线路可以最优分布，同时电源线与讯号线同管分布，有效减少管线材料的使用；
• 逐灯控制，满足多功能场景（图 5-159 1-10V 解决方案与 DALI 解决方案示意图）。

节能效益
• 高效 LED 灯具较普通 LED 灯具节约用能 75%，年节电量可达 220 万 kWh。

房间照明功率密度 Room lighting power density	标准值（W/m²） standard value	设计值（W/m²） design value
20K展厅 20K exhibition hall	≤9	3.36
50K展厅 50K exhibition hall	≤9	4.08

表 5-27 房间照明标准值与设计值表
Table 5-27 Standard Value and Designed Value of Room Lighting

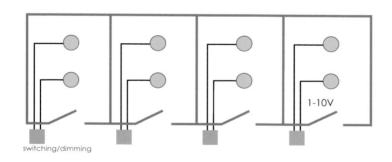

1-10V解决方案：每个房间至少需要一个2路开关/调光执行器
1-10Vsolution:at least one 2-way switching/dimming actuator is needed per room

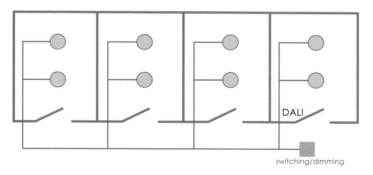

DALI解决方案：更多功能-更少的组件。
在输出端只有两根线，该接口具有与16个开关/调光功能相同的功能
DALI solution:more functions -fewer components.
with just two wires at the output the interface has the same funtionality as 16 switching/dimming

图 5-159 1-10V 解决方案与 DALI 解决方案示意图
Figure 5-159 1-10V Solution and DALI Solution

③ Indoor Environment

Light Efficiency
• Full use of efficient LED lamps, visual comfort is 140 lm/W, 56% higher than traditional LED lamps;
• The design value of the lighting power density is more than 60%. lower than the standard value.

Dali Dimmer
• The combination of control signal line and control equipment is adopted for intelligent addressing, so that the line can be optimally distributed, and the power line and signal line can be distributed in the same pipe at the same time, effectively reducing the use of pipeline materials;
• Lamps control for multi-functional scene(Figure 5-159).

Energy-saving Benefit
• Efficient LED lamps can save 75% of energy consumption compared with ordinary LED lamps, and the annual energy saving can reach 2.2 million kWh.

④ 建造运营

全钢结构
• 高强钢筋比例 100%，高强钢材比例 98.5%，均为高耐久性材料；
• 可再循环材料比例 14.93%。

预制构件
• 钢构件（图 5 - 160）、幕墙、屋面构建全部采用"BIM 下单，工厂预制"模式；
• 预制清水外挂墙板，预制外墙 2474 件，预制内墙 2528 件；
• 减少现场湿作业，节费 213 万元，节约工时 7098h（图 5-161）。

地上综合管廊
• 减少挖方 21 万平方米，便于安装，节省工期；
• 明装敷设，综合集中排布，为后期维护提供极大便利。

绿色施工
• 太阳能 +LED 照明，安全节能；
• 办公区雨水收集、基坑降水回用；
• 混凝土内支撑回收加工，用于地平回填，实现材料循环利用；
• 全场自动喷洒防尘，裸土覆盖。

精明运行策略
• 巧妙运用水蓄冷量，实现不同冷负荷比例下制冷机组平稳运行；
•25% 冷负荷下，无需开启制冷机组，蓄冷承担全部空调任务。

图 5- 160 钢结构实景图
Figure 5-160 Steel Structure Reality Image

图 5-161 预制构件实景图
Figure 5-161 prefabricated Components

Construction and Operation

Full-steel Structure
• High strength reinforcement ratio 100%, high strength steel ratio 98.5%, all high durability materials;
• Recyclable materials ratio 14.93%.

Prefabricated Components
• The construction of steel components(Figure 5-160), curtain wall and roof adopts the mode of "BIM ordering and factory prefabrication";
• Prefabricated bare concrete exterior wallboard consists of 2474 pieces of prefabricated outer wall, 2528 pieces of prefabricated inner wall;
• Reduced on-site wet construction, saving 2.13 million yuan and 7098 working hours(Figure 5-161).

Above-ground Utility Tunnel
• Reduce cutting excavation by 210,000 cubic meters, facilitate installation and save construction period;
• Plain installation, integrated and centralized arrangement, providing great convenience for later maintenance.

Green Construction
• Solar energy +LED lighting, safe and energy-saving;
• Rainwater collection in office area, reuse of foundation pit dewatering;
• Recycling and processing the internal concrete support for ground backfilling;
• Automatic dust-proof spraying and bare soil covering.

Smart Operation Strategy
• Smart use of water storage capacity to achieve the stable operation of refrigerating unit under different cooling load ratio;
• There is no need to start the refrigerating Unit under 25% cooling load.

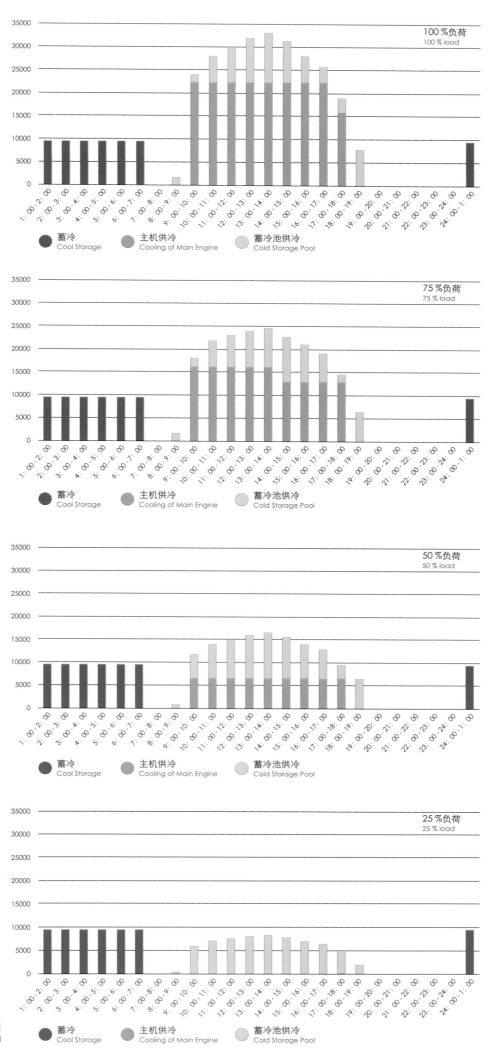

图 5-162　不同负荷状态下供冷分析图
Figure 5-162　Analysis Diagram Under Different Cooling Load

整体设计
Overall Design

编号 Number	技术名称 Technology
1	地下空间综合开发 comprehensive development of underground space
2	管网防渗漏 leakage prevention of pipe network
3	太阳能系统建筑一体化 Solar system architecture integration
4	水蓄冷 chilled water storage
5	透水铺装 permeable pavement
6	雨水花园 rainwater garden
7	雨水收集与处理 rainwater collection and treatment
8	再生水回用 recycled water reuse
9	海绵屋顶 sponge roof
10	本土植物及复层绿化优化配置 optimized allocation of native plants and multi-layer greening
11	场地风环境优化 wind environment optimization of the site
12	场地噪声污染防治 prevention and control of site noise pollution
13	场地光污染防治 prevention of light pollution on site
14	垃圾分类收集 garbage collection

编号 Number	技术名称 Technology
1	地上综合管廊技术 overground comprehensive pipe corridor technology
2	BIM技术 BIM technology
3	可周转模块应用 returnable module application
4	混凝土、砂浆回用 reuse of concrete and mortar
5	太阳能+LED照明 Solar +LED lighting
6	基坑降水回用 reuse of foundation pit dewatering
7	办公区雨水收集 rainwater collection in office area
8	自动喷洒防尘技术 automatic spraying dust-proof technology
9	分类分项计量技术 classification and item measurement technology
10	能耗水耗监测平台 monitoring platform for energy consumption and water consumption
11	对外开放展示平台 open platform
12	云会展 cloud

建造运营
Construction Operation

建筑本体
Architecture Noumenon

编号 Number	技术名称 Technology
1	建筑外遮阳 exterior shading
2	围护结构节能分析 energy saving analysis of envelope structure
3	隔热设计 heat insulation design
4	局部空间溶液除湿 solution dehumidification for part of the space
5	空调末端独立控制 air conditioning terminal independent control
6	大温差送风送冷 large temperature difference of cooling air supply
7	高强度钢材利用 Use of high strength steel
8	节水器具使用 use of water-saving appliances
9	综合能耗分析 comprehensive energy analysis
10	结构体系优化 structural system optimization
11	电梯节能（DC-DC电源、无齿轮牵引机）elevator energy saving (DC-DC power supply, gearless tractor)

编号 Number	技术名称 Technical name
1	室内空调气流组织分析 air distribution analysis of indoor air conditioning
2	座椅送风 ventilation from seat
3	中央廊道物理环境优化 physical environment optimization of central corridor
4	空调部分负荷高效运行控制 efficient operation control of partial load of air conditioning
5	墙体、门窗、室内构造隔声 indoor structures sound insulation of walls, doors and windows
6	吸音降噪、声掩蔽 sound absorption and noise reduction, sound masking
7	声学模拟分析 acoustic simulation analysis
8	设备降噪 equipment noise reduction
9	室内自然通风优化 optimization of indoor natural ventilation
10	室内污染物控制 indoor pollutant control
11	采光优化设计 optimal lighting design
12	室内环境质量在线监测 online monitoring of indoor environmental quality
13	高效LED照明 LED lighting of high efficiency
14	DALI调光技术 DALI dimming technology
15	建筑智能化集成 Intelligent building integration

室内环境
Indoor Environment

表 5-28 52 项绿建技术护航精明会展
Table 5-28 52 Green Construction Technology

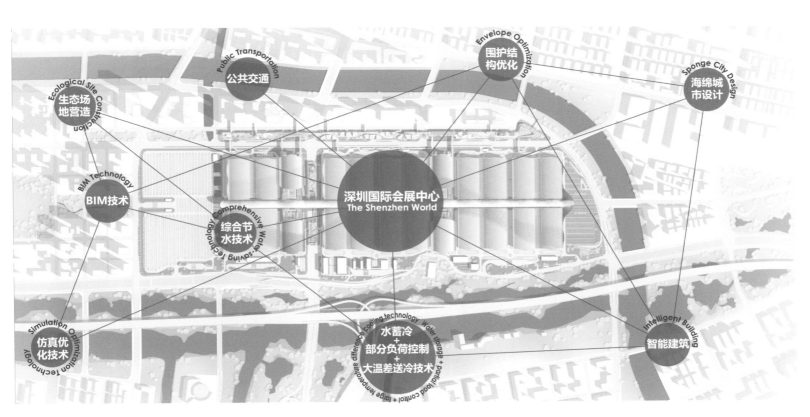

图 5-163 九大技术亮点锁定国际认证
Figure 5-163 Nine Technical Highlights with International Certification

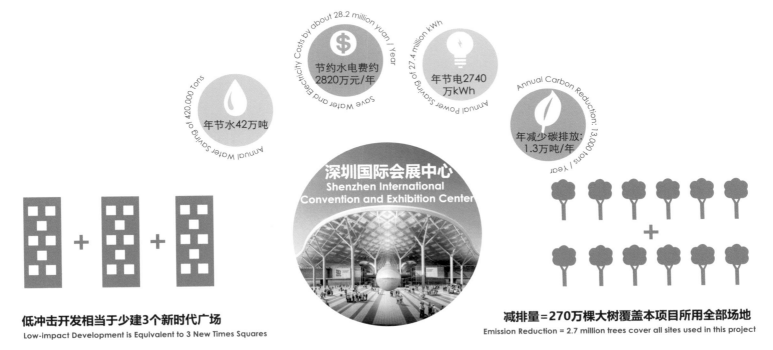

低冲击开发相当于少建3个新时代广场
Low-impact Development is Equivalent to 3 New Times Squares

减排量=270万棵大树覆盖本项目所用全部场地
Emission Reduction = 2.7 million trees cover all sites used in this project

图 5-164 未来城市名片——"绿色生态先锋"
Figure 5-164 Business Card of future city - "Green Ecological Pioneer"

"远" 景

新时代的起点
The Start of A New Era

450 用户体验
User Experience

458 开局之年
The Year Ahead

459 未来图景
Visions for The Future

Far Sight

用户体验
User Experience

深圳国际会展中心，是深圳会展业的重要城市基础设施，是关系到深圳未来百年发展大计的标志性工程，是深圳市委市政府的一项重大决策，项目采用全新的"建设、运营＋综合开发"（BO＋D）的一体化运作模式来进行新场馆的规划、建设和运营。通过邀请招标方式，确定了招商蛇口与华侨城联合体为政府代建新会展中心一期，并由招商蛇口和华侨城联合发起成立的深圳市招华国际会展运营有限公司主导后续展馆运营工作。

深圳国际会展中心自 2019 年 11 月正式开放以来，截止目前已经承接了几场展览活动：宝安产业博览会、智能装备博览会、DMP 大湾区工业博览会、2020 央视新年音乐会录制、VIVO 集团年会，以及 AUBE 欧博设计年会。

期间，AUBE 欧博设计有机会采访了深圳市招华国际会展运营有限公司总经理毛大奔先生。作为会展运营方，他从运营挑战、用户体验、未来蓝图等几个方面，以长远的全局观和完整的用户视角——解读了正式拉开帷幕、登上时代舞台的深圳国际会展中心的全貌，展示了一个真实、立体，充满生命力的会展景象。

The Shenzhen World is an important urban infrastructure of exhibition industry, as well as a landmark project for the future goal of Shenzhen's one hundred year development. Therefore, the Shenzhen municipal government determined to adopt a brand-new integrated operation mode of BO+D for its construction, operation and overall development. Through invitation for bids , the consortium of China Merchants Shekou and OCT are selected to take charge of the development of the Shenzhen World, while the Shenzhen Zhaohua international conference and exhibition operation Co., LTD. established by the consortium is responsible for the subsequent operation.

Since opened in November 2019, there has been several exhibitions and events held in the Shenzhen World, for example: DMP Industrial Expo, 2020 New Year Music Concert, VIVO Annual Meeting.

Recently, we've got the opportunity to interview the deputy general manager Mr. Mao Daben from Zhaohua International Exhibition Operation co., LTD.,. For the perspective of exhibition operator, He talked about the operational challenges, user experience, and future visions of the Shenzhen World, revealing us a real, three-dimensional, emerging exhibition center.

运营挑战："三个一流与国际化"
Challenge of Operation: Three First-class and Internationalization

站在视野开阔的中央廊道上，毛大奔再次称赞了这个新展馆的鱼骨式布局，"整齐大气，流线清晰，辨识性强"，优秀的硬件设施固然能给运营带来各种优势，但毛大奔表示真正的挑战或压力来自于对"三个一流与国际化"理念的坚持和执行（图 6-1），"在长久的、多层面的工作中，我们需要将国际体系和本地展馆运营管理经验结合，确保所有来办展、参展、参观的客户能够享受到国际标准的运营服务，为全球客户打造世界一流的绿色智慧展馆和全球性的商贸平台，持续为全球客户创造价值。"

图 6-1　大湾区工业博览会现场
Figure 6-1　Greater Bay Area Industrial Expo

Standing in the central concourse with a great view, Mao praised again about the herringbone layout of the building, "magnificent, clear, and distinctive". The excellent hardware facilities are definite an advantage for the operator. However, Mao said that the real challenge or pressure lies in the persistence and execution of "three first class and internationalization " idea.(Figure 6-1) "For the long-term, multifaceted work ahead, we need to combine the international system to local operation management experience, to guarantee that all visitors, exhibitors, customers can enjoy the international standard of service, creating a world-class green and smart exhibition center and a global business platform for global customers, and continue to create value for global customers."

图 6-2　展会现场采访（主办方）
Figure 6-2　Exhibition Interview (sponsor)

一流的绿色智慧展馆

深圳国际会展中心致力于打造规模宏大、节能环保的一流绿色展馆和全球最具代表性的新一代智慧展馆。深圳国际会展中心秉承绿色理念，在建中应用了 52 项全球领先的绿建技术，力争成为全球首个获得三个绿色认证的展馆，即国家绿色建筑二星级（同类型，国内首例），美国 LEED-NC 金级（会展类，世界首例），英国 BREEAM（会展类，世界首例）。

而在打造智慧展馆层面，深圳国际会展中心以"数字化、智慧化、智能化"为目标，与华为、腾讯、中国电信、埃森哲等领先的企业合作，通过智慧业务体系和智慧平台建设，笃志将深圳市国际会展中心全力打造成为全球最具代表性的新一代智慧展馆（图 6-2）。

World-class Green and Smart Exhibition Center

The Shenzhen World is committed to build as a world-class green exhibition center with large scale, energy efficiency and environmental protection, as well as a new smart exhibition center leading the world standard. It has applied 52 world's leading green building technologies, and strive to be the first exhibition building that holds three green certifications, namely the National Green Building Two-star (first in China), the American LEED - NC gold (first among the world)and British BREEAM (first among the world).

In regard of creating a smart exhibition center, the Shenzhen World is achieving the goal of "digital, wisdom, intelligence" by cooperating with world leading companies such as Huawei, Tencent, China Telecom, Accenture, etc. By building a smart business system and smart platform, the Shenzhen World is determined to become the new generation of smart exhibition center around the world(Figure 6-2).

图 6-3　会展中心展会现场
Figure 6-3　Exhibition in the Shenzhen World

图 6-4　会展中心展会现场
Figure 6-4　Exhibition in the Shenzhen World

国内首创多功能展厅复合模式

深圳国际会展中心 1-16 号馆为标准展馆，每个面积均为 2 万平方米。全部采用单层、无柱、大跨度空间设计，可灵活组合，既满足各类大型展会的需求，也能为小型展会提供个性化服务（图 6-3）。

北区的 17、18、20 这 3 个特殊展厅具备展览、会议、宴会、体育赛事、大型文艺活动和高端商务活动等一系列功能，这种功能的复合利用来自于国际运营方及国际运营顾问的经验传授，是国内首创的展厅多功能复合模式，将极大提升场馆的综合服务能力（图 6-4，图 6-5）。

The First Multi-functional Exhibition Hall in China

There are 16 standard exhibition halls within the Shenzhen World, each of which is 20,000 square meters. All of them are designed with single-story, column-free and long-span space, which can be combined flexibly to meet the needs of various large exhibitions and provide personalized services for small exhibitions(Figure 6-3).

No. 17, No.18 and No. 20 exhibition halls in the north are special exhibition halls used for ultra-large exhibitions, as well as various conferences, banquets, sporting events, large-scale cultural activities and high-end business events. This multi-functional mode is inspired from the experience of international operators and international operations consultants, which is first in China.(Figure 6-4, 6-5)

图 6-5　央视 2020 年扬帆远航大湾区新年音乐会
Figure 6-5　2020 CCTV New Year's Concert

五维一体的立体交通
Transportation with Five Advantages

深圳国际会展中心位于未来深圳经济和城市发展重点区域和港澳大湾区经济核心区，在未来将依托深圳机场、快船码头、高铁枢纽、深中通道、沿江高速、多条地铁线等，拥有五维一体的立体交通条件。

The Shenzhen Wolrd is located in the future economic and, urban development key area of Shenzhen, and the economic core area of the GBA. Therefore, it will have five advantages relate to the Shenzhen Airport, dock, high-speed rail hub, ShenZhong highway, riverside highway, and multiple subway lines, etc.

图 6-6　登陆大厅屋顶花园
Figure 6-6　Roof Garden of ArrivaHall

地段优势

深圳国际会展中心选址于深圳宝安机场 T3 航站楼以北约 7km 处，客流和物流无需经过深圳市区就可直达馆内，又紧靠沿江高速和两条地铁线，并邻近福永码头，同时兼备海陆空立体化交通优势。

Location Advantage

The Shenzhen World is located about 7km away from the north side of the Shenzhen Bao'an Airport Terminal 3. Passenger flow and logistics can connect to the Shenzhen World directly without passing through the city downtown area. Besides, it is also close to the riverside expressway, two subway lines, and the Fuyong wharf, which is form a three-dimensional transportation advantage.

自驾

目前展馆已直接接驳 2 条高速公路（沿江高速和广深高速）、1 条快速路（海滨大道）。展馆周边交通方面，展馆周边 12 条新建道路、2 条改扩建道路及国展立交建设、完善工作进展顺利，目前已几近尾声。

Self-driving

At present, there are two expressways (Yanjiang and Guangshen expressways) and one fast road (Haibin Avenue)connected with the Shenzhen World. Right now, people can take line 11 to the Fuyong or Qiaotou Station, which is 3-3.5km away from the Shenzhen World. Besides, 12 new roads, 2 expanded roads and Guozhan interchange are close to complete.

公交

会展专线智慧公交（起点：桥头站、塘尾站，终点：展馆登录大厅）、机场专线和常规公交接驳线现已开通。
地铁设施建设在有序地推进中，20 号线预计 2020 年开通，12 号线预计 2022 年开通，目前还需依赖已开通运行的 11 号线（福永站和桥头站距展馆 3-3.5 公里）。

Public Transportation

Smart bus (starts Qiaotu station, Tangwei station, Destination: Arrival Hall), airport special line and shuttle buses are under operation.
Subway is under construction, line 20 is expected to open in 2020, and line 12 in 2022. Right now, people can take line 11 to the Fuyong or Qiaotou Station, which is 3-3.5km away from the Shenzhen World.

用户体验：加快优化完善

作为用户、使用者、体验者，来自社会各界的参展商、活动主办方、配套服务商、专业观众、普通观众等对深圳国际会展中心的整体印象普遍反映较好：建筑造型美观大气，彰显现代化和国际范；结构布局整齐合理，展厅划分清晰；空间视野开阔，硬件设施优良；交通流线清晰，各区分流有序合理；货运通道、卸货区域宽敞阴凉、使用高效。

但对于一个刚刚投入使用的全新展馆而言，仍存在许多尚未完善的不足之处，有待加强和改良。例如公共交通尚未完善、餐饮配套及种类较少、部分区域标识不清晰、周边城市配套缺乏等。对此，毛大奔一一介绍了他们在现有技术和条件允许基础上做的一些优化和完善，其他急待完善的工作也正在加快推进中，而且会做得更加细致。

图 6-7 会展中心中央廊道内景
Figure 6-7 Inner view of the Central Concourse of the Shenzhen World

Improvement of User Experience

People from all sectors of society, such as exhibitors, event organizers, supporting service providers, professional visitors, ordinary visitors, etc., were all impressed by this new grand exhibition center. According to their words of descriptions: the building is magnificent and beautiful, presenting an modern and international image; the layout is neat and clear, and the exhibition halls are easy to find; it has great broad view, and equipped with excellent hardware facilities; the traffic flow is clear with reasonable and orderly divisions, the area for freight is spacious, shady and efficient, etc.

However, as a new exhibition center, there are still some deficiencies and shortcomings that need to be stressed and improved. For example, public transportation is not yet completed, the catering is insufficient, the signs in some areas are not clear enough, and other commercial services are scarce, etc. In this regard, Mao said they have optimized everything they could under the circumstance of existing technology and current conditions. The further improvement is also speeding up with more refined details according to the feedbacks.

图 6-8 展会现场采访 -- 参展商
Figure 6-8 Exhitition Interview (Exhibitor)

智慧停车

展馆的智慧停车系统能够做到车位预订、实时动态调整、后期统计和分析，可以根据趋势分析，为后续展会预估预留展位，提升资源管理能力及资源利用率。停车场内设置各类的车位摄像机、车位查询屏和车位引导屏等可有效解决传统展馆常见的排队长、距离远、支付烦等一系列痛点：

- 停车引导：观众可通过手机快速引导直达展会最近的停车区域；

- 无感支付：对于驾车者，出场无需排队等候，不停车即可出入停车场；

- 车位引导：停车场出入口系统联动，由停车场管理平台统一管理，支持余位显示以及反向寻车等功能（图6-9）。

图 6-9　展厅货运通道
Figure 6-9　Freight Routes of Exhibition Hall

Smart Parking System

The smart parking system can be used for parking reservation, real-time dynamic adjustment, statistics and analysis. It can reserve booths for subsequent exhibitions according to trend analysis, so as to improve resource management and utilization. Various cameras, query screens and guidance screens in the parking space can effectively solve a series of pain points such as long queues, long distances and annoying payment:

- Parking Guidance: visitors can be quickly guided to the nearest parking area hear the exhibition hall through mobile phones;

- Non-inductive Payment: drivers no need to wait in line, and can enter and leave the parking space without stopping;

-Parking Guidance: entrance and exit system of parking space are linked, which is managed by the parking management platform. It can display how many parking spots left and support reverse car searching(Figure 6-9).

图 6-10　展会现场采访 -- 货运工人
Figure 6-10　Exhitition Interview (Freight worker)

货运场地

鱼骨式双边布局，人车分流的立体交通体系，宽阔平坦且充足阴凉的货运场地，能够最大化满足灵活组织、管理布撤展车辆的需求，并且保证人流和货流分离不交叉，安全快速，真正做到布展、观展、撤展组织高效清晰（图6-10，图6-11）。

Freight space

The fish-bone layout and the traffic system that separates pedestrian and vehicles, and the wide, flat and sufficient shady freight space ensures flexible organization and management for vehicles and freight trunks, which also guarantee the safe and fast move-in and move-out of all vehicles with clear instructions and high efficiency.(Figure 6-10, 6-11)

图 6-11　展厅货运场地
Figure 6-11　Freight Yard of Exhibition Hall

5G 全覆盖网络

深圳国际会展中心是全球首家实现 5G 移动通信全覆盖的展馆。用户可以采用手机流量、场馆 Wifi 等多种途径满足个人上网需求；展商在此上网基础上，还可提供专线接入，通过接入机房直接上网，满足快速上网的需求。

Full Coverage of 5G Network

The Shenzhen World is the first exhibition center in the world that achieves a full coverage of 5G mobile communications. Users can use mobile phone, WIFI, and other ways to connect to the internet; Besides, special line assess is also provided for exhibitors, through which to connect the computer room directly for fast Internet access.

图 6-12　会展中心展会现场
Figure 6-12　Exhibition in the Shenzhen World

就餐

深圳国际会展中心除了提供超大面积的就餐场地外，还提供多元的订餐取餐服务。用户可以通过小程序、会展 APP 等新一代移动互联网工具进行订餐服务，并选择配送时间或取餐方式，达到用科技手段错峰就餐，进一步解决就餐难的问题（图 6-13）。

Dining

Except large dining space, there also has a variety of food delivery service in the Shenzhen World.
People can order food through the mini programs on Wechat app or the Shenzhen World app. They can choose time or spot for food delivery freely to avoid peak hours(Figure 6-13).

图 6-13　展会现场采访（服务人员）
Figure 6-13　Exhitition Interview (Service Staff)

智慧安防

智慧安保为智能安防、智能监测、指挥调度提供一体化保障。展馆运用人脸识别、人工智能等新技术，融合多种物联感知数据的实时监测及预警分析，从"人、车、场馆、环境"实现事件的"可视、可控、可测、可查"，实现平安会展（图 6-14）。

Smart Security

Smart Security provides integrated guarantee for smart security, smart monitoring, command and dispatch. Face recognition, artificial intelligence and other new technologies are integrated with real-time monitoring and early warning analysis of a variety of connected object perception data, so as to achieve a "visual, controllable, measurable and searchable" record for "people, vehicles, venues and environment", thus guarantee the safety and security of the Shenzhen World(Figure 6-14).

图 6-14　会展中心安保队伍
Figure 6-14　Security team of Shenzhen World

城市配套

周边配套的皇冠假日酒店预计 2020 年 2 月开业，希尔顿和希尔顿花园酒店预计 2020 年 8 月开业，洲际酒店预计 2022 年开业。同时，忙碌的会展之余，客户也可以享受到优美的环境和极其便利的综合商业服务（图 6-15）。

Urban Supporting Facilities

Crowne Plaza Hotel nearby is expected to open in February 2020, the Hilton and Hilton Garden Hotel will open in August 2020, and Intercontinental Hotel will open in 2022. For exhibitors and visitors, the beautiful environment and convenient surrounding commercial services will be a good choice for resting and leisure time after busy exhibition hours(Figure 6-15).

图 6-15　会展中心展会现场
Figure 6-15　Exhibition in the Shenzhen World

室内精准导航

深圳国际会展中心室内定位导航系统主要提供实时动态导航服务，包括室内 3D 高精地图展示、室内定位、室内布局及公共设施分类列表、导航路径智能规划、紧急逃生、周边交通等一系列的便民惠民功能。用户可通过深圳国际会展中心 APP 或微信小程序实现精准导航，也可在大屏上进行查看。

Indoor Precise Navigation

The indoor position and navigation system mainly provide real-time dynamic navigation services, including indoor 3D high-precision map display, indoor positioning, indoor layout and category listings of public facilities, smart path guidance, emergency escape, surrounding traffic, and many others. Users can get accurate navigation through the Shenzhen World app or mini program on Wechat, or check and view on the large screen placed inside the exhibition center.

图 6-16　展会现场采访（观展买家）
Figure 6-16　Exhitition Interview (Trade Buyers)

小程序

深圳国际会展中心 Shenzhen World 小程序已上线。小程序功能涵盖展馆导航、餐饮、商旅、快递、商务中心、WIFI 等服务。它应用轻巧、便捷、十分易用，在展会期间为众多参展、观展人群提供全方位的智慧化服务，连通线上与线下场景，高效解决用户需求（图 6-17）。

Shenzhen World Mini Program

The Shenzhen World mini program on Wechat app has been launched, which people can use for navigation, food order and delivery, business travel, express delivery, business center, Wifi connection and many other services. It's smart and convenient, easy to use, providing all kinds for smart services for both exhibitors and visitors. It connects online and offline scenes, offer solutions for people's need effectively(Figure 6-17).

图 6-17　展会现场采访（访客）
Figure 6-17　Exhitition Interview (Visitor)

开局之年
The Year Ahead

图 6-18　央视 2020 年扬帆远航大湾区新年音乐会
Figure 6-18　2020 CCTV New Year's Concert

2020 年，是深圳国际会展中心全面运营的开局之年。截至采访日期，毛大奔表示 2020 年的展会排期和签约基本接近尾声，"2020 年全年，将举办 35 个展会活动，其中展览 28 个，会议活动 7 个。展览将覆盖机械电子，纺织服装，家居用品，汽车，玩具，广告标识等多个行业，其中两个最大的展览将达到 32 万平方米的规模（图 6-18，图 6-19）。"

在已经进入排期的展览中，一部分来自深圳的本土知名品牌展会，如深圳机械展，家具展，礼品展，光博会；也有来自广州、上海和北京的成熟展会品牌如 LED 广告标识展，埃森焊接展，法兰克福玩具展；还有国际国内知名展会公司在深圳开创的新展，如深圳建材家居展，华南工博会等。
毛大奔认为，这些行业内的知名展会汇聚全球各行业的优质资源，落户深圳国际会展中心，必将对周边片区、深圳乃至整个粤港澳大湾区的行业发展起到带动和引领的作用。深圳国际会展中心卓越的区位优势、一流的软硬件设施和服务，将全力赋能择址于此的展会，助力展会提质、升级。同时，深圳国际会展中心的国际化品牌和选址于此的展会的发展战略很多时候是彼此融合的，有利于展会与展馆运营双赢。

The year 2020 will be the first year of full operation of the Shenzhen World. Up to the interview, Mao said all the schedule and contract signing work of 2020 are almost done. There will be approximately 35 exhibitions and events in the year ahead, including 26 exhibitions and 2 conferences and events, involving many industries such as machinery and electronics, textile and clothing, household goods, automobiles, toys and advertising signs, among which the area of two largest exhibitions will reach 320,000 square meters(Figure 6-18, 6-19).

Among the exhibitions scheduled for 2020, some are local well-known brands, such as the Shenzhen Machinery Xxhibition, Furniture Exhibition, Present and Gift Exhibition and Optical expo; Some are mature exhibition brands from Guangzhoǔ, Shanghai and Beijing, such as the LED Advertising Sign Exhibition, Essen Welding Exhibition and Frankfurt Toy Exhibition; Some are new exhibitions that held by international and domestic well-known exhibition companies, such as the Shenzhen building Materials Exhibition, South China Industrial Fair and so on.

From Mao's perspective, these well-known exhibitions gather the high-quality resources of all industries in the world, which will surely drive and lead the industry development of the surrounding area, of Shenzhen and even of the whole GBA. With the advantage of location, first-class hardware and software facilities and services, the Shenzhen World will benefit the exhibitions held, and help them improve and upgrade to a new level, achieving a "win-win" goal.

图 6-19　央视 2020 年扬帆远航大湾区新年音乐会
Figure 6-19　2020 CCTV New Year's Concert

未来图景
Visions for The Future

会展经济一直被喻为行业发展的晴雨表和风向标。会展对产业发展有着"汇聚、引领、带动和落地"的核心功能效应。 对于深圳国际会展中心而言，毛大奔表示。"它要打造的不仅是一个一流的展馆，更是一个超大型会展综合体、一艘会展航母。"

深圳国际会展中心的投入运营，将直接带动深圳会展业规模大幅上升。根据统计和预测，5 年后，深圳市年度总展览面积将达到 500 到 700 万平方米，预计 10 万平方米大展可达到 30 到 40 个左右，其中 30 至 40 万平方米的超大型展览将达到 5 到 8 个。毛大奔认为，"届时深圳的会展业规模将有望成为全国第二。"

深圳国际会展中心的投入运营，将吸引大批国内外知名展览、会议、大型活动在深圳举办，从而吸引大批国内外知名会展企业在深圳设立总部、区域总部或开设分支机构，以及大批国内外会展产业上下游服务商进入深圳。 除此之外，大批的星级酒店、餐饮、商旅服务等也将向展馆附近及紧密联系的交通干线附近集聚，形成新的会展旅游产业集聚。

毋庸置疑，深圳国际会展中心将极大提高深圳市展览场馆基础设施水平，引领现代服务业高质量发展，为深圳抢占会展业未来发展制高点，打造比肩汉诺威、上海等城市的国际会展之都，成为新的经济发展增长极奠定坚实基础。对于深圳构筑开放层次更高、辐射作用更强的全面开放高地，推动建设粤港澳大湾区，建设中国特色社会主义先行示范区，具有十分重要的意义（图 6-20 ）。

图 6-20　国际会展中心远景
Figure 6-20　Distant View of the Shenzhen World Exhibition Center

Exhibition economy has been regarded as the barometer and weathervane of industry development, and it has the core effect of "gathering, leading, driving and landing" industrial development. For the Shenzhen World, Mao said in confidence: "Its goal is not just a world-class exhibition center , but also an ultra-large exhibition complex and an exhibition aircraft carrier."

As the Shenzhen World opens and operates, it will upgrade the scale of Shenzhen exhibition industry directly and significantly. According to statistics and forecasts, the total annual exhibition area of Shenzhen will reach 5-7 million square meters in five years, the number of large exhibitions of 100,000 square meters will be up to 30-50, and the number of held super large exhibitions of 300,000 to 400,000 square meters will be 5-8. By then, in Mao's view: " the scale of the exhibition industry in Shenzhen will reach or surpass that in Guangzhou and will be the second largest in China."

In addition, the Shenzhen World will attract a large number of domestic and foreign well-known exhibitions, conferences and large-scale events and activities, which means that more and more domestic and foreign well-known exhibition enterprises will set up headquarters, regional headquarters or branch offices in Shenzhen, bring numerous domestic and foreign exhibition industry upstream and downstream service providers. Meanwhile, a lot of star-level hotels, restaurants, business travel agencies will gather near the exhibition center closely and form a new tourism industrial agglomeration.

Beyond all question, the Shenzhen World will improve the level of the Shenzhen exhibition infrastructure, and lead the development of the modern high quality service industry, boost the prosperity of Shenzhen exhibition industry in the future, making Shenzhen an international exhibition city like Hanover and Shanghai, laying a solid foundation as new growth pole of economic development. It is of great significance for Shenzhen to build a more open city with great and broad influence, it will highly promote the construction of the GBA and will also play an important role on Shenzhen's building of the demonstration pilot zone for socialism with Chinese characteristics(Figure 6-20).

刘鹏

许芳

艾小雪

顾问天

Nicholas Green

石春阶

陈益明

张玉

Khaled Khamallah

Stephane Gautier

王波

赵敏

何志力

李爽

吴昌伟

苏飞

宋红

何浏

梁振亚

郭晓黎

肖

张素禄

孙乐刚

蒋宪新

陈纪兵

蒋金勇

胡海萍

陈磊

许振

刘俊跃

任春磊

彭志鹏

黄俊杰

李年长

马晓雯

欧天祺

Jean Pistre

归

廖晓华

张昌蓉

赵锐

蔡戈锋

IrinaValentina Du

刘鹏

梁铨捷

冯德瑜

徐舟

林建军

孟松

何迎军

李静

李俊

Michel Cova

白宇西

龚沁华

周同迅

廖林涛

周游

曾繁生

骆年红

周晓光

董孜

Negre Alain

李维博

祁孜威

邹礼滨

林先文

冯梓阳

Dimitri Frank

郝东翔

黄用军

张海军

钱忠美

陈诚

梁威

马军

吴大农

余安琪

张浩

蔡亮

胡涛

胡平川

陈虎

李威

刘硕

郭文波

龙漫

许少良

周欣

王超

刘慧敏

何远明

曾一鸣

罗凡

叶锦州

叶翠景

陈忠卿

谭绍斌

毛大奔

刘源

王潇楠

Yannick Denis

冯百如

沈奕君

叶清

林天鑫

王金铸

谢莹

于军

张静雯

任慧

曹

陈启明

黄展春
王超
Gabriel Pistre Isabelle Serr
峰 陈玉保
李屹
David de la FuenteBurguera
饶伟 陈玉莲 陈建勇
朝 王硕 聂云飞 刘毅
智慧 沈小锋 齐海英 徐玉胜
李宇星 沙军 屈伟萍 廖晓华 梁燕明 门云
赵成 张栋梁 田智华 肖毓培
舒龙 邝英武 孔维良 钟艳 王甜 常冬冬
延超 叶瑞迪 李嘉璇 向辉 袁巍
王卓 麻友博 张明泽 吴蓉 蔡文珊
吴箫 李泽鹏 卢东晴 张盛红
王石 刘喜文 温怀海 黄煜
杨龙 向国焘 黄茗 洪史章
杨光伟 陈珺 吴凡 胡振涛 陈赟然 蔡宏智
张发智 林天鑫 宋玲清 李时瑞 覃思
祝捷 洪祖光 辛伟 黄帅
贤莫 冯越强 李宗钢 李俊星 黄婷婷
岑士沛 郭甲英 汤衡
林文波 金运丰 罗文锦 宋国鸿 周全
黄柳 邹蔚 谢建伟 李瑾
rie Poli 卢婷婉 Grégory Aldéa
Michel PERISSE 罗文辉 Leonardo Mariani
蒲饶卿 丁荣 Lucie Leger
小全利 沈映红 钱宏周
李媛琴 张西利 李佳睿 潘成科
庄业东 邹积成 Aldo De Sousa
伟 凌礼 Anelyse Duperrier
马恒懿 劳玉明
文
陈曼文 刘友德
黄淑婷

附 录

Appendix

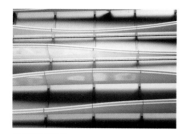

登录厅外墙装饰
Landing Hall Exterior Wall Decoration

2018.05.15 外立面施工
Facade construction

2018 年 8 月 1 日，标准展厅的样板区建设和装修进度非常顺利，即将精彩绽放：一是样板区的幕墙、PC 墙样板施工已经完成；二是地面铺装样板即将完成；三是卫生间天花，墙面施工完成；地面，墙面石材样板施工完成；四是样板段桥架搭设、管道安装已完成。

On August 1st, 2018, the construction and decoration of the model area of the standard exhibition hall is progressing very smoothly and will be wonderfully blooming: first, the curtain wall and PC wall model construction in the model area have been completed; The construction of the ground and wall stone samples is completed. Fourth, the bridge erection and pipeline installation of the sample sections have been completed.

终版施工图
Final construction drawing

2018 年 8 月 8 日，工程现场迎来了一场施工图交底大会。此次会议上，作为项目设计总牵头单位及深化设计和施工图设计方，AUBE 欧博设计为该项目总图、建筑、结构、给排水、电气、电讯、通风空调、绿建、交通等专业设计内容进行了现场答疑与研讨。

On August 8, 2018, the construction site ushered in a construction drawing submission meeting. At this meeting, as the general contractor of the project design and the design and construction design of the deepening design. AUBE design for the project general design、 architecture、 structure、 water supply and drainage、 electrical、 telecommunications、 ventilation and air conditioning、 green construction, transportation and other professional design Contents were answered on-site and discussed.

2018.09.07 山竹台风过后
2018.09.07 After the super typhoon Mangkhut

2019.01.28 钢结构全面封顶
Steel structure is fully capped

南登录大厅网壳整体提升
The overall improvement of the reticulated shell in the south login hall

2018 年钢结构建筑技术交流会"在深圳国际会展中心项目现场举行，来自 AUBE 欧博设计、深圳设计总院等几十家设计院近 70 余位结构设计专业的专家人员与会进行钢结构建筑设计与施工技术交流。

The "2018 Steel Structure Building Technology Exchange Conference" was held at the Shenzhen World project site. Nearly 70 experts from AUBE , Shenzhen Design Institute and other dozens of design institutes participated in the steel structure architectural design and construction technology exchange.

2019.09.30 竣工验收
Completion acceptance

2018 ———————————— 2019

2018.02.08
方案设计报建审批
Construction Approvals.

2018.06.20
地下室全面封顶
The construction of basement was completed.

2018.07.19
南登录大厅网壳整体提升
The overall improvement of the reticulated shell in the south landing hall.

2018.08.01
精装修招标完成，施工进场
Bidding of refined decoration was done, construction started.

2019.01.28
钢结构全面封顶
Steel structure was fully completed.

2019.06.14
获得二星绿色建筑设计标识证书
Recieved a certificate of two star green building design.

2019.09.30
竣工落成
Completion acceptance.

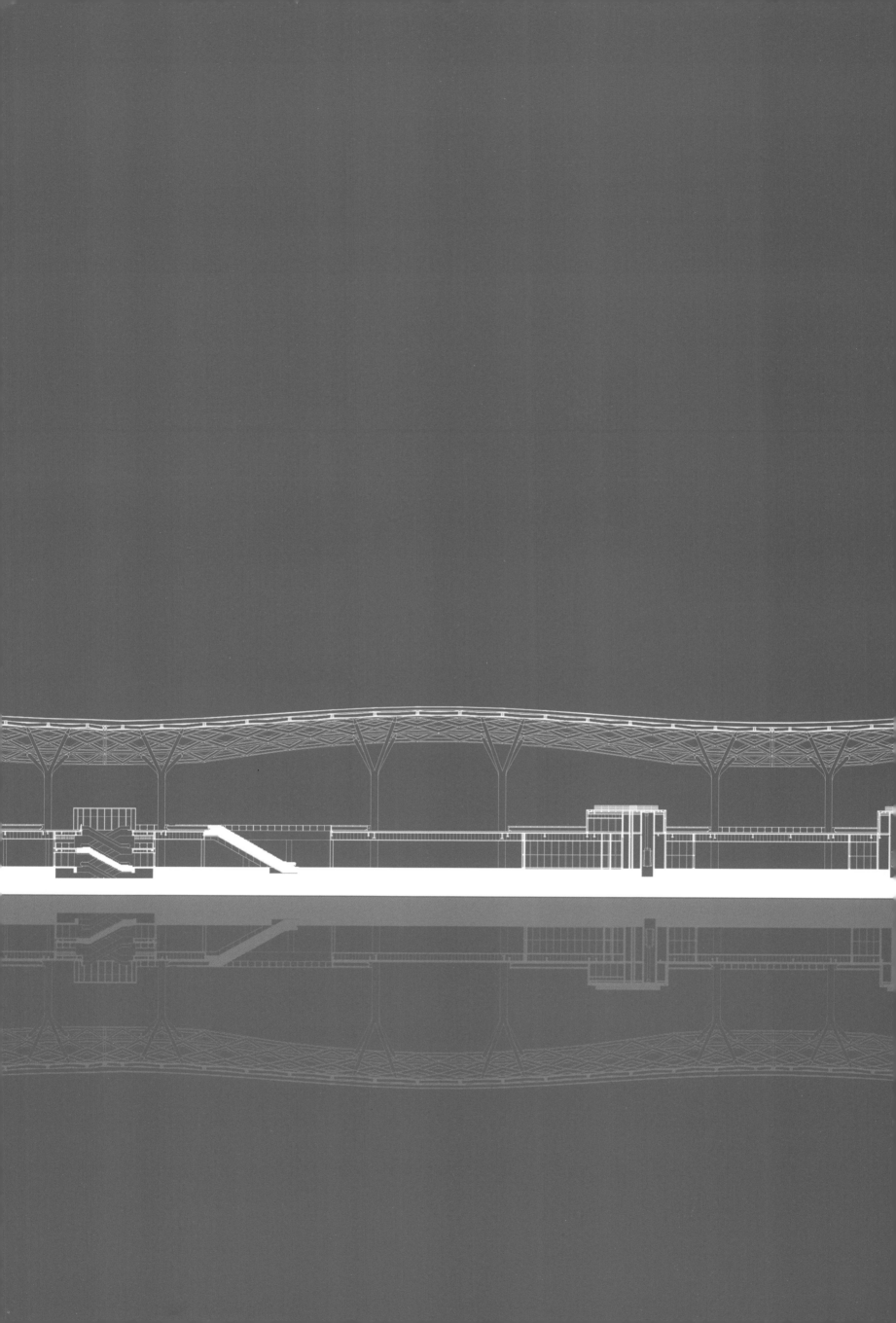

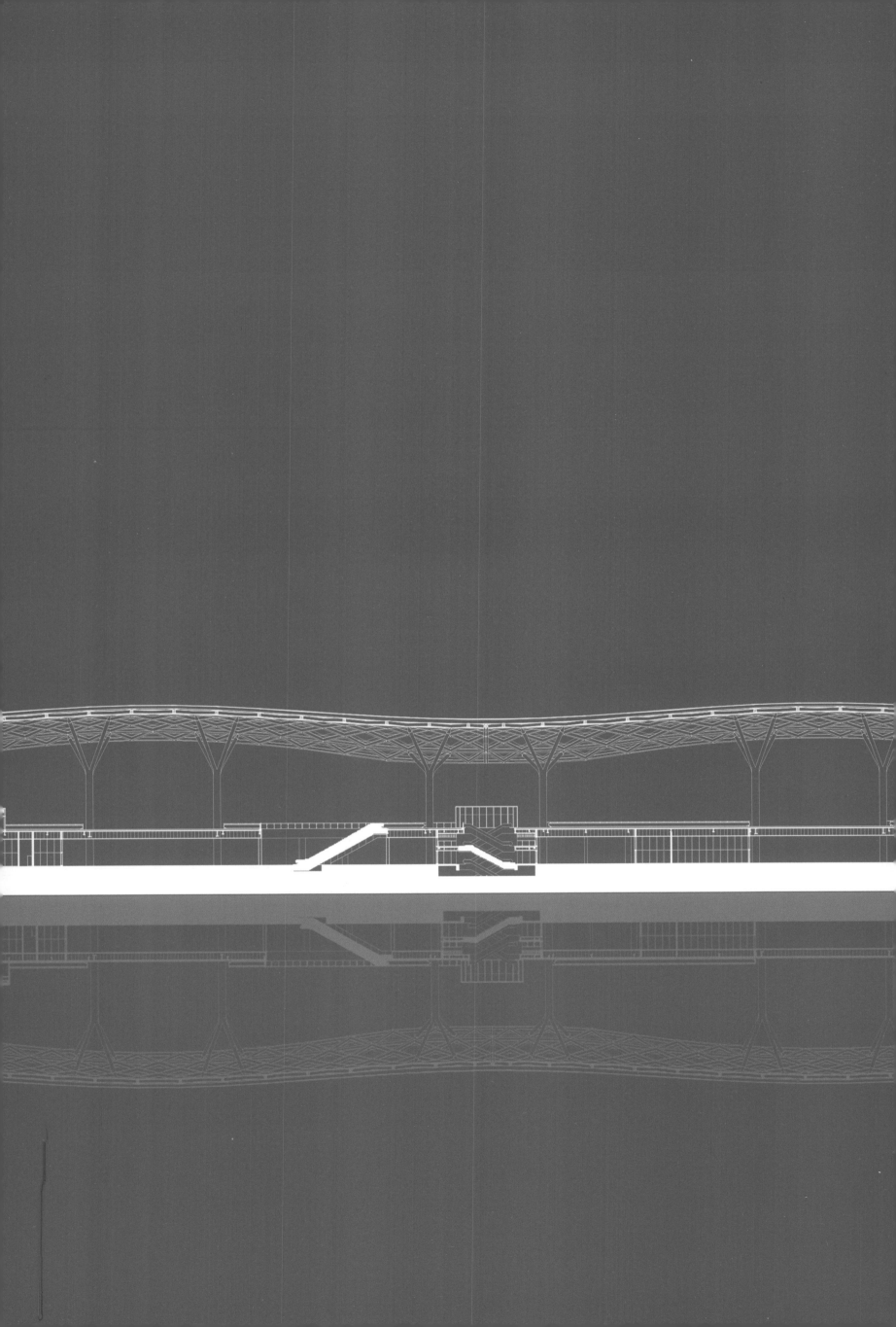

2018 年 6 月 20 日，在深圳国际会展□□
施工现场，10000 余名工人，500 余台□
员，48 台塔吊，300 余台汽车吊、履□□
个大区，100 余小块，分区分块在 1.□
的区域内组织平行施工。历经 1 个半月□
奋战，伴随着 80 万立方米混凝土的浇□□
57 万平方米的地下室结构实现全面封顶□

On June 20, 2018, at the Shenzhen□
project construction site, more than □
workers, more than 500 manag□
personnel, 48 tower cranes, more ti□
automobile cranes, crawler cranes, 25□
more than 100 small blocks The suł.□
organize parallel construction within □
of 1.7 kilometers. After one and a half□
of efforts day and night, along w□
pouring of 800,000 cubic meters of c□
the 570,000-square-meter basement s□
of the project was fully capped.

施工现场全景
construction site panorama

2018.01.05 AUBE 欧博设计团队成员□
2018.01.05 AUBE inspected the constru□

休闲带
Leisure belt

作为会展中心的配套设施，由欧博设计全专业全程打造的会展休闲带是会展与东侧商业地块之间重要的过渡链接带，也是满足复合功能需求的交通综合体，担负着柔化城市界面，提升城市空间品质的作用。

As the supporting facility of the Shenzhen World, the leisure belt designed by AUBE is an important transitional link between the exhibition center and the eastern commercial plot, and it is also a transportation complex that meets the needs of multi functions, and is responsible for softening the city interface to enhance the quality of urban space.

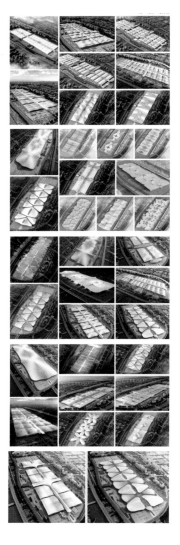

第五立面方案比选
Fifth façade plan comparison

AUBE 欧博设计与 VP 以及所有参与其中的伙伴们近十个月的努力和奋战，至少 103 个版本的方案比较，各方专家多次研讨，就是要为深圳这座城市献上一份珍贵稀有的礼物。

AUBE and the VP and all the partners involved in the efforts and struggles in the past ten months, at least 103 versions of the scheme comparison, experts from various parties have repeatedly discussed, to provide The city presented a precious and rare gift.

2017.12.22 展厅第一品桁架吊装
The first product truss hoisting
in the exhibition hall

na Merchants

□T signing ceremony

软基处理完成
Soft foundation processing completed

JWC，SMG 运营团队参与会展设计，设计联合体向指挥部汇报优化结果，并获得肯定。

The JWC and SMG operation team participated in the design of the exhibition, and the design consortium reported the optimization results to the headquarters, and obtained affirmation.

2017

2016.09.28
正式开工建设（软
基处理工程）

Construction started (soft
foundation treatment
project).

2016.09.30
获得规划部门方案设计
核查的复函

Obtained the verification of
planning design from the planning
department.

2017.01.12
JWC，SMG 运营团队参与会展
设计，设计联合体向指挥部汇报
优化结果，并获得肯定。

The JWC and SMG operation team participated
in the design of the exhibition, and the design
consortium reported the optimization results to
the headquarters, and obtained affirmation.

2017.05.01
施工总承包全面进场

General construction contractor
started working.

2017.06.01
市委常委会议，确
定建筑形态

The standing committee
meeting of municipal
committee determined
the architectural form of
Shenzhen World.

2017.09.30
广东省厅消防
专家评审

Review by Guangdong
Province Department
of Fire Fighting Experts.

2017.
展厅第

The first p□
in the exhi□

大事记年表
Chronicle of Events

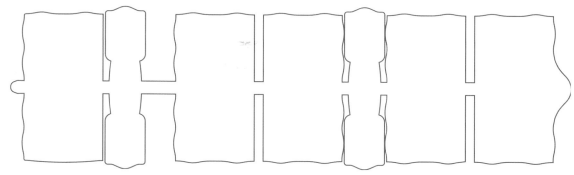

160 万平方米

全球建筑面积最大的单体建筑，总建筑面积达 160 万平方米。

The largest single building in the world, with a total construction area of 1.6 million square meters.

5 万平方米

全球最大单个展厅，展览面积达 5 万平方米。

The largest single exhibition hall in the world, with an exhibition area of 50000 square meters.

400 万立方米

基坑土方挖运量为全球房屋建筑领域之最，总土方挖运量 400 万立方米。

The excavation volume of foundation pit is the largest in the field of housing construction in the world, with a total excavation volume of 4 million cubic meters.

27 万吨

全球房屋建筑领域总用钢量最大的单体建筑，达 27 万吨。

The single building with the large total steel consumption in the globe housing construction field is 27000 tons.

58 万平方米

全球房屋建筑领域面积最大，长度最的无缝钢筋混凝土结构地下室，地下室面积达 58 万平方米，最大长度 1700 米。

The basement of seamless reinforced concrete structure with the large area and the largest length in the world's housing construction field ha an area of 580000 square meters an a maximum length of 1700 meters.

目前福田深圳会展中心能容纳的展会数量已近饱和，高交会，家具展等展会受限于展馆规模，无法扩大展览面积。深圳会展与许多其他城市相比，已不具备竞争优势。

At present, the number of exhibitions that can be accommodated in the Shenzhen Convention and Exhibition Center is nearly saturated, and the problem of the exhibition area is becoming increasingly prominent. Due to the constraints of the large-scale exhibitions, other cities were eventually selected.

川会展中心总建筑面积约 28 万展览面积 10 万平方米，于正式投入使用。据 2018 年数会展中心举办的展览近 100 场，力 1800 多场，年展览面积超过方米。

enzhen Convention and ion Center integrated on, conference, business, g, entertainment and other ns into one, and formally put in 2004. Since its opening in here have been about 100 ions held in the Shenzhen ntion and Exhibition Center ear, and more than 1,800 nce activities.

ruary 2, 2016, the "Shenzhen ational Convention and ion Center Construction ering Design Calibration ittee" voted on-site and etermined:

sign proposal submitted by Design by AUBE) and Valode Architects (VP) won the bid!

Valode & Pistre + AUBE

2016 年 2 月 2 日，经"深圳国际会展中心建筑工程设计定标委员会"现场投票，最终确定：
由 AUBE 欧博设计和 Valode&Pistre Architects（VP）提交的设计方案中标！

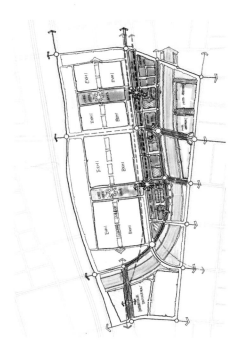

功能定案会
Meeting of function confirming

第一次与招商碰面会
The first meeting with Ch

招商 / 华侨城签约仪式
Investment invitation /

OCT 华侨

招商蛇

规划形成"一河、两带、三片区"的规划结构。一河为福永海河，是串联片区空间的主要线索；会展中心为核心功能带；会展休闲带是片区重要的线性公共开放空间和交通服务空间；西岸片区是酒店、办公、餐饮、购物、娱乐、旅游等功能为一体的大型城市综合片区；东岸片区环境相对独立，受会展干扰较少，主要功能为公寓，可以提供商务居住需求；南岸片区设置大型都市娱乐综合体，形成独特的旅游目的地。

The plan forms a planning structure of "one river, two belts, and three areas". One River is the Fuyonghai River, which is the main clue of the space in the series; the Convention and Exhibition Center is the core functional zone; the Leisure Zone is an important linear public open space and transportation service space in the zone; and the West Bank Zone is the hotel, office, catering, shopping, entertainment, A large-scale urban integrated area with integrated tourism and other functions; the east bank area is relatively independent and less affected by people's; and its main function is apartment, which can provide business residential needs; the south bank area sets up a large-scale urban entertainment complex to form a unique tourist destination .

2016

.06.01
信委提出尽快建设第展中心，拟选址宝安区域

an Economic and Information ission proposed to build the exhibition center as soon as

2014.05.01
市政府确定项目选址、规模，提出"三个一流"目标

In the international competition, general design contractor(AUBE+VP) won the bidding.

2016.02.02
AUBE 欧博设计和 Valode &Pistre Architects（VP）提交的设计方案中标

The design proposal submitted by AUBE and Valode & Pistre Architects (VP) won the bid.

2016.04.01
会展中心城市设计详细蓝图获批

The detailed blueprint for the urban design of Shenzhen World was approved.

2016.06.21
最终拟定设计任务书

Finalized design task.

2016.08.29
引入投资建设运营机构（招商 + 华侨城）

The companies (China Merchants + OCT) of investment, construction and operation took part in.

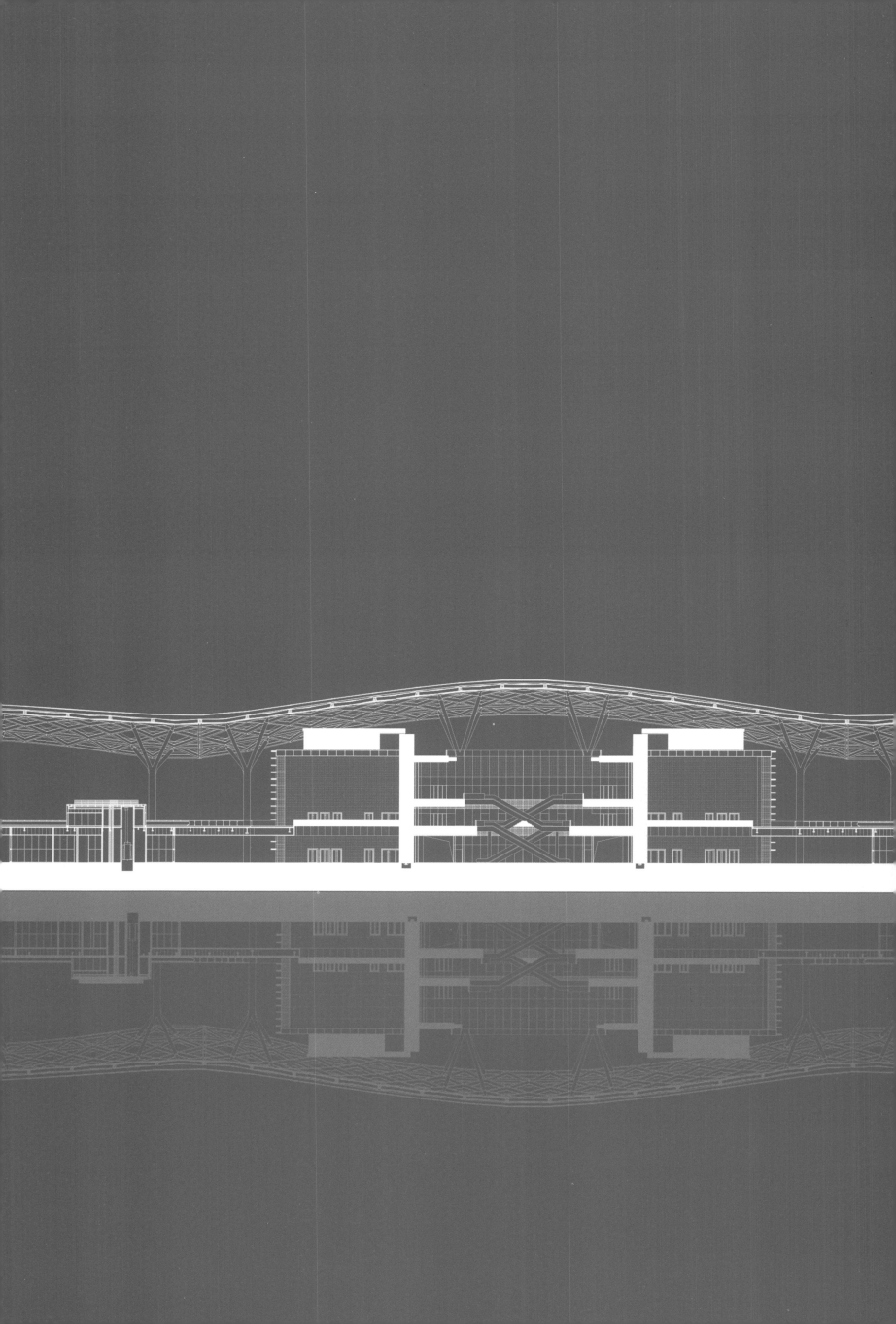

464 大事记年表
Chronicle of Events

470 设计单位
Project Team

476. 建设单位
Construction Team

478 联合编辑单位
Co-editorial Team

深圳欧博工程设计顾问有限公司
Shenzhen AUBE Architectural Engineering Design Co., Ltd.

AUBE 欧博设计 1997 年创立于法国巴黎，是一家国际的、领先的、持续的专业型公司，由法国欧博建筑与城市规划设计公司、其派驻中国的深圳代表处，以及深圳市欧博工程设计顾问有限公司等实体共同组成。20 余年来，欧博以"国际化经验、地域化实践、设计提升价值"为设计理念、"国际水准提供专业化服务"为客户宗旨，以"集成一体化"设计方法，致力构建城市人居美好生活。

AUBE 欧博设计，作为集建筑、景观、城市规划于一体的国际化设计院品牌，旗下包括 2 大核心产品（会展核心功能区、街区生活综合体）、3 大特色产品（建筑设计、城市设计、景观设计）、4 大服务专项（实验性设计、BIM | VR | 装配式、设计管理、设计总承包）、5 大合力专业（建筑、规划、景观、结构、机电）。AUBE 欧博设计同时拥有欧盟和法语系国家、地区资质（法国国家建筑师行会、法国大巴黎地区建筑行会），以及中国建筑行业建筑工程甲级、城乡规划编制乙级、风景园林工程设计专项甲级资质。截止目前，AUBE 欧博设计共参与和完成包括深圳国际会展中心、华侨城生态广场、深圳人才公园、深圳半岛城邦、贵阳 201 大厦、绍兴国际会展中心、深圳 CFC 长富中心在内的重大设计项目 1600 余个。

公司愿景
在我们未来的规划中
把我们的房子建造成人人都喜欢的景观
并且能进入历史

公司理念
以"建筑"为业者　　应对一块石保持基本的歉意
以"规划"为业者　　应对一个人保持基本的善意
以"景观"为业者　　应对一棵树保持基本的敬意
在某种范围内　　　不必要分先后
木——林——森　　这三个汉字
可以解释　　　　　建筑——规划——景观这三个观念

AUBE Design was founded in 1997 in Paris, France, is an international, leading, continuous professional company, composed of AUBE Architecture and Urban Planning Design, its representative office in Shenzhen, China, and Shenzhen AUBE Engineering Design Consultants Co., Ltd. For more than 20 years, AUBE has taken "international experience, regionalization practice, design to enhance value" as the design concept and "international standard to provide professional service" as the client's tenet, and "integrated" design approach to build a better life in urban living.

AUBE Design, as an international design institute brand integrating architecture, landscape and urban planning, includes 2 core products (exhibition core function area, street life complex), 3 specialty products (architectural design, urban design, landscape design), 4 service specialties (experimental design, BIM | VR | assembly, design management, design general contracting), 5 combined expertise (architecture, planning, landscape, structure, mechanical and electrical). AUBE design has both EU and Francophone national and regional qualifications (French National Institute of Architects, French Regional Chamber of Architecture of Greater Paris), as well as the Chinese construction industry construction engineering grade A, urban and rural planning compilation grade B, landscape engineering design special grade A.
Up to now, AUBE Design has participated in and completed more than 1,600 major design projects, including Shenzhen International Convention and Exhibition Center, Overseas Chinese Town Eco Plaza, Shenzhen Talent Park, Shenzhen Peninsula City State, Guiyang 201 Building, Shaoxing International Convention and Exhibition Center, and Shenzhen CFC Changfu Center.

Company vision
Our planning for the future is,
To build our house into a landscape that everyone will love,
And to be written in history.

Company philosophy
Those who are in the business of "construction"should maintain a basic apology for a stone.
Those who are in the business of "planning" should maintain a basic goodwill towards a person.
Those who are in the business of "landscape" should maintain a basic respect for a tree.
To a certain extent, no need to prioritize, wood-tree-forest , these three Chinese characters can explain the three concepts of architecture - planning - landscape.

主编
Editor in chief

冯越强
FENG YUEQIANG

创始人 / 董事长 / 首席设计师 / 高级工程师
Founder/ Chairman / Chief Designer / Senior Engineer

工作经历
Work experience

1987-1989 上海同济大学建筑城规学院教师
1990-1993 法国巴黎 ABC 建筑事务所主任建筑师
1993-1997 法国巴黎 SCAU 国际公司主任建筑师
1997 至今 深圳市欧博工程设计顾问有限公司

1987-1989 Teacher, College of Architecture and Urban Planning, Tongji University, Shanghai
1990-1993 Director Architect, ABC Architecture, Paris, France
1993-1997 Director Architect, SCAU International, Paris, France
1997-Present, Shenzhen AUBE Architectural Engineering Design co., Ltd.

主要作品
Major works

深圳市"半岛城邦"规划、建筑、景观设计和一、三、四期建筑设计（已建成）方案主创
深圳市南山商业文化中心区核心区景观设计（已建成）方案主创
贵阳生态国际会议展览中心（100 万平方米，最高建筑 201 米）（已建成）方案主创及部分施工图
深圳市龙华文体中心（建造中）方案主创
深圳市国际会展中心（建造中）主持人
深圳市南山区九年制央校（已建成）主持人

Planning, construction, landscape design and architectural design of Phase I, III and IV of Shenzhen "Peninsula City State" (completed)
Landscape design for the core area of Nanshan Commercial and Cultural Center, Shenzhen (completed)
Design and construction drawings for the Guiyang Ecological International Convention and Exhibition Center (1 million square meters, the highest building is 201 meters) (completed)
Shenzhen Longhua Cultural and Sports Centre (under construction)
Director of Shenzhen International Convention and Exhibition Center (under construction)
Director of Shenzhen Nanshan 9-year Central School (completed)

获奖情况
Awards

深圳市首届杰出建筑师
深圳市 2018 年行业领军人物
深圳鹏基商务时空获深圳市十三届优秀工程公共建筑三等奖
贵阳生态国际会议展览中心获深圳市首届建筑工程质量金奖和 Leed 铂金奖及中国三星级奖

The First Outstanding Architect in Shenzhen
Shenzhen 2018 Industry Leaders
Shenzhen Pengji Commercial Space was awarded the third prize of Shenzhen's 13 outstanding engineering public building
Guiyang Ecological International Convention and Exhibition Centre was awarded the first Gold Award and Leed Platinum Award for Construction Quality in Shenzhen and the China Three Star Award.

主要成员
Main members

Michel Perisse
董事合伙人、设计总监、
法国国家建筑师

Director partner, design
director
French national architect

林建军 LIN JIANJUN
董事总经理

Managing director

杨光伟 YANG GUANGWEI
董事、执行总设计师

Director, executive chief
designer

白宇西 BAI YUXI
董事副总经理、执行总设计师

Deputy general manager
executive chief designer

丁 荣 DING RONG
董事副总经理、执行总建筑师
一级注册建筑师、高级工程师

Deputy general manager,
executive chief architect,
first class registered
architect, senior engineer

祝 捷 ZHU JIE
董事、执行总景观师
高级工程师

Director, executive
landscape architect,
senior engineer

郭晓黎 GUO XIAOLI
董事副总经理、执行总规划师
一级注册建筑师、注册规划师

Deputy general manager
and executive general
planner, First class
registered architect,
registered planner

黄用军 HUANG YONGJUN
董事副总经理、总工程师、
教授级高工、一级注册结构工
程师

Deputy general manager,
chief engineer, professor
level senior engineer, first
level registered structural
engineer

沙 军 SHA JUN
董事副总经理

Deputy General Manager

黄 煜 HUANG YU
MEC 中心总经理、机电总工
程师注册电气工程师、高级
工程师

MEC center general
manager, chief
mechanical and electrical
engineer, registered
electrical engineer, senior
engineer

卢东晴 LU DONGQING
ARC 建筑创意中心副总经理

Deputy general manager
of ARC architectural
creative center

编辑团队成员
Editorial Team Member

编辑团队
Editorial Team
统筹：董孜孜
Planning Editor: Dong Zizi
编辑：董孜孜 廖林涛 邹蔚 张发智 黄柳 岑土沛 冯梓阳
Editorial Design Team: Dong Zizi, Liao Lintao, Zouwei, Zhang Fazhi, Huangliu, Cen Tupei, Feng Ziyang
翻译：邹梦璇
Translator: Zou Mengxuan

法国 VP 建筑设计事务所
Valode & Pistre Architects

法国 VP 建筑设计事务所是欧洲十大著名设计事务所之一，也是法国最大的建筑设计事务所。事务所由 Denis Valode 先生和 Jean Pistre 先生于 1980 年创办的。至今，VP 已有 300 名来自 15 个不同国家的建筑师、规划师和室内设计师。

法国 VP 建筑设计事务所总部设在法国巴黎，内部设有四个分公司：建筑设计公司、城市规划公司、建筑结构技术公司和室内设计公司。VP 的主要设计作品分布在全球 16 个国家和地区，主要以大型公共建筑、商业中心和居住区为主，如写字楼、宾馆、国际展览会议中心、科研中心、博物馆、大学、医院、综合商业中心和高档住宅等。

公司是从以下一系列工程逐渐享有声誉的：波尔多的现代艺术博物馆、巴黎近郊 Aulnay sous Bois 的欧莱雅工厂、雷诺技术中心、壳牌石油总部、法国航空公司总部和航空城、拉德方斯的达芬奇大学、巴黎市内的贝西村商业街以及拉德方斯的 T1 大厦。

自 2005 年法国 VP 建筑设计事务所进入中国市场以来，经过 9 年的发展，中国团队已相应成熟。在酒店、办公、医院、住宅、室内设计等领域均已取得了不俗的成绩。现诚邀英才，希望能提供一个共同发展的平台。

VP architectural design firm in France is one of the top ten famous design firms in Europe and the largest architectural design firm in France. The firm was founded in 1980 by Mr Denis Valode and Mr Jean pistre. So far, VP has 300 architects, planners and interior designers from 15 different countries.

VP architectural design office, headquartered in Paris, France, has four branches: architectural design company, urban planning company, architectural structure technology company and interior design company. VP's main design works are distributed in 16 countries and regions around the world, mainly including large-scale public buildings, business centers and residential areas, such as office buildings, hotels, international exhibition and conference centers, scientific research centers, museums, universities, hospitals, comprehensive business centers and high-end residences.

The company has gradually gained its reputation from a series of projects: the Museum of modern art in Bordeaux, the L'Oreal factory in the suburb of aulnay sous Bois, Renault technology center, shell oil headquarters, Air France headquarters and aviation city, Da Vinci University in radfangs, the business street of Bessie village in Paris and T1 building in radfangs.

Since the French VP architectural design firm entered the Chinese market in 2005, after 9 years of development, the Chinese team has matured accordingly. In hotel, office, hospital, residential, interior design and other fields have made remarkable achievements. We sincerely invite talents and hope to provide a platform for common development.

丹尼斯·瓦洛德
DENIS VALODE

丹尼斯·瓦洛德于 1969 年获得法国国立高等美术学院 (Ecole Nationale Supérieure des Beaux Arts) (教育单元 1) 建筑师文凭。1970 年，他成为了法国国立高等美术学院的教师，负责建筑学和城市规划工作室。1971 年，由于一个采用合成材料的工业化公共住房设计方案，他成为了新建筑学项目 (programme Architecture Nouvelle) 的获奖人。1980 年，他与让·皮斯特尔创办瓦洛德与皮斯特尔建筑师事务所并成为了数个项目，如波尔多当代艺术博物馆 (Musée d'Art Contemporain de Bordeaux) 和波尔多贝利埃铸造厂 (Fonderie du Bélier) 的设计师。

他于 1988 年获得了法国建筑学会 (Académie d'Architecture) 银奖，并于 1990 年获得了法国国家工业促进金奖 (médaille de vermeil d'encouragement pour l'Industrie Nationale)。位于欧奈 (Aulnay) 的欧莱雅 (Oréal) 项目分别于 1992 年和 1993 年获得银角尺 (Equerre d'Argent) 奖和 Quaternario 大奖。1997 年，他获得艺术与文学骑士勋章，再于 1999 年，因其与让·皮斯特尔完成的全部设计而获得了法国建筑学会银奖后，他于 2001 年成为了该学会的成员。2002 年获得荣誉骑士勋章 (Chevalier de la Légion d'Honneur)。2007 年获得艺术与文学荣誉勋位。

Denis VALODE received, in 1969, his architectural degree from the Ecole Nationale Supérieure des Beaux-Arts (UP1), where he taught architecture and urban design beginning in 1971. In 1971, he won first prize in the Architecture Nouvelle series with a social housing design to be built from synthetic materials. In 1980, with Jean PISTRE, he co-founded the architectural firm VALODE et PISTRE, building upon early projects such as the Musée d'Art Contemporain at Bordeaux and the Fonderie du Bélier, an industrial project also located in the same city. He was awarded the Médaille d'Argent from the Académie d'Architecture in 1988 and the Médaille de Vermeil d'Encouragement from the Industrie Nationale in 1990. The L'Oréal Factory design was rewarded with the Equerre d'Argent in 1992 and the Quaternario Prize in 1993.

Denis VALODE was named Chevalier des Arts et des Lettres in 1997 and following his award of the Grande Médaille d'Argent from the Académie d'Architecture in 1999 for his carreer achievements was named, with Jean PISTRE, a member of the Académie of Architecture in 2001. He was named Officier des Arts et des Lettres in 2007.

让·皮斯特尔
JEAN PISTRE

让·皮斯特尔于 1974 年获得法国国立高等美术学院 (教育单元 1) 建筑师文凭，当仍为大学生时，因纸板娱乐住宅设计方案而成为新建筑学项目的获奖人。

1980 年，他与丹尼斯·瓦洛德创办瓦洛德与皮斯特尔建筑师事务所并成为了数个项目，如波尔多当代艺术博物馆 (Musée d'Art Contemporain de Bordeaux) 和波尔多贝利埃铸造厂 (Fonderie du Bélier) 的设计师。他于 1988 年获得了法国建筑学会 (Académie d'Architecture) 银奖，并于 1990 年获得了法国国家工业促进金奖 (médaille de vermeil d'encouragement pour l'Industrie Nationale)。位于欧奈 (Aulnay) 的欧莱雅 (Oréal) 项目分别于 1992 年和 1993 年获得银角尺 (Equerre d'Argent) 奖和 Quaternario 大奖。1997 年，他获得艺术与文学骑士勋章，再于 1999 年，因其与丹尼斯·瓦洛德完成的全部设计而获得了法国建筑学会银奖后，他于 2001 年成为该学会的成员。2011 年，让·皮斯特尔被任命为斯科尔科沃 (Skolkovo) 基金城市规划委员会主席，该项目是位于莫斯科附近的国际性科研中心项目。

Jean PISTRE received, in 1974, his architectural degree from the Ecole Nationale Supérieure des Beaux-Arts (UP1), winning as a student first prize in the Architecture Nouvelle series with a housing project conceived in cardboard. In 1980, with Denis VALODE, he co-founded the architectural firm VALODE et PISTRE, building upon early projects such as the Musée d'Art Contemporain at Bordeaux and the Fonderie du Bélier, an industrial project also located in the same city. He was awarded the Médaille d'Argent from the Académie d'Architecture in 1988 and the Médaille de Vermeil d'Ecouragement from the Industrie Nationale in 1990. The L'Oréal design was rewarded with the Equerre d'Argent in 1992 and the Quaternario Prize in 1993. Jean PISTRE was named Chevalier des Arts et des Lettres in 1997, and following his award of the Grande Médaille d'Argent from the Académie d'Architecture in 1999 for his career achievements was named, with Denis VALODE, a member of the Académie of Architecture in 2001. In 2011, Jean Pistre was named President of the Skolkovo Urban Council, a project for the creation of a scientific pole near Moscow.

宋国鸿
合伙建筑师
法国 VP 建筑设计事务所中国区总经理
PARTNER ARCHITECT
VP CHINA DIRECTOR

宋国鸿先生于 2003 年毕业于法国波尔多建筑景观设计学院 (Ecole d'Architecture et de paysage de Bordeaux) 获得法国政府建筑师文凭和法国木建筑硕士 DESS 证书。他曾工作于多家波尔多和巴黎的设计事务所并于 2004 年加入法国 VP 建筑设计事务所。自 2005 年担任法国 VP 建筑设计事务所中国区总经理。他工作经验广泛，涉足众多领域的建筑设计如超高层以及大型综合商业建筑、文教建筑、酒店、医疗等，至今，VP 中国公司发展到 50 多人设计团队并在北京和上海拥有办事处。他是法国大巴黎建筑师协会会员，自 2012 年成为法国 VP 建筑设计事务所集团合伙人。

Guohong Song received his architectural degree and degree DESS Bois from the Bordeaux School of Architecture and Landscape Architecture in 2003. He worked in several architectural offices in Bordeaux and Paris, before joining Valode & Pistre Architectes in 2004. In 2005 he was named director of Valode & Pistre's operations in China. Under his direction, Valode & Pistre China has expanded to over 50 employees located in offices in Beijing and Shanghai, and is currently working on several large scale projects, including mixed use towers, cultural buildings, hotels, and commercial centers.

In 2012 he joined the Order of Architectes in Ile de France (Greater Paris) and he became an Associate in the office of Valode & Pistre.

深圳市招华国际会展发展有限公司
Shenzhen Zhaohua International Conference and Exhibition Operation Co., Ltd

深圳市招华国际会展发展有限公司由两大央企招商局蛇口工业区控股股份有限公司和深圳华侨城股份有限公司联合发起成立，主要负责深圳全新展馆——深圳国际会展中心及其周边商业业态的建设。

招商局蛇口工业区控股股份有限公司（简称"招商蛇口"，股票代码001979）是招商局集团（中央直接管理的国有重要骨干企业）旗下城市综合开发运营板块的旗舰企业，也是集团核心资产整合平台及重要的业务协同平台。

招商蛇口创立于1979年。40年前建设开发运营的深圳蛇口片区是中国改革开放的发源地，为中国经济发展做出了重要的历史贡献，孵化并培育了以招商银行、平安保险、中集集团、招商港口等为代表的一批知名企业。2015年12月30日，招商蛇口吸收合并招商地产实现无先例重组上市，打造了国企改革的典范和中国资本市场创新标杆。

招商蛇口聚合了原招商地产和蛇口工业区两大平台的独特优势，以"中国领先的城市和园区综合开发运营服务商"为战略定位，聚焦园区开发与运营、社区开发与运营、邮轮产业建设与运营三大业务板块，以"前港-中区-后城"独特的发展经营模式，参与中国以及"一带一路"重要节点的城市化建设。

招商蛇口致力于成为"美好生活的承载者"，从城市功能升级、生产方式升级、生活方式升级三个角度入手，为城市发展与产业升级提供综合性的解决方案，配套提供多元化的、覆盖客户全生命周期的产品与服务。

Shenzhen Zhahua International Convention and Exhibition Development Co., Ltd. was jointly established by two central enterprises, China Merchants Shekou Industrial Zone Holdings Co.,Ltd and Shenzhen Overseas Chinese Town Co.,Ltd. It's mainly responsible for the construction of the new Shenzhen International Convention and Exhibition Center and its surrounding commercial properties.

China Merchants Shekou Industrial Zone Holdings Co., Ltd. ("China Merchants Shekou", stock code: 001979) is the flagship enterprise of China Merchants Group, an important state-owned enterprise directly managed by the central government, and a platform for the integration of the Group's core assets and important business collaboration.

Founded in 1979, China Merchants Shekou is the birthplace of China's reform and opening-up and has made important historical contributions to China's economic development, incubating and nurturing a number of well-known enterprises represented by China Merchants Bank, Ping An Insurance, CIMC, China Merchants Port, etc. On December 30, 2015, China Merchants Shekou absorbed and merged China Merchants Real Estate to achieve unprecedented restructuring and listing, creating a model for reform of state-owned enterprises and a benchmark for innovation in China's capital market.

China Merchants Shekou combines the unique advantages of the former China Merchants Real Estate and Shekou Industrial Zone, and is strategically positioned as a "leading integrated urban and park development and operation service provider in China", focusing on three business segments: park development and operation, community development and operation, and cruise industry construction and operation.

China Merchants Shekou is committed to becoming the "bearer of a better life", providing comprehensive solutions for urban development and industrial upgrading from three perspectives: urban function upgrading, production mode upgrading and lifestyle upgrading, and providing diversified products and services that cover the entire life cycle of customers.

华侨城集团有限公司

Overseas Chinese Town Group Co., Ltd

华侨城集团有限公司，是国务院国资委直接管理的大型中央企业，1985 年诞生于改革开放的前沿阵地——深圳，是国家首批文化产业示范园区、中国文化企业 30 强、中国旅游集团 20 强。集团连续 8 年获得国务院国资委年度业绩考核 A 级评价，2018 年实现营业收入突破千亿，利润总额突破 200 亿元，资产总额近 4500 亿元，效益位列央企前 20 位，收入和利润增速位列央企前 10 名。

多年来，华侨城集团通过独特的创想文化，致力于提升中国人的生活品质。其中康佳、欢乐谷连锁主题公园、锦绣中华·中国民俗文化村、世界之窗、东部华侨城、欢乐海岸、波托菲诺小镇、纯水岸、OCT-LOFT 华侨城创意文化园、OCAT 华侨城当代艺术中心、华侨城大酒店、威尼斯睿途酒店等均为行业领先品牌。

集团早期在实践中摸索出"旅游＋地产"创业模式，用旅游扩大社会影响，用地产收益反哺旅游，实现良性互动，取得一定成绩。进入新时代，华侨城紧跟国家新型城镇化战略，在"创新、协调、绿色、开放、共享"五大发展理念的指导下，提出"文化＋旅游＋城镇化"和"旅游＋互联网＋金融"的创新发展模式，支持康佳积极推进"科技＋产业＋城镇化"的战略落地，围绕"中国文化产业领跑者、中国新型城镇化引领者、中国全域旅游示范者"的战略定位，通过产业"文化化"和文化"产业化"，基于当前主业，形成了文化产业、旅游产业、新型城镇化、电子产业及相关业务投资五大发展方向，探索"文化＋"模式，从战略、体制、技术、产品、商业模式等维度全面创新，形成特色发展之路，实现跨越式发展。目前已布局全国 50 余座城市，形成了新型城镇化、全域旅游、文化创意产业、产融平台、科技及产业园区、城市综合体开发运营、主题酒店开发运营、智慧管理输出等业务板块。

Overseas Chinese Town Group Co., Ltd is a large centralized enterprise directly managed by the SASAC of the State Council. It was born in 1985 in Shenzhen, the frontier of reform and opening up, and is the first batch of national cultural industry demonstration park, the top 30 cultural enterprises in China, and the top 20 China Tourism Group. In 2018, the Group achieved operating income of more than 100 billion yuan, total profit of more than 20 billion yuan and total assets of nearly 450 billion yuan, ranking among the top 20 central enterprises in terms of efficiency and top 10 central enterprises in terms of revenue and profit growth rate.

Over the years, OCT Group has been committed to improving the quality of life of Chinese people through its unique creative culture. Among them, Konka, Happy Valley Chain Theme Park, Fairview China - Chinese Folk Culture Village, Window of the World, Eastern OCT, Happy Coast, Portofino Town, Pure Water Coast, OCT-LOFT OCT Creative Culture Park, OCAT OCT Contemporary Art Center, OCT Hotel, Venice Ritual Hotel, etc. are all leading brands in the industry.

In its early years, the Group has developed the "tourism + real estate" entrepreneurial model in practice, using tourism to expand social impact, and using real estate proceeds to feed tourism and achieve positive interaction, and has achieved certain achievements. Under the guidance of the five development concepts of "Innovation, Coordination, Green, Openness and Sharing", OCT has proposed the innovative development models of "Culture + Tourism + Urbanization" and "Tourism + Internet + Finance" to support Konka in actively promoting the strategy of "Technology + Industry + Urbanization", and based on the strategic positioning of "China Cultural Industry Leader, China New Urbanization Leader and China Tourism Demonstrator", Konka has formed five development directions of cultural industry, tourism industry, new urbanization, electronics industry and related business investment based on the current main business through "culturalization" and "industrialization" of industry, and explored the "cultural +" model to fully innovate in terms of strategy, system, technology, products, business model and other dimensions to form a characteristic development path and achieve leapfrog development. At present, it has laid out more than 50 cities nationwide, forming business segments such as new urbanization, regional tourism, cultural and creative industries, production and financing platform, technology and industrial parks, urban complex development and operation, theme hotel development and operation, intelligent management output.

黄展春剧场建筑设计顾问（北京）有限公司
Huang zhanchun Theater Architectural Design Consultant (Beijing) Co., Ltd

本公司于 2006 年在北京成立，是一家以观演空间、艺文空间及娱乐空间为工作范畴的专业设计顾问公司，业务主要针对剧院、音乐厅、演艺厅、报告厅、演播室、录音室、体育馆、会议空间及影院等场所进行规划设计或顾问咨询，我们对于前述空间拥有丰富的设计顾问经验，并与其他不同专业设计单位间有着良好的整合经验，因此无论项目规模的大小及功能形态的多元，我们均能针对业主需求提出最佳的规划方案，提供高水平的服务质量。

本公司最大的资产是拥有高素质的专业设计顾问人才，其涵盖多种专业领域，公司业务包括剧场顾问、观演建筑建筑设计、观演建筑室内设计、建筑声学设计（厅堂音质与噪声控制）及剧场专业设备系统设计，包括舞台机械、舞台专业灯光、音视频设计等等，配合项目良好管理的机制，辅以完善的行政支持，可确保每一项目从前期顾问咨询、设计规划、施工监督到完工验收之过程都获得良好的质量管控。

本公司提供的服务形式包括顾问咨询、规划设计及现场指导。透过专业的设计顾问、丰富的经验与良好的管理，必能共创一优秀作品。

The company was established in Beijing in 2006, and is a professional design consultancy firm with the scope of work in theatre, concert hall, auditorium, lecture hall, studio, recording studio, gymnasium, conference space, cinema, etc. We have rich experience in the design of the aforementioned spaces and good integration experience with other professional design units, so no matter the size and function of the project is diversified, we can plan and design or consult. We have rich experience in the design of the aforementioned spaces and have good integration experience with other professional design units, so no matter the size of the project and the diversity of functions, we can propose the best planning solutions to meet the needs of the owners and provide high quality services.

The company's biggest asset is its highly qualified professional design consultants, which covers a variety of professional fields, including theatre consultancy, architectural design for performing architecture, interior design for performing architecture, architectural acoustic design (hall sound and noise control) and theatre professional equipment system design, including stage machinery, stage professional lighting, audio-visual design... and so on, with the project good management mechanism, supplemented by perfect administrative support, to ensure that each project from the pre-consultation, design planning, construction supervision to the completion of the acceptance process of good quality control. With the mechanism of good project management and perfect administrative support, each project can be ensured good quality control from pre-consultation, design planning, construction supervision to completion acceptance.

The services include consulting, planning and design, and on-site guidance. Through professional design consultants, rich experience and good management, we can create an excellent work together.

黄展春
HUANG ZHANCHUN

陈建勇
CHEN JIANYONG

梁振亚
LIANG ZHENYA

项目主持人
Project Director

工作经历：
黄展春剧场建筑设计顾问（北京）有限公司
文化设施工艺专家（中华人民共和国住房和城乡建设部）
台湾大誉设计顾问有限公司创始人

Work Experience:
Huang Zhanchun Theatre Architectural Design Consultant (Beijing) Co.,Ltd
Expert in cultural facilities and crafts (Ministry of Housing and Urban-Rural Development of the People's Republic of China)
Founder of Taiwan Dayu Design Consultants Ltd,

项目负责人
Project Leader

工作经历：
易和乐道建筑设计工作室
北京实创博威建筑设计院

Work Experience:
Yiheledao Architectural Design Studio
Beijing Shichuangbowei Architectural Design Institute

声学工程师
Acoustical Engineer

工作经历：
黄展春剧场建筑设计顾问（北京）有限公司
辽宁山水装饰公司

Work Experience:
Huang Zhanchun Theatre Architectural Design Consultant (Beijing) Co.,Ltd
Liaoning Shanshui Decoration Company

深圳市城市交通规划设计研究中心
Shenzhen Urban Transport Planning Center

深圳市城市交通规划设计研究中心（以下简称"中心"）创建于 1996 年，一直以来是深圳市委市政府最重要的交通决策咨询机构，是深圳市交通综合治理和轨道交通建设重要的技术支撑单位。中心致力于提供最先进的城市交通技术与服务，提供以大数据分析为基础、以协同规划为引领、以品质设计为支撑、以集成建设为实践、以智慧运维为反馈的城市交通整体解决方案。

中心已成为国内资质最全、等级最高、人员规模最大、专业类别最多的城市交通专业研究机构之一。目前具备城乡规划甲级、工程咨询甲级、工程设计甲级以及信息系统集成等多项资质；拥有一支多学科、多专业、高学历、高素质、业务经验丰富的专业团队，人员规模超过 1500 人，其中博、硕士占比超过 50%，中、高级职称占比超过 40%，专业涵盖交通规划、城规建筑、工程设计、景观园林、系统集成、软件开发等类别。

目前业务范围已覆盖全国 30 多个省市，100 余座城市，累计完成各类项目千余项，先后获得全国优秀规划设计一等奖、詹天佑奖、省部级科学技术一等奖等各级奖项百余项。

Shenzhen Urban Transport Planning Center (hereinafter referred to as "SUTPC") was founded in 1996, and has been the most important transportation decision-making and advisory body of the Shenzhen Municipal Party Committee and Government, and is an important technical support unit for comprehensive transportation management and rail transportation construction in Shenzhen.
SUTPC is committed to providing the most advanced urban transportation technology and services, providing integrated urban transportation solutions based on big data analysis, led by collaborative planning, supported by quality design, integrated construction as a practice, intelligent operation and maintenance as feedback.

SUTPC has become one of the most fully qualified, highest-ranking, largest staff and most specialized urban transportation research institutions in China. It has a multidisciplinary, multi-disciplinary, highly educated, highly qualified and experienced professional team with a staff size of more than 1,500, of which more than 50% are PhD and Master, and more than 40% are middle and senior titles.

At present, the business scope has covered more than 30 provinces and 100 cities in China, and have completed more than 1,000 projects of various types, and have won more than 100 awards at all levels, such as the First Prize of National Excellent Planning and Design, the Zhan Tianyou Award, the First Prize of Science and Technology at the provincial and ministerial level.

张贻生
ZHANG YISHENG

项目主管 / 副总工程师 / 高级工程师 / 注册咨询工程师

Project Director/ Deputy Chief Engineer/ Senior Engineer/ Registered Consulting Engineer

徐惠农
XU HUINONG

项目总师 / 副总工程师 / 注册城市规划师 / 注册咨询工程师

Project Chief/ Deputy Chief Engineer/ Registered Urban Planners/ Registered Consulting Engineer

蒋金勇
JIANG JINYONG

项目负责人 / 博士 / 高级工程师 / 注册咨询工程师

Project manager/Ph.D./ Senior Engineer/ Registered Consulting Engineer

李方卫
LI FANGWEI

交通规划负责人 / 资深工程师 / 擅长片区交通规划设计

Transport planning manager/ experienced engineer/Good at transport planning & design of districts

陈 磊
CHEN LEI

交通设计负责人 / 主任工程师 / 擅长复杂工程交通规划设计

Transport design manager/ Chief Engineer/ Good at transport planning & design of complex projects

张素禄
ZHANG SULU

交通模型负责人 / 资深工程师 / 擅长交通仿真模拟分析

Transport model manager/ experienced engineer/Good at traffic simulation analysis

深圳市广田建筑装饰设计研究院
Shenzhen Grandland Construction Decoration Design Institute

深圳市广田建筑装饰设计研究院是广田股份（股票代码：002482）的下属全资公司，经过 20 年的磨砺积淀，已发展为目前兼具设计师规模及行业号召力的综合型设计机构之一。目前拥有国内外设计专家学者 2000 余名。由国际创意设计中心、国内设计领域专家、高级顾问团队、七个综合分院、四个专业分院（轨道交通、园林景观、陈设艺术、幕墙）以及机电所组成。服务范围涵盖：酒店空间设计、商业综合体设计、办公空间设计、别墅样板房设计、公共空间设计，配套专业有：幕墙设计、机电设计、智能化系统设计、园林景观设计、软装设计、灯光设计。广田设计院代表工程：深圳国际会展中心、香港亚洲电视台办公楼、阿里巴巴北京望京总部办公楼、华大基因中心、厦门国际中心主塔楼、深圳华为软件中心、宝能第一空间、西安印力诺富特酒店、青岛鑫江温德姆酒店、黄山北海宾馆、海口希尔顿、遵义逸林希尔顿酒店、北京地铁新机场线、天津地铁 10 号线一期工程。

广田设计院自成立以来，已成功与美国 WATG、美国 ABC、澳大利亚 HASSELL、德国 HID 等多家世界著名设计机构密切合作，打造了众多国际知名五星级品牌酒店及地标性项目，设计师及设计作品荣获国际国内大奖五百余项，公司与科研院校建立合作，已与新加坡南洋理工大学、中央美院、广州美院、四川美院、天津美院等高校签约并开展深度合作，在行业内起到示范作用。

Shenzhen Grandland Construction Decoration Design Institute, a wholly-owned subsidiary of Grandland Group(stock code:002482), has developed into one of the comprehensive design institutions in China with the largest designer scale and the most influence in the industry after nearly 20 years of accumulation. At present, there are more than 2,000 design experts and scholars at home and abroad. It is composed of the International Creative Design Center, domestic design experts, senior consultant team, seven comprehensive branches, four professional branches (rail transportation, landscape, furnishing art, curtain wall) and the Institute of Electrical and Mechanical. The service scope includes: hotel space design, commercial complex design, office space design, villa model room design, public space design, supporting specialties include: curtain wall design, mechanical and electrical design, intelligent system design, landscape design, soft decoration design, lighting design. Representative projects of Guangtian Design Institute: Shenzhen International Convention and Exhibition Center, Hong Kong Asia Television Office Building, Alibaba Beijing Wangjing Headquarters Office Building, Huada Genetic Center, Xiamen International Center Main Tower, Shenzhen Huawei Software Center, Baoneng First Space, Xi'an Innofot Hotel, Qingdao Xinjiang Wyndham Hotel, Huangshan Beihai Hotel, Haikou Hilton, Zunyi Yilin Hilton Hotel, Beijing Metro New Airport Line, Tianjin Metro Line 10 Phase I Project.

Ever since its inception, the institute has successfully cooperated with many world famous design institutes including WATG, ABC, HASSELL, HID , etc, accomplished numerous international well-known five star brand hotels and landmark projects. The company also set a model of enterprise-academic institution cooperation by signing contracts and carrying out deep teamwork with Singapore Nangyang Technological University, Central Academy of Fine Arts, Guangzhou Academy of Fine Arts, Sichuan Academy of Fine Arts, Tianjin Academy of Fine Arts, etc.

孙乐刚
SUN LEGANG

麻友博
MA YOUBO

门云
MEN YUN

深圳广田集团股份有限公司，深圳广田装饰设计研究院总院院长、董事、设计总监。中国四川美院研究生导师，广州美术学院研究生导师。

President, director and design director of Shenzhen Grandland Group Co., Ltd. and Shenzhen Grandland Construction Decoration Design Institute. Graduate Supervisor of Sichuan Academy of Fine Arts, China. Graduate Supervisor of Guangzhou Academy of Fine Arts.

深圳广田设计研究院第一分院设计副总监，方案六组组长。负责大型公共空间，商场办公及大型高端办公等方向的设计主导工作。

Deputy director of the first branch of Shenzhen Grandland Construction Decoration Design Institute, head of the scheme six team. Responsible for the design of large public spaces, shopping mall offices and large high-end offices.

深圳广田集团股份有限公司，深圳广田设计研究院项目七部总监，主要从事大型项目管理，整体深化设计等。

Director of Project Department 7 of Shenzhen Grandland Group Co., Ltd. and Shenzhen Grandland Construction Decoration Design Institute. Mainly engaged in large-scale project management and overall design deepening.

卓展工程顾问（北京）有限公司
Zhuozhan Engineering Consultants (Beijing) Co., Ltd.

卓展于 2004 年创建，是一家独立的、跨行业的工程顾问公司，我们的服务覆盖商业、教育、医疗、养老、工业等不同领域，为客户提供从咨询、设计到项目管理以及全方位工程顾问服务。我们在全国多个大中城市设有分公司，拥有超过 200 名的境内外资深顾问和专业人员，骨干成员都是建筑行业的资深人员，拥有超过 20 年的工作经验。卓展致力于培养人才，营造良好工作环境，鼓励创新科技，追求卓越成就，目标成为国际级的工程顾问公司。

我们的项目从超高超大型综合体、高星级酒店、甲级办公楼、别墅住宅、服务式公寓到学校、医院、养老、绿色建筑项目等。我们关注建筑环境，致力为工程建筑提供可持续的、绿色高能效环保的解决方案。我们深信，创造并提供更好的解决方案，是满足客户需求的最佳方式。

卓展作为一家全方位的机电顾问服务提供商，为客户提供全过程的优质服务，包括机电系统的方案设计、初步设计、招投标图纸及文件，投标分析，施工图设计、施工监管以及竣工验收等。我们也可以根据客户的需求，按项目的实际情况为客户提供阶段性或全过程的服务。

Founded in 2004, China-team is an independent, cross-industry engineering consulting company. Our services cover the building of variety aspect, such as commercial, education, medical care, pension and industrial, with the range from design, project management to total engineering consultant services.

We set up office in cities across the country, with more than 200 senior consultants and professionals, with core members with 20+ years working experience.

With the goal of becoming an international engineering consulting company, China-team is committed to cultivating talents, creating productive working environment, encouraging innovation and technology, and pursuing excellence.
Our projects range from shopping mall, luxury hotels, Grade-A office, villa residences, serviced apartments, campus, hospitals and green building. We focus on building environment and strive for sustainable and environmental friendly solutions to meet our customer needs.

蔡宏智
Johnnie Choi

从 2010 年至今，亲自担任七十个以上大型综合项目垂直交通顾问，熟练掌握各种建筑功能的垂直交通设计、系统审核、招投标、施工配合等工作，曾服务多家大型开发商及酒店管理公司，配合多家世界知名建筑师事务所及建筑设计院，完成多个标志性建筑方案设计。

擅长垂直交通动线及运能分析，了解国内外垂直交通行业最新技术及设计标准，深入剖析垂直交通对建筑空间布局、客流货流动线影响，为项目实现最大价值。
通过十年项目管理经验，熟悉项目规划流程、项目进度计划、建立设计团队及各机电专业设计工作配合等范围，具备较全面的项目协调及管理能力。

Since 2010, Johnnie Choi has taken part as vertical transportation consultant in over 70+mega-scale projects; he is proficient in vertical transportation design, system performance audit, tender and construction management for variety type of building. Johnnie helped some world-renowned architects and developer completed their iconic design.

With his strength in traffic analysis, he provides In-depth analysis on the impact of vertical traffic to the design of buildings, as the result of maximum commercial value for the project.

深圳市西利标识设计制作有限公司
Shenzhen Xili Sign Design & Production Co., Ltd

深圳市西利标识设计制作有限公司于 1996 年在中国深圳成立，是一家专业从事城市公共环境、公用建筑、房地产项目、商业地产、旅游景区、酒店等项目，致力于标识系统规划·设计·制作以及城市公共环境标识产品开发的专业性标识公司。深圳市西利标识研究院成立于 2010 年 7 月，是一家经深圳市民政局批准的专业培养标识行业高级技术人才的综合性学术机构。

西利标识经过 20 多年的沉淀，已经形成了一套完整规范的标识系统规划·设计·制作体系。西利标识运用关联行业的、专业的、先进的科学技术，将新颖、明快、醒目、富有个性的设计理念贯彻到标识系统规划设计中。用标准国际化、规划全面化、功能突出化、设计人性化、风格个性化来把握标识系统规划设计的主题， 是西利标识一直坚持的方向。

代表项目有：深圳国际会展中心、西藏鲁朗国际旅游小镇、云南丽江雪山艺术小镇、北京十三陵旅游景区、无锡清名桥历史文化街 区、北京中关村软件园、北京国际机场一号航站楼、海南博鳌亚洲论坛会议中心、重庆国际博览中心、长沙德思勤城市广场、上海复旦大学附属中山医院等。

Shenzhen Xili Sign Design and Production Co., Ltd. was established in 1996 in Shenzhen, China. The company is a professional sign company that specializes in the production and development of urban public environment signs, which is committed to the planning, design and implementation of signage systems for the environment, public buildings, real estate projects, commercial real estate, tourist attractions, hotels and other projects. Shenzhen Xili Sign Research Institute was established in July 2010, which is a comprehensive academic institution for senior technical talents in industry, and has been approved by Shenzhen Civil Affairs Bureau to cultivate the professional signage and product development.

After more than 20 years of experience, Xili has formed a complete and standardized system of planning, design and production. Xili uses professional and advanced science and technology of related industries to implement the novel, bright, eye-catching and personalized design concept into the logo system planning and design. The company has always insisted on the direction of internationalization of standards, comprehensive planning, functional highlighting, humanization of design, and individualization of style to grasp the theme of logo system planning and design.

Representative projects include: Shenzhen International Convention and Exhibition Center, Lulang International Tourism Town in Tibet, Lijiang Xueshan Art Town in Yunnan Province, Beijing 13th Mausoleum Tourist Attractions, Qingmingqiao Historical and Cultural Street in Wuxi, Zhongguancun Software Park in Beijing, Beijing International Airport Terminal 1, Hainan Boao Forum for Asia Conference Center, Chongqing International Expo Center, Desqin City Plaza in Changsha, Zhongshan Hospital affiliated with Fudan University in Shanghai, etc.

张西利
ZHANG XILI

深圳市西利标识研究院院长
深圳市西利标识设计制作有限公司董事长
西利标识设计中心总设计师
深圳国际会展中心标识项目总顾问

President of Shenzhen Xili Sign Research Institute
Chairman of Shenzhen Xili Sign Design and Production Co., Ltd
Chief Designer of Xili Sign Design Center
General Consultant of Shenzhen World Exhibition & Convention Center Sign Design Project

周游
ZHOU YOU

深圳市西利标识设计制作有限公司项目二部总经理
深圳市西利标识研究院认证注册标识项目管理师
美国项目管理协会（PMI）认证 PMP 项目经理
深圳国际会展中心标识项目经理

General Manager of Project 2 Department, Shenzhen Xili Sign Design and Production Co., Ltd
Registered Sign Project Manager of Shenzhen Xili Sign Research Institute
Project Management Institute (PMI) Certified PMP Project Manager
Project Manager of Shenzhen World Exhibition & Convention Center Sign Design Project

曾繁生
ZENG FANSHENG

深圳市西利标识设计制作有限公司设计经理
标识系统规划设计师、标识深化设计师
深圳国际会展中心标识项目主创设计

Design Manager, Shenzhen Xili Sign Design and Production Co., Ltd
Sign System Planner, Sign Design Deepening Designer
Master Designer of Shenzhen World Exhibition & Convention Center Sign Design Project

深圳市建筑科学研究院股份有限公司
Shenzhen Institute of Building Research Co., Ltd

深圳市建筑科学研究院股份有限公司为国有控股上市企业，前身是 1982 年成立的事业单位深圳市建筑科学研究中心，2017 年在深圳证券交易所 A 股创业板上市（股票代码 300675）。公司主要从事绿色城市建设技术的研究、开发和应用，集科研攻关、建筑设计、绿色咨询、生态规划、检测评估、项目管理等多重服务于一体。经过多年的发展，公司已成为我国建设领域内提供绿色建筑策划、设计、咨询、项目管理、检测评估等全过程综合性技术服务的品牌机构。公司现有员工 600 多人，其中研究与技术人员比例在 90% 以上，专业范围涵盖规划、建筑学、建筑技术、结构、岩土、建材、暖通空调、给排水、污水处理、水利、环境、景观、电气、建筑智能、太阳能研究、地理、地质、造价、质控等在内的建筑全领域。公司主编、参编绿色建筑和建筑节能相关的国家、省、市各级标准、规范共计 96 项，直接参与和推动了全国建筑节能强制实施和绿色建筑的普及。从 2003 年起率先面向行业开展绿色建筑和建筑节能设计与咨询服务，至今累计已完成 1500 多个项目，约 1.2 亿平米建筑的技术咨询，支持工程获得绿色建筑设计标识 300 余项、运营标识 30 余项，在绿色建筑技术服务领域走在全国前列。

Shenzhen Institute of Building Research Co., Ltd., a state-owned listed enterprise, was formerly Shenzhen construction science research center, a public institution established in 1982. In 2017, it was listed on the gem board of Ahenzhen stock exchange (stock code 300675). The company is mainly engaged in the research, development and application of green city construction technology, integrating scientific research, architectural design, green consulting, ecological planning, inspection and evaluation, project management and other services. After years of development, the company has become a brand organization providing whole-process comprehensive technical services such as green building planning, design, consulting, project management, testing and evaluation in the field of construction in China. Company has more than 600 staff, of which the proportion of research and technical personnel more than 90%, professional planning, architecture, construction techniques and range, structural, geotechnical, construction materials, HVAC, water supply and drainage, sewage treatment, water conservancy, environment, landscape, electrical, building intelligent, solar energy research, geography, geology, cost, quality, construction areas. The company has participated and edited in compiling 96 national, provincial and municipal standards and codes related to green building and building energy conservation, directly participating in and promoting the mandatory implementation of building energy conservation and the popularization of green building in China. Since 2003, it has taken the lead in providing green building and building energy-saving design and consulting services to the industry. So far, it has completed more than 1,500 projects, technical consulting for 120 million square meters of buildings, and more than 300 green building design marks and 30 operating marks for supporting projects, making it a leader in the field of green building technical services in China.

牛润卓
NIU RUNZHUO

陈益明
CHEN YIMING

肖雅静
XIAO YAJING

高级工程师，负责绿建咨询和物理环境分析优化。从前期规划阶段开始介入，提出会展中心绿建体系，对项目围护结构及机电节能方案进行系统优化分析，提出优化建议。

Senior engineer, in charge of green construction consulting and physical environment analysis and optimization. Participating in from the early planning stage, establish the green construction system of the Shenzhen World, made a comparative analysis of the project's energy saving scheme, and put forward optimization Suggestions.

高级工程师，负责 LEED 及 BREEAM 国际绿建认证咨询。为本项目国际咨询提供专业服务，配合开展认证工作，助力深圳会展中心完成 LEED 金级、BREEAM 三星的目标。

Senior engineer, in charge of consulting of LEED and BREEAM international green construction certification. Provide professional services for the international consultation of this project, cooperate with the certification work, and help Shenzhen World to achieve the goal of LEED gold and BREEAM three stars certification.

工程师，负责仿真模拟。综合评估会展中心的场地风环境、声环境、能耗、水耗、生态环境影响因子等。给出合理优化建议，助力会展中心的绿建目标达成。

Engineer, in charge of simulation, comprehensively evaluated the wind environment, sound environment, energy consumption, water consumption, ecological environment impact factors, etc. Give reasonable optimization Suggestions to help achieve the green construction goal of the Shenzhen World.

华纳工程咨询（北京）有限公司

Huana Ingenieure Consulting (Beijing) Co., Ltd

华纳工程咨询有限公司是一家从事专业幕墙顾问服务的公司，德国公司总部 SuP Ingenieure GmbH 位于德国达姆施塔特，是德国建筑幕墙工程界著名和重要的设计及设计审核机构之一，有着极强的专业技术实力和丰富的工作经验。

华纳工程咨询（北京）有限公司作为外资企业于 2007 年 7 月注册成立。2017 年入选首届全国建筑幕墙顾问行业联盟副理事长单位，2018 年评委全国幕墙顾问咨询行业 20 强之一，目前，中国公司除在北京外，同时在上海、深圳等地均设有分支机构。在建筑幕墙全生命周期的技术服务上，华纳（北京）公司拥有成熟而丰富的经验：不论建筑幕墙概念设计、深化设计、招投标文件编制和评标，还是设计文件审核以及现场检查，华纳（北京）公司都能为业主和建筑师提供专业、细致、一流的幕墙顾问服务。

华纳 SuP 不仅与众多国内外知名建筑师及设计院进行合作，也为很多房地产企业奉上了诸多精品项目。在华纳（北京）公司成功进行顾问服务的多个标志性建筑项目中，其建筑业态涵盖广泛：大型场馆、交通枢纽、超高层办公楼及酒店、商业综合体、科技园区、高档别墅等，都不乏华纳的经典之作。同时，在其参与过的 500 多个项目中，很多经典项目均获得中国建筑界顶级奖项"鲁班奖"和节能建筑"LEED"金奖的认证。

SuP is a professional façade consultant with technical expertise. Its headquarter SuP Ingenieure GmbH is located in Darmstadt, Germany and is one of the leading design and design review agency in the German façade design industry, we have strong technical expertise and extensive experiences.

SuP Construction Consultant (Beijing) Co., Ltd. was established in July 2007 as a foreign-funded enterprise. SuP was selected as vice-chairman corporation in the first national facade consulting industry union in 2017, and one of the top 20 façade consultants in China at 2018. So Far, SuP China have set up branches in Shanghai, Shenzhen except headquarter in Beijing. In terms of the technical services for the full lifecycle of building facade, SuP (Beijing) has mature and rich experience in providing professional, meticulous and first-class facade consultation services to owners and architects, no matter it is conceptual design,phase, deepening design phase, bidding document preparation and evaluation phase, design document review phase or site inspection phase.

SuP not only collaborates with many famous local architects and design institutes but also worked with many famous foreign architect and design institutes, and SuP accomplished many projects for the famous real estate companies. Among the landmark projects that SuP (Beijing) has successfully accomplished, the firm's construction portfolio covers a wide range of different functions: like large venues, transportation hubs, super tall office buildings and hotels, commercial complexes, technology parks, high-end villas, etc.. SuP accomplished many masterpiece together with all parties amount those projects. At the same time, among the more than 500 projects completed, many of the classic projects have been certified with the "Luban Award" and the "LEED" Gold Award, which are the top awards in the China construction industry.

苏州奥体中心
Suzhou Olympic Sports Center

深圳当代艺术馆及城市规划展览馆
Shenzhen Museum of Modern Art and City Planning Exhibition Hall

杭州华为
Huawei Hangzhou

北京中国人寿总部
Beijing China Life
Insurance Headquarters

成都中海财富中心
Chengdu Zhonghai
Wealth Center

深圳汉京
Shenzhen Hangking
Center

中建不二幕墙装饰有限公司

Zhongjian Bu'er Curtain Wall Decoration Co. Ltd

中建不二幕墙装饰有限公司是全球最大投资建设集团——中国建筑股份有限公司的全资骨干成员企业中建五局（湖南三强、中建三甲）的全资子公司，是一家集建筑幕墙、装饰装修、光电幕墙设计和施工于一体的高新技术企业，综合实力位居中国建筑幕墙行业前八强，拥有建筑幕墙、装饰装修专业承包壹级、设计专项甲级资质，具备华中地区最先进的单元式幕墙、节能系统门窗生产基地，年生产能力高达 1000 万平米。中建不二现有职工逾千人，其中一级建造师、注册造价师、高级工程师等高技能人才近 300 人，市场布局覆盖环渤海、长三角、珠三角和长江经济带、雄安新区、港珠澳大湾区等国家经济热点区域并踏足"一带一路"沿线国家。作为全国最早起步的幕墙企业之一，中建不二连续 12 年通过高新技术企业认证，创造了全国"四个第一"的业绩，将科技创新、BIM 技术等融入施工，先后承建了一大批地标项目。

深圳国际会展中心便是其中之一，为了完成深圳国际会展中心的幕墙建设，中建不二的建设者们无畏艰难，跑步进场，明确"五个一"的工作目标和思路。建设中秉承"六严"准则有序开展工作：严控质量、严把技术、严查材料、严开工序、严抓检查、严举竞赛。在精挑"布料"上，中建不二精准把控 5 万余块铝板，通过为每块铝板编号，确保所有铝板的加强筋都在同一水平线上，保证了"布料"的质量与美观；在登录厅屋顶为每个格栅板块制作一个"支架"，实现板块准确定位，将安装误差控制在 1 毫米内，做到严丝缝合；幕墙内部搭建一套"防水神经系统"，形成防水神经网络，并通过反复测试确保滴水不漏，以满足抗台防雨的性能。

除此之外，中建不二还参建了全球最大的机场单体卫星厅项目——上海浦东国际机场；世界唯一悬浮于矿坑之上的冰雪乐园——大王山冰雪世界；世界级高档中央商务区——埃及新首都 CBD 幕墙装饰工程等标志性工程。

The company is a wholly owned subsidiary of China Construction V Bureau (Hunan Top 3, China Construction Top 3), the world's largest investment and construction group, a high-tech enterprise integrating the design and construction of architectural curtain wall, decorative decoration, photoelectric curtain wall. The company now has more than 1,000 employees, including nearly 300 high-skilled personnel such as first-class builders, registered architects and senior engineers, and the market layout covers the Bohai Sea Rim, Yangtze River Delta, Pearl River Delta and Yangtze River Economic Belt, Xiong'an New District, Hong Kong-Zhuhai-Macao Greater Bay Area, and other national economic hotspots and countries along the "One Belt, One Road" route. As one of the earliest start-up curtain wall enterprises in China, the company has passed the high-tech enterprise certification for 12 years in a row, creating the national "four firsts" performance, integrating scientific and technological innovation and BIM technology into the construction, and has undertaken a large number of landmark projects.

In order to complete the construction of the curtain wall of Shenzhen World Exhibition & Convention Center, the builders of the company were fearless and hardworking, jogging into the site to clarify the work objectives and ideas of "five one". In the construction of the "six strict" guidelines to carry out work in an orderly manner: strict quality control, strict technology, strict investigation of materials, strict opening procedures, strict inspection, strict competition. By numbering each aluminum plate to ensure that the reinforcement of all aluminum plates are on the same horizontal line, the quality and beauty of the "fabric" is guaranteed; a "bracket" is made for each grille plate on the roof of the registration hall to achieve accurate positioning of the plates, the installation error is controlled to within 1 mm, so as to achieve strict stitching; a set of "waterproof nerve system" is built inside the curtain wall to form a waterproof nerve network, and through repeated tests to ensure that no water drips, in order to meet the performance of anti-standard and rainproof.

In addition, the company also participated in the construction of the world's largest airport monolithic satellite hall project - Shanghai Pudong International Airport; the world's only ice and snow park suspended above the pit - Dawang Mountain Ice World; the world-class high-end central business district - the new capital of Egypt's CBD curtain wall decoration project and other landmark projects.

中国建筑五局办公大楼

陶剑

TAO JIAN

深圳会展项目执行指挥长

Executive Commander of Shenzhen World
Exhibition & Convention Center

钟勇

ZHONG YONG

深圳会展项目设计负责人

Design director of Shenzhen World
Exhibition & Convention Center

江河幕墙
Jangho Group Company Limited

江河幕墙是全球高端幕墙第一品牌。江河幕墙成立于 1999 年，是集产品研发、工程设计、精密制造、安装施工、咨询服务、成品出口于一体的幕墙系统整体解决方案提供商，是全球幕墙行业领导者，在北京、上海、广州、成都、武汉等地建有一流的研发设计中心和生产基地，业务遍布全球二十多个国家和地区，拥有国家级企业技术中心、国家认定博士后科研工作站、中国幕墙行业首家国际认可的 CNAS 出口企业检测中心，是国家高新技术企业、国家技术创新示范企业、首批国家级知识产权优势企业。

近年来江河幕墙在全球各地承建了数百项难度大、规模大、影响大的地标建筑，荣获了包括中国建设工程鲁班奖在内的国内外顶级荣誉逾百项。其中，承建200 米以上摩天大楼逾百项，包括世界第一高楼沙特王国塔（1007 米）、中国第一高楼武汉绿地中心（636 米）、北京第一高楼中国尊（528 米）以及天津周大福金融中心、上海中心大厦、广州东塔、深圳华润国际商业中心、迪拜无限塔、阿布扎比天空塔等地标建筑；承建大型文化、金融、商业综合体逾千项，包括"全球十大最强悍工程"之一中央电视台新址、上海世博文化中心、澳门梦幻城、新加坡金沙娱乐城、阿布扎比金融中心、卡塔尔巴瓦金融中心等地标建筑；承建大型交通枢纽逾 40 项，包括中国四大直辖市五大机场、北上广三大枢纽火车站、阿布扎比国际机场等地标建筑。

江河幕墙秉承绿色建筑理念，依托技术领先、品质领先、服务领先、成本领先之竞争优势，以全球精品工程，铸就幕墙行业典范，持续缔造城市建筑传奇。

Jangho Group Company Limited (hereinafter referred to as "Jangho"), founded in 1999, is a large-scale multinational enterprise listed on the Main Board A shares of Shanghai Stock Exchange (Stock Code: 601886), headquartered in Beijing, previously named as Beijing Jangho Curtain Wall Company Limited.

Taking "for people's living environment, health and benefits" as its mission, Jangho actively explores business in the fields of green architectural decoration and health care by sticking to the strategy of "dual-core business, diversified development". Up till now, with many top brands, namely JANGHO, Sundart, Gangyuan, SLD, Vision and Zeming, the business of Jangho has extended to more than 20 countries and regions worldwide, making it play a leader role in the stage of curtain wall, interior decoration, health care etc. In addition, Jangho has been honored with many qualifications and glories, such as National High-tech Enterprise, the member of the first batch of National Intellectual Property Enterprises, the National Model Enterprise in Technology Innovation, National Technical Center of Enterprise and Post-doctoral Research Center with National Recognition, etc., besides, it often appears on the lists of China Top 500 Listed Enterprises and China Top 500 Private Enterprises.

Jangho adheres to the concept of green building, relying on the competitive advantages of leading technology, leading quality, leading service and leading cost, with global quality projects, casts a model of curtain wall industry and continues to create urban building legend.

闫忠云
YAN ZHONGYUN

北京江河幕墙系统工程有限公司副总工程师

Deputy Chief Engineer of Beijing Jianghe
Curtain Wall System Engineering Co., Ltd.

王顺雨
WANG SHUNYU

北京江河幕墙系统工程有限公司副总

Deputy General Manager of Beijing Jianghe
Curtain Wall System Engineering Co., Ltd.

张绍敏
ZHANG SHAOMIN

北京江河幕墙系统工程有限公司设计经理

Design Manager of Beijing Jianghe
Curtain Wall System Engineering Co., Ltd.

深圳市方大建科集团有限公司
Shenzhen Fangda Building Technology Group Co., Ltd.

深圳市方大建科集团有限公司创立于 1991 年，是我国第一批 A+B 股股票同时上市的民营企业，作为深圳国际会展中心主要参建单位，我们以打造精品工程、实现建筑师理念为目标，致力于为客户提供优质、精细化的服务，专筑高品质、高效率的幕墙工程。

方大建科总部位于深圳，下设北京、上海、成都、香港、澳洲、东南亚等区域公司和广州、珠海、重庆、南京、厦门、布里斯班、悉尼、胡志明等 20 多个国内和海外办事处，在东莞、上海、成都、南昌等地建有大型幕墙研发制造基地。目前，方大建科有数千名员工，业务遍及 170 多个国家和地区，建造了超过 4600 万平米建筑的幕墙，是全球高端幕墙系统的精品供应商和服务商。

方大建科在超高层、机场、大型场馆、企业总部等高端幕墙领域为客户提供有竞争力、高品质、高效率的产品、解决方案与服务，创造了我国建筑幕墙行业多项第一，包括全球最大的会展中心——深圳·国际会展中心、全球最高的纯钢结构建筑——深圳·汉京中心（350m）、全球最大的穿孔不锈钢双层装饰幕墙——深圳市当代艺术馆与城市规划展览馆、全球最大的青花瓷建筑群——南昌·万达茂等大批标志性项目。其中，方大建科承建的深圳国际会展中心中央廊道屋面幕墙工程，南北向长约 1800m，东西向宽为 42.5m，幕墙面积约 15.6 万㎡，造价超 3 亿元人民币。

Founded in Year 1991, Shenzhen Fangda Building Technology Group Co., Ltd. is one of the first listed private companies in both B and A share market in China. As a main contractor for Shenzhen International Convention & Exhibition Center, Fangda is dedicated to providing excellent service and high quality façadesolutions to our customers.

Headquartered in Shenzhen, Fangda has regional offices in Beijing, Shanghai, Chengdu, Hong Kong, Australia, Southeast Asia and more than 20 representative offices in domestic and overseas cities like Guangzhou, Zhuhai, Chongqing, Nanjing, Xiamen, Brisbane, Sydney, Ho Chi Minh and others with thousands of employers. In Dongguan, Shanghai, Chengdu and Nanchang, Fangda has its own curtain wall R&D and production base. Up till now, Fangda has provided products and services to more than 170 countries&areas and built over 46,000,000 sqmfaçade construction. Nowit is a leading supplier for high-end façade systems in the world.

Fangda provides competitive, high efficiency and high-performance façade solutions and its projects coverhigh-rising buildings, airports, stadiums, and enterprise headquarters. It creates multiply No.1s in curtain wall industry in China, including the largest convention center in the world-Shenzhen International Convention and Exhibition Center, the tallest full steel structure in the world- Shenzhen Hanking Center (350m in height), the largest double-skin perforated stainless steel curtain wall – Shenzhen Museum of Contemporary Art and Urban Planning, the largest chinese ceramic building cluster in the world- Nanchang Wanda Mall and others. Among them, the façade area ofShenzhen International Convention and Exhibition Center undertaken by Fangda covers a total area of about 156,000 square meters. The roofing façade ofthe central corridor, which is about 1800m in length (from south to north) and 42.5 meters in width(from east to west), values 300,000,000 RMB.

刘云飞
LIU YUNFEI

方大建科深圳国际会展中心项目项目经理

Fangda Project Manager of Shenzhen World Exhibition & Convention Center

方大建科深圳国际会展中心项目团队

Fangda Team of Shenzhen World Exhibition & Convention Center

徐强
XU QIANG

方大建科深圳国际会展中心项目主设

Fangda Chief Designer of Shenzhen World Exhibition & Convention Center

方大建科承办深圳国际会展中心幕墙技术研讨活动

Photo taken during Shenzhen World Exhibition & Convention Center Façade Technology Seminar organized by Fangda

中建深圳装饰有限公司
China Construction Shenzhen Decoration Co., Ltd

中建深圳装饰有限公司（以下简称"中建深装"），是世界 500 强企业中国建筑集团有限公司所属中建装饰集团的龙头子公司，是建设部首批批准成立的 8 家国家大型建筑装饰公司之一，在近几年中国建筑装饰行业百强、中国建筑幕墙行业百强上榜央企中，两项指标均高居第一，被誉为"中国建筑装饰行业的先锋"。

中建深装于 1985 年在深圳成立，原名中国建筑第三工程局深圳装饰设计公司，2007 年，伴随着中国建筑的整体上市需要，企业更名为中建三局装饰有限公司，并将总部迁往北京。2015 年，由于"一带一路"战略的市场需要，企业将总部重新迁回深圳，并于 2017 年 4 月正式更名为中建深圳装饰有限公司。

中建深装现有员工 2000 多人，企业资产 20 多亿元，年施工产值 50 亿元，实现了在全国三十多个省市区，以及巴哈马、阿尔及利亚、斯里兰卡等国外市场多领域的经营开拓，承建了众多地标性工程，鲁班奖数量位居行业前列。

China Construction Shenzhen Decoration Co., Ltd (hereinafter referred to as "CCSZD"), CCSZD is the leading subsidiary of "China Construction Decoration Group Co., ltd (CCDG)" and belongs to "China State Construction Engineering Corporation (CSCEC)". CSCEC is one of the 500 Fortune Companies in the world.Under the direct leadership of the CSCEC, CCSZD is one of the first large-scale building decoration companies approved by the Ministry of Construction in China. In recent years, CCSZD is top 100 enterprises in China's architectural decoration, and top 100 enterprises in China's curtain wall.

CCSZD was established in 1985 in Shenzhen and the initial name is "China Construction Third Engineering Bureau Shenzhen Decoration Design Engineering Company."In 2007, with the "China Construction" going public, the company changed its name to "China Construction Three Bureau Decoration Co., Ltd.". Its headquarters moved Beijing, capital of China.In 2015, with the strategic policy during "The Belt and Road", the company's headquarter moved back to Shenzhen, and in 2017 changed its name to "China Construction Shenzhen Decoration Co., Ltd (CCSZD)".

CCSZD currently has more than 1,500 employees and nearly 20 billion RMB corporate assets. The annual output value is nearly 50 billion yuan. Its business scope has distributed nearly every provincial capitals in China and some foreign countries (Bahamas, Algeria, Sri Lanka ...). The construction of many regional signs projects has won a large number of Chinese Luban Award (the highest honor in China construction).

何林武
HE LINWU

中建深圳装饰有限公司设计研究院（幕墙）广东分院副院长，长期从事于幕墙设计并主持了多个地标性幕墙项目的设计工作，多次获得全国优秀幕墙设计师等称号，近年来致力于装配式幕墙系统的研究。

Vice President of Guangdong Branch, has been engaged in curtain wall design for a long time and presided over the design of several landmark curtain wall projects, and has won the title of National Excellent Curtain Wall Designer many times. In recent years, he has been focusing on on assembled curtain wall systems.

森特士兴集团股份有限公司
Center Int Group Co., Ltd.

森特士兴集团股份有限公司（以下简称：森特股份）创建于 2001 年，注册资本金 4.8 亿元人民币。集团员工 1200 余人，25 家分支机构分布全国，业务覆盖建筑金属围护系统和生态治理两大领域，是国内第一家在主板上市提供金属围护系统解决方案的专业公司。

从金属围护系统专业咨询、科技研发、工程设计、材料加工到施工安装，森特股份为客户提供工业建筑及公共建筑高端金属建筑围护一体化解决方案。作为国内金属建筑围护行业的领军企业，连续多年荣登"全国金属围护系统行业十强企业"榜单。工程业绩累计超过 2100 个，建筑面积累计达 10000 万平方米，涵盖了机场、火车站、会展中心、商业展馆、文体场馆等公共建筑，以及航空航天、汽车、电子电器、物流、机械、电力、化工冶金等工业建筑。屡获"鲁班奖""金禹奖"等行业大奖。

近年来，集团参建了一系列重大、重点公共建筑类项目。深圳国际会展中心以其独特的造型、庞大的体量，强大的功能性及不可忽视的经济作用蜚声国内。会展中心标准展厅屋面板长 110 米，超大展厅长达 260 米，屋面最顶部的装饰层由 23 万多块不同规格颜色的铝单板组成，同时隐藏了很多滑移天窗和气动开启天窗。跨度大、构造层复杂，对屋面系统的抗风揭和防渗漏性能要求极为严苛。作为深圳国际会展中心最大的屋面承建商，森特股份项目团队综合了项目临海的自然环境因素和功能性需求，以及国内外大量的屋面案例分析和多次专家论证，最终采用了 66H 抗风揭防渗漏屋面系统。该系统屋面板固定座与板肋一起参与咬合；不滑动固定座，将屋面板的热胀冷缩量在每个跨内以轻微的形变予以抵消；板型的断面特征，提高了板型的截面惯性矩，使该系统具有更强的抗风揭和防水性能，能够更好的应对深圳地区多台风暴雨的气候挑战。目前，66H 抗风揭防渗漏屋面系统在北京大兴国际机场、海口美兰机场、杭州萧山国际机场、珠海长隆海洋科学馆等大型公共建筑项目上也得到了广泛运用。

Center INT Group (hereinafter referred to as: Center INT) was founded in 2001, with a registered capital of RMB 480 million. With more than 1200 employees and 25 branches all over China, the Group is the first professional company in China to provide metal enclosure system solutions listed on the main board.

From metal enclosure system professional consulting, R & D, engineering, material processing to construction and installation, Cent shares to provide customers with industrial buildings and public buildings high-end metal building enclosure integrated solutions. As a leading company in the metal envelope industry in China, we have been listed as one of the "Top Ten Enterprises in the Metal Envelope System Industry" in China for many consecutive years. The company has over 2,100 projects with a total construction area of 10 million square meters, covering public buildings such as airports, railway stations, convention centers, commercial halls, cultural and sports venues, as well as industrial buildings such as aerospace, automotive, electrical and electronic, logistics, machinery, electricity, chemical and metallurgical industries. We have won many industry awards such as "Luban Award" and "Jinyu Award".

In recent years, the Group has participated in a series of major and key public construction projects. Shenzhen International Convention and Exhibition Center is famous in China for its unique shape, huge volume, strong functionality and economic role that cannot be ignored. The standard showroom roof is 110 metres long and the oversized showroom is 260 metres long, with the topmost decorative layer consisting of more than 230,000 aluminium veneers in various colours and concealing many sliding skylights and pneumatic opening skylights. The large span and complex construction layers require the roofing system to be extremely demanding in terms of wind and leakage resistance. As the largest roofing contractor of the Shenzhen International Convention and Exhibition Center, the Centec project team integrated the natural environmental factors and functional requirements of the project waterfront, as well as a large number of domestic and international roofing case analysis and many expert arguments; and finally adopted the 66H windproof and leakproof roofing system. The system's roof mounts participate in the galling together with the ribs; the non-sliding mounts counteract the thermal expansion and contraction of the roof by a slight deformation within each span; the cross-sectional characteristics of the plate type improve the cross-sectional moment of inertia of the plate type, making the system more windproof and waterproof, which can better meet the climate challenges of many typhoons and storms in Shenzhen.At present, 66H windproof and leakproof roofing system is widely used in large public building projects such as Beijing Daxing International Airport, Haikou Meilan Airport, Hangzhou Xiaoshan International Airport, Zhuhai Changlong Marine Science Center, etc.

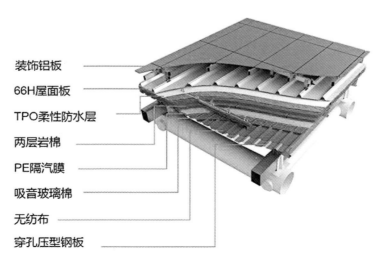

装饰铝板
66H屋面板
TPO柔性防水层
两层岩棉
PE隔汽膜
吸音玻璃棉
无纺布
穿孔压型钢板

抗风揭防渗漏屋面系统
Anti-wind exposed roof system

浙江中南建设集团有限公司

Zhejiang Zhongnan Construction Group Co., Ltd.

浙江中南建设集团有限公司成立于 1984 年，总部坐落于杭州市高新滨江区。集团公司注册资金 11.58 亿，集团公司占地近 660 亩，建筑面积逾 15 万平米，拥有占地面积 8 万平方米的现代化幕墙门窗生产加工基地，连续多年业内排名前三甲，公司现有技术、行政管理人员 1000 余人，年产值 50 多亿元，公司设计院下设 22 个设计所，拥有设计师 500 余人，国家级幕墙门窗专家 5 人，国家光伏专家 2 人，省级专家 6 人，高级工程师 30 余人。全国设有重庆、湖南、云南、东北、江西、西安等多个分院，并在广州、江苏、河南、北京、上海、天津等二十余个省三十多个城市设有分公司或办事处，覆盖全国市场，并在马来西亚、泰国、缅甸等境外地区设有分公司或办事处，在迪拜、安哥拉等中东地区承接完成了多个项目。

近年来，公司承建的幕墙工程荣获二十多项"鲁班奖"、三十多项"中国建筑工程装饰奖"、多项"全国金奖"、上百项"钱江杯"、"省优工程奖"、"省优设计奖"及外省市多项优质工程奖等，系长三角建筑幕墙龙头企业，综合实力进入全国建筑幕墙专业承包企业前茅。在中南，"铸造长城般的品质"是孜孜不倦的追求，更是兢兢业业的实践。怀着对幕墙事业的满腔热忱，中南幕墙，誓以"四个一流"打造核心竞争力，为振兴民族幕墙产业、让一座座中南幕墙长城，永远傲立在幕墙产业的最前沿！

Zhejiang Zhongnan Construction Group Co., Ltd. was established in 1984 and is headquartered in Hangzhou's High-tech Binjiang District. The group company has a registered capital of 1,158 million, covers an area of 660 mu, construction area of more than 150,000 square meters, has a modern curtain wall doors and windows production and processing base covering an area of 80,000 square meters, ranked the top three in the industry for many years, the company has more than 1,000 technical and administrative personnel, the annual output value of more than 5 billion yuan, the company has 22 design institutes under the design institute, has more than 500 designers, 5 national curtain wall doors and windows experts, 2 national PV experts, 6 provincial experts, senior engineers more than 30 people. The company has many branches in Chongqing,
Hunan, Yunnan, Northeast China, Jiangxi, Xi'an and more than 30 cities in more than 20 provinces, including Guangzhou, Jiangsu, Henan, Beijing, Shanghai, Tianjin, etc., covering the national market, as well as branches or offices in Malaysia, Thailand, Myanmar and other overseas regions, and has completed many projects in Dubai, Angola and other Middle East regions.

In recent years, the curtain wall projects constructed by the company have won more than 20 "Luban Award", more than 30 "China Construction Engineering Decoration Award", many "National Gold Award", hundreds of "Qianjiang Cup", "Provincial Excellent Project Award", "Provincial Excellent Design Award" and many quality project awards from other provinces and cities, etc. The company is a leading enterprise of curtain wall construction in the Yangtze River Delta, and its comprehensive strength is among the top professional curtain wall construction contracting enterprises in China. In Zhongnan, "casting the Great Wall-like quality" is a tireless pursuit, but also a conscientious practice. With full enthusiasm for the curtain wall business, Zhongnan Curtain Wall vows to build its core competitiveness with "four first-class", in order to revitalize the national curtain wall industry and make the Great Wall of Zhongnan Curtain Wall stand at the forefront of the curtain wall industry forever!

工厂风貌
Factory style

办公环境
Office environment

深圳国际会展中心场地勘察
Site survey of Shenzhen World Exhibition & Convention Center

深圳市水务规划设计院有限公司
Shenzhen water planning and Design Institute Co., Ltd

深圳国际会展中心（一期）全过程造价咨询
Whole process cost consultation of Shenzhen World Exhibition & Convention Center (phase I)

深圳市航建工程造价咨询有限公司
深圳市永达信工程造价咨询有限公司
Shenzhen aviation Construction Engineering Cost Consulting Co., Ltd
Shenzhen yongdaxin Engineering Cost Consulting Co., Ltd

深圳国际会展中心（一期）软基处理工程
Soft foundation treatment project of Shenzhen World Exhibition & Convention Center (phase I)

中交第四航务工程局有限公司
中铁二十五局集团有限公司
深圳市工勘岩土集团有限公司
CCCC fourth Harbor Engineering Co., Ltd
China Railway 25th Bureau Group Co., Ltd
Shenzhen engineering exploration geotechnical Group Co., Ltd

深圳国际会展中心（一期）质量安全文明施工评估服务合同
Shenzhen World Exhibition & Convention Center (phase I) quality, safety and civilized construction evaluation service contract

深圳市瑞捷建筑工程咨询有些公司
Shenzhen Ruijie Construction Engineering Consulting Co., Ltd

深圳国际会展中心（一期）基坑支护和桩基础工程
Foundation pit support and pile foundation engineering of Shenzhen World Exhibition & Convention Center (phase I)

广东潮通建筑园林工程有限公司
深圳市工勘岩土集团有限公司
深圳市建筑工程股份有限公司
深圳市基础工程有限公司
Guangdong CHAOTONG Architectural Landscape Engineering Co., Ltd
Shenzhen engineering exploration geotechnical Group Co., Ltd
Shenzhen Construction Engineering Co., Ltd
Shenzhen Foundation Engineering Co., Ltd

深圳市工程监理与相关服务
Shenzhen engineering supervision and related services
广州珠江工程建设监理有限公司
Guangzhou Pearl River Engineering Construction Supervision Co., Ltd

深圳国际会展中心（一期）施工总承包工程施工
Construction of EPC project of Shenzhen World Exhibition & Convention Center (phase I)
中国建筑股份有限公司
China Construction Corporation

深圳国际会展中心（一期）幕墙工程
Curtain wall project of Shenzhen World Exhibition & Convention Center (phase I)
北京江河幕墙系统工程有限公司
苏州金螳螂幕墙有限公司
浙江宝业幕墙装饰有限公司
浙江中南建设集团有限公司
深圳市方大建科集团有限公司

中建不二幕墙装饰有限公司
中建深圳装饰有限公司
Beijing Jianghe curtain wall system engineering Co., Ltd
Suzhou Golden Mantis curtain wall Co., Ltd
Zhejiang Baoye curtain wall decoration Co., Ltd
Zhejiang Zhongnan Construction Group Co., Ltd
Shenzhen fangdajianke Group Co., Ltd
CSCEC Buer curtain wall decoration Co., Ltd
CSCEC Shenzhen Decoration Co., Ltd

深圳国际会展中心（一期）机电承包工程施工工程
Shenzhen World Exhibition & Convention Center (phase I) mechanical and electrical contract project construction project

中建安装工程有限公司 // 中国建筑股份有限公司（联合体）
CSCEC Installation Engineering Co., Ltd. // CSCEC (joint venture)

深圳国际会展中心（一期）装修工程
Decoration project of Shenzhen World Exhibition & Convention Center (phase I)

深圳市洪涛装饰股份有限公司
深圳市晶宫设计装饰工程有限公司
深圳市奇信建设集团股份有限公司
深圳文业装饰设计工程股份有限公司
中建深圳装饰有限公司

Shenzhen Hongtao Decoration Co., Ltd
Shenzhen Jinggong design and Decoration Engineering Co., Ltd
Shenzhen Qixin Construction Group Co., Ltd
Shenzhen Wenye decoration design Engineering Co., Ltd
CSCEC Shenzhen Decoration Co., Ltd

深圳国际会展中心（一期）智能化工程施工合同
Intelligent project construction contract of Shenzhen World Exhibition & Convention Center (phase I)

中通服咨询设计研究院有限公司 // 中国电信股份有限公司
China Telecom Consulting Design & Research Institute Co., Ltd. // China Telecom Co., Ltd

深圳国际会展中心（一期）园林绿化工程
Shenzhen World Exhibition & Convention Center (phase I) Landscaping project

棕榈生态城镇发展股份有限公司
深圳市万年青园林绿化有限公司
Palm Ecological Town Development Co., Ltd
Shenzhen evergreen Landscape Co., Ltd

深圳国际会展中心（一期）标识导示系统施工及相关服务
Construction and related services of signage system of Shenzhen World Exhibition & Convention Center (phase I)

常州奥兰通标识公司
深圳市自由美标识有限公司
Changzhou Orlando logo Co., Ltd.
Shenzhen free beauty logo Co., Ltd

深圳国际会展中心（一期）艺术品设计及制作采购安装
Art design, production, purchase and installation of Shenzhen World Exhibition & Convention Center (phase I)
尤艾普（上海）艺术设计咨询有限公司
UEP (Shanghai) Art Design Consulting Co., Ltd

凤塘大道会展段施工总承包工程施工
General contracting project construction of Fengtang Avenue exhibition section
中国建筑股份有限公司
China Construction Corporation

深圳国际会展中心

深圳市欧博工程设计顾问有限公司　编著

出品人：冯越强

编著：冯越强　林建军　杨光伟　卢东晴　林文波　董孜孜

主编：冯越强

执行总编：杨光伟　丁荣

副主编：林建军　白宇西　祝捷　郭晓黎　卢东晴　李媛琴　林文波　王石　董孜孜

撰稿人：冯越强　林建军　杨光伟　丁荣　祝捷　黄用军　黄煜　卢东晴　李媛琴　王石　蒋宪新　许少良　廖晓华　邝英武　赵峰　王晓洁　黄丹　梁燕明　黄帅　罗凡　常冬冬　巫志强　何志力　何远明　骆年红　张浩　胡海萍　罗荣灿　吴少光　许振裕　冯明　舒龙　聂云飞　张昌蓉　叶清　梁威

统筹：董孜孜

排版：董孜孜

编辑：廖林涛　邹蔚　张发智　黄柳　岑土沛　冯梓阳

特约编辑：邹梦璇

翻译：邹梦璇

责任校对：卢东晴　林文波　王石　吴凡

装帧设计：邹蔚　岑土沛

开本：377mmx260mm 1/8

建设单位：深圳市招华国际会展发展有限公司

马军　李年长　吕涛　李时端　代君正　毛大奔　凌礼　王卓　黄俊杰　周欣　蔡亮　陈虎　邓碧有　胡振涛　李俊星　李威　李宗钢　刘硕　刘思源　刘源　梁铨捷　屈伟萍　任春磊　谭绍斌　王超　谢莹　辛伟　袁巍　张海军　赵锐　郑阳

Shenzhen World Exhibition & Convention Center

Presented by SHENZHEN AUBE ARCHITECTURAL ENGINEERING DESIGN CO., LTD.

Producer: Feng Yueqiang

Chief Planner: Feng Yueqiang, Lin Jianjun, Yang Guangwei, Lu Dongqing, Lin Wenbo, Dong Zizi

Planning Editor: Feng Yueqiang

Executive Chief Editor:Yang Guangwei, Ding Rong

Associate Editor: Lin Jianjun, Bai Yuxi, Zhu Jie, Guo Xiaoli, Lu Dongqing, Li Yuanqin, Lin Wenbo, Wang Shi, Dong Zizi

Contributor: Feng Yueqiang, Lin Jianjun, Yang Guangwei, Ding Rong, Zhu Jie, Huang Yongjun, Huang Yu, Lu Dongqing, Li Yuanqin, Wang Shi, Jiang Xianxin, Xu Shaoliang, Liao Xiaohua, Kuang Yingwu, Zhao Feng, Wang Xiaojie, Huang Dan, Liang Yanming, Huang Shuai, Luo Fan, Chang Dongdong, Wu Zhiqiang, He Zhili, He Yuanming, Luo Nianhong, Zhang Hao, Hu Haiping, Luo Rongcan, Wu Shaoguang, Xu Zhenyu, Feng Ming, Shu Long, Nie Yunfei, Zhang Changrong, Ye Qing, Liang Wei

Coordinator: Dong Zizi

Layout Design: Dong Zizi

Editorial Design Team: Liao Lintao, Zou Wei, Zhang Fazhi,Huang Liu,Cen Tupei, Feng Ziyang,

Contributing Editor : Zou Mengxuan

Translator:Zou Mengxuan

Proofreading:Lu Dongqing, Lin Wenbo, Wang Shi, Wu Fang

Binding design:Zou Wei,Cen Tupei

Size: 377mmx260mm 1/8

Construction institution： Shenzhen Zhaohua international conference and exhibition operation Co., Ltd

Ma Jun, Li Nianchang, Lv Tao, Li Shiduan, Dai Jun, Mao Daben, Ling Li, Wang Zhuo, Huang Junjie, Zhou Xin, Cai Liang, Chen Hu, Deng Biyou, Hu Zhentao, Li Junxing, Li Wei, Li Zonggang, Liu Shuo, Liu Siyuan, Liu Yuan, Liang Quanjie, Qu Weiping, Ren Chunlei, Tan Shaobin, Wang Chao, Xie Ying, Xin Wei, Yuan Wei, Zhang Haijun, Zhao Rui, Zheng Yang

AUBE 欧博设计 / 深圳市欧博工程设计顾问有限公司

地址：中国 深圳 南山 华侨城创意文化园北区 B1 栋 501

联系电话：86-755-26930800，26930801 传真：86-755-26918376

邮政编码：518053

AUBE CONCEPTION / SHENZHEN AUBE ARCHITECTURAL ENGINEERING DESIGN CO., LTD.

Add: 501,building B-1,OCT LOFT North,NanShan District,Shenzhen

Tel: 86-755-26930800, 26930801/ Fax: 86-755-26918376

Web: www.aube-archi.com